W9-DFQ-392

Gasoline-Engine Management

Robert Bosch GmbH

Robert Bosch GmbH

Gasoline-Engine Management

2nd
Edition, completely revised and extended

Published by:

Postfach 1129,
D-73201 Plochingen.
Automotive Aftermarket Division,
Product Marketing – Diagnostics &
Test Equipment (AA/PDT5)

Editor in Chief:
Dipl.-Ing. (FH) Horst Bauer.

Editorial team:
Dipl.-Ing. Karl-Heinz Dietsche,
Dipl.-Ing. (FH) Thomas Jäger,
Sven A. Hutter.

Translated by:
STAR Deutschland GmbH
Member of STAR Group

1st edition 1999
2nd edition, completely revised and extended,
January 2004

Printed in Germany. Imprimé en Allemagne.

Bentley Publishers
1734 Massachusetts Avenue
Cambridge, MA 02138
USA
Tel: (800) 423-4595
Fax: (617) 876-9235
Internet: www.bentleypublishers.com
E-Mail: sales@bentleypublishers.com

ISBN 0-8376-1052-4

ND1

With his invention of the four-stroke engine, Nikolaus August Otto developed an internal-combustion engine which was still suitable for use in the first automobiles at the end of the 19th century. This engine then went from strength to strength and promoted the spread of the automobile. The gasoline engines now fitted to modern cars are still working on the same principle which is known by the inventor's name: the "Otto cycle".

One problem at the beginning of the automobile era was the ignition. This remained a problem until Robert Bosch came up with his reliable solution. He improved on Otto's sluggish magneto-electric impulse ignition, which Otto used on his fixed engines, and produced a lighter make-and-break magneto ignition that made Otto-cycle engines rotate at higher speeds. It also made it possible to fit an engine into an automobile. The high-voltage magneto ignition (also developed by the Bosch corporation), the spark plug and battery ignition finally satisfied increased requirements. They now guarantee reliable ignition of the air-fuel mixture in the gasoline engine. Incidentally, the spark plug celebrated its centenary in 2002.

Starting in the 1970s, advances in engine construction became necessary owing to the tightening-up of legislation on emission control. Electronic systems have played a key role in endeavors to boost performance and comfort while reducing pollutant emission at the same time. The Bosch corporation has developed systems that have always kept pace with new demands. Gasoline fuel injection and the ignition system have played vital roles. They were combined in the Bosch engine management system called Motronic. Additional subsystems have been continually added to Motronic, such as cylinder-charge control and the electronically controlled throttle valve (electronic throttle control). Bosch engineers faced another major challenge: to re-engineer Motronic to control a spark-ignition engine which has gasoline directly injected into the combustion chamber.

This book combines several booklets from the "Expert Know-How on Automotive Technology" series. It describes how different engine-management systems are designed and how they and their components work. A historical section also reviews earlier engine-control systems, gasoline-injection systems, and ignition systems. Ultimately, these systems still exist on older car models.

Another topic in this book is emissions-control legislation, which has a major impact on features in Motronic systems. This chapter explains the complexities of legal regulations.

It requires an enormous development effort before a Motronic system can be fitted to a new vehicle model. One chapter is this book is therefore devoted to an overview of how these systems are developed to production maturity at Bosch.

An extensive Table of Contents and Index of Technical Terms, as well as an Index of Abbreviations, make this book an ideal reference work on the subject of "Gasoline Engine Management".

The Editorial Team

Contents

10 History of the automobile
10 Development history
12 Pioneers of automotive technology
14 Robert Bosch's life's work

16 Basics of the gasoline (SI) engine
16 Operating concept
20 Combustion knock
23 Torque and power
24 Engine efficiency
26 Specific fuel consumption
28 Fuels for spark-ignition engines (gasoline)

34 Gasoline-engine management
34 Technical requirements
36 Cylinder-charge control
39 A/F-mixture formation
42 Ignition

44 History of gasoline-engine management system development
44 Overview
45 Mechanical systems
46 Electronic systems
49 Motronic engine-management system

50 Systems for cylinder-charge control
50 Air-charge control
52 Variable valve timing
55 Controlled charge flow
56 Exhaust-gas recirculation (EGR)
57 Dynamic supercharging
60 Mechanical supercharging
62 Exhaust-gas turbocharging
65 Intercooling

66 Gasoline fuel injection: An overview
66 Overview
68 Beginnings of mixture formation
76 Evolution of gasoline injection systems
76 D-Jetronic
82 K-Ketronic
92 KE-Jetronic
94 L-Jetronic
108 Mono-Jetronic

128 Fuel supply
128 Overview
129 Fuel supply on manifold-injection systems
130 Fuel supply for gasoline direct injection
133 Evaporative-emissions control system
134 Electric fuel pump
136 Fuel filter
137 Fuel rail
137 Fuel-pressure regulator
138 Fuel-pressure damper
138 Fuel tank
138 Fuel lines

140 Manifold fuel injection
140 Overview
141 Operating concept
142 Electromagnetic fuel injectors
144 Types of fuel injection

146 Gasoline direct injection
146 Overview
147 Operating concept
148 Rail
148 High-pressure pump
150 Pressure-control valve
152 High-pressure injector
154 Combustion process
155 A/F-mixture formation
156 Operating modes

158 Ignition-system overview
158 Overview
160 Early ignition evolution
170 Battery ignition systems over the years

176 Inductive ignition system
176 Design
177 Operating concept
177 Applications
178 Ignition driver stage
178 Ignition coil
180 Voltage distribution
181 Spark plug
182 Connecting devices and interference-suppressor equipment
182 Ignition voltage
183 Ignition energy
185 Ignition timing

186 Ignition coils
186 Function
187 Requirements
188 Design concept and operation
193 Design variations
198 Ignition-coil electronics
200 Electrical parameters
202 Ignition-coil development using simulation tools
203 Manufacturing ignition coils

204 Spark plugs
204 Function
205 Application
206 Requirements
207 Design
210 Electrode materials
211 Spark-plug concepts
212 Electrode gap
213 Spark position
214 Spark-plug heat ranges
216 Spark-plug selection
220 Spark-plug performance
222 Versions
228 Spark-plug type designations
229 Simulation-based spark-plug development
230 Spark-plug manufacture

232 Motronic engine management
232 System overview
238 M-Motronic
246 ME-Motronic
250 MED-Motronic

256 Sensors
256 Automotive applications
258 Temperature sensors
259 Fuel-level sensor
260 Sensor-plate potentiometer
261 Throttle-valve sensor
262 Accelerator-pedal sensors
264 Hall-effect phase sensors
266 Induction-type sensors for transistorized ignition
267 Hall-effect sensors for transistorized ignition
268 Inductive engine-speed sensors
269 Piezoelectric knock sensors
270 Micromechanical pressure sensors
273 Thick-film pressure sensors
275 High-pressure sensors
276 Sensor-flap (impact-pressure) air-flow sensor LMM
278 Hot-wire air-mass meter HLM
279 Hot-film air-mass meter HFM2
280 Hot-film air-mass meter HFM5
282 Two-step Lambda oxygen sensors
286 LSU4 planar broad-band Lambda oxygen sensors

288 Electronic control unit (ECU)
288 Operating conditions
288 Design
288 Data processing

294 Electronic control systems
294 Overview
296 Subsystems and main functions

304 Electronic diagnosis
304 Self-diagnosis
307 On-Board Diagnosis (OBD)
308 OBD – General requirements
311 OBD – Diagnosis System Management (DSM)
312 OBD – Individual diagnoses

324 Data transfer between automotive electronic systems
324 System overview
324 Serial data transfer (CAN)
329 Prospects

330 Exhaust emissions
330 Overview
331 Major components
332 Combustion by-products
334 Factors affecting raw emissions

338 Reducing emissions
338 Overview
339 Post-combustion thermal treatment

340 Catalytic emissions control
340 Overview
340 Oxidation-type catalytic converter
341 Three-way catalytic converter
344 NO_X accumulator-type catalytic converter
346 Lambda control loop
348 Catalytic-converter heating

350 Emissions-control legislation
350 Overview
352 CARB legislation
356 EPA regulations
358 EU regulations
362 US test cycles
364 European test cycle
365 Japanese test cycles
366 Emissions testing
368 Evaporative-emissions testing

370 Service technology
370 Overview
372 Testing on-board control units
376 Testing the ignition system
377 Ignition coils in service
380 Handling spark plugs
386 Emissions inspections (AU)
388 Emissions measurement concept

390 ECU development
390 Overview
394 Hardware development
398 Function development
400 Software development
404 Application-related adaptation

412 Index of technical terms
412 Technical terms
417 Abbreviations

Background information

22 Real-world fuel economy
61 History of compressor-engine automobiles
81 The story of fuel injection
132 Integration in the vehicle: In-tank unit
151 Exploiting GDI to reduce fuel consumption
164 Presentation of the Bosch low-tension ignition magneto
166 Magneto ignition applications
169 Bosch battery ignition
182 Accident hazard
197 Ignition coil retrospective: the asphalt coil
219 100 years of spark plugs – product variety
237 Motronic systems in motorsport
257 Miniaturization
274 Micromechanics
293 Performance of electronic control units
333 Ozone and smog
349 Patents
360 Greenhouse effect
375 Global service
411 The Bosch Boxberg Test Center

Authors

Basics of the Gasoline Engine
Dipl.-Ing. Michael Oder,
Dr.-Ing. Rainer Ortmann,
Dipl.-Ing. Werner Häming,
Dipl.-Ing. Werner Hess.

Gasoline-Engine Control System
Dipl.-Ing. Michael Oder,
Dipl.-Ing. Walter Gollin.

Cylinder-Charge Control Systems
Dipl.-Ing. Georg Mallebrein,
Dipl.-Ing. Christian Köhler,
Dipl.-Ing. Oliver Schlesiger,
Dipl.-Ing. Michael Bäuerle.

Fuel Supply
Dr. Thomas Frenz,
Dr.-Ing. Dieter Lederer,
Dipl.-Ing. Albert Gerhard,
Dipl.-Betriebsw. Michael Ziegler,
Dipl.-Ing. (FH) Eckhard Bodenhausen,
Dipl.-Ing. (FH) Annette Wittke,
Dipl.-Ing. (FH) Bernd Kudicke.

Intake-Manifold Fuel Injection
Dipl.-Ing. (FH) Klaus Joos.

Gasoline Direct Injection
Dipl.-Ing. Michael Oder,
Dipl.-Ing. Helmut Rembold.

Inductive Ignition System
Dipl.-Ing. Walter Gollin.

Ignition Coils
Dipl.-Ing. (FH) Klaus Lerchenmüller,
Dipl.-Ing. (FH) Markus Weimert,
Dipl.-Ing. Tim Skowronek.

Spark Plugs
Dipl.-Ing. Erich Breuser.

Motronic Engine Management
Dipl.-Ing. Bernhard Mencher,
Dipl.-Ing. Christian Köhler,
Dipl.-Red. Ulrich Michelt.

Electronic Control Unit
Dipl.-Ing. Adolf Fritz,
Dipl.-Ing. Martin Kaiser.

Electronic Control and Regulation
Dr.-Ing. Jürgen Haag.

Electronic Diagnostics
Dr.-Ing. Matthias Knirsch,
Dr.-Ing. Matthias Tappe,
Dr. Michael Eggers.

Exhaust-Gas Emissions
Dipl.-Ing. Eberhard Schnaibel,
Dipl.-Ing. Christian Köhler.

Pollutant Reduction
Dipl.-Ing. Christian Köhler.

Catalytic Emission-Control Systems
Dipl.-Ing. Eberhard Schnaibel.

Emissions-Control Legislation
Dr. Michael Eggers,
Dr.-Ing. Gerold König.

Workshop Technology
Dipl.-Ing. Rainer Rehage,
Dipl.-Ing. (FH) Volker Engel,
Dipl.-Ing. (FH) Hans-Günther Weißhaar,
Rainer Heinzmann,
Karl-Heinz Vocke.

Electronic Control Unit Development
Dipl.-Ing. Adolf Fritz,
Dipl.-Phys. Lutz Reuschenbach,
Dipl.-Ing. (FH) Bert Scheible,
Dipl.-Ing. Eberhard Frech.

Other authors
Dipl.-Phys. Patentanwalt Ralf-Holger Behrens
(patents),
Dietrich Kuhlgatz
(Robert Bosch's Life's Work)

and the Editorial Team in collaboration with the relevant specialist departments at Bosch.

Unless otherwise stated, the above are employees of Robert Bosch GmbH, Stuttgart.

History of the automobile

Mobility has always played a crucial role in the course of human development. In almost every era, man has attempted to find the means to allow him to transport people over long distances at the highest possible speed. It took the development of reliable internal-combustion engines that were operated on liquid fuels to turn the vision of a self-propelling "automobile" into reality (combination of Greek: autos = self and Latin: mobilis = mobile).

Development history

It would be hard to imagine life in our modern day without the motor car. Its emergence required the existence of many conditions without which an undertaking of this kind would not have been possible. At this point, some development landmarks may be worthy of note. They represent an essential contribution to the development of the automobile:

- About 3500 B.C.
  The development of the wheel is attributed to the Sumerians
- About 1300 AD
  Further refinement of the carriage with elements such as steering, wheel suspension and carriage springs
- 1770
  Steam buggy by Joseph Cugnot
- 1801
  Étienne Lenoir develops the gas engine
- 1870
  Nikolaus Otto builds the first four-stroke internal-combustion engine

The first journey with an engine-powered vehicle is attributed to Joseph Cugnot (in 1770). His lumbering, steam-powered, wooden three-wheeled vehicle was able to travel for all of 12 minutes on a single tankful of water.

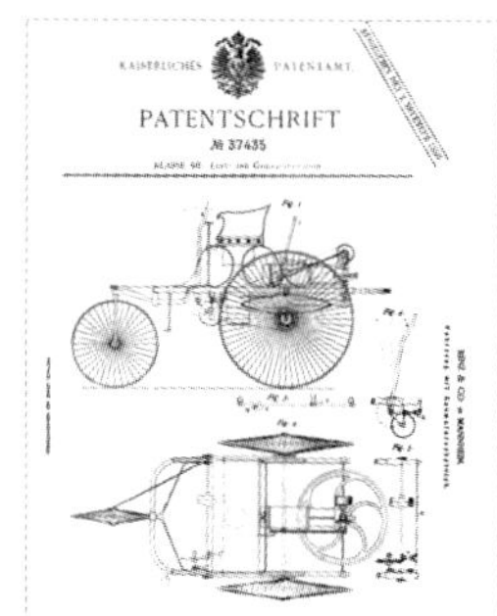

The patent issued to Benz on January 29, 1886 was not based on a converted carriage. Instead, it was a totally new, independent construction (Source: DaimlerChrysler Classic, Corporate Archives)

In 1885 Carl Benz enters the annals of history as the inventor of the first automobile. His patent marks the beginning of the rapid development of the automobile powered by the internal-combustion engine. Public opinion remained divided, however. While the proponents of the new age lauded the automobile as the epitome of progress, the majority of the population protested against the increasing annoyances of dust, noise, accident hazard, and inconsiderate motorists. Despite of all this, the progress of the automobile proved unstoppable.

Daimler Motorized Carriage, 1894 (Source: DaimlerChrysler Classic, Corporate Archives)

In the beginning, the acquisition of an automobile represented a serious challenge. A road network was virtually nonexistent, repair shops were unknown, fuel was purchased at the drugstore, and spare parts were produced on demand by the local blacksmith. The prevailing circumstances made the first long-distance journey by Bertha Benz in 1888 an even more astonishing accomplishment. She is thought to have been the first woman behind the wheel of a motorized vehicle. She also demonstrated the reliability of the automobile by journeying the then enormous distance of more than 100 kilometers (about 60 miles) between Mannheim and Pforzheim in south-western Germany.

In the early days, however, few entrepreneurs – with the exception of Benz – considered the significance of the engine-powered vehicle on a worldwide scale. It was the French who were to help the automobile to greatness. The firm of Panhard et Levassor purchased the industrial rights for Daimler engines and built their own automobiles in France. Panhard pioneered construction features such as the steering wheel, inclined steering column, clutch pedal, pneumatic tires, and the tube-type radiator.

In the years that followed, the industry mushroomed with the arrival of companies such as Peugeot, Citroën, Renault, Fiat, Ford, Rolls-Royce, Austin, and others. The influence of Gottlieb Daimler, who was selling his engines almost all over the world, added significant impetus to these developments.

Stemming from horse-drawn carriages, the motor car would soon evolve to become the car as we know it today. However, at that time, every motor car was the product of manual labor. This changed dramatically when Henry Ford introduced the assembly line in 1913. His Model T revolutionized the car industry in the US. As a result, the motor car ceased to be a luxury. Mass production dropped the price of cars. This meant that large sectors of the general public could finally afford to buy one. Although Citroën and Opel were the first to bring the assembly line to Europe, owning a car only become popular in the 1920s.

Car makers were quick to realize that, if they were to be successful on the market, they would have to satisfy customer wishes more. Motor racing wins were exploited for commerical advertising. As speeds advanced to ever higher records, professional racing drivers and the names of car brands became for ever inscribed on spectators' memories. Product ranges were also widened. As a result, the following decades saw a variety of car designs that mirrored the prevailing zeitgeist. For example, streamlined cars were unable to gain a footing before the Second World War as demand called for large cars. Manufacturers of the time produced memorable cars such as the Mercedes-Benz 500 K, the Rolls-Royce Phantom III, the Horch 855, and the Bugatti Royale.

WWII had a significant influence on the development of smaller cars. The Volkswagen car, later known as the "Beetle", was designed by Ferdinand Porsche and was made in Wolfsburg. At the end of the war the main demand was for small, affordable cars. The industry therefore produced cars such as the Goliath GP 700, Lloyd 300, Citroën 2CV, Trabant, Isetta, and the Fiat 500 C. Car production also began to evolve new standards. There was a greater emphasis on technology, integrated accessories and a reasonable price.

Today the emphasis is on occupant safety. The larger volume of traffic and higher speeds are making airbags, ABS, TCS, ESP and intelligent sensors virtually indispensable. Ongoing development has been powered by innovative engineering by the car industry and market demands. But there are sectors that continue to present future challenges. They include further reductions in pollution by using alternative energy such as fuel cells.

One thing, however, is not expected to change in the near future – it is the one concept that has been associated with the automobile for more than a century, and which had inspired its original creators – it is the enduring ideal of individual mobility.

More than 15 million units were produced of the Model T, affectionately called "Tin Lizzie". This record would be topped only by the Volkswagen Beetle in the 1970s

(Photos: Ford, Volkswagen AG)

Contemporary studies indicate what automobiles of tomorrow might look like (Photo: Peugeot)

In 1899 the Belgian Camille Jenatzy was the first human to break the 100 km/h barrier. Today the speed record stands at 1227.9 km/h.

Mercedes-Benz 500 K Convertible C, 1934 (Source: DaimlerChrysler Classic, Corporate Archives)

Pioneers of automotive technology

Owing to the large number of people who contributed to the development of the automobile, this list makes no claim to completeness

1866: Nikolaus August Otto (Photo: Deutz AG) acquires the patent for the atmospheric gas machine

Nikolaus August Otto (1832–1891), born in Holzhausen (Germany), developed an interest in technical matters at an early age. Beside his employment as a traveling salesman for food wholesalers, his main interest was devoted to the functioning of gas-powered engines.

From 1862 onward he dedicated himself totally to engine construction. He managed to make improvements to the gas engine invented by the French engineer, Étienne Lenoir. For this work, Otto was awarded the gold medal at the 1867 Paris World Fair. Together with Daimler and Maybach, he developed an internal-combustion engine based on the four-stroke principle he had formulated in 1861. The resulting engine is known as the "Otto engine" to this day. In 1884 Otto invented the magneto ignition, which allowed engines to be powered by gasoline. This innovation would form the basis for the main part of Robert Bosch's life's work.

Otto's achievement was that he was the first to build the four-stroke internal-combustion engine and demonstrate its superiority over all its predecessors.

Gottlieb Daimler (Photo: DaimlerChrysler Classic, Corporate Archives)

Gottlieb Daimler (1834–1900) hailed from Schorndorf (Germany). He studied mechanical engineering at the Polytechnikum engineering college in Stuttgart. In 1865 he met the highly talented engineer Wilhelm Maybach. From that moment on, the two men would be joined in a lasting cooperation. Besides inventing the first motorcycle, Daimler mainly worked on developing a gasoline engine suitable for use in road vehicles. In 1889 Daimler and Maybach introduced the first "steel-wheeled vehicle" in Paris featuring a two-cylinder V-engine. Scarcely one year later, Daimler was marketing his fast-running Daimler engine on an international scale. In 1891, for example, Armand Peugeot successfully entered a vehicle he had engineered himself in the Paris-Brest-Paris long-distance trial. It proved both the worth of his design and the dependability of the Daimler engine he was using.

Daimler's merits lie in the systematic development of the gasoline engine, and in the international distribution of his engines.

Wilhelm Maybach (Photo: MTU Friedrichshafen GmbH)

Wilhelm Maybach (1846–1929), a native of Heilbronn (Germany), completed his apprenticeship as a technical draftsman. Soon after, he worked as a design engineer. Among his employers was the firm of Gasmotoren Deutz AG (founded by Otto). He already earned the nickname of "king of engineers" during his own lifetime.

Maybach revised the gasoline engine and brought it to production maturity. He also developed water cooling, the carburetor, and the dual-ignition system. In 1900 Maybach built a revolutionary, alloy-based racing car. This vehicle was developed in response to a suggestion by an Austrian businessman named Jellinek. His order for 36 of these cars was given on condition that the model was to be named after his daughter Mercedes.

Maybach's virtuosity as a design engineer pointed the way to the future of the contemporary automobile industry.

His death signaled the end of the grand age of the automotive pioneers.

1886: As inventor of the first automobile fitted with an internal-combustion engine, Benz enters the annals of world history (Photo: DaimlerChrysler Classic, Corporate Archives)

Carl Friedrich Benz (1844–1929), born in Karlsruhe (Germany), studied mechanical engineering at the Polytechnikum engineering college in his hometown. In 1871 he founded his first company, a factory for iron-foundry products and industrial components in Mannheim.

Independently of Daimler and Maybach, he also pursued the means of fitting an engine in a vehicle. When the essential claims stemming from Otto's four-stroke engine patent had been declared null and void, Benz also developed the surface carburetor, the electrical ignition, the clutch, water cooling, and a type of gearshift, besides his own four-stroke engine. In 1886 he applied for his patent, and presented his motor carriage to the public. In the period until the year 1900, Benz was able to offer more than 600 models for sale. In the period between 1894 and 1901 the factory of Benz & Co. produced the "Velo" which, with a total output of about 1200 units, may be called the first mass-produced automobile. In 1926 Benz merged with Daimler to form the "Daimler-Benz AG".

The legacy of Carl Friedrich Benz may be seen in the construction of the first automobile and in its industrial manufacture.

Henry Ford (Photo: Ford)

Henry Ford (1863–1947) hailed from Dearborn, Michigan (USA). Although Ford had found secure employment as an engineer with the Edison Illuminating Company in 1891, his personal interests were dedicated to the advancement of the gasoline engine.

In 1893 the Duryea Brothers built the first American automobile. Ford managed to even the score in 1896 by introducing his own car, the "Quadricycle Runabout", which was to serve as the basis for numerous additional designs. In 1908 Ford introduced the legendary "Model T", which was mass-produced on assembly lines from 1913 onward. Beginning in 1921, with a 55-percent share in the country's industrial production, Ford dominated the domestic automobile market in the USA.

The name Henry Ford is synonymous with the motorization of the United States. It was his ideas that made the automobile accessible to a broad segment of the population.

Rudolf C. K. Diesel (Photo: Historical Archives of MAN AG)

Rudolf Christian Karl Diesel (1858–1929), born in Paris (France), decided to become an engineer at the age of 14. He graduated from the Polytechnikum engineering college in Munich with the best marks the institution had ever given in its entire existence.

In 1892 Diesel was issued the patent for the "Diesel engine" that was later to bear his name. The engine was quickly adopted as a stationary power plant and marine engine. In 1908 the first commercial truck was powered by a diesel engine. However, its entrance in the world of passenger cars would take several decades. The diesel engine became the power plant for the serial-produced Mercedes 260 D as late as in 1936. Today's diesel engine has reached a level of development that places it on a par with the gasoline engine.

With the high efficiency rating of his engine, Diesel has made a major contribution to a more economical utilization of the internal-combustion engine. Although Diesel became active internationally by granting production licenses, he failed to earn due recognition for his achievements during his lifetime.

"It has always been an unbearable thought to me that someone could inspect one of my products and find it inferior in any way. For this reason, I have constantly endeavored to produce work that withstands the closest scrutiny, in other words, the very best of the best."

Robert Bosch

Robert Bosch's life's work (1861–1942)

(Photos: Bosch Archives)

Robert Bosch, born September 23, 1861 in Albeck near Ulm (Germany), was the scion of a wealthy farmer's family. After completing his apprenticeship as a precision fitter, he worked temporarily for a number of enterprises, where he continued to hone his technical skills and expand his merchandising abilities and experience. After six months as an auditor studying electrical engineering at Stuttgart technical university, he traveled to the United States to work for "Edison Illuminating". He was later employed by "Siemens Brothers" in England.

In 1886 he decided to open a "Workshop for Precision Mechanics and Electrical Engineering" in the back of a house in Stuttgart's west end. He employed another mechanic and an apprentice. His first responsibility was the installation and repair of telephones, telegraphs, lightning rods, and other services involving precision engineering. His dedication in finding rapid solutions to new problems also helped him gain a competitive lead in his later activities.

To the automobile industry, the low-tension voltage magneto ignition developed by Bosch in 1897 represented – much unlike its unreliable predecessors – a true breakthrough. This product became the starting point for the rapid expansion of the enterprise run by Robert Bosch. He always managed to bring the purposefulness of the world of technology and economics into harmony with the needs of humanity. Bosch was a trailblazer in many aspects of social care.

ROB. BOSCH
Rothebühlstr. 75 B.
Telephone, Haustelegraphen.
Fachmännische Prüfung und Anlegung von Blitzableitern. (37
Anlegung und Reparatur elektr. Apparate, sowie aller Arbeiten der Feinmechanik.

First ad in the Stuttgart daily "Der Beobachter" (The Observer), 1887

Robert Bosch performed technological pioneering work in developing and bringing the following products to maturity:

- Low-voltage magneto ignition
- High-voltage magneto ignition for higher engine speeds (engineered by his colleague Gottlob Honold)
- Spark plug
- Ignition distributor
- Battery (passenger vehicles and motorcycles)
- Electrical starter
- Generator (alternator)
- Lighting system with first electric headlamp
- Car radio (manufactured by "Ideal-Werke", renamed "Blaupunkt" in 1938)
- First lighting system for bicycles
- Bosch horn
- Battery ignition
- Bosch semaphore turn signal (initially ridiculed as being typical of German sense of organization – now the indispensable direction indicator)

At this point, many other achievements, also in the area of social engagement, would be worthy of note. They are clear indicators that Bosch was veritably ahead of his time.

His visionary mind prepared the way for many advances in car development. The growing number of motorists gave rise to a need for repair workshops. In the 1920s Robert Bosch started a customer service organization. In 1926 the organization was renamed the "Bosch-Dienst" (Bosch Service) and the name was registered.

Bosch has similarly high ambitions in social welfare. After introducing the 8-hour da in 1906, he started paying his workers high wages. In 1920 he donated one million Reichsmarks to support technical education. When the 500 thousandth magneto ignation was produced, he abolished Saturday afternoon work at his factories. Other Bosch improvements included pensions, jobs for the handicapped, and a vacation scheme. In 1913 the Bosch credo resulted in the start of a training workshop for 104 apprentices.

By mid-1914 the name of Bosch was known all over the world. But the period between the two world wars would be a time of great expansion. By 1914 about 88 percent of products made in Stuttgart were shipped abroad. This included major defense contracts. However, the atrocities of the time made him spurn the profits he had made. He therefore donated 13 million Reichsmarks to social causes.

After the end of WWI it was difficult to regain a foothold on foreign markets. In the United States, for example, Bosch factories, sales offices, and the corporate logo and symbol had been confiscated and sold to an American company. One of the consequences was that products appeared under the "Bosch" brand name that were not truly Bosch-made. It would take until the end of the 1920s before Bosch had reclaimed all of his former rights and was able to reestablish himself in the United States. The Founder's unyielding determination to overcome any and all obstacles returned the company to the markets of the world and, at the same time, imbued the minds of Bosch employees with the international significance of Bosch as an enterprise.

First offices in London's Store Street (Photo: Bosch Archives)

"To each his own automobile"
Such was the Bosch claim in a 1931 issue of the Bosch employee magazine "Bosch Zünder" (Bosch Ignitor)·

Two notable events are further examples of Robert Bosch's social commitment.

In 1936 he donated funds to construct a hospital that was officially opened in 1940. In his inaugural speech, Robert Bosch emphasized his personal dedication in terms of social engagement: "Every job is important, even the lowliest. Let no man delude himself that his work is more important than that of a colleague."

When Robert Bosch died in 1942, the world mourned an entrepreneur who not only pioneered technology and electrical engineering, but who also promoted social welfare. Robert Bosch today is an example of progressive zeitgeist, untiring hard work, social care, business acumen, and promoting education. He formulated his vision of progress in the words: "Knowledge, ability, and will are important, but success only comes from their harmonious interaction."

In 1964 the Robert Bosch Foundation was inaugurated. Its activities include the promotion and support of health care, welfare, education, as well as sponsoring the arts and culture, humanities and social sciences. The Foundation continues to nurture the founder's ideals to this day.

Basics of the gasoline (SI) engine

The gasoline or spark-ignition (SI) internal-combustion engine uses the Otto cycle[1]) and externally supplied ignition. It burns an air/fuel mixture and in the process converts the chemical energy in the fuel into kinetic energy.

For many years, the carburetor was responsible for providing an A/F mixture in the intake manifold which was then drawn into the cylinder by the downgoing piston.

The breakthrough of gasoline fuel injection, which permits extremely precise metering of the fuel, was the result of the legislation governing exhaust-gas emission limits. Similar to the carburetor process, with manifold fuel injection the A/F mixture is formed in the intake manifold.

Even more advantages resulted from the development of gasoline direct injection, in particular with regard to fuel economy and increases in power output. Direct injection injects the fuel directly into the engine cylinder at exactly the right instant in time.

Operating concept

The combustion of the A/F mixture causes the piston (Fig. 1, Pos. 8) to perform a reciprocating movement in the cylinder (9). The name reciprocating-piston engine, or better still reciprocating engine, stems from this principle of functioning.

The conrod (10) converts the piston's reciprocating movement into a crankshaft (11) rotational movement which is maintained by a flywheel (11) at the end of the crankshaft. Crankshaft speed is also referred to as engine speed or engine rpm.

Four-stroke principle

Today, the majority of the internal-combustion engines used as vehicle power plants are of the four-stroke type.

The four-stroke principle employs gas-exchange valves (5 and 6) to control the exhaust-and-refill cycle. These valves open and close the cylinder's intake and exhaust passages, and in the process control the supply of fresh A/F mixture and the forcing out of the burnt exhaust gases.

1st stroke: Induction

Referred to top dead center (TDC), the piston is moving downwards and increases the volume of the combustion chamber (7) so that fresh air (gasoline direct injection) or fresh A/F mixture (manifold injection) is drawn into the combustion chamber past the opened intake valve (5).

The combustion chamber reaches maximum volume (V_h+V_c) at bottom dead center (BDC).

2nd stroke: Compression

The gas-exchange valves are closed, and the piston is moving upwards in the cylinder. In doing so it reduces the combustion-chamber volume and compresses the A/F mixture. On manifold-injection engines the A/F mixture has already entered the combustion chamber at the end of the induction stroke. With a direct-injection engine on the other hand, depending upon the operating mode, the fuel is first injected towards the end of the compression stroke.

At top dead center (TDC) the combustion-chamber volume is at minimum (compression volume V_c).

[1]) Named after Nikolaus Otto (1832-1891) who presented the first gas engine with compression using the 4-stroke principle at the Paris World Fair in 1878.

3rd stroke: Power (or combustion)
Before the piston reaches top dead center (TDC), the spark plug (2) initiates the combustion of the A/F mixture at a given ignition point (ignition angle). This form of ignition is known as externally supplied ignition. The piston has already passed its TDC point before the mixture has combusted completely.

The gas-exchange valves remain closed and the combustion heat increases the pressure in the cylinder to such an extent that the piston is forced downward.

4th stroke: Exhaust
The exhaust valve (6) opens shortly before bottom dead center (BDC). The hot (exhaust) gases are under high pressure and leave the cylinder through the exhaust valve. The remaining exhaust gas is forced out by the upwards-moving piston.

A new operating cycle starts again with the induction stroke after every two revolutions of the crankshaft.

Valve timing

The gas-exchange valves are opened and closed by the cams on the intake and exhaust camshafts (3 and 1 respectively). On engines with only 1 camshaft, a lever mechanism transfers the cam lift to the gas-exchange valves.

The valve timing defines the opening and closing times of the gas-exchange valves. Since it is referred to the crankshaft position, timing is given in "degrees crankshaft". Gas flow and gas-column vibration effects are applied to improve the filling of the combustion chamber with A/F mixture and to remove the exhaust gases. This is the reason for the valve opening and closing times overlapping in a given crankshaft angular-position range.

The camshaft is driven from the crankshaft through a toothed belt (or a chain or gear pair). On 4-stroke engines, a complete working cycle takes two rotations of the crankshaft. In other words, the camshaft only turns at half crankshaft speed, so that the step-down ratio between crankshaft and camshaft is 2:1.

1 Complete working cycle of the 4-stroke spark-ignition (SI) gasoline engine (example shows a manifold-injection engine with separate intake and exhaust camshafts)

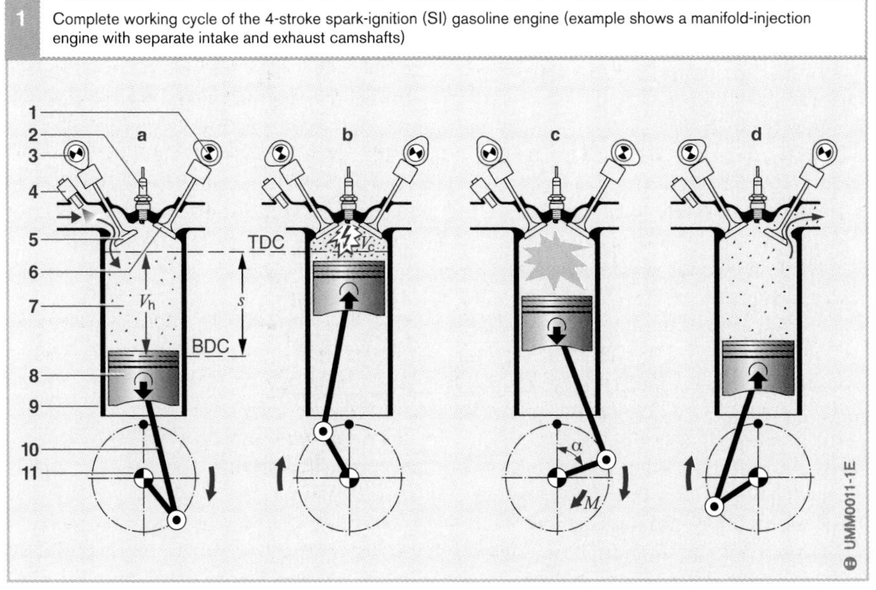

Fig. 1
a Induction stroke
b Compression stroke
c Power (combustion) stroke
d Exhaust stroke

1 Exhaust camshaft
2 Spark plug
3 Intake camshaft
4 Injector
5 Intake valve
6 Exhaust valve
7 Combustion chamber
8 Piston
9 Cylinder
10 Conrod
11 Crankshaft

M Torque
α Crankshaft angle
s Piston stroke
V_h Piston displacement
V_c Compression volume

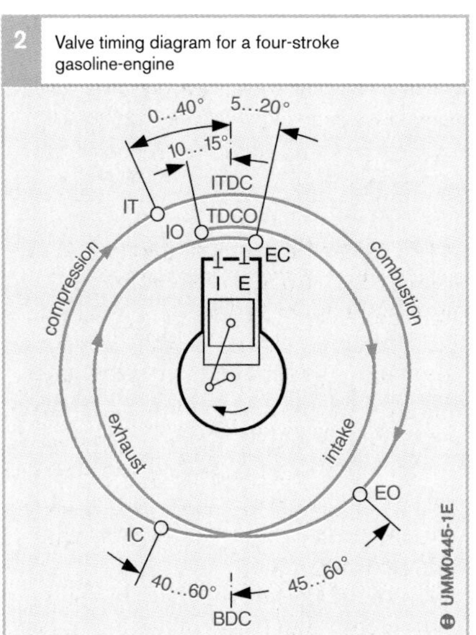

2 Valve timing diagram for a four-stroke gasoline-engine

Fig. 2

I	Intake valve
IO	Intake valve opens
IC	Intake valve closes
E	Exhaust valve
EO	Exhaust valve opens
EC	Exhaust valve closes
TDC	Top Dead Center
TDCO	Overlap at TDC
ITDC	Ignition at TDC
BDC	Bottom Dead Center
IT	Ignition

The valve timing diagram (Figure 2) illustrates opening and closing durations for intake and exhaust valves.

Compression

The difference between the maximum swept volume V_h and the compressed volume V_c is the compression ratio $\varepsilon = (V_h + V_c)/V_c$.

The engine's compression ratio is a vital factor in determining

- Torque generation
- Power generation
- Fuel economy and
- Emissions of harmful pollutants

The gasoline-engine's compression ratio ε varies according to design configuration and the selected form of fuel injection (manifold or direct injection $\varepsilon = 7 \ldots 13$). Extreme compression ratios of the kind employed in diesel powerplants ($\varepsilon = 14 \ldots 24$) are not suitable for use in gasoline engines. Because the knock resistance of the fuel is limited, the extreme compression pressures and the high combustion-chamber temperatures resulting from such compression ratios must be avoided in order to prevent spontaneous and uncontrolled detonation of the air/fuel mixture. The resulting knock can damage the engine.

Air/fuel ratio

Complete combustion of the air/fuel mixture relies on a stoichiometric mixture ratio. A stoichiometric ratio is defined as 14.7 kg of air for 1 kg of fuel, that is, a 14.7 to 1 mixture ratio.

The air/fuel ratio λ (lambda) indicates the extent to which the instantaneous monitored air/fuel ratio deviates from the theoretical ideal:

$$\lambda = \frac{\text{induction air mass}}{\text{theoretical air requirement}}$$

The lambda factor for a stoichiometric ratio is λ 1.0. λ is also referred to as the excess-air factor.

Richer fuel mixtures result in λ figures of less than 1. Leaning out the fuel produces mixtures with excess air: λ then exceeds 1. Beyond a certain point the mixture encounters the lean-burn limit, beyond which ignition is no longer possible.

Induction-mixture distribution in the combustion chamber

Homogeneous distribution

The induction systems on engines with manifold injection distribute a homogeneous air/fuel mixture throughout the combustion chamber. The entire induction charge has a single excess-air factor λ (Figure 3a). Lean-burn engines, which operate on excess air under specific operating conditions, also rely on homogeneous mixture distribution.

Stratified-charge concept

A combustible mixture cloud with $\lambda \approx 1$ surrounds the tip of the spark plug at the instant ignition is triggered. At this point the remainder of the combustion chamber contains either non-combustible gas with no fuel, or an extremely lean air/fuel charge. The corresponding strategy, in which the ignitable mixture cloud is present only in one portion of the combustion chamber, is the stratified-charge concept (Figure 3b). With this concept, the overall mixture – meaning the average mixture ratio within the entire combustion chamber – is extremely lean (up to $\lambda \approx 10$). This type of lean operation fosters extremely high levels of fuel economy.

3 Induction-mixture distribution in the combustion chamber

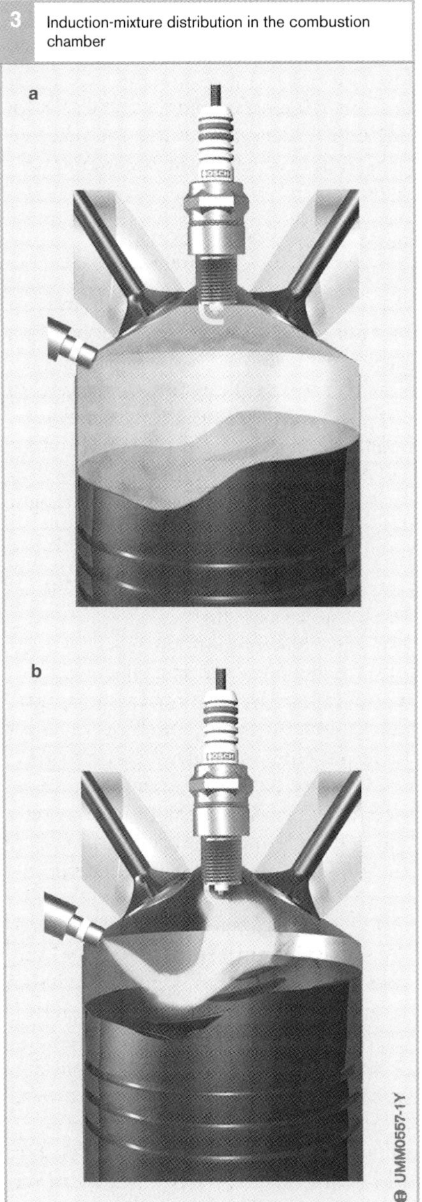

Fig. 3
a Homogeneous mixture distribution
b Stratified charge

Efficient implementation of the stratified-charge concept is impossible without direct fuel injection, as the entire induction strategy depends on the ability to inject fuel directly into the combustion chamber just before ignition.

Ignition and flame propagation

The spark plug ignites the air/fuel mixture by discharging a spark across a gap. The extent to which ignition will result in reliable flame propagation and secure combustion depends in large part on the air/fuel mixture λ, which should be in a range extending from $\lambda = 0.75...1.3$. Suitable flow patterns in the area immediately adjacent to the spark-plug electrodes can be employed to ignite mixtures as lean as $\lambda \leq 1.7$.

The initial ignition event is followed by formation of a flame-front. The flame front's propagation rate rises as a function of combustion pressure before dropping off again toward the end of the combustion process. The mean flame front propagation rate is on the order of 15...25 m/s.

The flame front's propagation rate is the combination of mixture transport and combustion rates, and one of its defining factors is the air/fuel ratio λ. The combustion rate peaks at slightly rich mixtures on the order of $\lambda = 0.8...0.9$. In this range it is possible to approach the conditions coinciding with an ideal constant-volume combustion process (refer to section on "Engine efficiency"). Rapid combustion rates provide highly satisfactory full-throttle, full-load performance at high engine speeds.

Good thermodynamic efficiency is produced by the high combustion temperatures achieved with air/fuel mixtures of $\lambda = 1.05...1.1$. However, high combustion temperatures and lean mixtures also promote generation of nitrous oxides (NO_x), which are subject to strict limitations under official emissions standards.

Combustion knock

Among the factors imposing limits on the latitude for enhancing an engine's thermodynamic efficiency and increasing power-plant performance are spontaneous pre-ignition and detonation. This highly undesirable phenomenon is frequently accompanied by an audible "pinging" noise, which is why the generally applicable term for this condition is "knock". Knock occurs when portions of the mixture ignite spontaneously before being reached by the flame front. The intense heat and immense pressure peaks produced by combustion knock subject pistons, bearings, cylinder head and head gasket to enormous mechanical and thermal loads. Extended periods of knock can produce blown head gaskets, holed piston crowns and engine seizure, and leads to destruction of the engine.

The sources of combustion knock

The spark plug ignites the air/fuel mixture toward the end of the compression stroke, just before the piston reaches top dead centre (TDC). Because several milliseconds can elapse until the entire air/fuel mixture can ignite (the precise ignition lag varies according to engine speed), the actual combustion peak occurs after TDC.

The flame front extends outward from the spark plug. After being compressed during the compression stroke, the induction mixture is heated and pressurized as it burns within the combustion chamber. This further compresses any unburned air/fuel mixture within the chamber. As a result, some portions of the compressed air/fuel mixture can attain temperatures high enough to induce spontaneous auto-ignition (Figure 1). Sudden detonation and uncontrolled combustion are the results.

When this type of detonation occurs it produces a flame front with a propagation rate 10 to 100 times that associated with the normal combustion triggered by the spark plug (approximately 20 m/s). This uncontrolled combustion generates pressure pulses which spread out in circular patterns from the core of the process. It is when these pulsations impact against the walls of the cylinder that they generate the metallic pinging sound typically associated with combustion knock.

Other flame fronts can be initiated at hot spots within the combustion chamber. Among the potential sources of this hot-spot ignition are spark plugs which during operation heat up excessively due to their heat range being too low. This type of pre-ignition produces engine knock by initiating combustion before the ignition spark is triggered.

Engine knock can occur throughout the engine's speed range. However, it is not possible to hear it at extremely high rpm, when its sound is obscured by the noise from general engine operation.

Factors affecting tendency to knock

Substantial ignition advance: Advancing the timing to ignite the mixture earlier produces progressively higher combustion-chamber temperatures and correspondingly extreme pressure rises.

High cylinder-charge density: The charge density must increase as torque demand rises (engine load factor). This leads to high temperatures during compression.

Fuel grade: Because fuels with low octane ratings furnish only limited resistance to knock, compliance with manufacturer's specifications for fuel grade(s) is vital.

Excessively high compression ratio: One potential source of excessively high compression would be a cylinder head gasket of less than the specified thickness. This leads to higher pressures and temperatures in the air/fuel mixture during compression. Deposits and residue in the combustion chamber (from ageing, etc.) can also produce a slight increase in the effective compression ratio.

Cooling: Ineffective heat dissipation within the engine can lead to high mixture temperatures within the combustion chamber.

Geometry: The engine's knock tendency can be aggravated by unfavorable combustion-chamber geometry. Poor turbulence and swirl characteristics caused by unsatisfactory intake-manifold tract configurations are yet another potential problem source.

Engine knock with direct gasoline injection

With regard to engine knock, when operating with homogeneous A/F mixtures direct-injection gasoline engines behave the same as manifold-injected power plants. One major difference is the cooling effect exerted by the evaporating fuel during direct injection, which reduces the temperature of the air within the cylinder to levels lower than those encountered with manifold injection.

During operation in the stratified-charge mode it is only in the area immediately adjacent to the spark plug tip that an ignitable mixture is present. When the remainder of the combustion chamber is filled with air or inert gases, there is no danger of spontaneous ignition and engine knock. Nor is there any danger of detonation when an extremely lean air/fuel mixture is present within these outlying sections of the combustion chamber. The ignition energy required to generate a flame in this kind of lean mixture would be substantially higher than that needed to spark a stoichiometric combustion mixture. This is why stratified-charge operation effectively banishes the danger of engine knock.

1 The sources of combustion knock

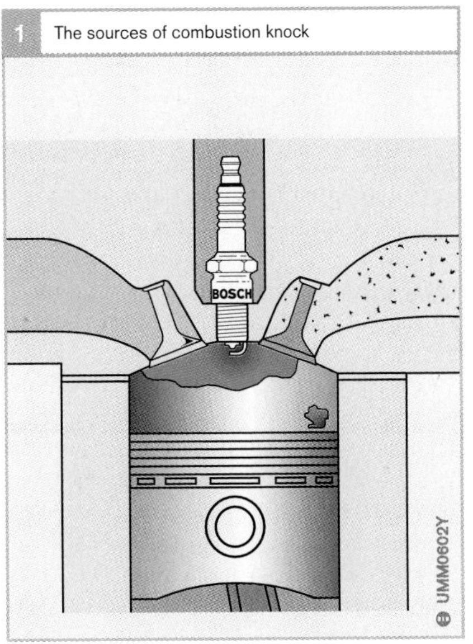

Avoiding consistent engine knock

To effectively avoid pre-ignition and detonation, ignition systems not equipped with knock detection rely on ignition timing with a safety margin of 5...8 degrees (crankshaft) relative to the knock limit.

Ignition systems featuring knock detection employ one or several knock sensors to monitor acoustic waves in the engine. The engine-management ECU detects knock in individual combustion cycles by analysing the electrical signals relayed by these sensors. The ECU then responds by retarding the ignition timing for the affected cylinder to prevent continuous knock. The system then gradually advances the ignition timing back toward its original position. This progressive advance process continues until the ignition timing is either back at the initial reference point programmed into the engine's software map, or until the system starts to detect knock again. The engine management regulates the timing advance for each cylinder individually.

The limited number of combustion events with mild knock of the kind that also occur with knock control are not injurious to the health of the engine. On the contrary: They help dissolve deposits formed by oil and fuel additives within the combustion chamber (on intake and exhaust valves, etc.), allowing them to be combusted and/or discharged with the exhaust gases.

Advantages of knock control

Thanks to reliable knock recognition, engines with knock control can use higher compression ratios. Co-ordinated control of the ignition's timing advance also makes it possible to do without the safety margin between the timing point and the knock threshold; the ignition timing can be selected for the "best case" instead of the "worst case" scenario. This provides benefits in terms of thermodynamic efficiency. Knock control

- reduces fuel consumption,
- enhances torque and power, and
- allows engine operation on different fuels within an extended range of octane ratings (both premium and regular unleaded, etc.).

Real-world fuel economy

Manufacturers must furnish fuel-consumption data for their vehicles. The official figures are calculated based on the composition of the exhaust gases monitored during emissions testing. Emissions testing is conducted based on a standardised test procedure, or driving cycle. The standardized procedure provides emissions figures suitable for comparison among vehicles.

Motorists can make a major contribution to improved fuel economy by adopting a suitable driving style. Potential fuel savings vary according to a variety of factors.

By adopting the practices listed below, the "economy-minded" motorist can achieve fuel savings of 20...30 % compared to the "average" driver. The latitude available for enhancing fuel economy depends upon a number of factors. Especially significant among these are operating environment (urban traffic or long-distance cruising, etc.) and general traffic conditions. This is why attempts to quantify the precise savings potential represented by each individual factor are not always logical.

Increasing fuel economy

- Tire pressures: Remember to increase inflation pressures when vehicle is loaded to capacity (savings: roughly 5 %)
- Accelerate at wide throttle openings and low engine speeds, upshift at 2,000 rpm
- Drive in the highest possible gear: even at engine speeds below 2,000 rpm it is possible to apply full throttle
- Plan ahead to avoid continuous alternation between braking and acceleration
- Exploit the potential of the trailing-throttle fuel cutoff
- Switch off engine during extended stops, such as at traffic lights with extended red phases and at railroad crossings, etc. (3 minutes of idling consumes as much fuel as driving 1 kilometer)
- Use full-synthetic engine oils (savings of approximately 2 % according to the manufacturer)

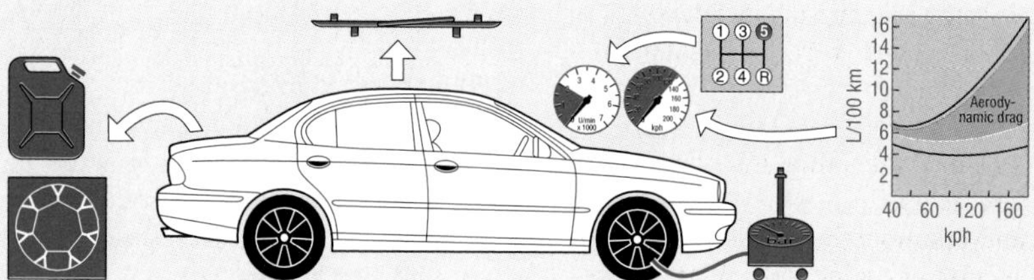

Negative influences on fuel economy

- Added vehicle weight caused by unnecessary ballast in the luggage compartment (adds roughly 0.3 litres/100 km)
- High driving speeds
- Increased aerodynamic drag from roof-mounted racks and luggage carriers
- Activation of supplementary electrical accessories such as rear-screen defroster, fog lamps (approximately 1 liter/1 kW load)
- Contaminated air filter and worn spark plugs (observe service intervals)

SMK1827E

Torque and power

Defined as a physical unit, torque M is the product of force F times leverage s:

$$M = F \cdot s$$

The connecting rods press against the crankshaft journals to convert the pistons' linear travel into rotary motion. As the air/fuel mixture expands it presses the piston downward. The force from the connecting rod is converted to torque at a rate defined by the radius, and thus the leverage, of the crankshaft journal arm.

The leverage relevant for torque generation is the component exerted at a right angle to the direction of force. The force and the leverage angle are parallel at top dead centre (TDC). This results in an effective leverage force of zero. The ignition timing must be selected to trigger ignition while the crankshaft is rotating through a phase of increasing leverage (0...90 °crankshaft). This allows the engine to generate maximum torque.

Each power plant's maximum torque-production potential M is defined by an array of factors stemming from the engine's basic design configuration (displacement, combustion-chamber geometry, etc.). Varying the quality and quantity of the air/fuel mixture represents the primary tool for adapting torque generation to reflect instantaneous vehicle demand.

Figure 1 shows the torque and power curves for a typical gasoline engine equipped with manifold injection; these curves are plotted as a function of engine speed. Many automotive periodicals provide these kinds of diagrams with their road tests. As engine speed rises, torque initially climbs to its maximum M_{max}. As engine speed moves past the torque peak, torque falls off again, reflecting the fact that the reduced opening periods available to the intake valves place limits on induction-charge density.

Engine designers focus on attempting to obtain maximum torque at low engine speeds of around 2,000 rpm. This rpm range coincides with optimal fuel economy. Engines with exhaust-gas turbochargers meet this demand catalogue.

The engine's power rating P rises as torque M and engine speed n increase. The relationship is defined as follows:

$$P = 2 \cdot \pi \cdot n \cdot M$$

The engine's power output continues to climb until the maximum output rpm $n_{spec.}$, where it peaks at $P_{spec.}$ Owing to the substantial decrease in torque, power generation drops again at extremely high engine speeds.

A transmission to vary conversion ratios is needed to adapt the engine's torque and power curves to meet the requirements of vehicle operation.

1 Sample power and torque curves for a manifold-injected gasoline-engine

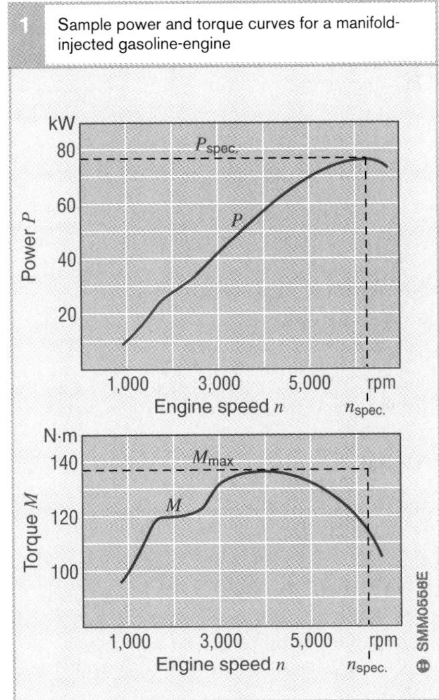

Fig. 1
M_{max} Torque peak
$P_{spec.}$ Rated max. power
$n_{spec.}$ Engine speed at max. power

Engine efficiency

Thermal efficiency

The internal-combustion engine does not convert all the energy which is chemically available in the fuel into mechanical work, and some of the added energy is lost. This means that an engine's efficiency is less than 100 % (Fig. 1). Thermal efficiency is one of the important links in the engine's efficiency chain.

Pressure-volume diagram (p-V diagram)

The p-V diagram is used to display the pressure and volume conditions during a complete working cycle of the 4-stroke IC engine.

The ideal constant-volume cycle

Fig. 2 (curve A) shows the compression and power strokes of an ideal process as defined by the laws of Boyle/Mariotte and Gay-Lussac. The piston travels from BDC to TDC (point 1 to point 2), and the A/F mixture is compressed without the addition of heat (Boyle/Mariotte). Subsequently, the mixture burns accompanied by a pressure rise (point 2 to point 3) while volume remains constant (Gay-Lussac).

From TDC (point 3), the piston travels towards BDC (point 4), and the combustion-chamber volume increases. The pressure of the burnt gases drops whereby no heat is released (Boyle/Mariotte). Finally, the burnt mixture cools off again with the volume remaining constant (Gay-Lussac) until the initial status (point 1) is reached again.

The area inside the points 1 – 2 – 3 – 4 shows the work gained during a complete working cycle. The exhaust valve opens at point 4 and the gas, which is still under pressure, escapes from the cylinder. If it were possible for the gas to expand completely by the time point 5 is reached, the area described by 1 – 4 – 5 would represent usable energy. On an exhaust-gas-turbocharged engine, the part above the atmospheric line (1 bar) can to some extent be utilized (1 – 4 – 5′).

Real p-V diagram

Since it is impossible during normal engine operation to maintain the basic conditions for the ideal constant-volume cycle, the actual p-V diagram (Fig. 2, curve B) differs from the ideal p-V diagram.

Measures for increasing thermal efficiency

The thermal efficiency rises along with increasing A/F-mixture compression. The higher the compression, the higher the pressure in the cylinder at the end of the compression phase, and the larger is the enclosed area in the p-V diagram. This area is an indication of the energy generated during the combustion process. When selecting the compression ratio, the fuel's antiknock qualities must be taken into account.

Manifold-injection engines inject the fuel into the intake manifold onto the closed intake valve, where it is stored until drawn into the cylinder. During the formation of the A/F mixture, the fine fuel droplets vaporize. The energy needed for this process is in the form of heat and is taken from the air and the intake-manifold walls. On direct-injection engines the fuel is injected into the combustion chamber, and the energy needed for fuel-droplet vaporization is taken from the air trapped in the cylinder which cools off as a result. This means that the compressed A/F mixture is at a lower temperature than is the case with a manifold-injection engine, so that a higher compression ratio can be chosen.

Thermal losses

The heat generated during combustion heats up the cylinder walls. Part of this thermal energy is radiated and lost. In the case of gasoline direct injection, the stratified-charge A/F mixture cloud is surrounded by a jacket of gases which do not participate in the combustion process. This gas jacket hinders the transfer of heat to the cylinder walls and therefore reduces the thermal losses.

Further losses stem from the incomplete combustion of the fuel which has condensed onto the cylinder walls. Thanks to the insulating effects of the gas jacket, these losses are reduced in stratified-charge operation. Further thermal losses result from the residual heat of the exhaust gases.

Losses at λ =1

The efficiency of the constant-volume cycle climbs along with increasing excess-air factor (λ). Due to the reduced flame-propagation velocity common to lean A/F mixtures, at $\lambda > 1.1$ combustion is increasingly sluggish, a fact which has a negative effect upon the SI engine's efficiency curve. In the final analysis, efficiency is the highest in the range $\lambda = 1.1...1.3$. Efficiency is therefore less for a homogeneous A/F-mixture formation with $\lambda = 1$ than it is for an A/F mixture featuring excess air. When a 3-way catalytic converter is used for emissions control, an A/F mixture with $\lambda = 1$ is absolutely imperative for efficient operation.

Pumping losses

During the exhaust and refill cycle, the engine draws in fresh gas during the 1st (induction) stroke. The desired quantity of gas is controlled by the throttle-valve opening. A vacuum is generated in the intake manifold which opposes engine operation (throttling losses). Since with a gasoline direct-injection engine the throttle valve is wide open at idle and part load, and the torque is determined by the injected fuel mass, the pumping losses (throttling losses) are lower.

In the 4th stroke, work is also involved in forcing the remaining exhaust gases out of the cylinder.

Frictional losses

The frictional losses are the total of all the friction between moving parts in the engine itself and in its auxiliary equipment. For instance, due to the piston-ring friction at the cylinder walls, the bearing friction, and the friction of the alternator drive.

1 Efficiency chain of an SI engine at $\lambda = 1$

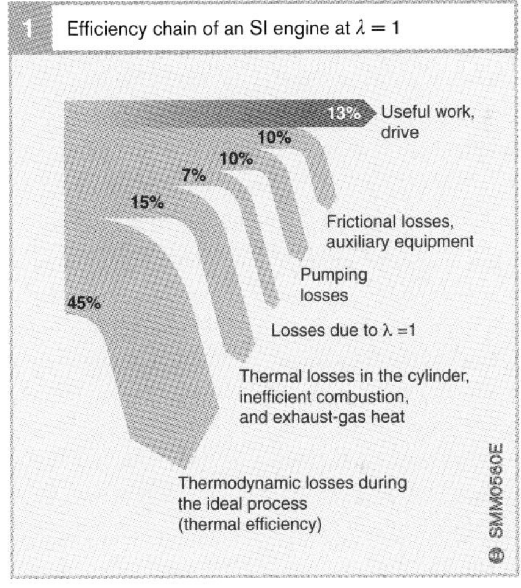

2 Sequence of the motive working process in the p-V diagram

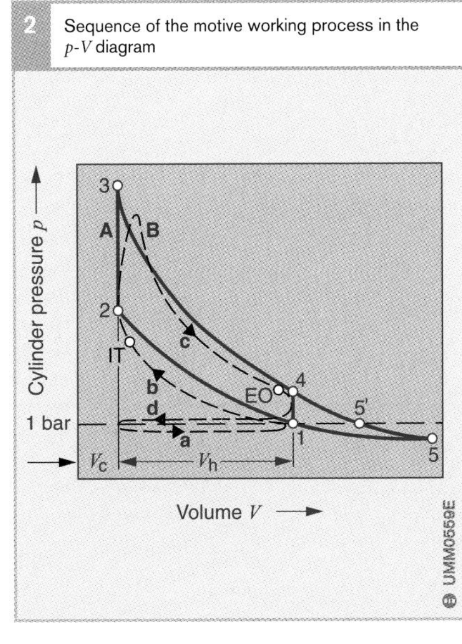

Fig. 2
A Ideal constant-volume cycle
B Real p-V diagram
a Induction
b Compression
c Work (combustion)
d Exhaust
IT Ignition point
EO Exhaust valve opens

Specific fuel consumption

Specific fuel consumption b_e is defined as the mass of the fuel (in grams) that the internal-combustion engine requires to perform a specified amount of work (kW · h, kilowatt hours). This parameter thus provides a more accurate measure of the energy extracted from each unit of fuel than the terms liters per hour, litres per 100 kilometers or miles per gallon.

Effects of excess-air factor

Homogeneous mixture distribution

When engines operate on homogeneous induction mixtures, specific fuel consumption initially responds to increases in excess-air factor λ by falling (Figure 1). The progressive reductions in the range extending to $\lambda = 1.0$ are explained by the incomplete combustion that results when a rich air/fuel mixture burns with inadequate air.

The throttle plate must be opened to wider apertures to obtain a given torque during operation in the lean range ($\lambda > 1$). The resulting reduction in throttling losses combines with enhanced thermodynamic efficiency to furnish lower rates of specific fuel consumption.

1 Effects of excess-air factor λ and ignition timing α_z on fuel consumption during operation with homogeneous mixture distribution

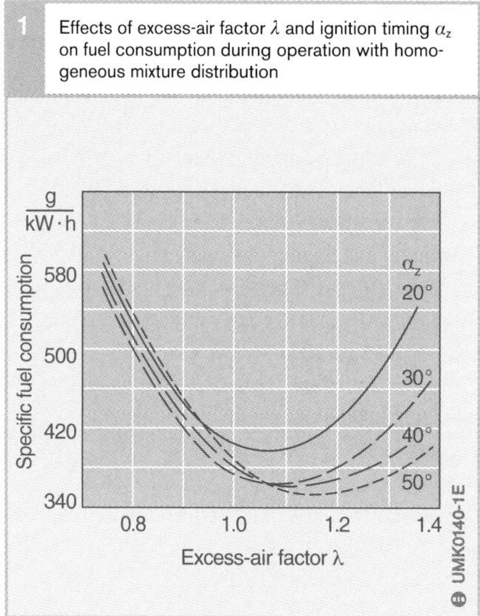

As the excess-air factor is increased, the flame front's propagation rate falls in the resulting, progressively leaner mixtures. The ignition timing must be further advanced to compensate for the resulting lag in ignition of the combustion mixture.

As the excess-air factor continues to rise, the engine approaches the lean-burn limit, where incomplete combustion takes place (combustion miss). This results in a radical increase in fuel consumption. The excess-air factor that coincides with the lean-burn limit varies according to engine design.

Stratified-charge concept

Engines featuring direct gasoline injection can operate with high excess-air factors in their stratified-charge mode. The only fuel in the combustion chamber is found in the stratification layer immediately adjacent to the tip of the spark plug. The excess-air factor within this layer is approximately $\lambda = 1$.

The remainder of the combustion chamber is filled with air and inert gases (exhaust-gas recirculation). The large throttle-plate apertures available in this mode lead to a reduction in pumping losses. This combines with the thermodynamic benefits to provide a substantial reduction in specific fuel consumption.

Effects of ignition timing

Homogeneous mixture distribution

Each point in the cycle corresponds to an optimal phase in the combustion process with its own defined ignition timing (Figure 1). Any deviation from this ignition timing will have negative effects on specific fuel consumption.

Stratified-charge concept

The range of possibilities for varying the ignition angle is limited on direct-injection gasoline engines operating in the stratified-charge mode. Because the ignition spark must be triggered as soon as the mixture cloud reaches the spark plug, the ideal ignition point is largely determined by injection timing.

2 Fuel-consumption map for gasoline engine with homogeneous induction mixture

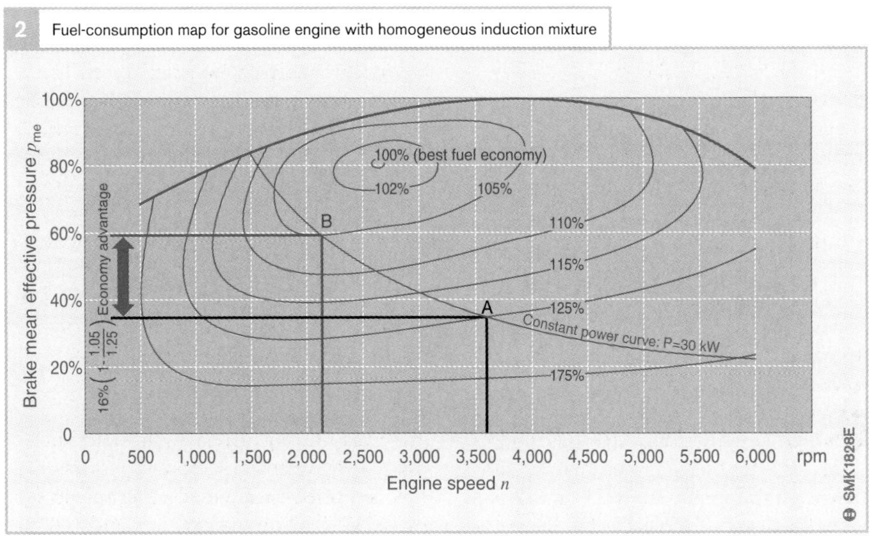

Fig. 2
Engine data:
4-cylinder gasoline engine
Displacement:
$V_H = 2.3$ litres
Power:
$P = 110$ kW at 5,400 rpm
Torque peak:
$M = 220$ N·m at 3,700...4,500 rpm
Brake mean effective pressure:
$p_{me} = 12$ bar (100%)

Calculating torque M and power P with numerical value equations:
$M = V_H \cdot p_{me}/0.12566$
$P = M \cdot n/9{,}549$

M in N·m
V_H in dm³
p_{me} in bar
n in rpm
P in kW

Achieving ideal fuel consumption

During operation on homogeneous induction mixtures, gasoline engines must operate on a stoichiometric air/fuel ratio of $\lambda = 1$ to create an optimal operating environment for the 3-way catalytic converter. Under these conditions using the excess-air factor to manipulate specific fuel consumption is not an option. Instead, the only available recourse is to vary the ignition timing. Defining ignition timing always equates with finding the best compromise between maximum fuel economy and minimal levels of raw exhaust emissions. Because the catalytic converter's treatment of toxic emissions is very effective once it is hot, the aspects related to fuel economy are the primary considerations once the engine has warmed to normal operating temperature.

Fuel-consumption map

Testing on an engine dynamometer can be used to determine specific fuel consumption in its relation to brake mean effective pressure and to engine speed. The monitored data are then entered in the fuel consumption chart (Figure 2). The points representing levels of specific fuel consumption are joined to form curves. Because the resulting graphic portrayal resembles a sea shell, the lines are also known as shell or conchoid curves.

As the diagram indicates, the point of minimum specific fuel consumption coincides with a high level of brake mean effective pressure p_{me} at an engine speed of roughly 2,600 rpm.

Because the brake mean effective pressure also serves as an index of torque generation M, curves representing power output P can also be entered in the chart. Each curve assumes the form of a hyperbola. Although the chart indicates identical power at different engine speeds and torques (operating points A and B), the specific fuel consumption rates at these operating points are not the same. At Point B the engine speed is lower and the torque is higher than at Point A. Engine operation can be shifted toward Point A by using the transmission to select a gear with a higher conversion ratio.

Fuels for spark-ignition engines (gasoline)

Overview

The most important energy source for use in production of automotive fuels is petroleum. Petroleum results from the decomposition of plants and animals over millions of years.

While petroleum contains four hydrocarbon compounds, not all are suitable for use in producing gasoline. This is why raw petroleum must be processed at the refinery. Two processing options are available:

- Distillation and filtration can be used to separate the individual components.
- Conversion processes (cracking, reforming, etc.) can be used to produce new hydrocarbon compounds from the existing components.

Figure 1 illustrates the structures of the most important hydrocarbon molecules. Linear molecule chains offer extremely good ignitability but only poor resistance to knock. Structures featuring molecule chains with side chains and molecular rings provide fuel components with better knock resistance.

In addition to the mineral fuels distilled from petroleum, other fuels may be used in specific individual applications. These include

- Alcohol (methanol, ethanol)
- Liquid petroleum gas and
- Natural gas

Four-stroke spark-ignition engines designed to operate on hydrogen are currently being tested.

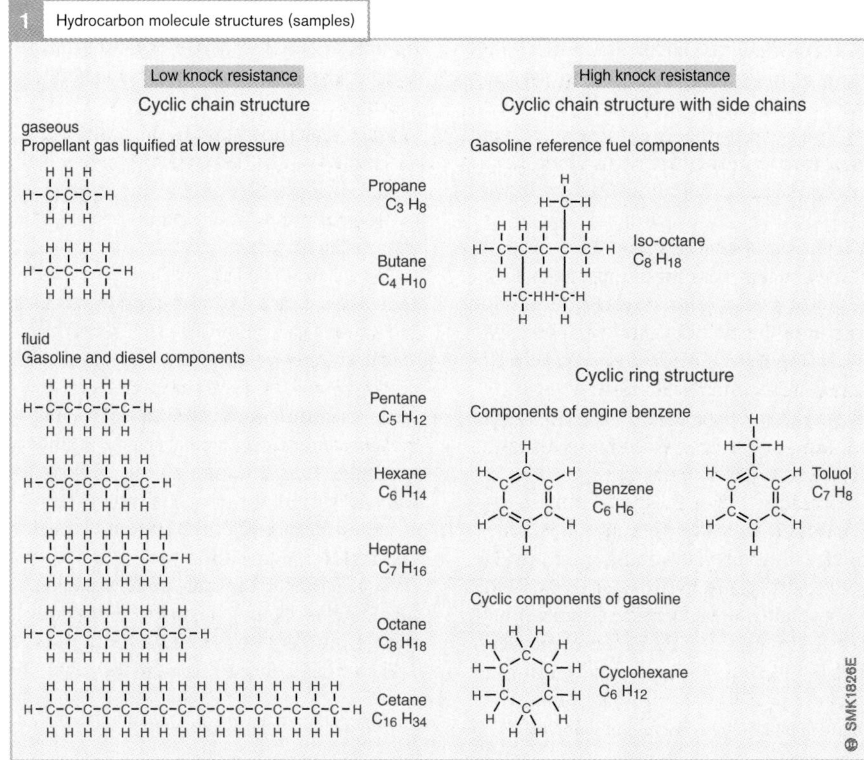

1 Hydrocarbon molecule structures (samples)

Fuel standards

The minimum performance requirements for gasolines are defined in various national standards. European Norm EN 228 defines the unleaded fuel marketed within Europe (Euro-Super). DIN EN 228 contains the corresponding German specifications along with supplementary descriptions of regular and super premium unleaded fuels.

The US specifications defining fuels for spark-ignition engines are contained in ASTM D4814 (American Society for Testing and Materials).

Components of gasoline

Fuels for spark-ignition engines are hydrocarbon compounds which may contain supplements in the form of oxygenous organic components or other additives to improve performance.

A distinction is made between regular and premium (super-grade) gasoline. Premium gasoline displays enhanced resistance to pre-ignition and detonation, and is formulated for use in high-compression engines. Volatility ratings vary according to region and whether the fuel is intended for summer or winter use.

Unleaded gasoline (EN 228)

Unleaded fuel is indispensable in vehicles that rely on catalytic converters for emissions control, as lead would damage the noble metals (e.g. platinum) in the catalytic converter and render it inoperative. It would also destroy the Lambda sensors that monitor exhaust-gas composition to support closed-loop mixture control.

Unleaded fuels are a special mixture of high-grade, high-octane components (e.g. platformates, alkylates and isomerisates). Non-metallic additives can be blended with the fuel to enhance knock resistance. Suitable additives include ether (methyltertiary-butylether, or MTBE) in concentrations of 3...15 % as well as alcohol-based mixtures (2...3 % methanol and higher alcohols).

Since the year 2000, maximum lead concentrations in gasoline have been officially limited to 5 mg/l.

Leaded gasoline

Gasolines containing lead were banned from the European market in 2000, with waivers being granted exclusively to bridge transition periods. While leaded fuel remains available in various individual countries throughout the world, its sales continue to decrease.

Fuels once relied on lead-alkyl compounds to discharge various functions, including lubrication of the exhaust valves. Engines manufactured since the 1980s no longer require fuels containing lead additives.

Characteristics

Net and gross calorific values

The specific values for the net (formerly: low) and gross (formerly: high, or combustion) calorific values, or H_u and H_o respectively, provide an index for the energy content of fuels. Only the net calorific value H_u (combustion vapor) is significant in dealing with fuels that produce water as one of their combustion products.

Oxygenates, fuel constituents which contain oxygen, such as alcohols, ether, and fatty-acid methyl ester have a lower calorific value than pure hydrocarbons, because their bound oxygen does not contribute to the combustion process. Power comparable to that available from conventional fuels can thus be attained only at the cost of higher fuel consumption.

Density

European standard EN 228 limits the approved fuel density range to 720...775 kg/m^3. Because premium fuels generally include a higher proportion of aromatic compounds, they are denser than regular gasoline, and also have a slightly higher calorific value.

Knock resistance

The octane rating defines the gasoline's anti-knock quality (resistance to pre-ignition). The higher the octane rating, the greater the resistance to engine knock. Two differing procedures are in international use for determining the octane rating; these are the Research Method and the Motor Method.

RON, MON

The number determined in testing using the Research Method is the Research Octane Number, or RON. It serves as the essential index of acceleration knock.

The Motor Octane Number, or MON, is derived from testing according to the Motor Method. The MON basically provides an indication of the tendency to knock at high speeds.

The Motor Method differs from the Research Method by using preheated mixtures, higher engine speeds and variable ignition timing, thereby placing more stringent thermal demands on the fuel under examination. MON figures are lower than those for RON.

Octane numbers up to 100 indicate the volumetric content in percent of C_8H_{18} iso-octane (trimethyl pentane) contained in a mixture with C_7H_{16} n-heptane at the point where the mixture's knock-resistance in a test engine is identical to that of the fuel being tested. Iso-octane, which is extremely knock-resistant, is assigned the octane number 100 (100 RON and MON), while n-heptane, with low-resistance to pre-ignition, is assigned the number 0.

Table 1

1 Essential properties of gasolines and fuels, EN 228 (valid since 1.1.2000)

Requirements	Unit	Parameter
Knock resistance		
Premium, min.	RON/MON	95/85
Regular, min. 1)	RON/MON	91/82.5
Super Plus 1)	RON/MON	98/88
Density	kg/m³	720...775
Sulphur, max.	mg/kg	150
Benzol, max.	Vol.-%	1
Lead, max.	mg/l	5
Volatility,		
Summer vapor pressure, min./max.	kPa	45/60
Winter vapor pressure, min./max.	kPa	60/90 1)
Min./max. summer evaporation % vol. at 70°C	%	20/48
Min./max. winter evaporation % vol. at 70°C	%	22/50
Min./max. evaporation in % volume at 100°C	46/71	
Min./max. evaporation in % volume at 150°C	75/	
Max. final boiling point	°C	210
VLI transition time 3), max. 2)		1,150 1)

1) National specifications for Germany

2) VLI = Vapor Lock Index

3) Spring and autumn

Enhancing knock resistance
Normal (untreated) straight-run gasoline has only limited resistance to knock. Various refinery components must be added to obtain a fuel with an adequate octane rating. It is also important to maintain the highest possible octane level throughout the entire boiling range.

Cyclic hydrocarbons (aromatics) and branched chains (iso-paraffins) provide greater knock resistance than straight-chain molecules (n-paraffins). While additives based on oxygenous components (methanol, ethanol, methyl tertiary butyl ether) have a positive effect on the octane number, they also raise volatility levels, and can cause problems by interacting with materials in fuel systems.

Volatility
Gasolines must satisfy stringent volatility requirements to ensure satisfactory operation. They must simultaneously contain an adequate proportion of volatile components to ensure reliable cold starting. At the same time volatility should not be so high as to lead to starting and performance problems during operation in high-temperature environments ("vapor lock"). Still another factor is environmental protection, which demands that evaporative losses be held low.

Various parameters are available for use in defining volatility. EN 228 defines 10 different volatility classes distinguished by various levels of vapor pressure, boiling-point curve and VLI (Vapor Lock Index). To meet special requirements stemming from variations in climatic conditions, countries can incorporate specific individual classes into their own national standards.

Boiling curve
Certain ranges on the boiling curve have a particularly pronounced effect on performance. These can be defined according to volumetric fuel evaporation at three specific temperatures. The volume that evaporates up to 70 °C must be large enough to promote good cold-start response, but not so large as to foster formation of vapor bubbles when the engine is hot. While the percentage of the fuel that vaporizes up to 100 °C determines the engine's warm-up characteristics, this factor's most pronounced effects are reflected in the acceleration and response provided by the power plant once it warms to normal operating temperature. The vaporized volume up to 150 °C should be high enough to minimize dilution of the engine's lubricating oil, especially with the engine cold.

Vapor pressure
Fuel vapor pressure as measured at 38 °C/ 100 °F in accordance with EN 13016-1 is primarily an index of the safety with which the fuel can be pumped into and out of the vehicle's tank. The Reid method is an alternative method of measuring fuel vapor pressure. All specifications place limits on this vapor pressure. Germany, for example, prescribes maxima of 60 kPa in summer and 90 kPa in winter.

On fuel-injection engines, when dealing with vapor-bubble problems it is more important to know the vapor pressure at higher temperatures (80...100 °C) is the most significant element in analysing drivability problems from vapor bubbles encountered in fuel-injected engines. Research procedures are in place, and a definitive standard is currently being prepared. Fuels to which methanol is added are characterized by a pronounced rise in vapor pressure at high temperatures.

Vapor/liquid ratio
This specification provides an index of a fuel's tendency to form vapor bubbles. It is based on the volume of vapor generated by a specific quantity of fuel at a defined temperature.

A drop in pressure (e.g. when driving over a mountain pass) accompanied by an increase in temperature will raise the vapor/ liquid ratio and with it the probability of operating problems. ASTM D4814 specifies vapor-liquid ratios for gasolines.

Vapor Lock Index (VLI)
This parameter is the sum of the Reid vapor pressure (in kPa × 10), and that proportion of the fuel that vaporizes up to a temperature of 70 °C. Both components in the equation are absolute data, the latter being derived from the boiling-point curve prior to multiplication by a factor of 7. The VLI provides more useful information on the fuel's influence on starting and operating a hot engine than that supplied by conventional data.

An accurate correlation exists between the VLI and the vapor/liquid ratio in fuels without alcohol-based additives.

Sulphur content
The NO_X accumulator-type catalytic converter is responsible for post-combustion treatment of emissions on gasoline direct-injection vehicles. Sulphur in the fuel forms sulphate deposits within these devices, chemically paralyzing the catalytic layer. While accumulations of sulphurous residue can be dispersed by heating the catalytic converter, the heating process has a negative effect on fuel economy.

For this reason, but also in order to comply with the scheduled reductions in SO_2 emissions, it is necessary to even further reduce the sulphur content of future fuels. Sulphur levels must fall to below 10 ppm.

Additives
Along with the structure of the hydrocarbons (refinery components), it is the additives which determine the ultimate quality of a fuel. The packages generally used combine individual components with various attributes.

Extreme care and precision are required both when testing additives and in determining their optimal concentrations. Undesirable side-effects must be avoided. This is why the fuel producer assumes responsibility for defining additive quantities and mixing the selected substances into the fuel. Each individual brand of fuel receives a specific additive package as the tanker is filled at the refinery. Vehicle operators should refrain from adding supplementary additives on their own.

Anti-ageing additives
These agents are added to fuels to improve their stability during storage, and are particularly important in fuels that also contain cracked components. They inhibit the oxidation that is otherwise promoted by reactions with atmospheric oxygen and prevent catalytic reactions with metal ions (metal deactivators).

Preventing contamination in the intake system
The entire intake system (injectors, intake valves) should remain free of contamination and deposits for several reasons. A clean intake tract is essential for maintaining the factory-defined air/fuel ratios, as well as for trouble-free operation and minimal exhaust emissions. To achieve this end, effective detergent agents should be added to the fuel.

Corrosion protection
Moisture transported in the fuel leads to corrosion in fuel-system components. An extremely effective remedy is afforded by anti-corrosion additives designed to form a protective layer below any moisture layers.

Environmentally compatible gasolines
Environmental authorities and legislative bodies are imposing increasingly stringent regulations on fuels to ensure low evaporation and pollutant emissions (ecologically sound fuels, reformulated gasoline). As defined in the regulations, these fuels' salient characteristics include reduced vapor pressure along with lower levels of aromatic components, benzene and sulphur, and special specifications for the final boiling point. Additives designed to prevent deposit formation within the intake tract are also mandatory in the US.

Alternate fuels for spark-ignition engines

Vehicles must be specifically adapted for operation with alternate fuels. Any vehicle powered by an internal-combustion engine can theoretically be converted for operation on liquified petroleum gas. The most common procedure is to equip the vehicles for dual use, allowing selection between gasoline and LPG.

Liquified petroleum gas

LPG, or **L**iquid **P**etroleum **G**as, consists primarily of butane and propane, and is in limited use as a fuel for motor vehicles. It is a by-product of the petroleum refining process and can be liquified under pressure (at 2...20 bar, depending upon the relative proportions of propane and butane, and the temperature). LPG is characterized by a high octane number (RON >100).

The high quality of the mixtures available using LPG and air limits toxic emissions (including CO_2 as well as other components such as polycyclic hydrocarbons) to substantially lower levels than those produced with gasoline . LPG is also free of lead and sulphur compounds.

Natural gas

Natural gas is more plentiful than petroleum, and reserves are also less intensively exploited. As a result, natural gas poses an interesting alternative as a fuel for automotive applications. The design of systems for operation on compressed natural gas is virtually identical to the configuration employed for liquified petroleum gas.

The primary component of natural gas is methane (CH_4), which is present in proportions of 80...99 %. The remainder consists of inert gases such as carbon dioxide, nitrogen and other low-grade hydrocarbons.

One option for transporting natural gas in automotive applications is as CNG (**C**ompressed **N**atural **G**as). Compressed to 160...200 bar, CNG provides only a relatively limited cruising range. The alternative is LNG (**L**iquid **N**atural **G**as), which is carried at –162 °C in a special tank which is resistant to cold. Owing to the not inconsiderable expense of storing the gas in liquid form, natural gas is used in compressed form in virtually all applications.

Natural-gas vehicles are characterized by low CO_2 emissions. This is due to the favorable hydrogen/carbon ratio of almost 4:1 (gasoline: 2.3:1), and the resulting shift in the primary combustion products CO_2 and H_2O. Apart from virtually particle-free combustion, they also provide practically regligible levels of NO_X, CO and NMHC ("non-methane hydrocarbons": the sum of all hydrocarbons minus methane) emissions when equipped with 3-way catalytic converters. Methane is classified as non-toxic, and is therefore not considered to be a pollutant.

Alcohol fuels

Among the alternate fuels available for use in spark-ignition engines are alcohols (primarily methanol and ethanol) and their derivatives (such as ether). Methanol can be manufactured from readily available natural hydrocarbons found in plentiful substances such as coal, natural gas, heavy oils, etc. In certain countries (such as Brazil, but also in the USA), biomass (sugar cane, wheat) is distilled to produce ethanol for use as an engine fuel and fuel additive.

In the absence of comprehensive methanol-distribution networks to ensure universal availability, engines and engine-management systems must be designed for flexible dual-fuel operation (ranging from pure gasoline to max. 85 % methanol).

Lambda oxygen sensors permit optimal conversion of emissions within the catalytic converter.

Vehicles must be specially adapted to accommodate differences in calorific value and in various other factors relative to conventional fuels.

Gasoline-engine management

In modern-day vehicles, closed and open-loop electronic control systems are becoming more and more important. Slowly but surely, they have superseded the purely mechanical systems (for instance, the ignition system). Without electronics it would be impossible to comply with the increasingly severe emissions-control legislation.

Technical requirements

One of the major objectives in the development of the automotive engine is to generate as high a power output as possible, while at the same time keeping fuel consumption and exhaust emissions down to a minimum in order to comply with the legal requirements of emissions-control legislation.

Fuel consumption can only be reduced by improving the engine's efficiency. Particularly in the idle and part-load ranges, in which the engine operates the majority of the time, the conventional manifold-injection SI engine is very inefficient. This is the reason for it being so necessary to improve the engine's efficiency at idle and part load without at the same time having a detrimental effect upon the normal engine's favorable efficiency in the upper load ranges. Gasoline direct injection is the solution to this problem.

A further demand made on the engine is that it develops high torque even at very low rotational speeds so that the driver has good acceleration at his disposal. This makes torque the most important quantity in the management of the SI engine.

SI-engine torque

The power *P* delivered by an SI engine is defined by the available clutch torque *M* and the engine rpm *n*. The clutch torque is the torque developed by the combustion process less friction torque (frictional torque in the engine), pumping losses, and the torque needed to drive the auxiliary equipment (Fig. 1).

1 Torque at the drivetrain

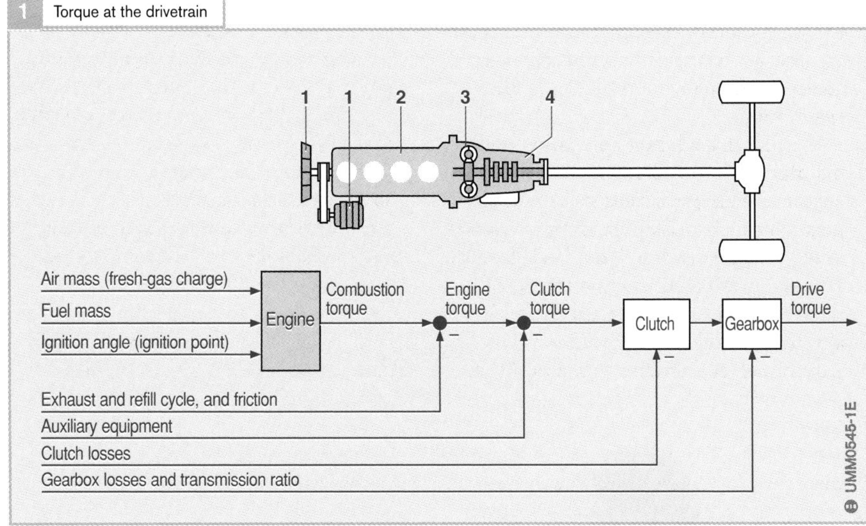

Fig. 1
1 Auxiliary equipment (alternator, A/C compressor etc.)
2 Engine
3 Clutch
4 Gearbox

The combustion torque is generated during the power stroke. In manifold-injection engines, which represent the majority of today's engines, it is determined by the following quantities:

- The air mass which is available for combustion when the intake valves close
- The fuel mass which is available at the same moment and
- The moment in time when the ignition spark initiates the combustion of the A/F mixture

The proportion of direct-injection SI engines will increase in the future. These engines run with excess air at certain operating points (lean-burn operation) which means that there is air in the cylinder which has no effect upon the generated torque. Here, it is the fuel mass which has the most effect.

Engine-management assignments

One of the engine management's jobs is to set the torque that is to be generated by the engine. To do so, in the various subsystems (ETC, A/F-mixture formation, ignition) all quantities that influence torque are controlled. It is the objective of this form of control to provide the torque demanded by the driver while at the same time complying with the severe demands regarding exhaust emissions, fuel consumption, power output, comfort and safety. It is impossible to satisfy all these requirements without the use of electronics.

In order that all these stipulations are maintained in long-term operation, the engine management continuously runs through a diagnosis program and indicates to the driver when a fault has been detected. This is one of the most important assignments of the engine management, and it also makes a valuable contribution to simplifying vehicle servicing in the workshop.

Subsystem: Cylinder-charge control

On conventional injection systems, the driver directly controls the throttle-valve opening through the accelerator pedal. In doing so, he/she defines the amount of fresh air drawn in by the engine.

Basically speaking, on engine-management systems with electronic accelerator pedal for cylinder-charge control (also known as EGAS or ETC/Electronic Throttle Control), the driver inputs a torque requirement through the position of the accelerator pedal, for instance when he/she wants to accelerate. Here, the accelerator-pedal sensor measures the pedal's setting, and the "ETC" subsystem uses the sensor signal to define the correct cylinder air charge corresponding to the driver's torque input, and opens the electronically controlled throttle valve accordingly.

Subsystem: A/F-mixture formation

During homogeneous operation and at a defined A/F ratio λ, the appropriate fuel mass for the air charge is calculated by the A/F-mixture subsystem, and from it the appropriate duration of injection and the best injection point. During lean-burn operation, and essentially stratified-charge operation can be classified as such, other conditions apply in the case of gasoline direct injection. Here, the torque-requirement input from the driver determines the injected fuel quantity, and not the air mass drawn in by the engine.

Subsystem: Ignition

The crankshaft angle at which the ignition spark is to ignite the A/F mixture is calculated in the "ignition" subsystem.

Cylinder-charge control

It is the job of the cylinder-charge control to coordinate all the systems that influence the proportion of gas in the cylinder.

Components of the cylinder charge

The gas mixture trapped in the combustion chamber when the intake valve closes is referred to as the cylinder charge. This is comprised of the fresh gas and the residual gas.

The term "relative air charge *rl*" has been introduced in order to have a quantity which is independent of the engine's displacement. It is defined as the ratio of the actual air charge to the air charge under standard conditions (p_0 = 1,013 hPa, T_0 = 273 K).

Fresh gas

The freshly introduced gas mixture in the cylinder is comprised of the fresh air drawn in and the fuel entrained with it (Fig. 1). On a manifold-injection engine, all the fuel has already been mixed with the fresh air upstream of the intake valve. On direct-injection systems, on the other hand, the fuel is injected directly into the combustion chamber.

1 Cylinder charge in the gasoline engine

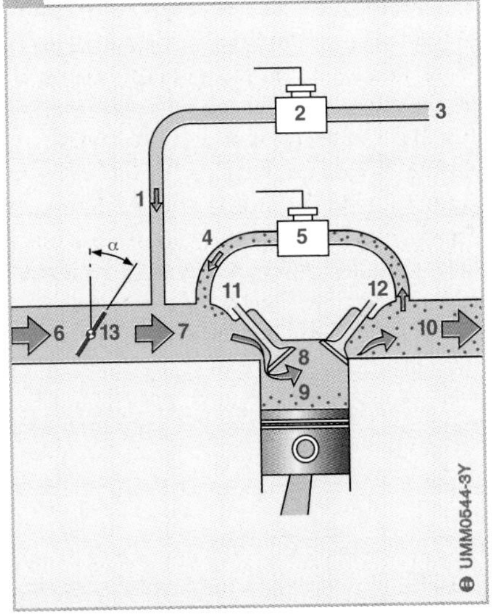

Fig. 1
1 Air and fuel vapor (from the evaporative-emissions control system)
2 Canister-purge valve with variable valve-opening cross-section
3 Connection to the evaporative-emissions control system
4 Returned exhaust gas
5 EGR valve with variable valve-opening cross-section
6 Air-mass flow (ambient pressure p_U)
7 Air-mass flow (manifold pressure p_S)
8 Fresh A/F-mixture charge (combustion-chamber pressure p_B)
9 Residual exhaust-gas charge (combustion-chamber pressure p_B)
10 Exhaust gas (exhaust-gas back pressure p_A)
11 Intake valve
12 Exhaust valve
13 Throttle valve

α Throttle valve-angle

The majority of the fresh air enters the cylinder with the air-mass flow (6, 7) via the throttle valve (13) and the intake valve (11). Additional fresh gas, comprising fresh air and fuel vapor, can be directed to the cylinder via the evaporative-emissions control system (3).

For homogeneous operation at $\lambda \leq 1$, the air in the cylinder after the intake valve (11) has closed is the decisive quantity for the work at the piston during the combustion stroke and therefore for the engine's output torque. In this case, the air charge corresponds to the torque and the engine load. During lean-burn operation (stratified charge) though, the torque (engine load) is a direct product of the injected fuel mass.

During lean-burn operation, the air mass can differ for the same torque. Almost always, measures aimed at increasing the engine's maximum torque and maximum output power necessitate an increase in the maximum possible charge. The theoretical maximum charge is defined by the displacement.

Residual gas

The residual-gas share of the cylinder charge comprises that portion of the cylinder charge which has already taken part in the combustion process. In principle, one differentiates between internal and external residual gas. The internal residual gas is that gas which remains in the cylinder's upper clearance volume following combustion, or that gas which is drawn out of the exhaust passage and back into the intake manifold when the intake and exhaust valves open together (that is, during valve overlap). External residual gas are the exhaust gases which enter the intake manifold through the EGR valve.

Residual exhaust gas comprises inert gas[1]) and, during excess-air operation, unburnt air. The inert gas in the residual exhaust gas does not participate in the combustion during the next power stroke, although it does have an influence on ignition and on the combustion curve.

The selective use of a given share of residual gas can reduce the NOx emissions.

In order to achieve the demanded torque, the fresh-gas charge displaced by the inert gas must be compensated for by a larger throttle-valve opening. This leads to a reduction in pumping losses which in turn results in a reduction in fuel consumption.

Controlling the fresh-gas charge

Manifold injection

The torque developed by a manifold-injection engine is proportional to the fresh-gas charge. The engine's torque is controlled via the throttle valve which regulates the flow of air drawn in by the engine. With the throttle valve less than fully open, the flow of air drawn in by the engine is throttled and the torque drops as a result. This throttling effect is a function of the throttle valve's setting, in other words its opened cross-section. Maximum torque is developed with the throttle wide open (Wide Open Throttle = WOT).

Fig. 2 shows the principal correlation between fresh-gas charge and engine speed as a function of throttle-valve opening.

2 Throttle characteristic-curve map for an SI engine
- - - Intermediate throttle-valve settings

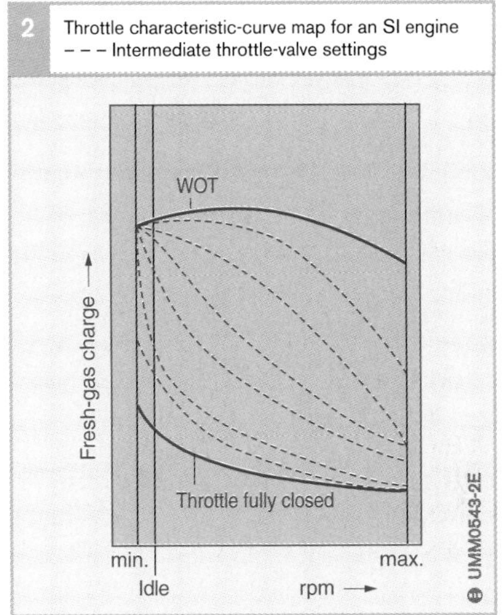

Direct injection

On direct-injection (DI) gasoline engines during homogeneous operation at $\lambda \leq 1$ (that is, not lean-burn operation), the same conditions apply as with manifold injection.

To reduce the throttling losses, the throttle valve is also opened wide in the part-load range. In the ideal case, there are no throttling losses with the throttle wide open (as it is during full-load operation). In order to limit the torque developed at part load, not all of the air mass entering the cylinder may participate in combustion. In lean-burn applications with excess air ($\lambda > 1$), some of the air drawn in remains as residual exhaust gas in the cylinder or is forced out during the exhaust stroke. In other words, it is not the air charge trapped in the cylinder which is decisive for the developed torque, but rather the fuel injected into the combustion chamber.

[1]) Components in the combustion chamber which behave inertly, that is, do not participate in the combustion process.

Exhaust and refill cycle

The replacement of the used/burnt cylinder charge (= exhaust gas) by a fresh-gas charge takes place using intake and exhaust valves which are opened and closed at precisely defined times by the cams on the camshaft (valve timing). These cams also define the valve-lift characteristic which influences the exhaust and refill cycle and with it the fresh-gas charge which is available for combustion.

Valve overlap, that is, the overlap of the opened times of the intake and exhaust valves, has a decisive influence on the exhaust-gas mass remaining in the cylinder. This exhaust-gas mass also defines the amount of inert gas in the fresh cylinder charge for the next power cycle. In such cases, one refers to "internal" EGR.

The inert-gas mass in the cylinder charge can be increased by "external" EGR. Exhaust pipe and intake manifold are connected by an EGR valve so that the percentage of inert gas in the cylinder charge can be varied as a function of the operating point.

Volumetric efficiency

For the air throughput, the total charge during a complete working cycle is referred to the theoretical charge as defined by the piston displacement. For the volumetric efficiency though, only the exhaust gas actually remaining in the cylinder is considered. Fresh gas drawn in during valve overlap, which is not available for the combustion process, is not considered.

The volumetric efficiency for naturally aspirated engines is 0.6...0.9. It depends upon the combustion-chamber shape, the opened cross-sections of the gas-exchange valves, and the valve timing.

Supercharging

The torque which can be achieved during homogenous operation at $\lambda \leq 1$ is proportional to the fresh gas charge. This means that maximum torque can be increased by compressing the air before it enters the cylinder (supercharging). This leads to an increase in volumetric efficiency to values above 1.

Dynamic supercharging

Supercharging can be achieved simply by taking advantage of the dynamic effects inside the intake manifold. The supercharging level depends on the intake manifold's design and on its operating point (for the most part, on engine speed, but also on cylinder charge). The possibility of changing the intake-manifold geometry while the engine is running (variable intake-manifold geometry) means that dynamic supercharging can be applied across a wide operating range to increase the maximum cylinder charge.

Mechanical supercharging

The intake-air density can be further increased by compressors which are driven mechanically from the engine's crankshaft. The compressed air is forced through the intake manifold and into the engine's cylinders.

Exhaust-gas turbocharging

In contrast to the mechanical supercharger, the exhaust-gas turbocharger is driven by an exhaust-gas turbine located in the exhaust-gas flow, and not by the engine's crankshaft. This enables recovery of some of the energy in the exhaust gas.

A/F-mixture formation

The A/F-mixture formation system is responsible for calculating the fuel mass appropriate to the amount of air drawn into the engine. This fuel is metered to the engine's cylinders through the fuel injectors.

A/F mixture

To run efficiently, the gasoline engine needs a given air/fuel (A/F) ratio. Ideal, theoretically complete combustion takes place at a mass ratio of 14.7:1, which is also referred to as the stoichiometric ratio. In other words, 14.7 kg of air are needed to burn 1 kg of fuel. Or, expressed in volumes, approx. 9,500 liters of air are needed to completely burn 1 liter of gasoline.

Excess-air factor λ

The excess-air factor λ has been chosen to indicate how far the actual A/F-mixture deviates from the theoretically ideal mass ratio (14.7:1). λ defines the ratio of the actually supplied air mass to the theoretical air mass required for complete (stoichiometric) combustion.

$\lambda = 1$: The inducted air mass corresponds to the theoretically required air mass.

$\lambda < 1$: This indicates air deficiency and therefore a rich A/F mixture. On a cold engine, it is necessary to enrich the A/F mixture by adding fuel to compensate for the fuel that has condensed on the cold manifold walls (manifold-injection engines) and cold cylinder walls and which, as a result, is not available for combustion.

$\lambda > 1$: This indicates excess air and therefore a lean A/F mixture. The maximum value for λ that can be achieved is defined by the so-called lean-misfire limit (LML), and is highly dependent upon the engine's design and construction, as well as upon the mixture-formation system used. At the lean-misfire limit the A/F mixture is no longer combustible, and this marks the point at which misfire starts. The engine begins to run very unevenly, fuel consumption increases dramatically, and power output drops.

Other combustion conditions prevail on direct-injection (DI) engines, and these are thus able to run with considerably higher λ figures.

Operating modes

Homogeneous ($\lambda \leq 1$): On manifold-injection engines, the A/F mixture in the manifold is drawn in past the open intake valve during the induction stroke. This leads to an essentially homogeneous mixture distribution in the combustion chamber.

This operating mode is also possible with DI gasoline engines, the fuel being injected into the combustion chamber during the induction stroke.

Homogeneous lean ($\lambda > 1$): The A/F mixture is distributed homogeneously in the combustion chamber with a defined level of excess air.

Stratified charge: This operating mode and those given below are only possible with direct-injection gasoline engines. Fuel is injected only shortly before the ignition point, and an A/F-mixture cloud forms in the vicinity of the spark plug.

Homogenous stratified charge: In addition to the stratified charge, there is a homogeneous lean A/F mixture throughout the complete combustion chamber. Dual injection is applied to achieve this form of A/F-mixture distribution.

Homogeneous anti-knock: Here, dual injection is also used to achieve an A/F-mixture distribution which to a great extent prevents combustion knock.

Stratified-charge/catalyst heating: Retarded (late) injection leads to the rapid warm-up of the catalytic converter.

1 Influence of the excess-air factor λ on the power P and on the specific fuel consumption b_e under conditions of homogeneous A/F-mixture distribution

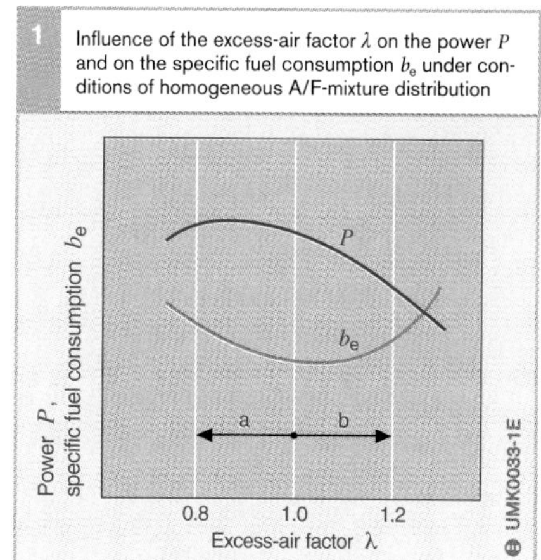

Fig. 1
a Rich A/F mixture (air deficiency)
b Lean A/F mixture (excess air)

2 Effect of the excess-air factor λ on the pollutant composition of untreated exhaust gas under conditions of homogeneous A/F-mixture distribution

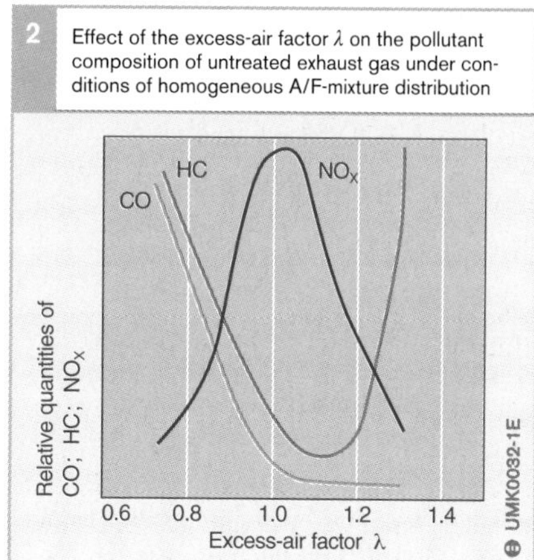

Specific fuel consumption, power and exhaust emissions

Manifold injection

Manifold-injection gasoline engines develop their maximum power output at 5...15% air deficiency (λ = 0.95...0.85), and their lowest fuel consumption at 10...20% excess air (λ = 1.1...1.2). Figs. 1 and 2 indicate the extent to which power output, fuel consumption, and exhaust emissions are all a function of the excess-air factor λ. It is immediately apparent that there is no excess-air factor at which all factors are at their "optimum". Best-possible fuel consumption together with best-possible power output are achieved with excess-air factors of λ = 0.9...1.1.

When a 3-way catalytic converter is used for the treatment of the exhaust gases, it is absolutely imperative that λ = 1 is maintained precisely when the engine has warmed-up. In order to comply with these requirements, the mass of the intake air must be measured exactly and a precisely metered fuel quantity injected.

An optimal combustion process though not only demands precision fuel injection, but also a homogeneous A/F mixture, which in turn necessitates efficient atomization of the fuel. If the fuel is not perfectly atomized, large fuel droplets are deposited on the walls of the manifold and/or combustion chamber. Since these fuel droplets cannot burn completely, they lead to increased HC emissions.

Gasoline direct injection

For gasoline direct injection, during homogeneous operation at $\lambda \leq 1$, the same conditions apply as with manifold injection. With stratified-charge operation though, a practically stoichiometric A/F mixture is only present in the stratified-charge mixture cloud near the spark plug. Outside this area, the combustion chamber is filled with fresh air and inert gas. Regarding the combustion chamber as a whole, the A/F mixture ratio is very high ($\lambda > 1$).

Since the complete combustion chamber is not filled with a combustible A/F mixture in this operating mode, torque output and power output both drop. Similar to manifold injection, maximum power can only be developed when the complete combustion chamber is filled with a homogeneous A/F mixture.

Depending upon the combustion process, and the A/F-mixture distribution in the combustion chamber, NO_x emissions are generated in the lean-burn mode which cannot be reduced by the 3-way catalytic converter. Here, for emissions control, it is necessary to take additional measures which call for a NO_x accumulator-type catalytic converter.

Engine operating modes

In some engine operating modes, the fuel requirement differs considerably from the steady-state requirements with the engine at operating temperature. This makes it necessary to take corrective measures in the A/F-mixture formation.

Start and warm-up

When starting with the engine cold, the inducted A/F-mixture leans-off. This is the due not only to inadequate mixing of the intake air with the fuel, but also to the fuel having less tendency to evaporate at low temperatures, and the pronounced wall wetting (condensation of the fuel) on the still-cold intake manifold (only on manifold-injection engines) and on the cylinder walls. To compensate for these negative effects, and to facilitate engine start, additional fuel must be provided during the cranking process.

Even after the engine has started, additional fuel must continue to be injected until it reaches operating temperature. This also applies to the gasoline direct-injection engine. Depending upon the engine's design and the combustion process, stratified-charge lean-burn operation is only possible with the engine at operating temperature.

Idle and part load

Once they have reached their operating temperature, conventional manifold-injection engines all run on a stoichiometric A/F mixture at idle and part load. On direct-injection gasoline engines though, the objective is to run the engine as often as possible with a stratified-charge. This is feasible at idle and at part load, the two operating modes with the highest potential for saving fuel, where fuel savings of as much as 40 % can be achieved with lean-burn operation.

Full load

Essentially, the conditions for manifold injection and gasoline direct injection are pretty much the same at full load. At WOT, it may be necessary to enrich the A/F mixture. As can be seen from Fig. 1, this permits the generation of maximum-possible torque and power.

Acceleration and deceleration

With manifold injection, the fuel's tendency to evaporate depends to a large extent upon the manifold pressure. This leads to the development of a fuel film (wall film) on the intake manifold in the vicinity of the intake valves. Rapid changes in manifold pressure, as occur when the throttle-valve opening changes suddenly, lead to changes in this wall film. Heavy acceleration causes the intake-manifold pressure to increase so that the fuel's evaporation tendency deteriorates, and the wall film thickens as a result. Being as a portion of the fuel has been deposited to form the wall film, the A/F mixture leans-off temporarily until the wall film has stabilized. Similarly, sudden deceleration leans to enrichment of the A/F mixture since the drop in manifold pressure causes a reduction in the wall film and the fuel from the wall film is drawn into the cylinder. A temperature-dependent correction function (transitional compensation) is used to correct the A/F mixture so as to ensure not only the best possible driveability, but also the constant A/F ratio as needed for the catalytic converter.

Wall-film effects are also encountered at the cylinder walls. With the engine at operating temperature though, they can be ignored on direct-injection gasoline engines.

Overrun

At overrun (trailing throttle), the fuel supply is interrupted (overrun fuel cutoff). Apart from saving fuel on downhill gradients, this protects the catalytic converter against overheating which could result from inefficient and incomplete combustion.

Ignition

It is the job of the ignition to ignite the compressed A/F-mixture at exactly the right moment in time and thus initiate its combustion.

Ignition system

In the gasoline (SI) engine, the A/F mixture is ignited by a spark between the electrodes of the spark plug. The inductive-type ignition systems used predominantly on gasoline engines store the electrical energy needed for the ignition spark in the ignition coil. This energy determines how long (dwell angle) the current must flow through the ignition coil to recharge it. The interruption of the coil current at a defined crankshaft angle (ignition angle) leads to the ignition spark and the A/F-mixture combustion.

In today's ignition systems, the processes behind the ignition of the A/F mixture are electronically controlled.

Ignition point

Changing the ignition point (ignition timing)

Following ignition, about 2 milliseconds are needed for the A/F mixture to burn completely. The ignition point must be selected so that main combustion, and the accompanying pressure peaks in the cylinder, takes place shortly after TDC. Along with increasing engine speed, therefore, the ignition angle must be shifted in the advance direction.

The cylinder charge (or fill) also has an effect upon the combustion curve. The lower the cylinder charge the slower is the flame front's propagation. For this reason, with a low cylinder charge, the ignition angle must also be advanced.

Influence of the ignition angle

The ignition angle has a decisive influence on engine operation. It determines

- The delivered torque
- The exhaust-gas emissions and
- The fuel consumption

The ignition angle is chosen so that all requirements are complied with as well as possible, whereby care must be taken that continued engine knock is avoided.

Ignition angle: Basic adaptation

On electronically controlled ignition systems, the ignition map (Fig. 1) takes into account the influence of engine speed and cylinder charge on the ignition angle. This map is stored in the engine-management data storage, and represents the basic adaptation of the ignition angle.

The x and y axes represent the engine speed and the relative air charge. The map's data points are formed by a given number of values, typically 16. A certain ignition angle is allocated to each pair of variates so that the map has 256 (16x16) adjustable ignition-angle values. By applying linear interpolation between two data points, the number of ignition-angle values is increased to 4,096.

Using the ignition-map principle for the electronic control of the ignition angle means that for every engine operating point it is possible to select the best-possible ignition angle. These ignition maps are generated by running the engine on the engine dynamometer.

1 Ignition map based on engine rpm n and relative air charge rl

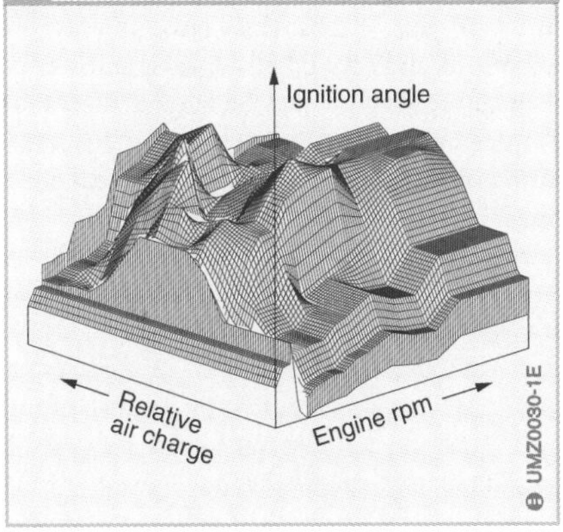

Additive ignition-angle adjustments

A lean A/F mixture is more difficult to ignite. This means that more time is needed before the main combustion point is reached. A lean A/F mixture must therefore be ignited sooner. The A/F ratio λ thus has an influence on the ignition angle.

The coolant temperature is a further variable which affects the choice of the ignition angle. Temperature-dependent ignition-angle corrections are therefore also necessary. Such corrections are stored in the data storage in the form of fixed values or characteristic curves (e.g. temperature-dependent correction). They shift the basic ignition angle by the stipulated amount in either the advance or retard direction.

Special ignition angle

There are certain operating modes, such as idle and overrun, which demand an ignition angle which deviates from those defined by the ignition map. In such cases, access is made to special ignition-angle curves stored in the data storage.

Knock control

Knock is a phenomenon which occurs when ignition takes place too early. Here, once regular combustion has started, the rapid pressure increase in the combustion chamber leads to the auto-ignition of the unburnt residual mixture which has not been reached by the flame front. The resulting abrupt combustion of the residual mixture leads to a considerable local pressure increase. This generates a pressure wave which propagates through the combustion chamber until it hits the cylinder wall. At low engine speeds and when the engine is not making too much noise, it is then audible as combustion knock. At high speeds, the engine noises blanket the combustion knock.

If knock continues over a longer period of time, the engine can be damaged by the pressure waves and the excessive thermal loading. To prevent knock on today's high-compression engines, no matter whether of the manifold-injection or direct-injection type, knock control belongs to the standard scope of the engine-management system. With this system, knock sensors detect the start of knock and the ignition angle is retarded at the cylinder concerned. To obtain the best-possible engine efficiency, therefore, the basic adaptation of the ignition angle (ignition map) can be located directly at the knock limit.

On direct-injection gasoline engines, combustion knock only takes place in homogeneous operation. There is no tendency for the engine to knock in the stratified-charge mode since there is no combustible mixture in the stratified charge at the combustion chamber's peripheral zones.

Dwell angle

The energy stored in the ignition coil is a function of the length of time current flows through the coil (energisation time). In order not to thermally overload the coil, the time required to generate the required ignition energy in the coil must be rigidly adhered to. The dwell angle refers to the crankshaft and is therefore speed-dependent.

The ignition-coil current is a function of the battery voltage, and for this reason the battery voltage must be taken into account when calculating the dwell angle.

The dwell-angle values are stored in a map, the x and y axes of which represent rpm and battery voltage.

History of gasoline-engine management system development

The functional principle of the gasoline engine has essentially remained unchanged since its initial deployment in a self-propelled vehicle more than 100 years ago. Nonetheless, a modern gasoline engine has very little in common with its ancestors. Engine technology has been subject to constant development. As engines have become increasingly more powerful and, in the past three decades, their emissions have gradually diminished, they also have become more economical in terms of fuel consumption. A major contribution to this development is owed to the changeover from mechanical to electronic engine-management systems.

Overview

Table 1 provides an overview of the development history of the major Bosch-developed management systems for gasoline engines. Fig. 1 depicts ongoing development of electronic control units, keeping pace with advances in microelectronics.

Fig. 1 Examples of Electronic Control Units (ECU)

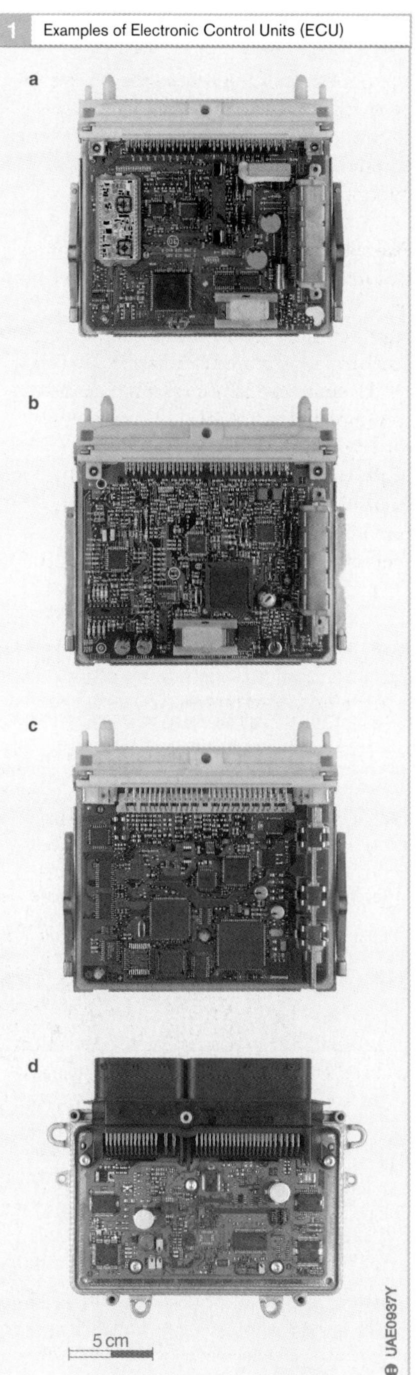

Fig. 1
- a Engine management with LH3.2-Jetronic fuel-injection system and
- b EZ129K electronic ignition system from the year 1992, each containing 8-bit microcontrollers
- c Motronic M4.4 from 1995, with 8-bit microcontroller. Integration of LH and electronic ignition system in a single electronic control unit plus additional functions (e.g. on-board diagnostics OBD)
- d Motronic ME9 from the year 2002, with 32-bit microcontroller. More compact size than M4.4 despite more functions (incl. electronic throttle control)

Table 1 Development of Bosch control systems for gasoline engines (examples)

Year	System
1965	Transistorized ignition (TI)
1967	D-Jetronic (electronic fuel-injection system, controlled by intake-manifold pressure)
1973	K-Jetronic (hydromechanical fuel-injection system)
1973	L-Jetronic (electronic fuel-injection system with air-flow sensing)
1979	M-Motronic (first engine-management system for fuel injection and ignition)
1981	LH-Jetronic (electronic fuel-injection system with air-flow sensing)
1982	KE-Jetronic (K-Jetronic with electronically controlled auxiliary functions)
1982	Electronic ignition system
1982	Knock control
1986	Electronic engine-performance control
1987	Mono-Jetronic (throttle-body injection)
1989	Mono-Motronic (throttle-body injection and ignition)
1994	ME-Motronic (Motronic with integrated electronic throttle control)
2000	MED-Motronic (Motronic for gasoline direct injection)

Mechanical systems

Up to the end of the 1960s the gasoline engine was managed by means of mechanical control devices. There were essentially two parameters that were adjustable, i.e. fuel quantity and ignition angle.

Fuel injection

Gasoline-injection systems initially failed to prevail over the more easily mastered carburetor principle. Controlled by depressing the accelerator pedal, the carburetor determines the fuel quantity to be vaporized, corresponding to the volume of air drawn into the engine. However, the mechanical control of the carburetor does not allow for the kind of precise fuel metering required today. This provided the needed breakthrough for the fuel-injection system.

K-Jetronic

This hydromechanical fuel-injection system continuously introduces fuel into the intake manifold via the fuel injectors (multipoint fuel injection) separately for each cylinder. The injected-fuel quantity is determined by the fuel pressure at the fuel injector.

The K-Jetronic meters fuel by purely mechanical means. The air-flow sensor measures the volumetric flow rate of air drawn into the engine. It also acts directly on the control plunger in the fuel distributor which influences fuel pressure and thus the metered fuel quantity.

Additional variables that impact on the injected-fuel quantity may also be considered (e.g. temperature).

The advantage of this mechanical fuel-injection system consisted in the cost advantage compared with electronic fuel-injection systems of the time.

2 K-Jetronic components

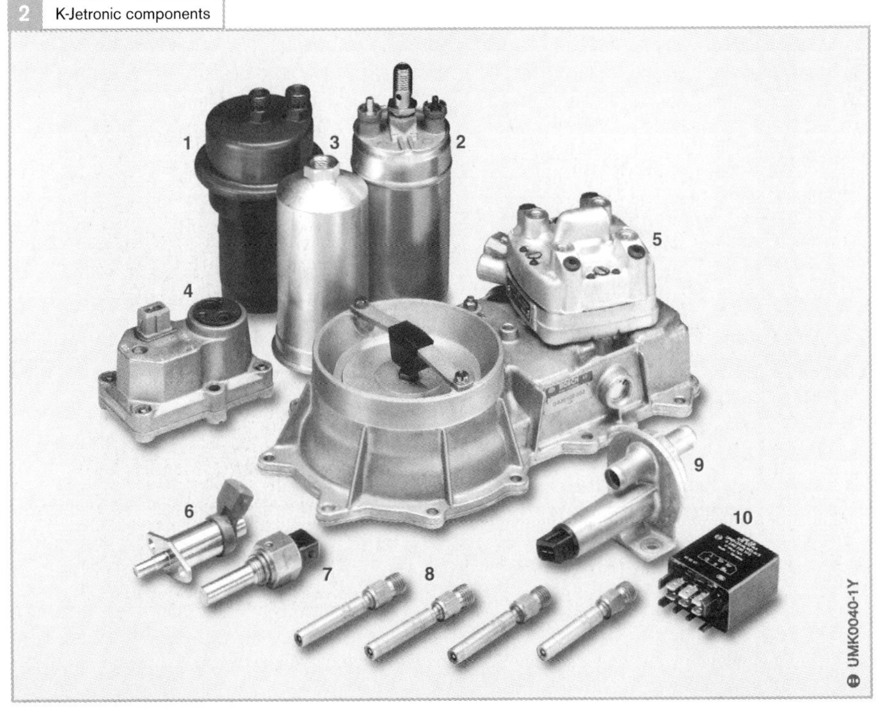

Fig. 2
1 Fuel accumulator
2 Electric fuel pump
3 Fuel filter
4 Warm-up regulator
5 Mixture-control unit with air-flow sensor and fuel distributor
6 Cold-start valve
7 Thermo-time switch
8 Fuel injectors
9 Auxiliary-air device
10 Electronic control relay

Ignition

Coil ignition system

The coil ignition system stores the ignition energy required to create a spark between the spark-plug electrodes in an ignition coil. The high arcing voltage is generated by transforming up the induction voltage that occurs when the coil current is cut off (mechanical distributor-contact points). The ignition distributor directs the ignition voltage to the individual spark plugs.

The control components of the coil ignition consist of the centrifugal advance mechanism, the vacuum unit, and the distributor contact points. Their interaction allows adjustment of the moment of ignition (ignition point) dependent on engine speed and engine load.

The described ignition system first went into mass production in 1934. In the years that followed, it quickly replaced the ignition magneto used until then to generate the high-tension ignition voltage.

Transistorized ignition

The first electronic component to be used in engine management was the transistor. It was able to replace the wear-prone distributor contact points. It also eliminated the influence of contact erosion on the ignition angle.

Transistorized ignition was introduced in 1965. However, the mechanical control of both ignition angle and dwell angle continued in existence for the next few years.

Electronic systems

Electronics gradually crept into the control devices handling fuel injection and ignition. Today there are no longer any systems featuring mechanical control.

Fuel injection

D-Jetronic

D-Jetronic is an electronically controlled multipoint fuel-injection system with intermittent fuel injection. The fuel pressure at the electromagnetic fuel injectors is constant, and the injected-fuel quantity is determined by the length of the injection pulses. This injection-pulse length is dependent on rotational speed and the back-pressure reaction in the intake manifold.

D-Jetronic obtains its rotational-speed signal by means of two maintenance-free contact points, which are located in the distributor at an offset of 180°. The intake-manifold pressure sensor supplies the engine-load data. Both engine and air temperature are measured by temperature sensors. They help to adjust the basic injection timing determined by load and rotational speed.

The D-Jetronic is designed using analog electronic components.

L-Jetronic

In contrast to D-Jetronic, the injection time of L-Jetronic is derived from the rotational speed and the rate of the intake-air flow. This is accomplished by an air-flow sensor located downstream of the throttle valve which sends an electrical signal dependent on the volumetric air flow to the electronic control unit. As the air-flow rate takes account all changes in the engine (e.g. wear, deposits in the combustion chamber), it allows a more accurate mixture composition than the method of measuring intake manifold pressure in the D-Jetronic.

L-Jetronic obtains its rotational speed signal by means of contact points in the ignition distributor of breaker-triggered ignition systems. In the case of breakerless ignition systems, this signal is obtained from terminal 1 of the ignition coil.

L-Jetronic uses analog technology. Its successor, the more advanced L3-Jetronic, uses digital technology for data processing. This allows the implementation of additional functions providing enhanced adjustment options.

KE-Jetronic

KE-Jetronic is based on the proven K-Jetronic which was enhanced by adding an electronic control unit (ECU), a primary-pressure regulator, and an electrohydraulic pressure actuator to control mixture composition. Electronic adjustment of fuel metering makes it easier to adjust mixture formation to both external conditions and engine operating status.

LH-Jetronic

LH-Jetronic essentially differs from L-Jetronic in the way it senses load. LH-Jetronic measures air mass instead of air-flow rate.

As a result, the measured value is independent of air density which is temperature- and pressure-dependent.

3 L-Jetronic components

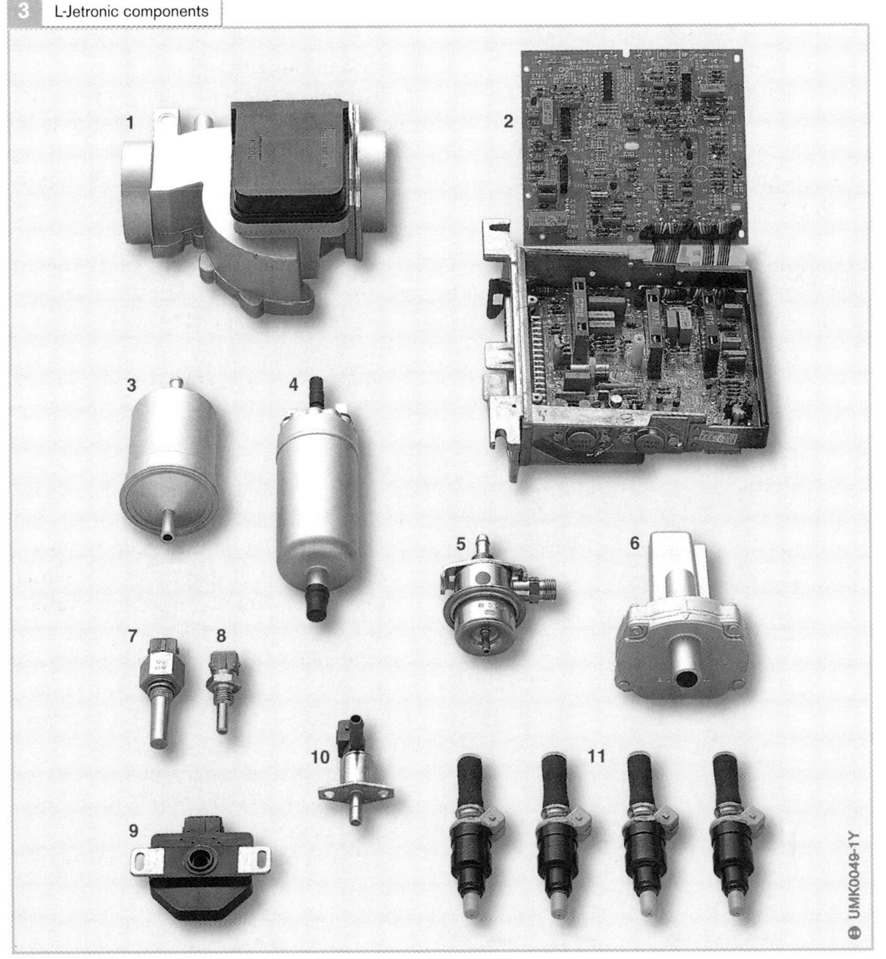

Fig. 3
1 Air-flow sensor
2 Electronic control unit
3 Fuel filter
4 Electric fuel pump
5 Fuel-pressure regulator
6 Auxiliary-air device
7 Thermo-time switch
8 Temperature sensor
9 Throttle-valve switch
10 Cold-start valve
11 Fuel injectors

Mono-Jetronic
Mono-Jetronic is a throttle-body injection system in which the fuel is injected at a central point by only a single electromagnetic fuel injector (throttle-body injection system, single-point fuel injection).

The cost advantage of this fuel-injection system over multipoint fuel-injection systems lead to the introduction of electronic gasoline injection in mid-range and compact cars.

Ignition

Mechanical adjustment of the ignition angle was only replaceable by an electronic adjustment after microcontrollers were introduced to the automobile. The load- and rotational-speed-dependent ignition angle is permanently stored in a map in program and data memories in the ignition-control unit. This stabilizes the ignition angle over the long term so that it is no longer impacted by wearing parts.

Electronic ignition systems are deployed in conjunction with electronic fuel-injection systems. These systems were installed on new car model launches up to 1998. In the meantime both ignition and fuel-injection systems are integrated in the Motronic system.

Electronic ignition
The electronic ignition system controls the ignition driver stage. The dwell angle and ignition angle are stored in program maps (map-controlled ignition). In calculating the ignition angle, additional influences, such as engine or intake-air temperature, are included to compute corrective adjustments.

Distributorless semiconductor ignition
The distributorless semiconductor ignition dispenses with the mechanical high-voltage distributor. Voltage distribution is performed electronically in the ignition control unit. The high voltage is generated in several ignition coils.

4 Overview of engine management systems

System designation		Injection type	Main controlled variable
D	Jetronic	Simultaneous (sequential on LH 4.1)	Intake manifold pressure
L			Air-flow rate/air mass
MONO		Throttle-body	Throttle valve position/engine speed
K		Continuous	Air-flow rate
KE		Continuous, with electronic adjustment	Air-flow rate
KE	Motronic		Air-flow rate
MONO		Throttle-body	Throttle valve position/engine speed
ML		Simultaneous/sequential	Air-flow rate
M			Air-flow rate/air mass
MP			Intake manifold pressure
ME		Sequential	Air mass
MED			Air mass

SMK1902E

Fig. 4
This diagram provides an overview of engine-management systems used by Bosch in the course of development.

Motronic engine-management system

Electronic fuel injection and electronic ignition helped to develop engines that delivered more power on the one hand, and that could comply with specified emission limits on the other. The increasing packaging density of electronic components produced more powerful microcontrollers and semiconductor memory chips with significantly greater storage capacity. As a result, the tasks handled by both electronic fuel injection and electronic map-controlled ignition could be assigned to a single microcontroller. This made it possible to combine both systems – electronic fuel injection and ignition – in a single electronic control unit. The result was the Motronic.

M-Motronic

The M-Motronic started production as early as 1979. It integrated the Jetronic multipoint injection system with an electronic map-controlled ignition. This made it possible to achieve the best possible balance between fuel metering and ignition control.

As a result of rapid advancements in semiconductor technology, the processing power of microcontrollers and the storage capacity of program and data memory chips increased at a steady pace. This made it possible to integrate a growing number of functions in the Motronic, such as knock control, and boost-pressure control for the exhaust-gas turbocharger. Emission-control legislation made functions such as exhaust-gas recirculation or tank ventilation, both of which reduce exhaust gases and evaporative emissions, indispensable requirements. In this way, this array of electronic control and regulation devices evolved into a complex engine-management system.

KE-Motronic

Due to the expense of electronics and fuel-injection components, M-Motronic was only fitted to luxury-performance vehicles at the beginning. However, the demand for compliance with emission limits led to the development of Motronic systems of lesser complexity. They were more suited to use in mid-range and compact cars.

KE-Motronic comprised a combination of KE-Jetronic, the proven electromechanical fuel-injection of its time, with an electronic map-controlled ignition in a single electronic control unit.

Mono-Motronic

The simplification of Mono-Motronic compared with M-Motronic consisted of the fact that there was only a single fuel injector placed at a central location to inject fuel into the intake manifold. Therefore, the fuel-injection system in the Mono-Motronic corresponded to the Mono-Jetronic.

ME-Motronic

ME-Motronic (start of production in 1994) is based on M-Motronic. As an additional component, electronic engine-performance control which started production as a separate system in 1986, is integrated in Motronic. In this Motronic system, also referred to as ETC (electronic throttle control), the conventional actuation of the throttle valve by a mechanical control cable was replaced by an electrically adjustable throttle valve plus an additional pedal travel sensor in the accelerator-pedal module.

MED-Motronic

Compared with ME-Motronic, MED-Motronic (start of production in 2000) features the additional function of gasoline direct injection. The high complexity of controlling and regulating tasks requires the use of microcontrollers featuring enormous computing power.

Systems for cylinder-charge control

On a gasoline engine running with a homogeneous A/F mixture, the intake air is the decisive quantity for the output torque and therefore for engine power. This means that not only is the fuel-metering system of special importance but also the systems which influence the cylinder charge. Some of these systems are able to influence the percentage of inert gas in the cylinder charge and thus also the exhaust emissions.

Air-charge control

For it to burn, fuel needs oxygen which the engine takes from the intake air. On engines with external A/F-mixture formation (manifold injection), as well as on direct-injection engines operating on a homogeneous A/F mixture with $\lambda = 1$, the output torque is directly dependent upon the intake-air mass. The throttle valve located in the induction tract controls the air flow drawn in by the engine and thus also the cylinder charge.

Conventional systems

Conventional systems (Fig. 1) feature a mechanically operated throttle valve (3). The accelerator-pedal (1) movement is transferred to the throttle valve by a linkage (2) or by a Bowden cable. The throttle valve's variable opening angle alters the opening cross-section of the intake passage (4) and in doing so regulates the air flow (5) drawn in by the engine, and with it the torque output.

To compensate for the higher levels of friction, the cold engine requires a larger air mass and extra fuel. And when, for instance, the A/C compressor is switched on more air is needed to compensate for the torque loss. This information is inputted to the ECU (8) in the form of an electrical signal (9), and the extra air is supplied by the air bypass actuator (7) directing the required extra air (6) around the throttle valve. Another method uses a throttle-valve actuator to adjust the throttle valve's minimum stop. In both cases though, it is only possible to electronically influence the air flow needed by the engine to a limited extent, for instance for idle-speed control.

1 Principle of the air control in a conventional system using a mechanically adjustable throttle valve and an air bypass actuator

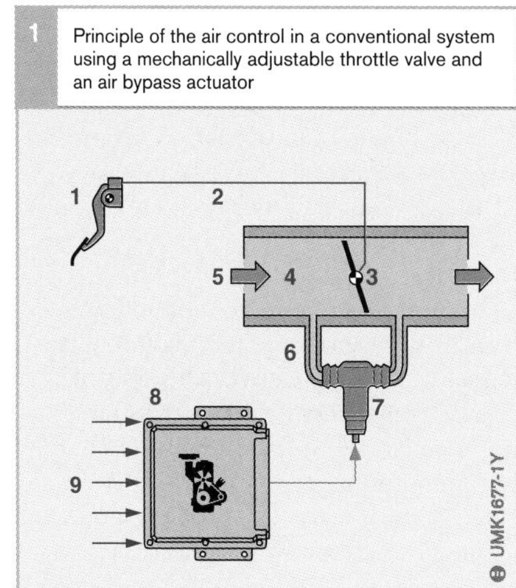

Fig. 1
1 Accelerator pedal
2 Bowden cable or linkage
3 Throttle valve
4 Induction passage
5 Intake air flow
6 Bypass air flow
7 Idle-speed actuator (air bypass actuator)
8 ECU
9 Input variables (electrical signals)

2 The ETC system (Electronic Throttle Control or EGAS)

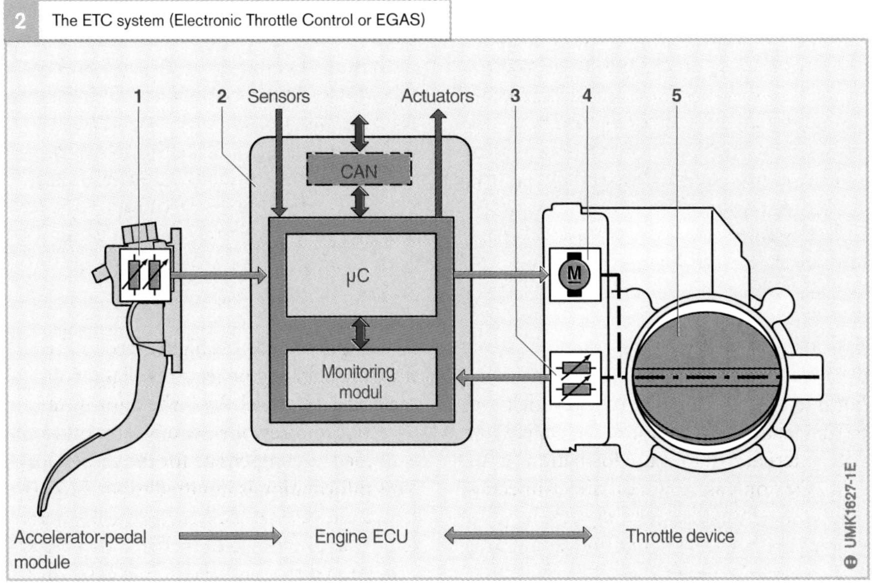

Fig. 2
1 Accelerator-pedal sensor
2 Engine ECU
3 Throttle-valve-angle sensor
4 Throttle-valve drive (DC motor)
5 Throttle valve

ETC systems

With ETC (Electronic Throttle Control, also known as EGAS), an ECU (Fig. 2, Pos. 2) is responsible for controlling the throttle valve (5). The DC-motor throttle-valve drive (4) and the throttle-valve-angle sensor (3) are combined with the throttle valve to form a unit, the so-called throttle device. To trigger the throttle device, the accelerator-pedal position, in other words the driver input, is registered by two potentiometers (accelerator-pedal sensor, 1). Taking into account the engine's actual operating status (engine speed, engine temperature, etc.) the engine ECU then calculates the throttle-valve opening which corresponds to the driver input and converts it into a triggering signal for the throttle-valve drive.

Using the feedback information from the throttle-valve-angle sensor regarding the current position of the throttle valve, it then becomes possible to precisely adjust the throttle valve to the required setting.

Two potentiometers on the accelerator-pedal and two on the throttle unit are a component part of the ETC monitoring concept. The potentiometers are duplicated for redundancy reasons. In case malfunctions are detected in that part of the system which is decisive for the engine's power output, the throttle valve is immediately shifted to a predetermined position (emergency or limp-home operation).

In the latest engine-management systems, the ETC control is integrated in the engine ECU which is also responsible for controlling ignition, fuel injection, and the auxiliary functions. There is no longer a separate ETC control unit.

The demands of emissions-control legislation are getting sharper from year to year. They can be complied with though thanks to ETC with its possibilities of further improving the A/F-mixture composition.

ETC is indispensable when complying with the demands made by gasoline direct injection on the overall vehicle system.

Variable valve timing

Apart from using the throttle-valve to throttle the flow of incoming fresh gas drawn in by the engine, there are several other possibilities for influencing the cylinder charge. The proportion of fresh gas and of residual gas can also be influenced by applying variable valve timing.

Of great importance for valve timing is the fact that the behaviour of the gas columns flowing into and out of the cylinders varies considerably as a function of engine speed or throttle-valve opening. With invariable valve timing, therefore, this means that the exhaust and refill cycle can only be ideal for one single engine operating range. Variable valve timing, on the other hand, permits adaptation to a variety of different engine speeds and cylinder charges. This has the following advantages:

- Higher engine outputs
- Favorable torque curve throughout a wide engine-speed range
- Reduction of toxic emissions
- Reduced fuel consumption
- Reduction of engine noise

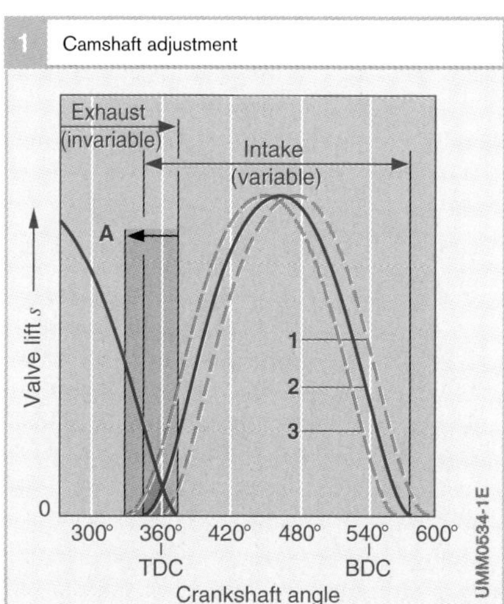

Fig. 1
1 Camshaft retarded
2 Camshaft normal
3 Camshaft advanced
A Valve overlap

Camshaft phase adjustment

In conventional IC engines, camshaft and crankshaft are mechanically coupled to each other through toothed belt or chain. This coupling is invariable.

On engines with camshaft adjustment, at least the intake camshaft, but to an increasing degree the exhaust camshaft as well, can be rotated referred to the crankshaft so that valve overlap changes. The valve opening period and lift are not affected by camshaft phase adjustment, which means that "intake opens" and "intake closes" remain invariably coupled with each other.

The camshaft is adjusted by means of electrical or electro-hydraulic actuators. On less sophisticated systems provision is only made for two camshaft settings. Variable camshaft adjustment on the other hand permits, within a given range, infinitely variable adjustment of the camshaft referred to the crankshaft.

Fig. 1 shows how the "position", or lift, of the open intake-valve changes (referred to TDC) when the intake camshaft is adjusted.

Retard adjustment of the intake camshaft

Retarding the intake camshaft leads to the intake valve opening later so that valve overlap is reduced, or there is no valve overlap at all. At low engine speeds (< 2,000 rpm), this results in only very little burnt exhaust gas flowing past the intake valve and into the intake manifold. At low engine speeds, the low residual exhaust-gas content in the intake of A/F mixture which then follows leads to a more efficient combustion process and a smoother idle. This means that the idle speed can be reduced, a step which is particularly favorable with respect to fuel consumption.

The camshaft is also retarded at higher engine speeds (> 5,000 rpm). Late closing of the intake valve, long after BDC, leads to a higher cylinder charge. This boost effect results from the high flow speed of the fresh gas through the intake valve which continues even after the piston has reversed its direction of travel and is moving upwards to compress the mixture. For this reason, the intake valve closes long after BDC.

Advance adjustment of the intake camshaft

In the medium speed range, the flow of fresh gas through the intake passage is much slower, and of course there is no high-speed boost effect. At medium engine speeds, closing the intake valve earlier, only shortly after BDC, prevents the ascending piston forcing the freshly drawn-in gas out past the intake valve again and back into the manifold. At such speeds, advancing the intake camshaft results in better cylinder charge and therefore a good torque curve.

At medium speeds, advanced opening of the intake camshaft leads to increased valve overlap. Opening the intake valve early means that shortly before TDC, the residual exhaust gas which has not already left the cylinder is forced out past the open intake valve and into the intake manifold by the ascending piston. These exhaust gases are then drawn into the cylinder again and serve to increase the residual-gas content of the cylinder charge. The increased residual gas content in the freshly drawn in A/F mixture caused by advancing the intake camshaft (internal EGR), affects the combustion process. The resulting lower peak temperatures lead to a reduction in NO_x.

The higher inert-gas content in the cylinder charge makes it necessary to open the throttle valve further, which in turn leads to a reduction of the throttling losses. This means that valve overlap can be applied to reduce fuel consumption.

Adjusting the exhaust camshaft

On systems which can also adjust the exhaust camshaft, not only the intake camshaft is used to vary the residual-gas content, but also the exhaust camshaft. Here, the total cylinder charge (defined by "intake closes") and the residual-gas content (influenced by "intake opens" and "exhaust closes") can be controlled independently of each other.

Camshaft changeover

Camshaft changeover (Fig. 2) involves switching the camshaft between two different cam contours. This changes both the valve lift and the valve timing (cam-contour changeover). The first cam defines the optimum timing and the valve lift for the intake and exhaust valves in the lower and medium speed ranges. The second cam controls the increased valve lift and longer valve-open times needed at higher speeds.

2 Camshaft changeover

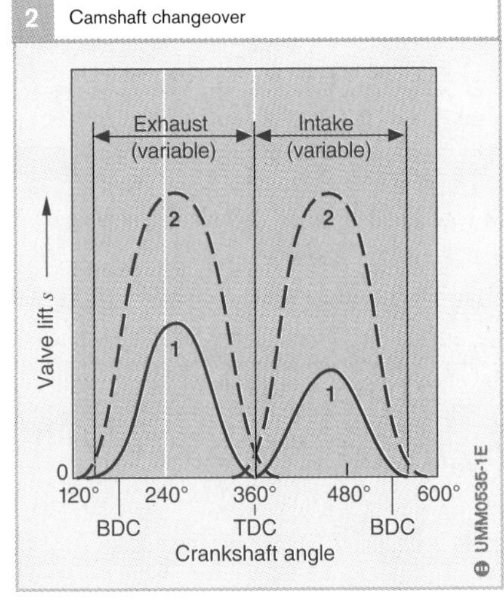

Fig. 2
1 Standard cam
2 Supplementary cam

3 Example of a system with fully variable adjustment of valve timing and of valve lift

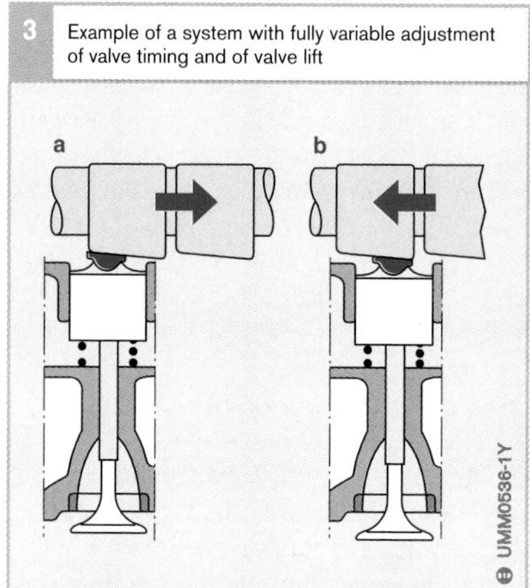

Fig. 3
a Minimum lift
b Maximum lift

At low and medium engine speeds, minimum valve lifts together with the associated small valve-opening cross-sections lead to a high inflow velocity and therefore to high levels of turbulence in the cylinder for the fresh air (gasoline direct injection) or for the fresh A/F mixture (manifold injection). This ensures excellent A/F mixture formation at part load. The high engine outputs required at higher engine speeds and torque demand (WOT) necessitate maximum cylinder charge. Here, the maximum valve lift is selected.

There are a variety of methods in use for switching-over between the different cam contours. One method, for instance, relies on a free-moving drag lever which engages with the standard rocking lever as a function of rotational speed. Another method uses changeover cup tappets.

Fully variable valve timing and valve lift using the camshaft

Valve control which incorporates both variable valve timing and variable valve lift is referred to as being fully variable. Even more freedom in engine operation is permitted by 3D cam contours and longitudinal-shift camshafts (Fig. 3). With this form of camshaft control, not only the valve lift (only on the intake side) and thus the opening angle of the valves can be infinitely varied, but also the phase position between camshaft and crankshaft.

Since the intake valve can be closed early with this fully variable camshaft control, this permits so-called charge control in which the intake-manifold throttling is considerably reduced. This enables fuel consumption to be slightly lowered in comparison with the simple camshaft phase adjustment.

Fully variable valve timing and valve lift without using the camshaft

For valve timing, maximum design freedom and maximum development potential are afforded by systems featuring valve-timing control which is independent of the camshaft. With this form of timing, the valves are opened and closed, for instance, by electromagnetic actuators. A supplementary ECU is responsible for triggering. This form of fully variable valve timing without camshaft aims at extensive reduction of the intake-manifold throttling, coupled with very low pumping losses. Further fuel savings can be achieved by incorporating cylinder and valve shutoff.

These fully variable valve-timing concepts not only permit the best-possible cylinder charge and with it a maximum of torque, but they also ensure improved A/F-mixture formation which results in lower toxic emissions in the exhaust gas.

Controlled charge flow

The airflow behavior in both intake manifold and cylinder is an essential factor in forming an ideal A/F mixture. Vigorous charge-flow movements make sure that the air-fuel mixture is well blended to achieve excellent, low-pollutant combustion.

There are different ways to generate a vigorous charge-flow movement.

Camshaft-lobe control

As described in the section entitled "Variable valve timing", selecting camshaft-lobe control allows for lower valve lift in the lower and medium engine-speed ranges. The resulting small cross-section of the intake-valve opening generates high flow velocity in the intake air or A/F mixture. For this reason, manifold injection makes sure that the air-fuel mixture is fully mixed.

Charge-flow control flap

A different process is used by systems with gasoline direct injection. In the intake-valve area, the intake manifold is split into two channels, one of which can be closed by means of a hinged flap – the charge-flow control flap – (Fig. 1). Directing the flow to specific parts of the intake valves, or only one of two intake valves, helps the gases in the combustion chamber to rotate or swirl.

The charge-flow control flap allows control the intensity of the charge flow. In stratified-charge mode, the flow moved in this way makes sure that the mixture is conveyed to the spark plug. It supports mixture formation at the same time.

In homogeneous-charge mode, the charge-flow control flap is normally open at high engine torque and rotational speeds. The full flow cross-section is required in this type of operating situation in order to obtain a high engine charge for high torques. The ideal mixture formation is achieved even without increased charge flow, simply by advancing fuel injection into the combustion chamber (as early as the induction stroke) at a high temperature.

1 Charge-flow control via the charge-flow control flap in the intake manifold

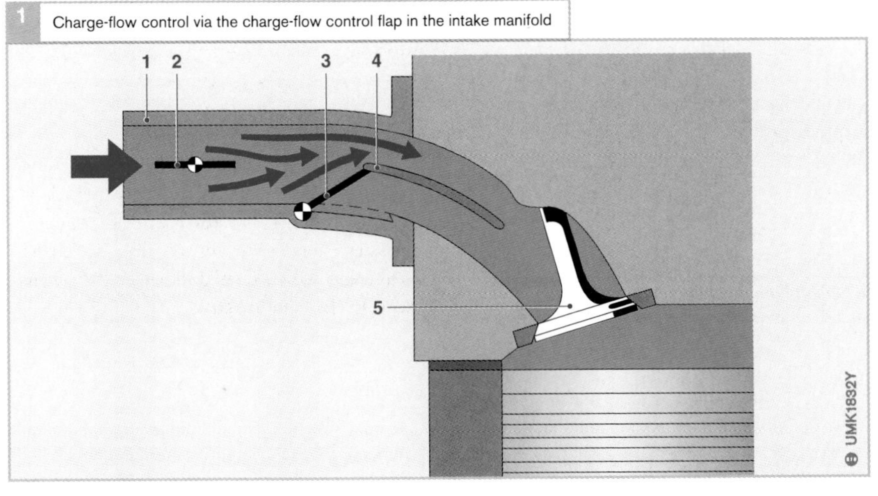

Fig. 1
1 Intake manifold
2 Throttle valve
3 Charge-flow control flap
4 Separating ridge
5 Intake valve

Exhaust-gas recirculation (EGR)

The mass of the residual gas remaining in the cylinder, and with it the inert-gas content of the cylinder charge, can be influenced by varying the valve timing. In this case, one refers to "internal" EGR. The inert-gas content can be influenced far more by applying "external" EGR with which part of the exhaust gas which has already left the cylinder is directed back into the intake manifold through a special line (Fig. 1, Pos. 3). EGR leads to a reduction of the NO_x emissions and to a slightly lower fuel-consumption figure.

Limiting the NO_x emissions

Since they are highly dependent upon temperature, EGR is highly effective in reducing NO_x emissions. When peak combustion temperature is lowered by introducing burnt exhaust gas to the A/F mixture, NO_x emissions drop accordingly.

Lowering fuel consumption

When EGR is applied, the overall cylinder charge increases while the charge of fresh air remains constant. This means that the throttle valve (2) must reduce the engine throttling if a given torque is to be achieved. Fuel consumption drops as a result.

EGR: Operating concept

Depending upon the engine's operating point, the engine ECU (4) triggers the EGR valve (5) and defines its opened cross-section. Part of the exhaust-gas (6) is diverted via this opened cross-section (3) and mixed with the incoming fresh air. This defines the exhaust-gas content of the cylinder charge.

EGR with gasoline direct injection

EGR is also used on gasoline direct-injection engines to reduce NO_x emissions and fuel consumption. In fact, it is absolutely essential since with it NO_x emissions can be lowered to such an extent in lean-burn operations that other emissions-reduction measures can be reduced accordingly (for instance, rich homogeneous operation for NO_x "Removal" from the NO_x accumulator-type catalytic converter). EGR also has a favorable effect on fuel consumption.

There must be a pressure gradient between the intake manifold and the exhaust-gas tract in order that exhaust gas can be drawn in via the EGR valve. At part load though, direct-injection engines are operated practically unthrottled. Furthermore a considerable amount of oxygen is drawn into the intake manifold via EGR during lean-burn operation.

Non-throttled operation and the introduction of oxygen into the intake manifold via the EGR therefore necessitate a control strategy which coordinates throttle valve and EGR valve. This results in severe demands being made on the EGR system with regard to precision and reliability, and it must be robust enough to withstand the deposits which accumulate in the exhaust-gas components as a result of the low exhaust-gas temperatures.

1 Exhaust-gas recirculation (EGR)

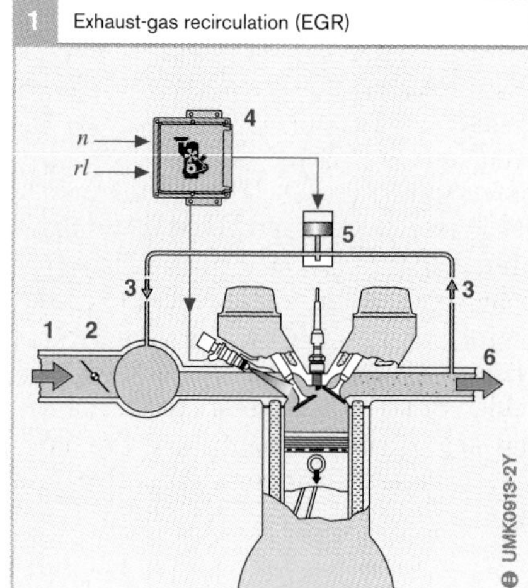

Fig. 1
1 Fresh-air intake
2 Throttle valve
3 Recirculated exhaust gas
4 Engine ECU
5 EGR valve
6 Exhaust gas

n Engine rpm
rl Relative air charge

Dynamic supercharging

Approximately speaking, the achievable engine torque is proportional to the fresh-gas content in the cylinder charge. This means that the maximum torque can be increased to a certain extent by compressing the air before it enters the cylinder.

The exhaust-and-refill processes are not only influenced by the valve timing, but also by the intake and exhaust lines. The piston's induction work causes the open intake valve to trigger a return pressure wave. At the open end of the intake manifold, the pressure wave encounters the quiescent ambient air from which it is reflected back again so that it returns in the direction of the intake valve. The resulting pressure fluctuations at the intake valve can be utilized to increase the fresh-gas charge and thus achieve the highest-possible torque.

This supercharging effect thus depends on utilization of the incoming air's dynamic response. In the intake manifold, the dynamic effects depend upon the geometrical relationships in the intake manifold and on the engine speed.

For the even distribution of the A/F mixture, the intake manifolds for carburetor engines and single-point injection (TBI) must have short pipes which as far as possible must be of the same length for all cylinders. In the case of multipoint injection (MPI), the fuel is either injected into the intake manifold onto the intake valve (manifold injection), or it is injected directly into the combustion chamber (gasoline direct injection). With MPI, since the intake manifolds transport mainly air and practically no fuel can deposit on the manifold walls, this provides wide-ranging possibilities for intake-manifold design. This is the reason for there being no problems with multipoint injection systems regarding the even distribution of fuel.

Ram-tube supercharging

The intake manifolds for multipoint injection systems are composed of the individual tubes or runners and the manifold chamber.

In the case of ram-tube supercharging (Fig. 1), each cylinder is allocated its own tube (2) of specific length which is usually attached to the manifold chamber (3). The pressure waves are able to propagate in the individual tubes independently.

The supercharging effect depends upon the intake-manifold geometry and the engine speed. For this reason, the length and diameter of the individual tubes is matched to the valve timing so that in the required speed range a pressure wave reflected at the end of the tube is able to enter the cylinder through the open intake valve (1) and improve the cylinder charge. Long, narrow tubes result in a marked supercharging effect at low engine speeds. On the other hand, short, large-diameter tubes have a positive effect on the torque curve at higher engine speeds.

1 Principle of ram-tube supercharging

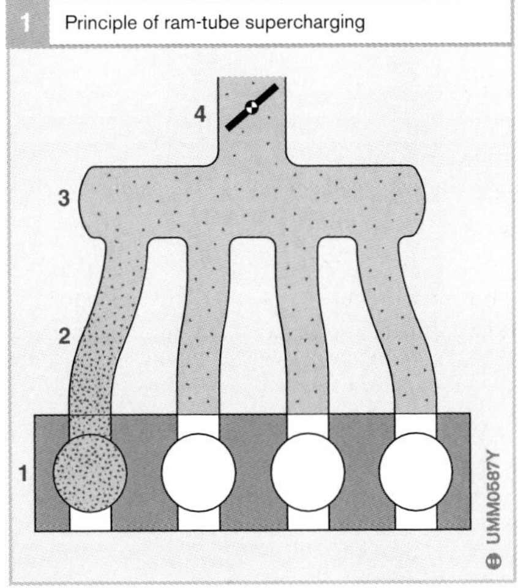

Fig. 1
1 Cylinder
2 Individual tube
3 Manifold chamber
4 Throttle valve

Tuned-intake-tube charging

At a given engine speed, the periodic piston movement causes the intake-manifold gas-column to vibrate at resonant frequency. This results in a further increase of pressure and leads to an additional supercharging effect. On the tuned intake-tube system (Fig. 2), groups of cylinders (1) with identical angular ignition spacing are each connected to a resonance chamber (3) through short tubes (2). The chambers, in turn, are connected through tuned intake tubes (4) with either the atmosphere or with the manifold chamber (5) and function as Helmholtz resonators.

The subdivision into two groups of cylinders each with its own tuned intake tube prevents the overlapping of the flow processes of two neighboring cylinders which are adjacent to each other in the firing sequence.

The length of the tuned intake tubes and the size of the resonance chamber are a function of the speed range in which the supercharging effect due to resonance is required to be at maximum. Due to the accumulator effect of the considerable chamber volumes which are sometimes needed, dynamic-response errors can occur in some cases when the load is changed abruptly.

Variable-geometry intake manifold

The supplementary cylinder charge resulting from dynamic supercharging depends upon the engine's working point. The two systems just dealt with increase the achievable maximum charge (volumetric efficiency), above all in the low engine-speed range (Fig. 3).

Practically ideal torque characteristics can be achieved with variable-geometry intake manifolds in which, as a function of the engine operating point, flaps are used to implement a variety of different adjustments such as:

- Adjustment of the intake-tube length
- Switch over between different intake-tube lengths or different tube diameters
- Selected switchoff of one of the cylinder's intake tubes on multiple-tube systems
- Switchover to different chamber volumes

Electrical or electropneumatically actuated flaps are used for change-over operations in these variable-geometry systems.

Fig. 2
1 Cylinder
2 Short tube
3 Resonance chamber
4 Tuned intake tube
5 Manifold chamber
6 Throttle valve

A Cylinder group A
B Cylinder group B

Fig. 3
1 System with tuned-intake-tube charging
2 System with conventional intake manifold

2 Principle of tuned-intake-tube charging

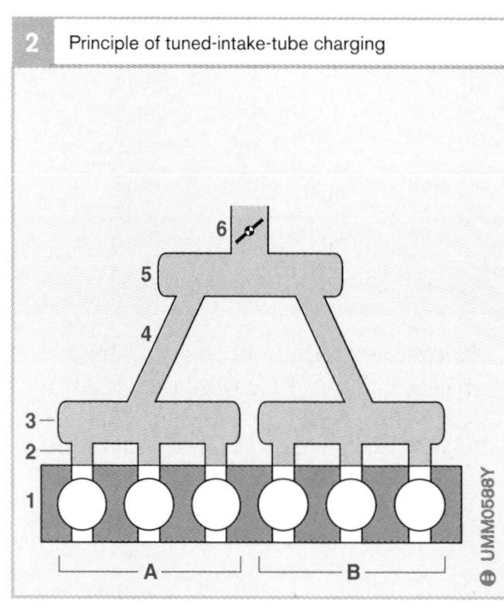

3 Increasing the maximum-possible cylinder air charge (volumetric efficiency) by means of dynamic supercharging

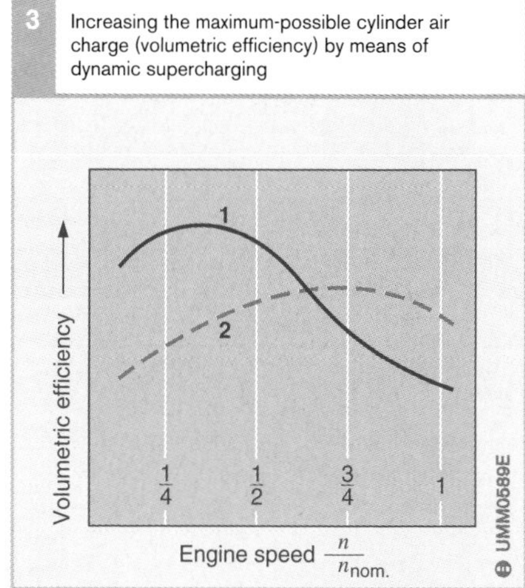

Ram-tube systems

The manifold system shown in Fig. 4 can switch between two different ram tubes. In the lower speed range, the changeover flap (1) is closed and the intake air flows to the cylinders through the long ram tube (3). At higher speeds and with the changeover flap open, the intake air flows through the short, wide diameter ram tube (4), and thus contibutes to improved cylinder charge at high engine revs.

5 Combined tuned-intake-tube and ram-tube system

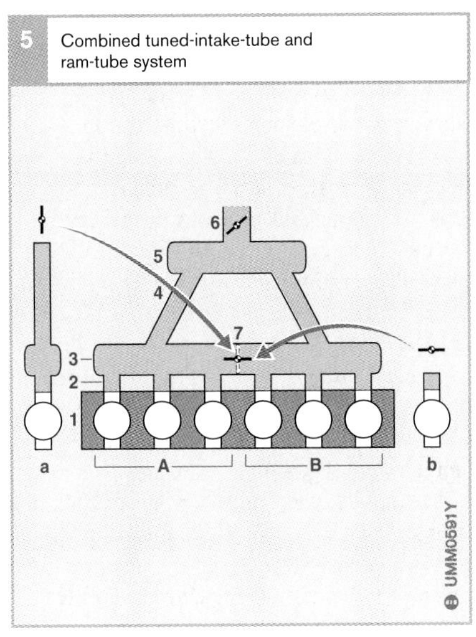

Fig. 5
1 Cylinder
2 Ram tube (short intake tube)
3 Resonance chamber
4 Tuned intake tube
5 Manifold chamber
6 Throttle valve
7 Changeover flap

A Cylinder group A
B Cylinder group B

a Intake-manifold conditions with changeover flap closed
b Intake-manifold conditions with changeover flap open

4 Ram-tube system

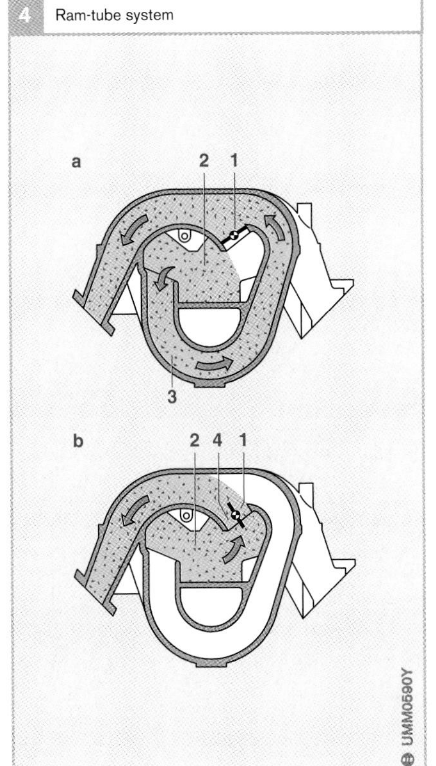

Fig. 4
a Manifold geometry with changeover flap closed
b Manifold geometry with changeover flap open

1 Changeover flap
2 Manifold chamber
3 Changeover flap closed: Long, narrow-diameter ram tube
4 Changeover flap opened: Short, wide-diameter ram tube

Tuned-intake-tube system

Opening the resonance flap switches in a second tuned intake tube. The changed geometry of this configuration has an effect upon the resonant frequency of the intake system. Cylinder charge in the lower speed range is improved by the higher effective volume resulting from the second tuned intake pipe.

Combined tuned-intake-tube and ram-tube system

When design permits the open changeover flap (Fig. 5, Pos. 7) to combine both the resonance chambers (3) to form a single volume, one speaks of a combined tuned-intake-tube and ram-tube system. A single intake-air chamber with a high resonant frequency is then formed for the short ram tubes (2).

At low and medium engine revs, the changeover flap is closed and the system functions as a tuned-intake-tube system. The low resonant frequency is then defined by the long tuned intake tube (4).

Mechanical supercharging

Design and operating concept

The application of supercharging units leads to increased cylinder charge and therefore to increased torque. Mechanical supercharging uses a compressor which is driven directly by the IC engine. Mechanically driven compressors are either positive-displacement superchargers with different types of construction (e.g. Roots supercharger, sliding-vane supercharger, spiral-type supercharger, screw-type supercharger), or they are centrifugal turbo-compressors (e.g. radial-flow compressor). Fig. 1 shows the principle of functioning of the rotary-screw supercharger with the two counter-rotating screw elements. As a rule, engine and compressor speeds are directly coupled to one another through a belt drive.

Boost-pressure control

On the mechanical supercharger, a bypass can be applied to control the boost pressure. A portion of the compressd air is directed into the cylinder and the remainder is returned to the supercharger input via the bypass. The engine management is responsible for controlling the bypass valve.

Advantages and disadvantages

On the mechanical supercharger, the direct coupling between compressor and engine crankshft means that when engine speed increases there is no delay in supercharger acceleration. This means therefore, that compared to exhaust-gas turbocharging engine torque is higher and dynamic response is better.

Since the power required to drive the compressor is not available as effective engine power, the above advantage is counteracted by a slightly higher fuel-consumption figure compared to the exhaust-gas turbocharger. This disadvantage though is somewhat alleviated when the engine management is able to switch off the compressor via a clutch at low engine loading.

1 Rotary-screw supercharger: Principle of functioning

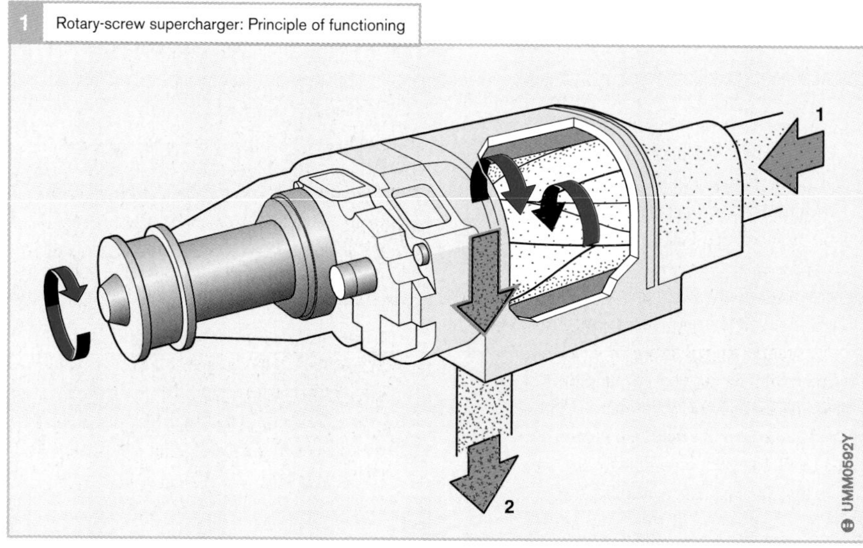

Fig. 1
1 Intake air
2 Compressed air

History of compressor-engine automobiles

At the 1921 Berlin Motor Show, the "Daimler Motoren Gesellschaft" introduced the passenger cars designated Type 6/20 HP and 10/35 HP, each featuring a four-cylinder compressor engine.

1922 marked the first time a vehicle powered by a compressor engine was entered in a car race. At the Targa Florio on the island of Sicily, Max Sailer drove a 28/95 HP Mercedes equipped with a compressor engine to victory in the production-car category.

In 1924 the new Mercedes 15/70/100 HP and 24/100/140 HP passenger cars with six-cylinder compressor engines were introduced at the Berlin Motor Show. On the Avus race track in Berlin that same year, NSU set spectacular speed records and victories with their 5/15 compressor racing car.

In 1927 the Type "S" Mercedes-Benz captured a triple victory at the inaugural race at the new Nuerburgring race track. The championship was won by Rudolf Caracciola. With its 6.8-litre displacement, the 6-cylinder compressor engine of the Type "S" delivered 180 HP and 120 HP without compressor. In 1932 Manfred von Brauchitsch, driving an "SSKL" (abbreviation for the German "super-sport-short-light") won the Avus race in Berlin, establishing a world record for this category at 200 km/h. The "SSKL" represented the final stage in the development of the "S" series.

The era of the Silver Arrows began in 1934. The very first time they were seen was at the International Eifel Race on the Nuerburgring. It ended with pilot von Brauchitsch winning and establishing a new speed record. In 1938 it was a streamlined Silver Arrow that established a speed record on public motorways that remains unbroken to this day: 426.6 km/h at 1 km, with a flying start. Silver Arrows with compressor engines were also built by Auto-Union; they engaged in riveting duels with the Mercedes-built Silver Arrows.

1934 saw the introduction of the Mercedes Type 500K featuring an 8-cylinder compressor engine. The successor model launched in 1936 featured an even more powerful 5.4-*l* engine. In 1938 the Type 770 "Grand Mercedes" (W150) was introduced at the Berlin Motor and Motorcycle Show. It featured a 7.7-liter, eight-cylinder in-line engine with compressor. This engine developed a power output of 230 HP.

Compared with today's compressor engines, the power output per engine size then was rather modest. The Mercedes SLK, launched in the late 1990s, develops 192 HP from a 2.3-l engine at 5,300 rpm.

1922
Mercedes 28/95 HP

1924
Mercedes Type 24/100
6 cylinders

1927
Mercedes Type "S"
6 cylinders, 6.8 *l*, 180 HP

1934
Mercedes Type 500K
8 cylinders

1938
"Grand Mercedes"
Type 770 (W150),
7.7 *l*, 230 HP

(All photos:
DaimlerChrysler Classic,
Corporate Archives)

Exhaust-gas turbocharging

Of all the possible methods for supercharging the IC engine, exhaust-gas turbocharging is the most widely used. Even on engines with low swept volumes, exhaust-gas supercharging leads to high torques and power outputs together with high levels of engine efficiency.

Whereas, in the past, exhaust-gas turbocharging was applied in the quest for increased power-weight ratio, it is today mostly used in order to increase the maximum torque at low and medium engine speeds. This holds true particularly in combination with electronic boost-pressure control.

Design and operating concept

The main components of the exhaust-gas turbocharger (Fig. 1) are the exhaust-gas turbine (3) and the compressor (1). The compressor impeller and the turbine rotor are mounted on a common shaft (2).

The energy needed to drive the exhaust-gas turbine is for the most part taken from the hot, pressurized exhaust gas. On the other hand, energy must be also used to "dam" the exhaust gas when it leaves the engine so as to generate the required compressor power.

The hot gases (Fig. 2, Pos. 7) are applied radially to the exhaust-gas turbine (4) and cause this to rotate at very high speed. The turbine-rotor blades are inclined towards the center and thus direct the gas to the inside from where it then exits axially.

1 Passenger-car exhaust-gas turbocharger (Shown: 3K-Warner, type K14)

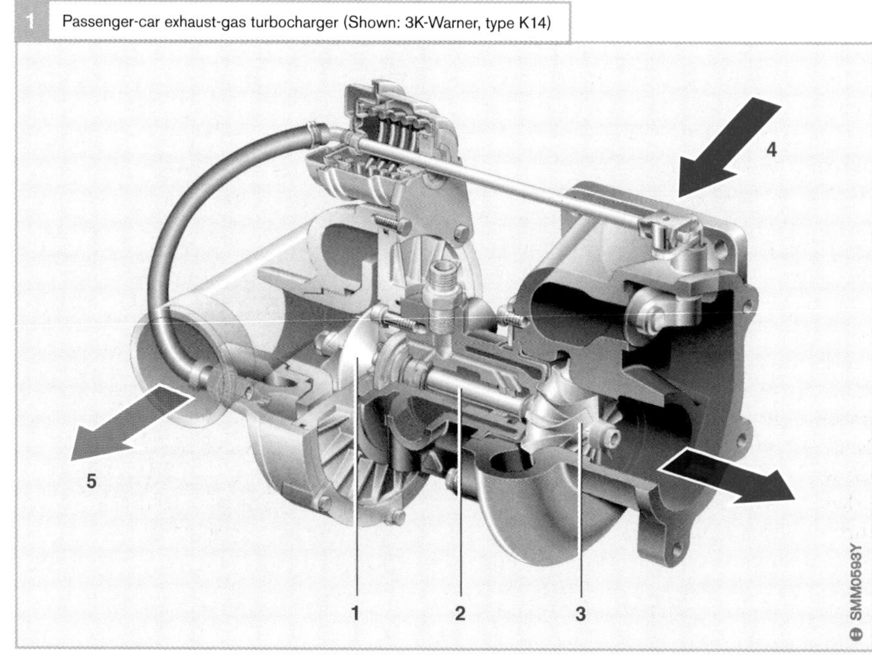

Fig. 1
1 Compressor impeller
2 Shaft
3 Exhaust-gas turbine
4 Intake for exhaust-gas mass flow
5 Outlet for compressed air

The compressor (3) also turns along with the turbine, but here the flow conditions are reversed. The fresh incoming gas (5) enters axially at the center of the compressor and is forced radially to the outside by the blades and compressed in the process.

Since the exhaust-gas turbocharger is located directly in the flow of hot exhaust gas it must be built of highly temperature-resistant materials.

Exhaust-gas turbochargers: Designs

Wastegate supercharger

The objective is for IC engines to develop high torques at low engine speeds. The turbine casing has therefore been designed for a low level of exhaust-gas mass flow, for instance WOT at ≤ 2,000 rpm. With high exhaust-gas mass flows in this range, part of the flow must be diverted around the turbine and into the exhaust system in order that the turbocharger is prevented from overcharging the engine. Diversion is via a bypass valve, the so-called wastegate (Fig. 2, Pos. 8). This flap-type bypass valve is usually integrated into the turbine casing.

The wastegate is actuated by the boost-pressure control valve (6). This valve is connected pneumatically to the pulse valve (1) through a control line (2). The pulse valve changes the boost pressure upon being triggered by an electrical signal from the engine ECU. This electrical signal is a function of the current boost pressure, information on which is provided by the boost-pressure sensor (BPS).

If the boost pressure is too low, the pulse valve is triggered so that a somewhat lower pressure prevails in the control line. The boost-pressure control valve then closes the wastegate and the proportion of the exhaust-gas mass flow used to power the turbine is increased.

If, on the other hand, the boost pressure is excessive, the pulse valve is triggered so that a somewhat higher pressure is built up in the control line. The boost-pressure control valve then opens the wastegate and the proportion of the exhaust-gas mass flow used to power the turbine is reduced.

VTG turbocharger

The VTG (Variable Turbine Geometry) is another method which can be applied to limit the exhaust-gas mass flow at higher engine speeds (Fig. 3, next page). The VTG supercharger is state-of-the-art on diesel engines, but has not yet become successful on gasoline engines due to the high thermal stressing resulting from the far hotter exhaust gases.

By varying the geometry, the adjustable guide vanes (3) adapt the flow cross-section, and with it the gas pressure at the turbine, to the required boost pressure. At low speeds, they open up a small cross-section so that the exhaust-gas mass flow in the turbine reaches a high speed and in doing so also brings the exhaust-gas turbine up to high speed (Fig. 3a).

2 Design and construction of an exhaust-gas turbocharger using a wastegate turbocharger as an example

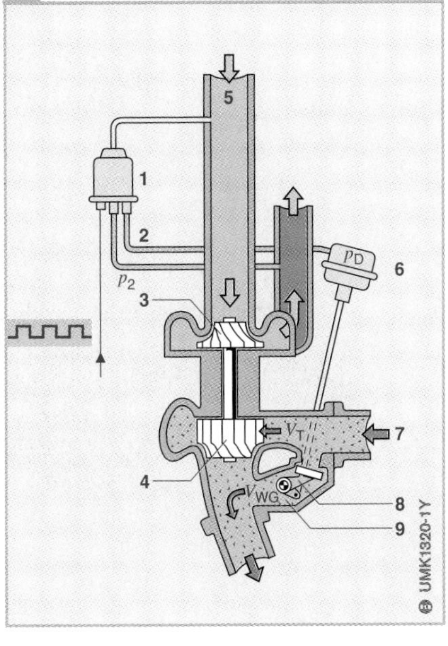

Fig. 2
1 Pulse valve
2 Pneumatic control line
3 Compressor
4 Exhaust-gas turbine
5 Fresh incoming air
6 Boost-pressure control valve
7 Exhaust gas
8 Wastegate
9 Bypass duct

⎍⎍ Triggering signal for pulse valve
V_T Volume flow through the turbine
V_{WG} Volume flow through the wastegate
p_2 Boost pressure
p_D Pressure on the valve diaphragm

At high engine speeds, the adjustable guide vanes (3) open up a larger cross-section so that more exhaust gas can enter without accelerating the exhaust-gas turbine to excessive speeds (Fig. 3b). This limits the boost pressure.

It is an easy matter to adjust the guide-vane angle by rotating the adjusting ring (2). Here, the guide vanes are adjusted to the desired angle either directly through individual adjusting levers (4) attached to the guide vanes, or by adjusting cam. The adjusting ring is rotated pneumatically via a barometric adjustment cell (5) using either vacuum or overpressure. This adjustment mechanism is triggered by the engine management so that the boost pressure can be set to the best-possible level in accordance with the engine's operating mode.

VST supercharger

On the VST (Variable Sleeve Turbine) supercharger, the "turbine size" is adapted by means of successively opening two flow passages (Fig. 4, Pos. 2 and 3) using a special control sleeve (4).

Initially, only one flow passage is opened, and the small opening cross-section results in high exhaust-gas flow speed and high turbine speeds (1). As soon as the permissible boost pressure is reached, the control sleeve successively opens the second flow passage, the exhaust-gas flow speed reduces accordingly, and with it the boost pressure.

Using the bypass channel (5) incorporated in the turbine casing, it is also possible to divert part of the exhaust-gas mass flow past the exhaust-gas turbine.

The control sleeve is adjusted by the engine management via a barometric cell.

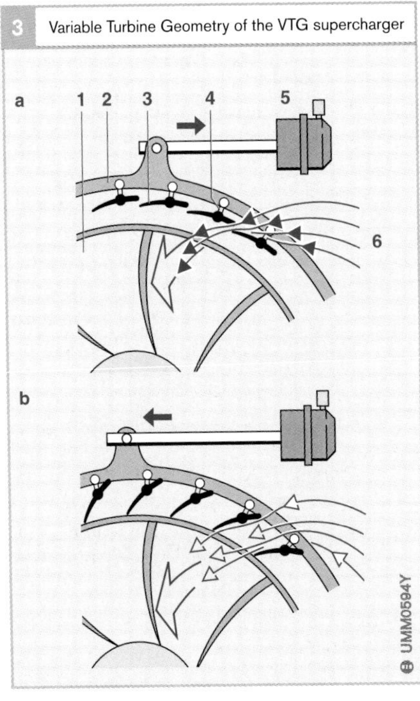

Fig. 3
- a Guide-vane setting for high boost pressure
- b Guide-vane setting for low boost pressure

1. Exhaust-gas turbine
2. Adjusting ring
3. Guide vanes
4. Adjusting lever
5. Barometric cell
6. Exhaust-gas flow

◄ High flow speed
◁ Low flow speed

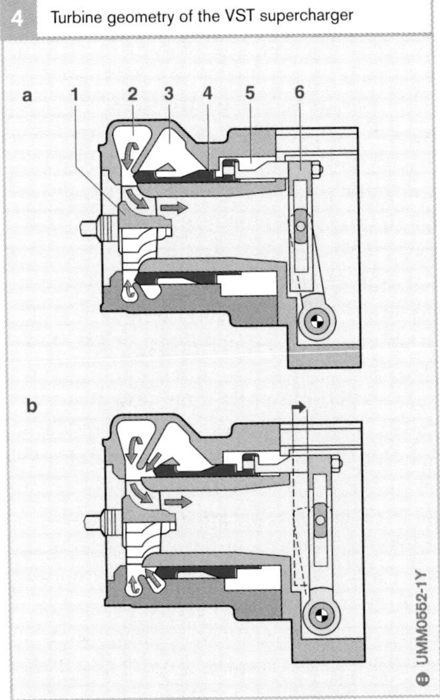

Fig. 4
- a Only 1 flow passage open
- b Both flow passages open

1. Exhaust-gas turbine
2. 1st flow passage
3. 2nd flow passage
4. Special control sleeve
5. Bypass duct
6. Adjustment fork

Exhaust-gas turbocharging: Advantages and disadvantages

Compared with a naturally-aspirated IC engine with the same output power, the major advantages are to be found in the turbocharged engine's lower weight and smaller size ("downsizing"). The turbocharged engine's torque characteristic is better throughout the usable speed range (Fig. 5, curve 4 compared to curve 3). All in all, at a given speed, this results in a higher output (A→B).

Due to its more favorable torque characteristic at WOT, the turbocharged engine generates the required power as shown in Fig. 5 (B or C) at lower engine speeds than the naturally aspirated engine. At part load, the throttle valve must be opened further, and the working point is shifted to an area with reduced frictional and throttling losses (C→B). This results in lower fuel-consumption figures even though turbocharged engines in fact feature less favorable efficiency figures due to their lower compression ratio.

The low torque that is available at very low engine speeds is a disadvantage of the turbocharger. In such speed ranges, there is not enough energy in the exhaust gas to drive the exhaust-gas turbine. In transient operation, even in the medium-speed range, the torque curve is less favorable than that of the natually aspirated engine (curve 5). This is due to the delay in building up the exhaust-gas mass flow. When accelerating from low engine speeds, this is evinced by the turbo flat spot.

The effects of this flat spot can be minimised by making full use of dynamic charge. This supports the supercharger's running-up characteristic. There are a number of other versions available, including a turbocharger with electric motor, or with an extra compressor driven by an electric motor. Independent of the exhaust-gas mass flow, these accelerate the compressor impeller and/or the air-mass flow, and in doing so avoid the turbo flat spot.

5 Power and torque characteristics of an exhaust-gas-turbocharged engine compared with those of a naturally aspirated engine

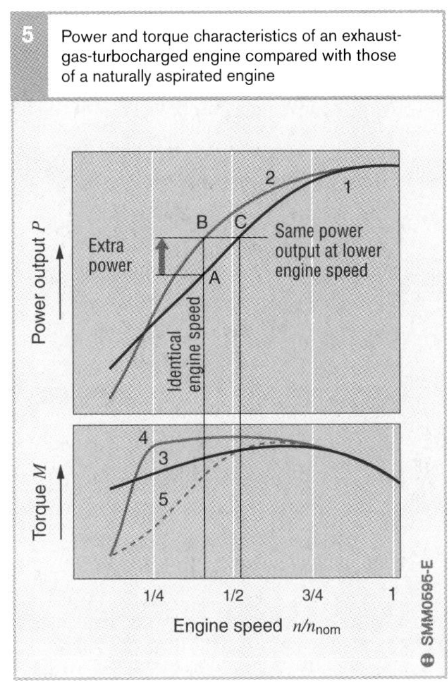

Fig. 5
1, 3 Naturally aspirated engine in steady-state operation
2, 4 Supercharged engine in steady-state operation
5 Torque curve of the supercharged engine in transient (dynamic) operation

Intercooling

The air warms up in the compressor during the compression process, but since warm air has a lower density than cold air, this temperature rise has a negative effect upon cylinder charge. The compressed, warmed air must therefore be cooled off again by the intercooler. Compared to supercharged engines with this facility, intercooling results in an increase in the cylinder charge so that it is possible to further increase torque and output power.

The drop in the combustion-air temperature also leads to a reduction in the temperature of the cylinder charge compressed during the compression cycle. This has the following advantages:

- Reduced tendency to knock
- Improved thermal efficiency resulting in lower fuel-consumption figures
- Reduced thermal loading of the pistons
- Lower NO_x emissions

Gasoline fuel injection: An overview

It is the job of the fuel-injection system, or carburetor, to meter to the engine the best-possible air/fuel mixture for the actual operating conditions.

Fuel-injection systems, particularly when they are electronically controlled, are far superior to carburetors in complying with the tight limits imposed on A/F-mixture composition. In addition, they are better from the point of view of fuel consumption, driveability, and power output. In the automotive sector, the demands imposed by increasingly severe emission-control legislation have led to the carburetor being completely superseded by electronic fuel injection.

At present, on the majority of these injection systems the A/F mixture is formed externally outside the combustion chamber (manifold injection). Systems based on internal A/F-mixture formation, that is with the fuel injected directly into the cylinder (gasoline direct injection), are coming more and more to the forefront though, since they have proved to be particularly suitable in the never-ending endeavours to reduce fuel consumption.

1 Multipoint fuel-injection system

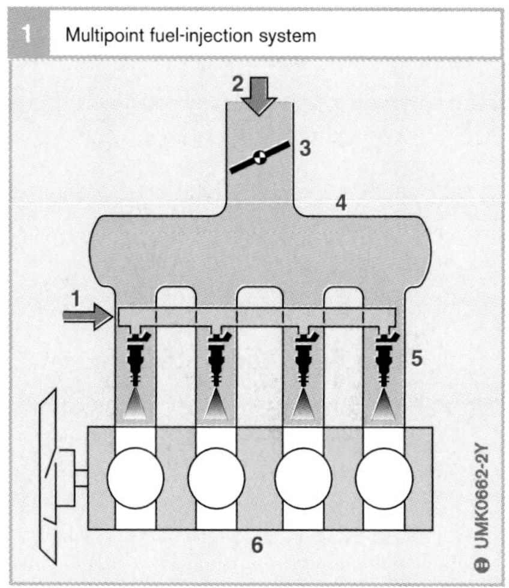

Fig. 1
1 Fuel
2 Air
3 Throttle valve
4 Intake manifold
5 Injector
6 Engine

Overview

External A/F-mixture formation

On gasoline injection systems with external A/F-mixture formation, the mixture is formed outside the combustion chamber, that is, in the intake manifold. Development of such systems was forced ahead to enable them to comply with increasingly severe demands. Today, only the electronically controlled multipoint injection systems are of any importance in this sector.

Multipoint fuel-injection systems

On a multipoint injection system, every cylinder is allocated its own injector which sprays the fuel directly onto the cylinder's intake valve (Fig. 1). Such injection systems are ideal for complying with the demands made on the A/F-mixture formation system.

Mechanical fuel-injection system

The K-Jetronic injection system operates without any form of drive from the engine, and injects fuel continuously. The injected fuel mass is not defined by the injector but by the system's fuel distributor.

Combined mechanical-electronic fuel-injection system

The KE-Jetronic is based on the basic mechanical system used for the K-Jetronic. Thanks to additional operational-data acquisition, this system features electronically controlled supplementary functions which permit the injected fuel quantity to be even more accurately adapted to changing engine operating conditions.

Electronic fuel-injection systems
Electronically controlled fuel-injection systems inject the fuel intermittently through electromagnetically operated injectors. The injected fuel quantity is defined by the injector opening time (for a given pressure drop across the injector).

Examples: L-Jetronic, LH-Jetronic, and Motronic in the form of an integrated engine-management system (M and ME-Motronic).

Single-point injection
Single-point injection (also known as throttle-body injection or TBI) features an electromagnetically operated injector located at a central point directly above the throttle valve. This injection system intermittently injects fuel into the intake manifold (Fig. 2). The Bosch single-point injection systems are designated Mono-Jetronic and Mono-Motronic.

Internal A/F-mixture formation

On direct-injection (DI) systems, the fuel is injected directly into the combustion chamber through electromagnetic injectors, one of which has been allocated to each cylinder (Fig. 3). A/F-mixture formation takes place inside the combustion chamber.

A/F-mixture formation inside the combustion chamber permits two completely different operating modes: In homogeneous operation, similar to external A/F-mixture formation, a homogeneous A/F mixture is present throughout the combustion chamber, and all the fresh air in the combustion chamber participates in the combustion process. This operating mode is therefore applied when high levels of torque are called for. In stratified-charge operation on the other hand, it is only necessary to have an ignitable A/F mixture around the spark plug. The remainder of the combustion chamber only contains fresh gas and residual gas without any unburnt-fuel content. This results in an extremely lean mixture at idle and part-load, with a corresponding drop in fuel consumption.

The MED-Motronic is used for the management of gasoline direct-injection engines.

2 Single-point injection (TBI) system

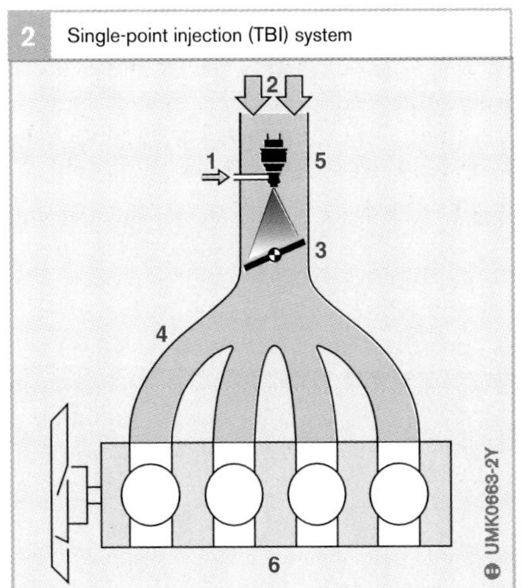

3 Direct-injection (DI) system

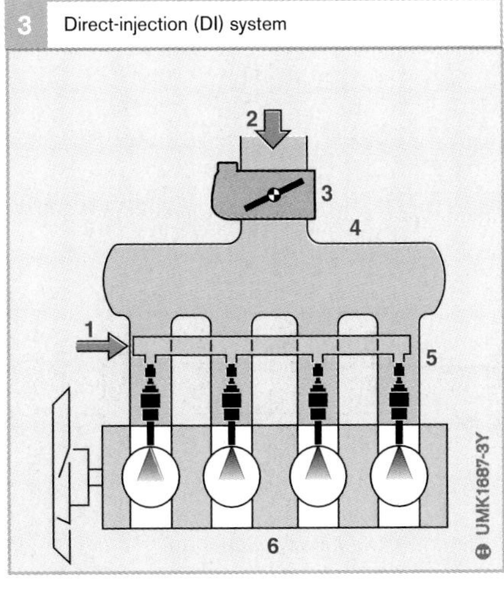

Fig. 2
1 Fuel
2 Air
3 Throttle valve
4 Intake manifold
5 Injector
6 Engine

Fig. 3
1 Fuel
2 Air
3 Throttle valve (ETC)
4 Intake manifold
5 Injectors
6 Engine

Beginnings of mixture formation

At a time when the first atmospheric engines were being designed by various inventors, a common problem was forming an ignitable mixture. The question of whether or not such an internal-combustion engine would be able to function at all. The solution was very much dependent on the ignition mechanism.

The basics of the carburetor were developed as early as the 18th Century. Inventors' efforts at the time were directed at vaporizing liquid fuels in such a way that they could be used to operate a lighting or heating device.

In 1795 Robert Street was the first to suggest to vaporize turpentine or creosote in an atmospheric engine. Around 1825 Samuel Morey and Eskine Hazard developed a two-cylinder engine for which they also designed the first carburetor. It was granted British patent no. 5402. Up to that time, mixture-formation systems were essentially fueled with turpentine or kerosene.

However, this situation changed in 1833, when Eilhardt Mitscherlich, professor of chemistry at the University of Berlin, managed to split benzoic acid by thermal cracking. The result was the so-called "Faraday's olefiant gas", which he called "benzene", the precursor of today's gasoline.

William Barnett designed the first carburetor for gasoline. He was awarded patent no. 7615 in 1838.

During this time, these designs were either wick carburetors (Fig. 1) or surface carburetors (Fig. 2). The first carburetor to be used in a vehicle was a wick carburetor. The wick drew up the fuel, similar to the oil-lamp principle. The wick was then exposed to an air stream in the engine, causing air and fuel to blend. By contrast, in the surface carburetor, in which the fuel was heated by the engine's exhaust gases, the result was a layer of vapor just above the fuel surface. This was then blended to form the required air-fuel mixture by introducing an air stream.

In Berlin in 1882 Siegfried Marcus applied for a patent for the brush carburetor (Fig. 3) that he had invented. This mixture generator used the interaction of a rapidly rotating cylindrical brush (3) driven by a drive pulley (1) and a fuel stripper (2) to form a mist of atomized fuel in the brush chamber (4). The fuel mist was then drawn into the engine via the inlet (5). The brush carburetor maintained its dominance for about 11 years.

1 Principle of a wick carburetor

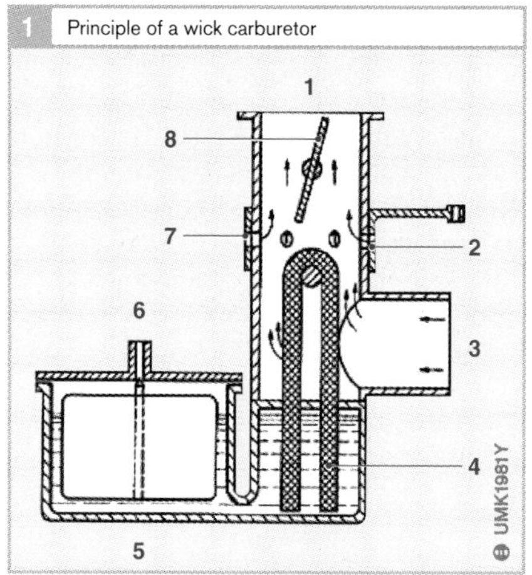

Fig. 1
1 Air/fuel mixture to engine
2 Annular slide valve
3 Air inlet
4 Wick
5 Float chamber with float
6 Fuel inlet
7 Auxiliary air
8 Throttle valve

2 Principle of a surface carburetor

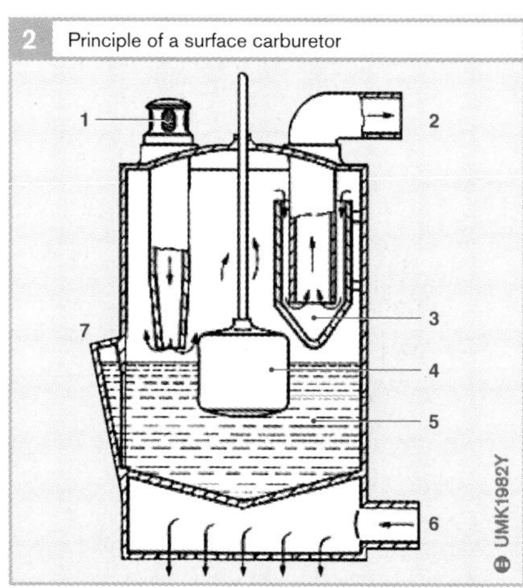

Fig. 2
1 Air inlet
2 A/F mixture to engine
3 Fuel separator
4 Float
5 Fuel
6 Engine exhaust gases
7 Fuel filler neck

3 Brush carburetor by Siegfried Marcus

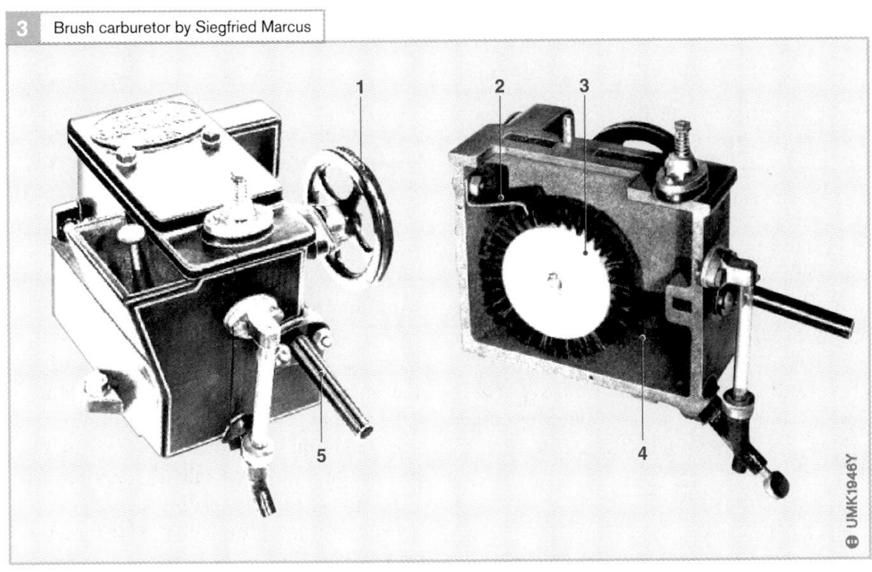

Fig. 3
1 Drive pulley
2 Fuel stripper
3 Rotating brush
4 Brush chamber
5 Intake fitting

In 1885 Nikolaus August Otto succeeded in his struggle to master the engine powered by hydrocarbon fuels (alcohol/gasoline); he had been working toward this goal since 1860. The first gasoline-engine, working on the four-stroke principle and equipped with a surface carburetor and an electrical ignition device of Otto's own construction, garnered the highest praise and profound recognition at the World Fair at Antwerp. This design was later built and sold in great numbers by the firm of Otto & Langen in Deutz over many years (Fig. 4).

4 Spark ignition engine by Nikolaus August Otto

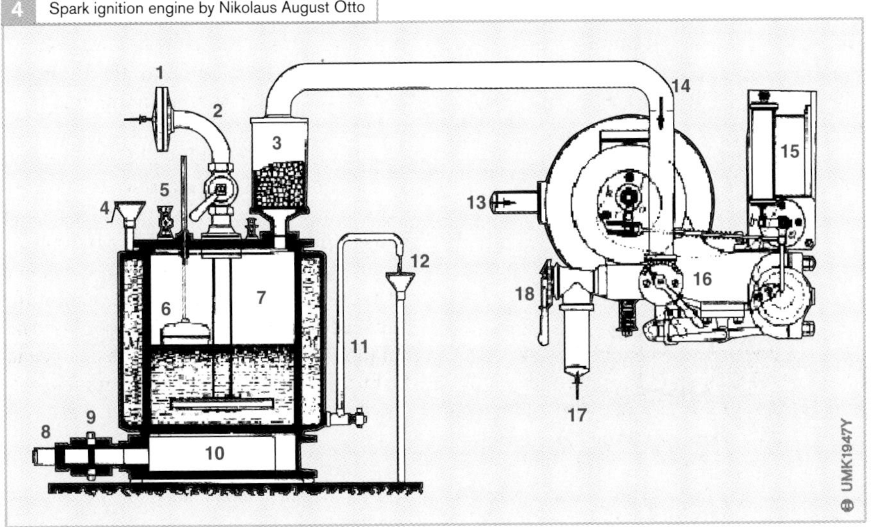

Fig. 4
A Carburetor
B Engine with ignition system

1 Air inlet
2 Air tube
3 Gravel canister (flame shield)
4 Water funnel
5 Fuel filler neck
6 Float
7 Fuel reservoir
8 Exhaust-gas inlet
9 Shutoff valve
10 Heating pad
11 Cooling-water jacket
12 Water tubing
13 Coolant inlet
14 Gas inlet
15 Ignition device
16 Gas shutoff valve
17 Air inlet
18 Air shutoff valve

In the same year Carl Benz installed a surface carburetor of his own design in his first "Patent Motor Carriage" (Fig. 5). A short while later, he improved the original design of his carburetor by adding a valve float, as he put it, "to always maintain the fuel level automatically at the same height".

In 1893 Wilhelm Maybach introduced his jet-nozzle carburetor (Fig. 7). In this device, the fuel was sprayed from a fuel nozzle onto a baffle surface, which caused the fuel to distribute in a cone-shaped pattern (Fig. 8).

5 Benz Motor Carriage with surface carburetor (vertically positioned)

6 Surface carburetor, 1885 (cutaway model)

7 Jet-nozzle carburetor by Wilhelm Maybach

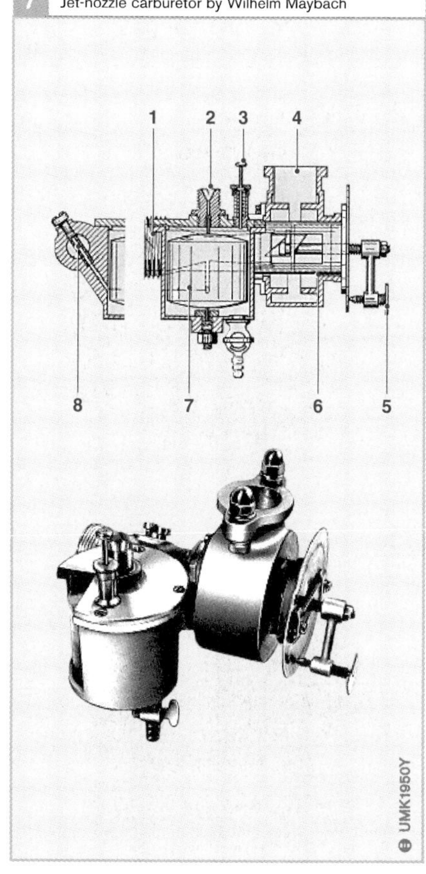

Fig. 7
1 Air inlet
2 Fuel inlet
3 Spring-loaded swab
4 A/F mixture outlet
5 Rotating-slide stop
6 Rotating-slide for mixture control
7 Float
8 Jet nozzle

1906/07 saw the introduction of the Claudel carburetor and the carburetor designs of François Bavery, both of which brought further advancement to carburetor design. In these carburetors, which were to become famous under the ZENITH brand name, the lean-fuel auxiliary or compensation jet delivers a virtually unchanged A/F mixture despite increasing air velocity (Fig. 9).

The same period also saw applications for the carburetor patents of Mennesson and Goudard. Their designs became world-famous under the SOLEX brand (Fig. 10).

The years that followed produced a proliferation of carburetor designs. In this context, some of their names, e.g. SUM, CUDELL, FAVORIT, ESCOMA, and GRAETZIN, deserve special mention. After the Haak carburetor was patented in 1906 and manufactured by the firm of PALLAS, Scüttler and Deutrich developed the PALLAS carburetor in 1912. It had a ring float and combination jet (Fig. 11).

In 1914 the Royal Prussian War Ministry sponsored a competition for benzol (benzene) carburetors. Even at that time the test specifications included the condition that the exhaust gases should be as clean as possible. Of the competing products bearing 14 different brand names, it was a ZENITH carburetor that won 1st prize. All the carburetors were examined at the test facility of the Technical University at Charlottenburg, and subjected to a demanding 800-km winter trial by the German army administration in automobiles of identical horsepower.

8 Principle of jet-nozzle carburetor

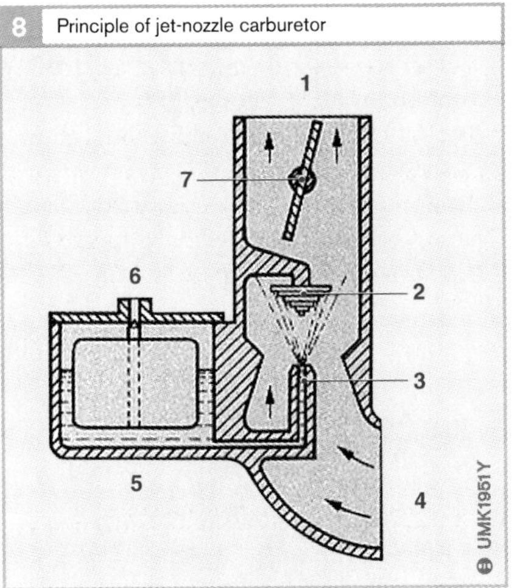

Fig. 8
1 Air/fuel mixture to engine
2 Baffle surface
3 Fuel jet (jet nozzle)
4 Air inlet
5 Float chamber with float
6 Fuel inlet
7 Throttle valve

9 ZENITH carburetor, type 22, 1910

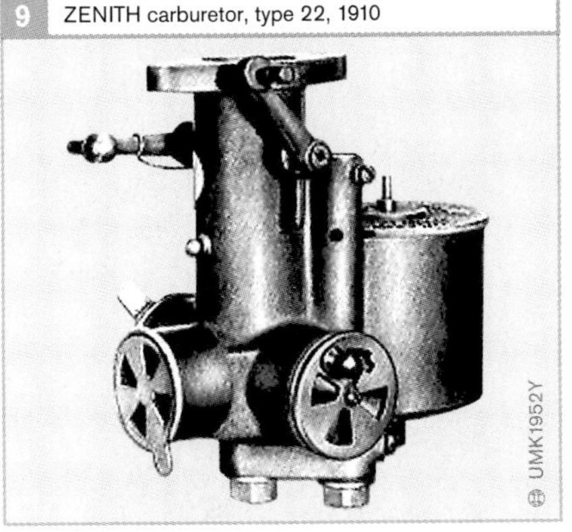

10 SOLEX carburetor, type DHR, 1912

11 PALLAS carburetor, type I, 1914

The period that followed marked the beginning of various attempts at detail work and specialization. A variety of model configurations and auxiliary devices was developed, e.g. rotating slides and preliminary throttles serving as start-assist measures, diaphragm systems replacing the float in aircraft carburetors, and pump systems providing acceleration aids. The diversity of these modifications was so extensive that any descriptive attempt would exceed the scope of this chapter.

In the 1920s, to obtain greater engine power, single and twin carburetors (carburetors featuring two throttle valves) were installed in the form of multiple carburetor systems (several single or twin carburetors with synchonized controls). In the decades that followed, the great variety of carburetor variants made by various manufacturers increased even further.

In parallel with the ongoing carburetor development for aircraft engines, the 1930s saw the development of the first gasoline-injection systems with direct injection (example in Fig. 12). This engine required two 12-cylinder in-line fuel-injection pumps, each of which was mounted on the crankcase between the cylinder banks (not visible in Fig. 12).
A pump of this type is shown in Fig. 13.
It has a total length of about 70 cm (27.5 in.).

In the late 1930s direct-injection systems were used in conjunction with the 9-cylinder radial BMW engines (Fig. 14) in the legendary three-engine Junkers Ju 52 aircraft. Especially noteworthy is the "boxer" (recip-

13 12-cylinder in-line fuel-injection pump (length approx. 70 cm)

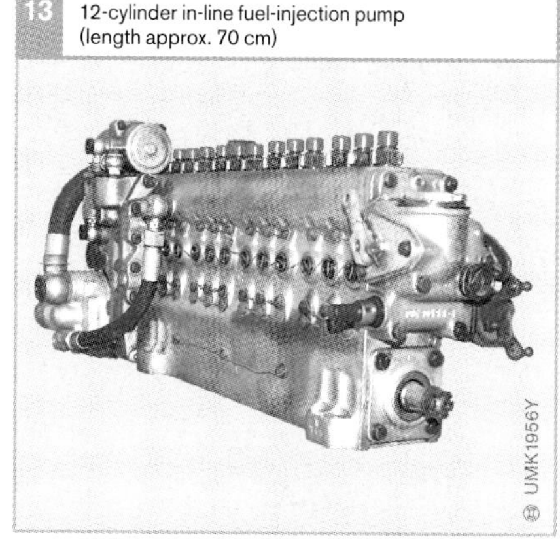

12 Aircraft engine by Daimler-Benz, with 24 cylinders in in-line-X configuration, type DB 604

UMK1955Y

Fig. 12
This aircraft engine was produced by the Daimler-Benz AG between 1939 and 1942.
There were models ranging from 48.5-liter displacement and 2,350 bhp (1,741 kW) to 50.0-liter displacement and 3,500 bhp (2,593 kW), all featuring gasoline direct injection by Bosch. The engine had a total length of 2.15 m.
(Photo: DaimlerChrysler Classic, Corporate Archives)

14 BMW radial engine with 9 cylinders

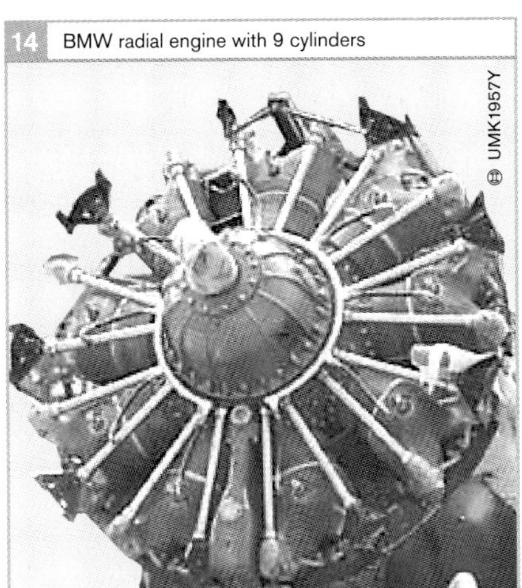

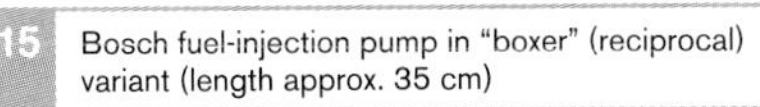

15 Bosch fuel-injection pump in "boxer" (reciprocal) variant (length approx. 35 cm)

rocal) configuration of the mechanical fuel-injection pump (Fig. 15) by Bosch.

In the 1950s this type of fuel-injection system working with direct fuel injection also made its debut in passenger cars. One example was the "Gutbrod Superior" of 1952 (Fig. 16), and the Goliath GP700E introduced in 1954 (Fig. 17). These two vehicles were compact cars powered by two-cylinder, two-stroke engines with a displacement of less than 1,000 cc. Their fuel-injection pumps were also compact in size (Fig. 18).

16 Gutbrod Superior 600 convertible (1950–1954; 1952 and later with direct injection)

17 Goliath GP700E (1951–1957; after 1954 with direct injection)

18 Two-cylinder fuel-injection pump (length approx. 15 cm)

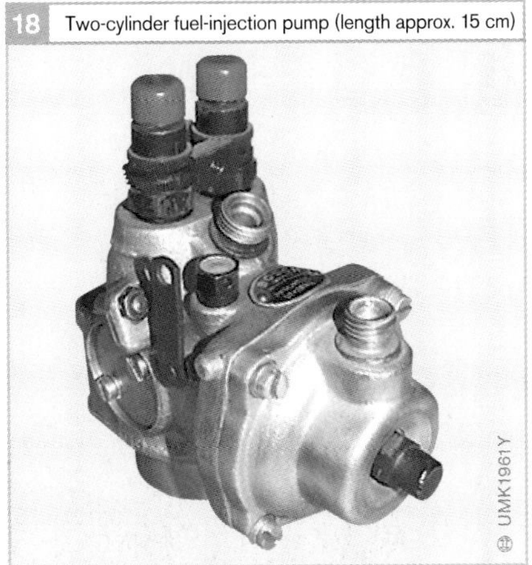

A component diagram of this two-cylinder fuel-injection system, which entered the annals of automotive history as the first gasoline direct-injection system in passenger cars, appears in Fig. 19 (next page).

19 Components of Bosch gasoline direct injection for the two-cycle engines in Gutbrod and Goliath cars

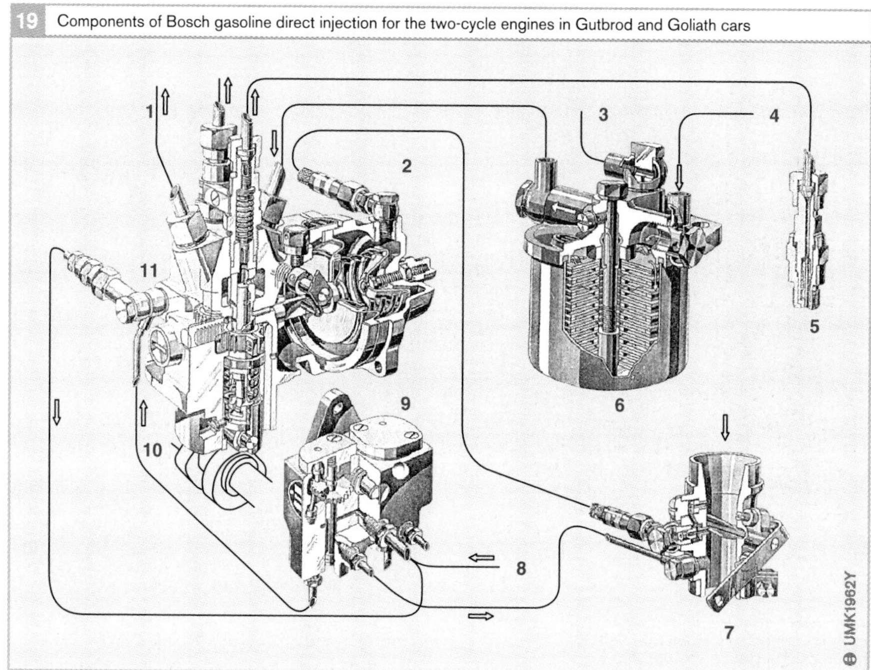

Fig. 19
1 Venting tube
2 Mixture-control-unit diaphragm block
3 Venting tube
4 From fuel tank
5 Fuel injector
6 Fuel filter
7 Mixture-control-unit flap supports
8 From oil reservoir
9 Lubricating-oil pump
10 Fuel-injection pump
11 Overflow valve

However, the Mercedes-Benz 300 SL sports car (Fig. 17) featured a Bosch-built gasoline direct-injection system. It was presented to the public on February 6, 1954 at the International Motor Sports Show in New York. Installed at a 50-degree slant, the car's 6-cylinder in-line engine (M198/11) had a displacement of 2996 cc, and delivered 215 bhp (159 kW).

A fundamentally different facet of A/F mixture formation for gasoline engines appeared during the latter part of WWII and for a while thereafter: the wood-gas generator.

The wood gas emitted by the glowing charcoal was used to form an ignitable A/F mixture (Fig. 21). However, it was hard to overlook the physical size of these wood gasifier systems (Fig. 22).

20 Mercedes-Benz 300 SL (1954)

Fig. 20
Photo:
DaimlerChrysler Classic, Corporate Archives

21 Schematic of a wood-gas generator system

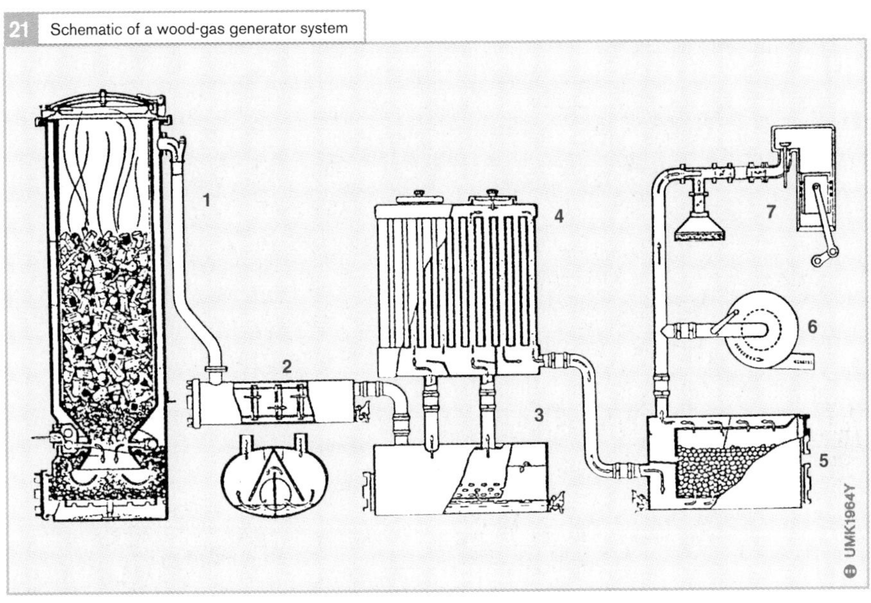

Fig. 21
1 Gas generator
2 Baffle plate cleaner
3 Settling tank
4 Gas cooler
5 Secondary cleaner
6 Bellows blower
7 Regulator group

22 Wood-gas generator system on a 1936 Adler Diplomat

Due to the gradual tightening-up of emission standards, the automotive industry has increasingly shunned the carburetor. In the early 1990s, however, there was a successful design by Bosch and Pierburg that equipped a modified conventional carburetor with modern-day actuators: the Ecotronic carburetor (Fig. 23). This carburetor made it possible to comply with the prevailing emission standards at the time, while at the same time ensuring economical fuel consumption.

At the conclusion of this brief review of the history of A/F mixture formation, it should be stated that the various carburetor types continued to be used well into the 1990s as standard equipment in passenger cars. In compact cars in particular, the carburetor enjoyed sustained popularity due to its lower cost.

23 Ecotronic (2EE) carburetor

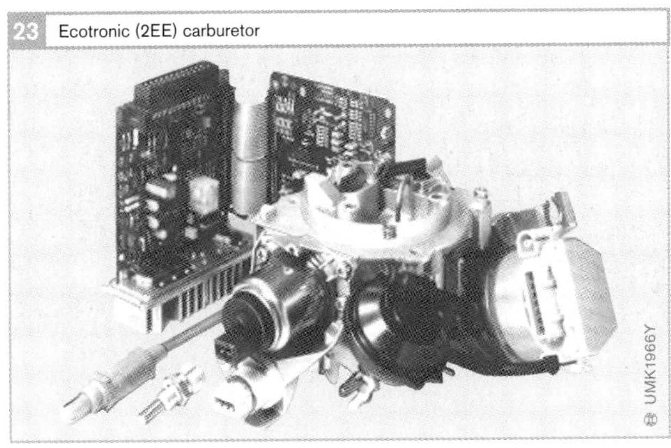

Evolution of gasoline injection systems

As early as 1885, when the first manifold injection system was introduced on a fixed industrial engine, many changes have occurred in the realm of gasoline-injection systems. Later attempts included a floatless carburetor with attached fuel-injection device installed in aircraft engines in 1925, and an electrically triggered fuel-injection device in a racing motorcycle in 1930. That was before Bosch finally developed a mechanically driven gasoline-injection pump for the Gutbrod Superior 600 and Goliath GP 700 vehicles. These were the first passenger cars with gasoline injection. The system was designed as a direct injection system. Even the legendary Mercedes 300 SL was equipped with gasoline direct injection featuring a mechanical in-line pump.

After passing though several development stages of manifold injection systems (described below), the current trend is again headed for direct-injection systems.

D-Jetronic

System overview

The pressure-controlled fuel-injection system, first introduced on the German market in the 1967 Volkswagen 1600 LE, was the first-ever system incorporating an electronic control unit (ECU). The pressure sensor (known as a "pressure probe" at the time) measures the pressure in the intake manifold and passes the obtained value on to the ECU as a representative engine-load variable. This is why it was given the name "D-Jetronic".

The electronic control unit (Fig. 1, Item 1) receives signals for intake-manifold pressure, intake-air temperature, cooling-water and/or cylinder-head temperature, throttle-valve position and movement, and starting, engine speed, and start of injection. The ECU processes this data, and sends electrical pulses to the fuel injectors (2). The electronic control unit is interconnected with the electrical components via a multiple connector and wiring harness. It contains approximately 300 components, about 70 of which are semiconductors.

1 D-Jetronic system schematic

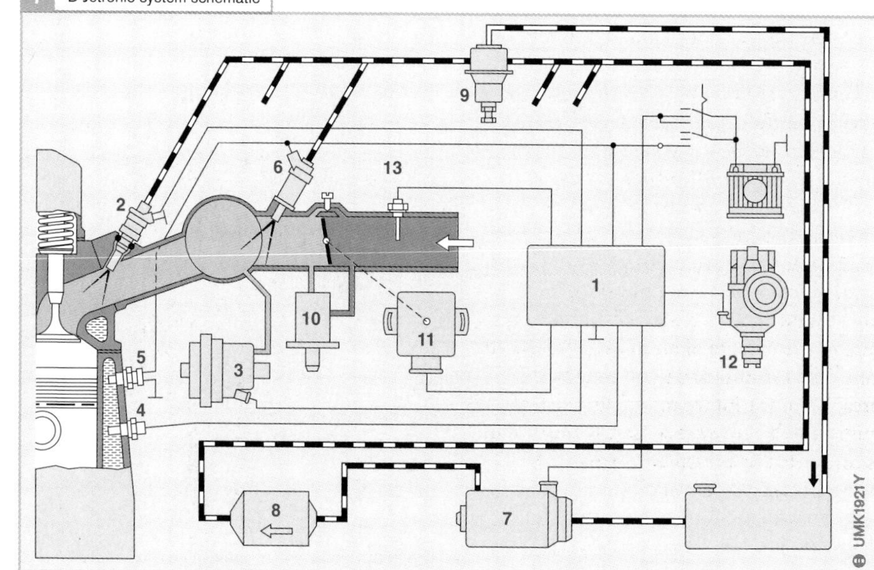

Fig. 1
1 Electronic control unit
2 Fuel injector
3 Pressure sensor
4 Temperature sensor, cooling water
5 Thermostatic switch/thermo-time switch
6 Electric start valve
7 Electric fuel pump
8 Fuel filter
9 Fuel-pressure regulator
10 Auxiliary-air valve
11 Throttle-valve switch
12 Injection trigger
13 Air-temperature sensor

The fuel injectors (2) spray the fuel into the intake manifolds of the cylinders. The pressure sensor (3) sends engine load data to the ECU. The temperature sensors (4) report the temperatures of air (13) and cooling water (4) to the electronic control unit. The thermostatic switch, also known as a thermotime switch (5), switches the electric cold-start valve (6), which injects additional fuel into the intake manifold during low-temperature starts. The electric fuel pump (7) continuously delivers fuel to the fuel injectors. The fuel filter (8) is integrated in the fuel line to remove contaminants. The fuel-pressure regulator (9) maintains a constant fuel pressure in the fuel lines. The temperature-dependent function of the auxiliary-air device (10) provides additional air during engine warm-up. The throttle-valve switch (11) sends engine idle, acceleration, and full-load states to the electronic control unit. The injection trigger (12) is located in the ignition distributor, and provides the ECU with control pulses signaling the start of injection and engine speed.

Mode of operation

The Bosch-engineered D-Jetronic comprises a gasoline injection system that is essentially controlled by intake-manifold pressure and engine speed.

Upstream of the throttle valve inside the intake manifold, the pressure is equal to the ambient atmospheric pressure. The pressure downstream of the throttle valve is lower, and changes with the throttle-valve position. This reduced intake-manifold pressure serves as the measured quantity for engine load, which is the most important information. Engine load is derived from the measure of the volume of air drawn into the engine. The information pertaining to the pressure in the intake manifold is determined by the pressure sensor.As the system is controlled by pressure (German "Druck"), it was named "D-Jetronic".

Start of injection

Special contact points in the ignition distributor determine – in accordance with camshaft control – the start of the pulse for opening the fuel injectors (Fig. 2). These contacts are located under the centrifugal adjusting device inside the ignition distributor, and are actuated by a cam on the distributor shaft. The ECU also derives engine speed data from the timed distance between the trigger pulses. Engine speed then becomes a factor in computing injection duration.

Injection duration

Injection duration is mainly dependent on two factors: engine load and engine speed. A pressure sensor and an injection trigger supply the required signals to the electronic control unit. After computing the required duration, the ECU sends electrical pulses to the fuel injectors to inject a greater or lesser quantity of fuel. This is the process for determining the "basic fuel quantity".

2 Ignition distributor with injection trigger

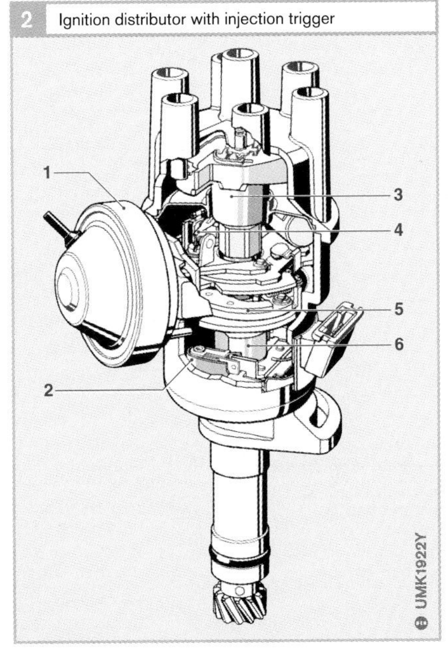

Fig. 2
1 Vacuum unit
2 Injection trigger
3 Distributor rotor
4 Distributor contact points
5 Centrifugal advance mechanism
6 Distributor cam

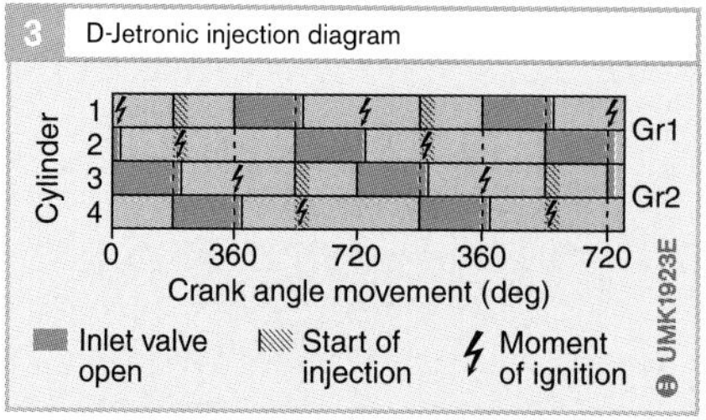

3 D-Jetronic injection diagram

It is instructive to note the following: The start of fuel injection is triggered by the injection trigger in the ignition distributor. The duration of the injection cycle – and thus the injected-fuel quantity – is determined by the pressure sensor via the electronic time switch on the electronic control unit. The injection-valve group (Fig. 3) is open throughout the pulse duration.

Pressure sensor

The measuring system of the pressure sensor (Fig. 4) is contained in a sealed metal housing. There is a line providing the connection to the engine intake manifold (Fig. 5).

The pressure sensor contains two aneroid capsules (Fig. 4, Items 2, 3) which shift the armature (6) of a coil (5). As the load increases, i.e. as pressure in the intake manifold increases, the aneroid capsules are compressed and the armature is drawn deeper into the coil. This changes its inductance.

The device is therefore a measuring transducer that converts a pneumatic pulse into an electrical signal. The induction-type pulse generator in the pressure sensor is connected to an electronic timer on the electronic control unit. It determines the duration of the electrical pulses required to trigger the fuel injectors. In this way, the intake-manifold pressure is directly converted to an injection duration.

When the throttle valve is closed, the intake-manifold pressure is low. The aneroid capsules are therefore compressed to a lesser degree, and push the armature out of the coil (Fig. 6). The coil inductance lowers, the pulse becomes shorter, and the valves inject a smaller amount of fuel.

Adaptation to operating conditions

Wide-open throttle

When the engine is in part-load range, fuel-quantity metering attempts to keep both fuel consumption and the ratio of unburned exhaust-gas components is as low as possible. At wide-open throttle (WOT), however, the fuel quantity is determined in accordance with the maximum engine output, which means that additional fuel must be injected when the engine is operating at wide-open throttle.

4 Pressure sensor with additional diaphragm for full-load enrichment

Fig. 4
1 Diaphragm
2 Aneroid capsule
3 Aneroid capsule
4 Leaf spring
5 Coil
6 Armature
7 Core
8 Part-load stop
9 Full-load stop
10 Valve

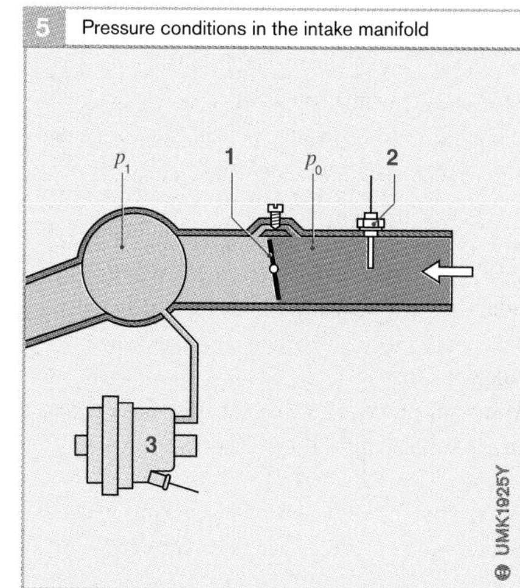

5 Pressure conditions in the intake manifold

Fig. 5
p_0 Atmospheric pressure
p_1 Pressure in the intake manifold

1 Throttle valve
2 Temperature sensor
3 Pressure sensor

The data for this full-load enrichment (pressure conditions in the intake manifold) is provided by the pressure sensor. The armature movement induced by the aneroid capsule changes the inductance of the coil. In the part-load range (Fig. 7), the atmospheric pressure p_0 is greater than the pressure in the intake manifold p_1. The diaphragm therefore contacts its part-load stop. Only aneroid capsules 1 and 2 act on the armature.

At wide-open throttle (Fig. 8), the intake-manifold pressure is roughly equivalent to atmospheric pressure. This allows the spring to press the diaphragm against its full-load stop. This additional movement signals the full-load status to the electronic control unit by superimposing itself on the movement caused by aneroid capsules 1 and 2. In systems that are subject to more stringent emission limits, full-load enrichment is controlled by an additional contact in the throttle-valve switch. As a result, the action of the diaphragm segment of the pressure sensor is not required for full-load enrichment.

Altitude compensation

In systems where full-load enrichment is performed by a throttle-valve switch, the pressure sensor does not contain a dual aneroid capsule. Instead, there is a closed aneroid capsule plus a second aneroid capsule that is open to ambient atmospheric pressure. In this way, not only the pressure in the intake manifold but also the differential pressure between ambient atmosphere and intake manifold are factored into the calculation. In practical terms, this means that, when the engine is operating in part-load range, the engine can be much better adapted to operation at varying altitudes.

Acceleration

When the throttle valve is opened – i.e. during acceleration – there is a slight delay in the pressure-rise signal that the pressure sensor sends to the electronic control unit. This brief response delay is a consequence of the interval required for the pressure to

6 Pressure sensor at idle $p_1 << p_0$

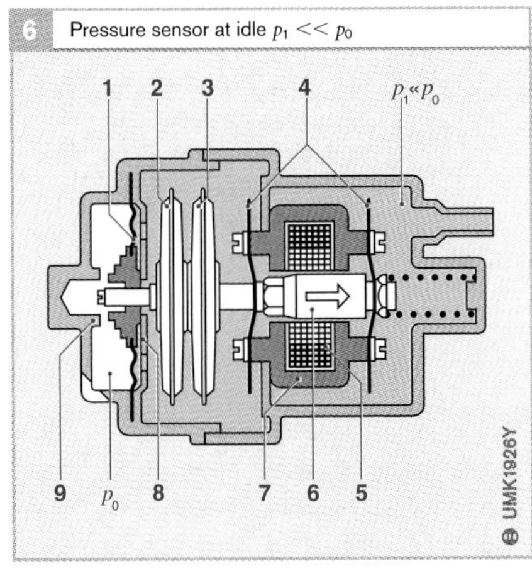

Fig. 6
Basic function:
Aneroid capsules 2 and 3 expanded
1 Diaphragm
2 Aneroid capsule
3 Aneroid capsule
4 Leaf spring
5 Coil
6 Armature
7 Core
8 Part-load stop
9 Full-load stop

p_0 Atmospheric pressure
p_1 Intake-manifold pressure

7 Pressure sensor at part-load $p_1 < p_0$

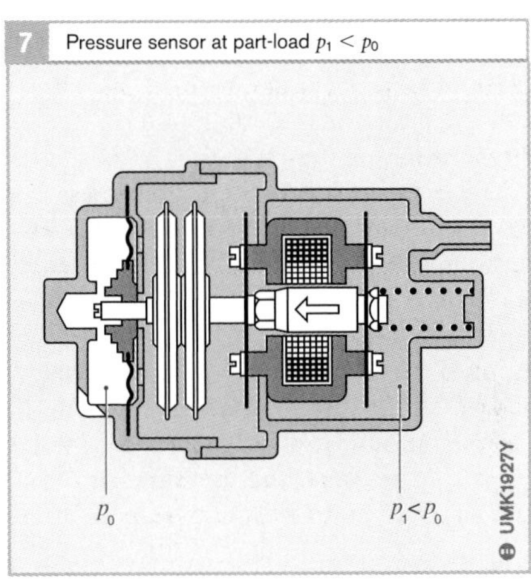

Fig. 7
Basic function:
Aneroid capsules 2 and 3 slightly compressed;
Auxiliary function:
Diaphragm 1 contacts part-load stop
p_0 Atmospheric pressure
p_1 Intake-manifold pressure

8 Pressure sensor at full load $p_1 \approx p_0$

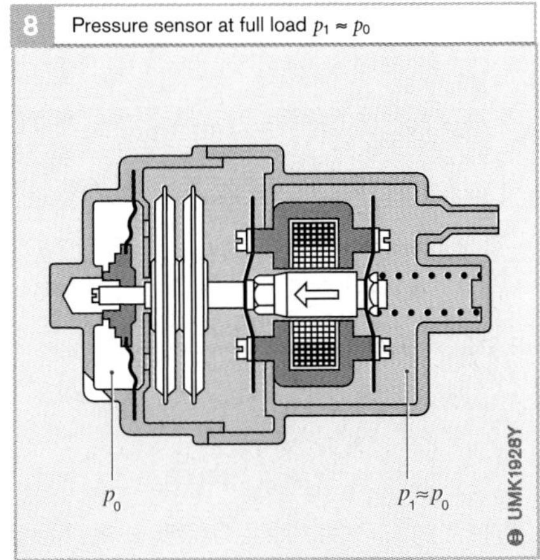

Fig. 8
Basic function:
Aneroid capsules 2 and 3 compressed;
Auxiliary function:
Diaphragm 1 contacts full-load stop
p_0 Atmospheric pressure
p_1 Intake-manifold pressure

9 Throttle-valve switch

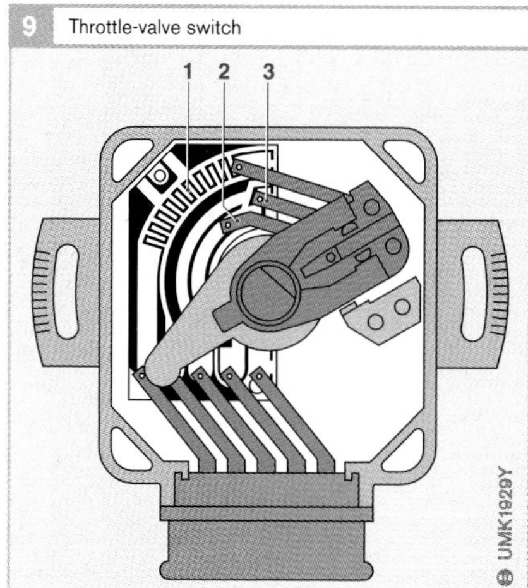

Fig. 9
1 Contact track for acceleration enrichment
2 Full-load contact
3 Idle contact

build up in the pressure sensor in response to a change in throttle-valve position. To bridge this brief response delay, a throttle-valve switch is used. This causes the electronic control unit to provide additional injection pulses as the throttle valve is opened.

Throttle-valve switch

The throttle-valve switch is equipped with sliding contacts and contact tracks. It is actuated directly by the throttle-valve shaft depending on throttle-valve movement.

During acceleration, the sliding contact runs along serrated contact tracks. The objective here is to extend the injection time and add injection pulses for acceleration enrichment.

Intake-air temperature

The injected-fuel quantity is essentially controlled by the intake-manifold pressure. However, the fuel quantity can be controlled only at a constant temperature since the density of the intake air is higher when the ambient temperature is lower. If the control unit does not take account of the air temperature, the A/F mixture would become too lean. This may cause combustion misses, especially at outside temperatures of 0 °C to –20 °C. To overcome this disadvantage, a temperature sensor is integrated. As air temperature drops (higher air density), the temperature sensor acts on the electronic control unit to increase the injected-fuel quantity as a factor of air density.

The temperature sensor consists of a temperature-sensitive resistor encased in a metal housing.

The story of fuel injection

The story of fuel injection extends back to cover a period of almost one hundred years. The Gasmotorenfabik Deutz was manufacturing plunger pumps for injecting fuel in a limited production series as early as 1898.

A short time later the uses of the venturi-effect for carburetor design were discovered, and fuel-injection systems based on the technology of the time ceased to be competitive.

Bosch started research on gasoline-injection pumps in 1912. The first aircraft engine featuring Bosch fuel injection, a 1,200-hp unit, entered series production in 1937; problems with carburetor icing and fire hazards had lent special impetus to fuel-injection development work for the aeronautics field. This development marks the beginning of the era of fuel injection at Bosch, but there was still a long path to travel on the way to fuel injection for passenger cars.

1952 saw a Bosch direct-injection unit being featured as standard equipment on a small car for the first time. A unit was then installed in the 300 SL, the legendary production sports car from Daimler-Benz. In the years that followed, development on mechanical injection pumps continued, and ...

In 1967 fuel injection took another giant step forward: The first electronic injection system: the intake-pressure-controlled D-Jetronic!

In 1973 the air-flow-controlled L-Jetronic appeared on the market, at the same time as the K-Jetronic, which featured mechanical-hydraulic control and was also an air-flow-controlled system.

In 1976, the K-Jetronic was the first automotive system to incorporate a Lambda closed-loop control.

1979 marked the introduction of a new system: Motronic, featuring digital processing for numerous engine functions. This system combined L-Jetronic with electronic program-map control for the ignition. The first automotive microprocessor!

In 1982, the K-Jetronic model became available in an expanded configuration, the KE-Jetronic, including an electronic closed-loop control circuit and a Lambda oxygen sensor.

These were joined by Bosch Mono-Jetronic in 1987: This particularly cost-efficient single-point injection unit made it feasible to equip small vehicles with Jetronic, and once and for all made the carburetor absolutely superfluous.

Bosch gasoline fuel injection from the year 1954

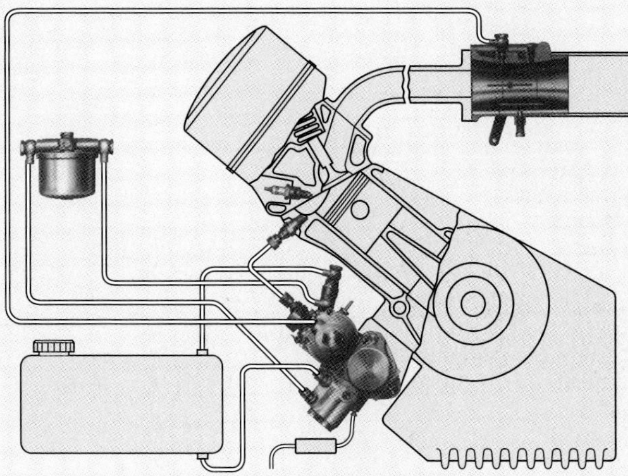

UMK1355Y

K-Jetronic

System overview

The K-Jetronic is a mechanically and hydraulically controlled fuel-injection system which needs no form of drive and which meters the fuel as a function of the intake air quantity and injects it continuously onto the engine intake valves. Specific operating conditions of the engine require corrective intervention in mixture formation and this is carried out by the K-Jetronic in order to optimize starting and driving performance, power output and exhaust composition. Owing to the direct air-flow sensing, the K-Jetronic system also allows for engine variations and permits the use of facilities for exhaust-gas aftertreatment for which precise metering of the intake air quantity is a prerequisite.

The K-Jetronic was originally designed as a purely mechanical injection system. Later, using auxiliary electronic equipment, the system also permitted the use of lambda closed-loop control.

The K-Jetronic fuel-injection system covers the following functional areas:

- Fuel supply
- Air-flow measurement and
- Fuel metering

Fuel supply

An electrically driven fuel pump (Figure 1, Pos. 2) delivers the fuel to the fuel distributor (9) via a fuel accumulator (3) and a filter (4). The fuel distributor allocates this fuel to the injection valves (6) of the individual cylinders.

Air-flow measurement

The amount of air drawn in by the engine is controlled by a throttle valve (16) and measured by an air-flow sensor (10).

Fuel metering

The amount of air, corresponding to the position of the throttle plate, drawn in by the engine serves as the criterion for metering of the fuel to the individual cylinders. The amount of air drawn in by the engine is measured by the air-flow sensor which, in turn, controls the fuel distributor. The air-flow sensor and the fuel distributor are assemblies which form part of the mixture control unit.

1 Schematic diagram of the K-Jetronic system with closed-loop lambda control

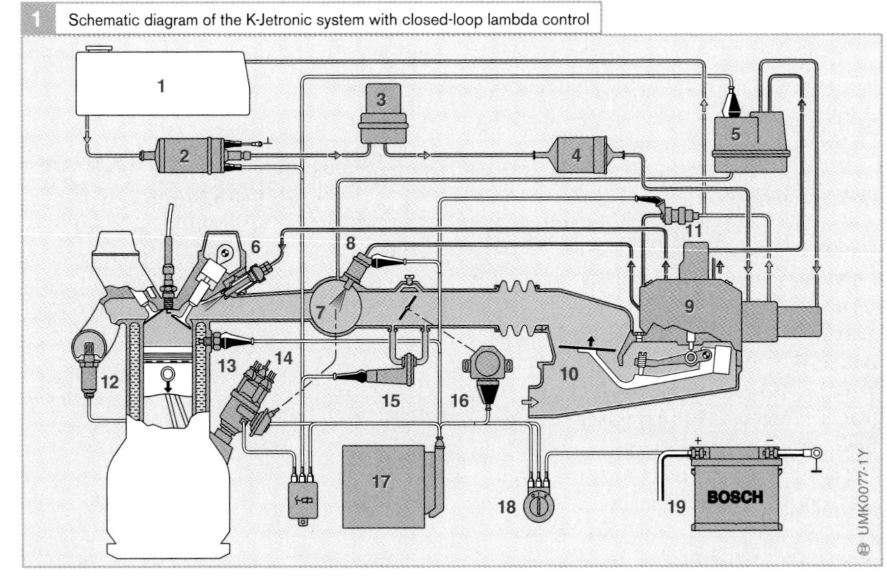

Fig. 1
1 Fuel tank
2 Electric fuel pump
3 Fuel accumulator
4 Fuel filter
5 Warm-up regulator
6 Injection valve
7 Intake manifold
8 Cold-start valve
9 Fuel distributor
10 Air-flow sensor
11 Timing valve
12 Lambda sensor
13 Thermo-time switch
14 Ignition distributor
15 Auxiliary-air device
16 Throttle-valve switch
17 ECU (for version with lambda closed-loop control)
18 Ignition and starting switch
19 Battery

Injection occurs continuously, i.e. without regard to the position of the intake valve. When the intake valve is closed, the fuel is "stored".

Mixture enrichment is controlled in order to adapt to various operating conditions such as start, warm-up, idle and full load. In addition, supplementary functions such as overrun fuel cut-off, engine-speed limiting and closed-loop lambda control are possible.

Fuel supply

The fuel supply system comprises

- Electric fuel pump
- Fuel accumulator
- Fine filter
- Primary-pressure regulator and
- Injection valves

An electrically driven roller-cell pump pumps the fuel from the fuel tank at a pressure of over 5 bar to a fuel accumulator and through a filter to the fuel distributor. From the fuel distributor the fuel flows to the injection valves. The injection valves inject the fuel continuously into the intake ports of the engine. Thus the system designation K-Jetronic (taken from the German for continuous). When the intake valves open, the mixture is drawn into the cylinder.

The fuel primary-pressure regulator maintains the supply pressure in the system constant and reroutes the excess fuel back to the fuel tank.

Electric fuel pump

The electric fuel pump is a roller-cell pump driven by a permanent-magnet electric motor. It delivers more fuel than the maximum requirement of the engine so that compression in the fuel system can be maintained under all operating conditions. A check valve in the pump decouples the fuel system from the fuel tank by preventing reverse flow of fuel to the fuel tank.

The electric fuel pump starts to operate immediately when the ignition and starting switches are operated and remains switched on continuously after the engine has started. A safety circuit is incorporated to stop the pump running and, thus, to prevent fuel being delivered if the ignition is switched on but the engine has stopped turning (for instance in the case of an accident).

The fuel pump is located in the immediate vicinity of the fuel tank and requires no maintenance.

Fuel accumulator

The fuel accumulator maintains the pressure in the fuel system for a certain time after the engine has been switched off in order to facilitate restarting, particularly when the engine is hot. The special design of the accumulator housing deadens the sound of the fuel pump when the engine is running.

The interior of the fuel accumulator is divided into two chambers by means of a diaphragm. One chamber serves as the accumulator for the fuel whilst the other represents the compensation volume and is connected to the atmosphere or to the fuel tank by means of a vent fitting. During operation, the accumulator chamber is filled with fuel and the diaphragm is caused to bend back against the force of the spring until it is halted by the stops in the spring chamber. The diaphragm remains in this position, which corresponds to the maximum accumulator volume, as long as the engine is running.

Fuel filter

The fuel filter retains particles of dirt which are present in the fuel and which would otherwise have an adverse effect on the functioning of the injection system. The fuel filter contains a paper element with a mean pore size of 10 μm backed up by a fluff trap.

Primary-pressure regulator

The primary-pressure regulator maintains the pressure in the fuel system constant. It is incorporated in the fuel distributor and holds the delivery pressure (system pressure) at about 5 bar. The fuel pump always delivers more fuel than is required by the vehicle engine, and this causes a plunger to shift in the pressure regulator and open a port through which excess fuel can return to the tank.

The pressure in the fuel system and the force exerted by the spring on the pressure-regulator plunger (Figure 2, Pos. 4) balance each other out. If, for instance, fuel-pump delivery drops slightly, the plunger is shifted by the spring (5) to a corresponding new position and in doing so closes off the port slightly through which the excess fuel returns to the tank. This means that less fuel is diverted off at this point and the system pressure is controlled to its specified level.

When the engine is switched off, the fuel pump also switches off and the primary pressure drops below the opening pressure of the injection valves. The pressure regulator then closes the return-flow port and thus prevents the pressure in the fuel system from sinking any further.

2 Primary-pressure regulator fitted to fuel distributor

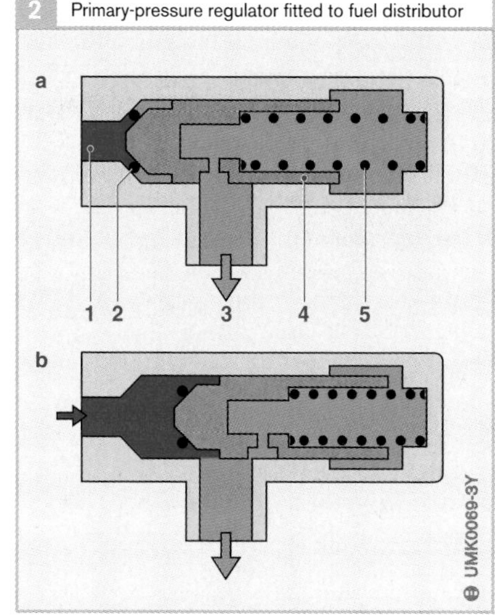

Fig. 2
a In rest position
b In actuated position

1 System-pressure entry
2 Seal
3 Return to fuel tank
4 Plunger
5 Spring

Fuel-injection valves

The injection valves open at a given pressure and inject the fuel metered to them into the intake passages and onto the intake valves. They are secured in special holders to insulate them against the heat radiated from the engine. The injection valves have no metering function themselves, and open of their own accord when the opening pressure of e.g. 3.5 bar is exceeded. The needle valve (Fig. 3), whose needle (3) oscillates ("chatters") audibly at high frequency when fuel is injected, ensures excellent atomization of the fuel even with the smallest of injection quantities. When the engine is switched off, the injection valves close tightly when the pressure in the fuel-supply system drops below their opening pressure. This means that no more fuel can enter the intake passages once the engine has stopped.

Air-shrouded injection valves improve the mixture formation particularly at idle. Using the pressure drop across the throttle valve, a portion of the air inducted by the engine is drawn into the cylinder through the injection valve: The result is excellent atomization of the fuel at the point of exit. Air-shrouded injection valves reduce fuel consumption and toxic emission constituents.

3 Fuel-injection valve

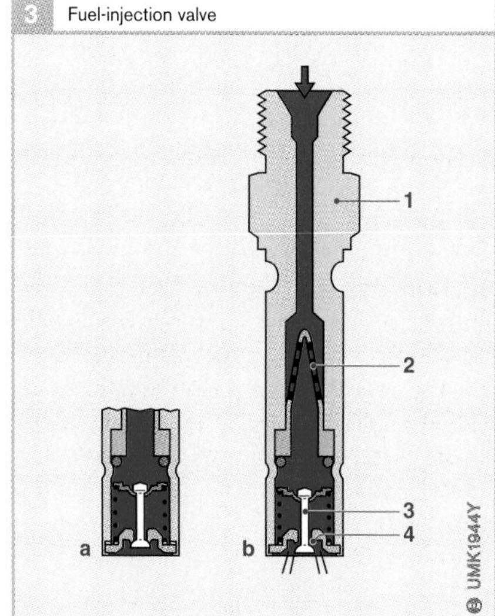

Fig. 3
a In rest position
b In actuated position

1 Valve housing
2 Filter
3 Valve needle
4 Valve seat

Fuel metering

The task of the fuel-management system is to meter a quantity of fuel corresponding to the intake air quantity.

Basically, fuel metering is carried out by the mixture control unit. This comprises the air-flow sensor and the fuel distributor. In a number of operating modes however, the amount of fuel required deviates greatly from the "standard" quantity and it becomes necessary to intervene in the mixture formation system (see section "Adaptation to operating conditions").

Air-flow sensor

The quantity of air drawn in by the engine is a precise measure of its operating load. The air-flow sensor operates according to the suspended-body principle, and measures the amount of air drawn in by the engine.

The intake air quantity serves as the main actuating variable for determining the basic injection quantity. It is the suitable variable for deriving the fuel requirement, and changes in the induction characteristics of the engine have no effect upon the formation of the air-fuel mixture.

Since the air drawn in by the engine must pass through the air-flow sensor before it reaches the engine, this means that it has been measured and the control signal generated before it actually enters the engine cylinders. The result is that, in addition to other measures described below, the correct mixture adaptation takes place at all times.

The air-flow sensor is located upstream of the throttle valve so that it measures all the air which enters the engine cylinders. It comprises an air funnel (Figure 4, Pos. 1) in which the sensor plate (suspended body, Pos. 2) is free to pivot. The air flowing through the funnel deflects the sensor plate by a given amount out of its zero position, and this movement is transmitted by a lever system to a control plunger which determines the basic injection quantity required for the basic functions.

4 Updraft air-flow sensor

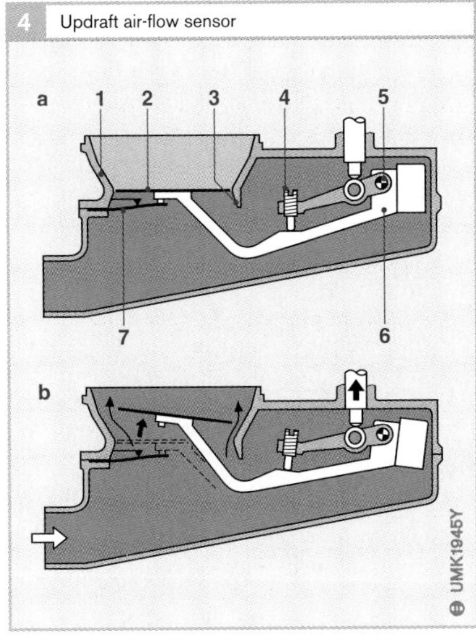

Fig. 4
a Sensor plate in its zero position
b Sensor plate in its operating position

1 Air funnel
2 Sensor plate
3 Relief cross-section
4 Idle-mixture adjusting screw
5 Pivot
6 Lever
7 Leaf spring

Considerable pressure shocks can occur in the intake system if backfiring takes place in the intake manifold. For this reason, the air-flow sensor is so designed that the sensor plate can swing back in the opposite direction in the event of misfire, and past its zero position to open a relief cross-section in the funnel (3). A rubber buffer limits the downward stroke (the upwards stroke on the downdraft air-flow sensor).

A counterweight compensates for the weight of the sensor plate and lever system (this is carried out by an extension spring on the downdraft air-flow sensor). A leaf spring (7) ensures the correct zero position in the switched-off phase.

Fuel distributor

Depending upon the position of the plate in the air-flow sensor, the fuel distributor meters the basic injection quantity to the individual engine cylinders. The position of the sensor plate is a measure of the amount of air drawn in by the engine. The position of the plate is transmitted to the control plunger by a lever (Figure 5, next page, Pos. 5).

5 Fuel distributor

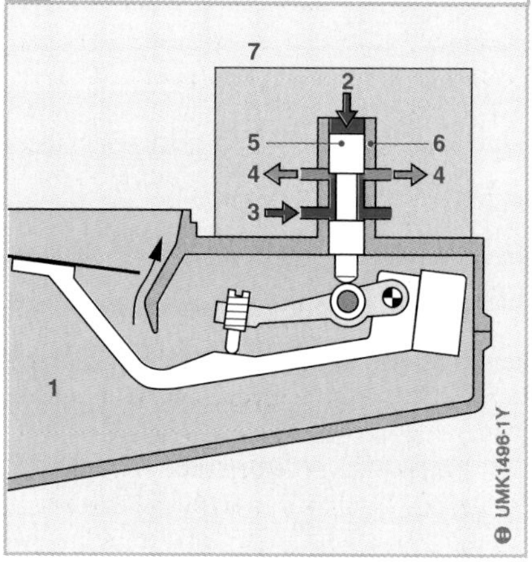

Fig. 5
1 Intake air
2 Control pressure
3 Fuel inlet
4 Metered quantity of fuel
5 Control plunger
6 Barrel with metering slits
7 Fuel distributor

Depending upon its position in the barrel with metering slits, the control plunger opens or closes the slits to a greater or lesser extent. The fuel flows through the open section of the slits to the differential pressure valves and then to the fuel injection valves.

If sensor-plate travel is only small, then the control plunger is lifted only slightly and, as a result, only a small section of the slit is opened for the passage of fuel. With larger plunger travel, the plunger opens a larger section of the slits and more fuel can flow. There is a linear relationship between sensor-plate travel and the slit section in the barrel which is opened for fuel flow.

6 Differential pressure valve with primary pressure and control pressure

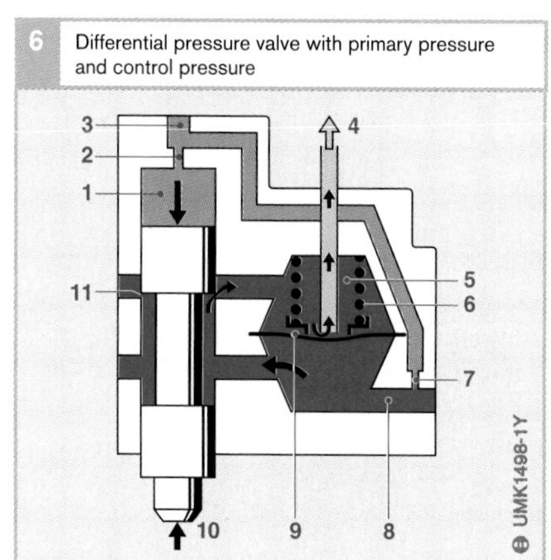

Fig. 6
1 Control-pressure effect (hydraulic force)
2 Damping restriction
3 Line to warm-up regulator
4 to the intake valve
5 Pressure in the upper chamber of the differential pressure valve (0.1 bar < primary pressure)
6 Governor spring
7 Decoupling restriction bore
8 Pressure in the lower chamber = primary pressure (delivery pressure)
9 Diaphragm
10 Effect of air pressure
11 Metering slits

A hydraulic force generated by the so-called control pressure is applied to the control plunger. It opposes the movement resulting from sensor-plate deflection. One of its functions is to ensure that the control plunger follows the sensor-plate movement immediately and does not, for instance, stick in the upper end position when the sensor plate moves down again. Further functions of the control pressure are discussed in the sections "Warm-up enrichment" and "Full-load enrichment".

Control pressure

The control pressure is tapped from the primary pressure through a restriction bore (Figure 6). This restriction bore serves to decouple the control-pressure circuit and the primary-pressure circuit from one another. A connection line joins the fuel distributor and the warm-up regulator (control-pressure regulator).

When starting the cold engine, the control pressure is about 0.5 bar. As the engine warms up, the warm-up regulator increases the control pressure to about 3.7 bar. The control pressure acts through a damping restriction on the control plunger and thereby develops the force which opposes the force of the air in the air-flow sensor. In doing so, the restriction dampens a possible oscillation of the sensor plate which could result due to pulsating air-intake flow.

The control pressure influences the fuel distribution. If the control pressure is low, the air drawn in by the engine can deflect the sensor plate further. This results in the control plunger opening the metering slits (11) further and the engine being allocated more fuel. On the other hand, if the control pressure is high, the air drawn in by the engine cannot deflect the sensor plate so far and, as a result, the engine receives less fuel.

In order to fully seal off the control-pressure circuit with absolute certainty when the engine has been switched off, and at the same time to maintain the pressure in the fuel circuit, the return line of the warm-up regulator is fitted with a check valve. This (push-up) valve is attached to the primary-pressure regulator and is held open during operation by the pressure-regulator plunger.

When the engine is switched off and the plunger of the primary-pressure regulator returns to its zero position, the check valve is closed by a spring.

Differential-pressure valves

The differential-pressure valves in the fuel distributor result in a specific pressure drop at the metering slits. The air-flow sensor has a linear characteristic. This means that if double the quantity of air is drawn in, the sensor-plate travel is also doubled. If this travel is to result in a change of delivered fuel in the same relationship, in this case double the travel equals double the quantity, then a constant drop in pressure must be guaranteed at the metering slits (Figure 6, Pos. 11), regardless of the amount of fuel flowing through them.

The differential-pressure valves maintain the differential pressure between the upper and lower chamber constant regardless of fuel throughflow. The differential-pressure is 0.1 bar. The differential-pressure valves enable high metering accuracy. Flat-seat type valves are used as differential pressure valves.

They are fitted in the fuel distributor and one such valve is allocated to each metering slit. A diaphragm separates the upper and lower chambers of the valve.

The lower chambers of all the valves are connected with one another by a ring main and are subjected to the primary pressure (delivery pressure). The valve seat is located in the upper chamber. Each upper chamber is connected to a metering slit and its corresponding connection to the fuel-injection line. The upper chambers are completely sealed off from each other. The diaphragms are spring-loaded and it is this helical spring that produces the pressure differential (Figure 6, Pos. 6).

If a large basic fuel quantity flows into the upper chamber through the metering slit, the diaphragm is bent downwards and enlarges the valve cross-section at the outlet leading to the injection valve until the set differential pressure once again prevails. If the fuel quantity drops, the valve cross-section is reduced owing to the equilibrium of forces at the diaphragm until the differential pressure of 0.1 bar is again present. This causes an equilibrium of forces to prevail at the diaphragm which can be maintained for every basic fuel quantity by controlling the valve cross-section.

Mixture formation

The formation of the air-fuel mixture takes place in the intake ports and cylinders of the engine. The continually injected fuel coming from the injection valves is "stored" in front of the intake valves. When the intake valve is opened, the air drawn in by the engine carries the waiting "cloud" of fuel with it into the cylinder. An ignitable air-fuel mixture is formed during the induction stroke due to the swirl effect.

Adaptation to operating conditions

In addition to the basic functions described up to now, the mixture has to be adapted during particular operating conditions. These adaptations (corrections) are necessary in order to optimize the power delivered, to improve the exhaust-gas composition and to improve the starting behavior and driveability.

Basic mixture adaptation

The basic adaptation of the air-fuel mixture to the operating modes of idle, part load and full load is by appropriately shaping the air funnel in the air-flow sensor (Figure 7).

If the funnel had a purely conical shape, the result would be a mixture with a constant air-fuel ratio throughout the whole of

7 Adaptation of the air-funnel shape

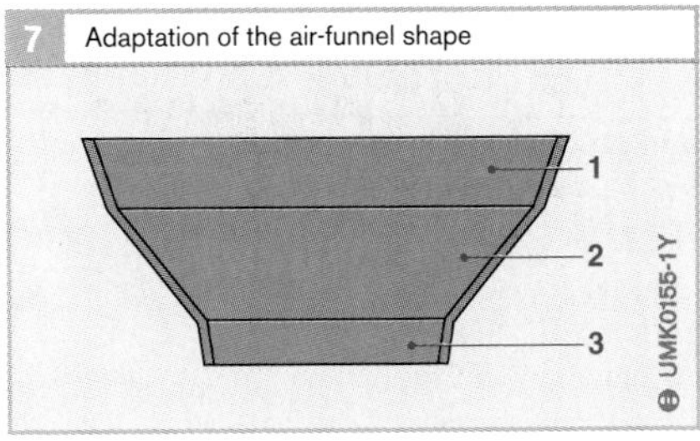

Fig. 7
1 For full load
2 For part load
3 For idle

the sensor plate range of travel (metering range). However, it is necessary to meter to the engine an air-fuel mixture which is optimal for particular operating modes such as idle, part load and full load. In practice, this means a richer mixture at idle and full load, and a leaner mixture in the part-load range. This adaptation is achieved by designing the air funnel so that it becomes wider in stages.

If the cone shape of the funnel is flatter than the basic cone shape (which was specified for a particular mixture, e.g. for $\lambda = 1$), this results in a leaner mixture. If the funnel walls are steeper than in the basic model, the sensor plate is lifted further for the same air throughput, more fuel is therefore metered by the control plunger and the mixture is richer. Consequently, this means that the air funnel can be shaped so that it is possible to meter mixtures to the engine which have different air-fuel ratios depending upon the sensor-plate position in the funnel (which in turn corresponds to the particular engine operating mode i.e. idle, part load and full load). This results in a richer mixture for idle and full load (idle and full-load enrichment) and, by contrast, a leaner mixture for part load.

Cold-start enrichment

Depending upon the engine temperature, the cold-start valve injects extra fuel into the intake manifold for a limited period during the starting process. In order to compensate for the condensation losses due to condensation on the cold cylinder walls, and in order to facilitate starting the cold engine during cold starting, extra fuel must be injected at the instant of start-up. This extra fuel is injected by the cold-start valve into the intake manifold. The injection period of the cold-start valve is limited by a thermo-time switch depending upon the engine temperature.

This process is known as cold-start enrichment and results in a “richer” air-fuel mixture, i.e. the excess-air factor λ is temporarily less than 1.

Cold-start valve

The cold-start valve is a solenoid-operated valve. An electromagnetic winding is fitted inside the valve. When unoperated, the movable electromagnet armature is forced against a seal by means of a spring and thus closes the valve. When the electro-magnet is energized, the armature which consequently has lifted from the valve seat opens the passage for the flow of fuel through the valve. From here, the fuel enters a special nozzle at a tangent and is caused to rotate or swirl. The result is that the fuel is atomized very finely and enriches the mixture in the manifold downstream of the throttle valve. The cold-start valve is so positioned in the intake manifold that good distribution of the mixture to all cylinders is ensured.

Thermo-time switch

The thermo-time switch limits the duration of cold-start valve operation, depending upon temperature. It consists of an electrically heated bimetal strip which, depending upon its temperature opens or closes a contact. It is brought into operation by the ignition/starter switch, and is mounted at a position which is representative of engine temperature. During a cold start, it limits the “on” period of the cold-start valve. In case of repeated start attempts, or when starting takes too long, the cold-start valve ceases to inject. Its “on” period is determined by the thermo-time switch which is heated by engine heat as well as by its own built-in heater. Both these heating effects are necessary in order to ensure that the “on” period of the cold-start valve is limited under all conditions, and engine over-enrichment is prevented. During an actual cold start, the heat generated by the built-in heater is mainly responsible for the “on” period (switch off, for instance, at –20 °C after 7.5 seconds). With a warm engine, the thermo-time switch has already been heated up so far by engine heat that it remains open and prevents the cold-start valve from going into action.

Warm-up enrichment

Warm-up enrichment is controlled by the warm-up regulator. When the engine is cold, the warm-up regulator reduces the control pressure to a degree dependent upon engine temperature and thus causes the metering slits to open further (Figure 8).

At the beginning of the warm-up period which directly follows the cold start, some of the injected fuel still condenses on the cylinder walls and in the intake ports. This can cause combustion misses to occur. For this reason, the air-fuel mixture must be enriched during the warm-up ($\lambda < 1$). This enrichment must be continuously reduced along with the rise in engine temperature in order to prevent the mixture being over-rich when higher engine temperatures have been reached. The warm-up regulator (control-pressure regulator) is the component which carries out this type of mixture control for the warm-up period by changing the control pressure.

Warm-up regulator

The change of the control pressure is effected by the warm-up regulator which is fitted to the engine in such a way that it ultimately adopts the engine temperature. An additional electrical heating system enables the regulator to be matched precisely to the engine characteristic.

The warm-up regulator comprises a spring-controlled flat seat (diaphragm-type) valve (Figure 8, Pos. 4) and an electrically heated bimetallic ring (2). Component 3 and components 8 to 12 are used for full-load enrichment; a basic warm-up regulator does not have these components.

In cold condition, the bimetal spring exerts an opposing force to that of the valve spring (7) and, as a result, reduces the effective pressure applied to the underside of the valve diaphragm (4). This means that the valve outlet cross-section is slightly increased at this point and more fuel is diverted out of the control-pressure circuit in order to achieve a low control pressure. Both the electrical heating system (1) and the engine heat the bimetal spring as soon as the engine is cranked. The spring bends, and in doing so reduces the force opposing the valve spring which, as a result, pushes up the diaphragm of the flat-seat valve. The valve outlet cross-section is reduced and the pressure in the control-pressure circuit rises.

Warm-up enrichment is completed when the bimetal spring has lifted fully from the valve spring. The control pressure is now solely controlled by the valve spring and maintained at its normal level. The control pressure is about 0.5 bar at cold start and about 3.7 bar with the engine at operating temperature.

Idle stabilization

In order to overcome the increased friction in cold condition and to guarantee smooth idling, the engine receives more air-fuel mixture during the warm-up phase due to the action of the auxiliary air device.

8 Warm-up regulator with full-load diaphragm

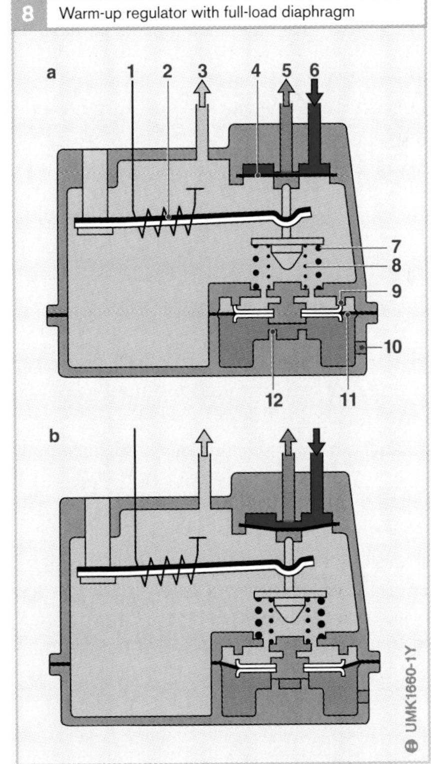

Fig. 8
- a During idle and part load
- b During full load

1. Electrical heating
2. Bimetal spring
3. Vacuum connection (from intake manifold)
4. Valve diaphragm
5. Return to fuel tank
6. Control pressure (from fuel distributor)
7. Outer valve spring
8. Inner valve spring
9. Upper stop
10. To atmospheric pressure
11. Diaphragm
12. Lower stop

When the engine is cold, the frictional resistances are higher than when it is at operating temperature and this friction must be overcome by the engine during idling. The engine draws in air additionally by means of the auxiliary-air device which bypasses the throttle valve. Due to the fact that this auxiliary air is measured by the air-flow sensor and taken into account for fuel metering, the engine is provided with more air-fuel mixture. This results in idle stabilization when the engine is cold.

Auxiliary-air device
In the auxiliary-air device, a perforated plate is pivoted by means of a bimetal spring and changes the open cross-section of a bypass line. This perforated plate thus opens a correspondingly large cross-section of the bypass line, as a function of the temperature, and this cross-section is reduced with increasing engine temperature and is ultimately closed. The bimetal spring also has an electrical heating system which permits the opening time to be restricted dependent upon the engine type. The installation location of the auxiliary-air device is selected such that it assumes the engine temperature. This guarantees that the auxiliary-air device only functions when the engine is cold.

Full-load enrichment
Engines operated in the part-load range with a very lean mixture require an enrichment during full-load operation, in addition to the mixture adaptation resulting from the shape of the air funnel.

This extra enrichment is carried out by a specially designed warm-up regulator. This regulates the control pressure depending upon the manifold pressure (Figure 8).

This model of the warm-up regulator uses two valve springs instead of one. The outer of the two springs (7) is supported on the housing as in the case with the normal-model warm-up regulator. The inner spring (8) however is supported on a diaphragm (11) which divides the regulator into an upper and a lower chamber. The manifold pressure (3) which is tapped via a hose connection from the intake manifold downstream of the throttle valve acts in the upper chamber. Depending upon the model, the lower chamber is subjected to atmospheric pressure either directly or by means of a second hose leading to the air filter.

Due to the low manifold pressure in the idle and part-load ranges, which is also present in the upper chamber, the diaphragm (11) lifts to its upper stop (9). The inner spring is then at maximum pretension. The pretension of both springs, as a result, determines the particular control pressure for these two ranges.

The throttle valve is opened further as full load approaches. The vacuum in the intake manifold decreases. The diaphragm leaves the upper stops and is pressed against the lower stops (12). The inner spring is relieved of tension and the control pressure reduced by the specified amount as a result. This results in mixture enrichment.

Acceleration response
Transitions from one operating condition to another produce changes in the mixture ratio which are utilized to improve driveability. The good acceleration response is a result of "overswing" of the air-flow sensor plate.

If, at constant engine speed, the throttle valve is suddenly opened, the amount of air which enters the combustion chamber, plus the amount of air which is needed to bring the manifold pressure up to the new level, flow through the airflow sensor. This causes the sensor plate to briefly "overswing" past the fully opened throttle point. This "overswing" results in more fuel being metered to the engine (acceleration enrichment) and ensures good acceleration response.

Supplementary functions

Overrun fuel cutoff

Smooth fuel cutoff effective during overrun responds as a function of the engine speed. The engine-speed information is provided by the ignition system. Intervention is via an air bypass around the sensor plate. A solenoid valve controlled by an electronic speed switch opens the bypass at a specific engine speed. The sensor plate then reverts to zero position and interrupts fuel metering. Cutoff of the fuel supply during overrun operation permits the fuel consumption to be reduced considerably not only when driving downhill but also in town traffic.

Engine speed limiting

The fuel supply can be cut off to limit the maximum permissible engine speed.

Lambda closed-loop control

Open-loop control of the air-fuel ratio is not adequate for observing the exhaust-gas limit values. The lambda closed-loop control system required for operation of a three-way catalytic converter necessitates the use of an electronic control unit on the K-Jetronic. The important input variable for this control unit is the signal supplied by the lambda sensor.

In order to adapt the injected fuel quantity to the required air-fuel ratio with $\lambda = 1$, the pressure in the lower chambers of the fuel distributor is varied. If, for instance, the pressure in the lower chambers is reduced, the differential pressure at the metering slits increases, whereby the injected fuel quantity is increased. In order to permit the pressure in the lower chambers to be varied, these chambers are decoupled from the primary pressure via a fixed restrictor, by comparison with the standard K-Jetronic fuel distributor.

A further restrictor connects the lower chambers and the fuel return line. This restrictor is variable: if it is open, the pressure in the lower chambers can drop (Figure 9). If it is closed, the primary pressure builds up in the lower chambers. If this restrictor is opened and closed in a fast rhythmic succession, the pressure in the lower chambers can be varied dependent upon the ratio of closing time to opening time. An electromagnetic valve, the frequency valve, is used as the variable restrictor. It is controlled by electrical pulses from the lambda closed-loop controller.

9 K-Jetronic with lambda closed-loop control

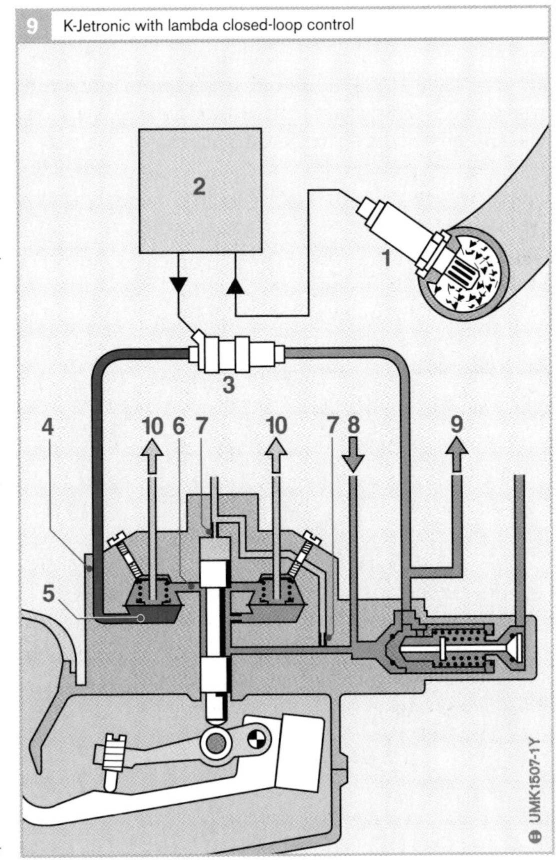

Fig. 9
1 Lambda sensor
2 Lambda closed-loop controller
3 Frequency valve (variable restrictor)
4 Fuel distributor
5 Lower chambers of the differential-pressure valves
6 Metering slits
7 Decoupling restrictor (fixed restrictor)
8 Fuel inlet
9 Fuel return line
10 To the injection valve

KE-Jetronic

The system design of the KE-Jetronic (Figure 1) is basically identical to that of the K-Jetronic. The most significant difference in the KE-Jetronic is the electronic mixture control, which influences the mixture formation via an actuator. This actuator is an electro-hydraulic pressure actuator.

Electro-hydraulic pressure actuator

Design

The electro-hydraulic pressure actuator (Figure 2) is mounted on the fuel distributor. The actuator is a differential-pressure controller which functions according to the nozzle/baffle-plate principle, and its pressure drop is controlled by the current input from the ECU. In a housing of non-magnetic material, an armature (Figure 3, Pos. 11) is suspended on a frictionless taut-band suspension element, between two double magnetic poles (5). The armature is in the form of a diaphragm plate made from resilient material.

Function

The magnetic flux of a permanent magnet (broken lines in Figure 3) and that of an electromagnet (unbroken lines) are superimposed upon each other in the magnetic poles and their air gaps. The permanent magnet is actually turned through 90 degrees referred to the focal plane. The paths taken by the magnetic fluxes through the two pairs of poles are symmetrical and of equal length, and flow from the poles, across the air gaps to the armature, and then through the armature. In the two diagonally opposed air gaps (Figure 3, L_2, L_3), the permanent-magnet flux, and the electromagnet flux resulting from the incoming ECU control signal are added, whereas in the other two air gaps (L_1, L_4) the fluxes are subtracted from each other. This means that, in each air gap, the armature, which moves the baffle plate, is subjected to a force of attraction proportional to the square of the magnetic flux. Since the permanent-magnet flux remains constant, and is proportional to the control current from the ECU flowing in the electromagnet coil, the resulting torque is proportional to this control current.

Fig. 1
1 Fuel tank
2 Electric fuel pump
3 Fuel accumulator
4 Fuel filter
5 Primary-pressure regulator
6 Fuel-injection valve (injector)
7 Intake manifold
8 Cold-start valve
9 Fuel distributor
10 Air-flow sensor
11 Electrohydraulic pressure actuator
12 Lambda sensor
13 Thermo-time switch
14 Engine-temperature sensor
15 Ignition distributor
16 Auxiliary-air device
17 Throttle-valve switch
18 ECU
19 Ignition and starting switch
20 Battery

1 Schematic diagram of a KE-Jetronic system with lambda closed-loop control

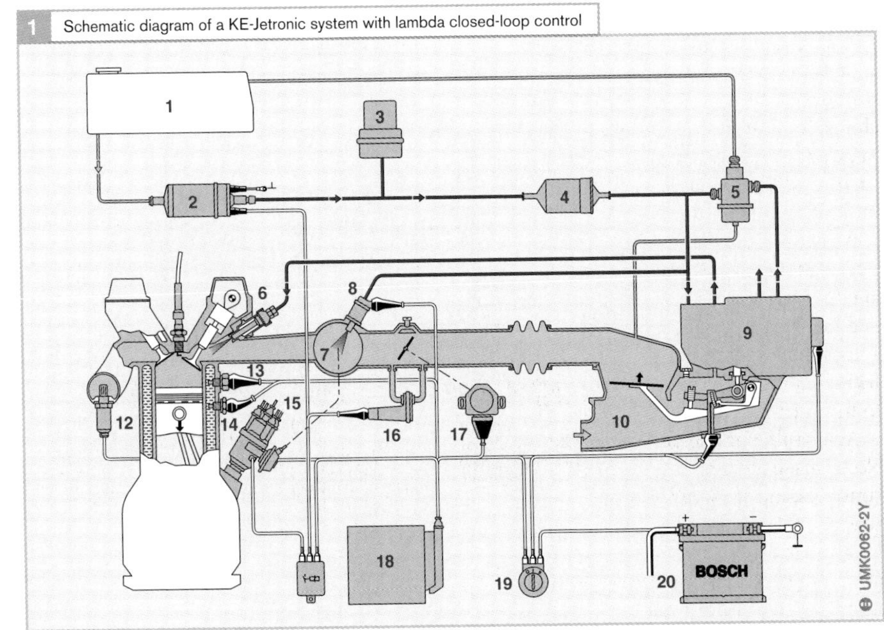

The basic moment of force applied to the armature has been selected so that, when no current is applied from the ECU, there results a basic differential pressure which corresponds preferably to $\lambda = 1$. This also means that, in the case of control current failure, limp-home facilities are available without any further correction measures being necessary. The jet of fuel which enters through the nozzle (Figure 3, Pos. 2) attempts to bend the baffle plate away against the prevailing mechanical and magnetic forces. Taking a fuel throughflow which is determined by a fixed restriction located in series with the pressure actuator, the difference in pressure between the inlet and outlet is proportional to the control current applied from the ECU. This means that the variable pressure drop at the nozzle is also proportional to the ECU control current, and results in a variable lower-chamber pressure.

At the same time, the pressure in the upper chambers changes by the same amount. This, in turn, results in a change in the difference at the metering slits between the upper-chamber pressure and the primary pressure and this is applied as a means for varying the fuel quantity delivered to the injection valves.

As a result of the small electromagnetic time constants, and the small masses which must be moved, the pressure actuator reacts extremely quickly to variations in the control current from the ECU. If the direction of the control current is reversed, the armature pulls the baffle plate away from the nozzle and a pressure drop of only a few hundredths of a bar occurs at the pressure actuator. This can be used for auxiliary functions such as overrun fuel cutoff and engine-speed limitation. The latter function is performed by interrupting the flow of fuel to the injection valves.

2 Electro-hydraulic pressure actuator fitted to the fuel distributor

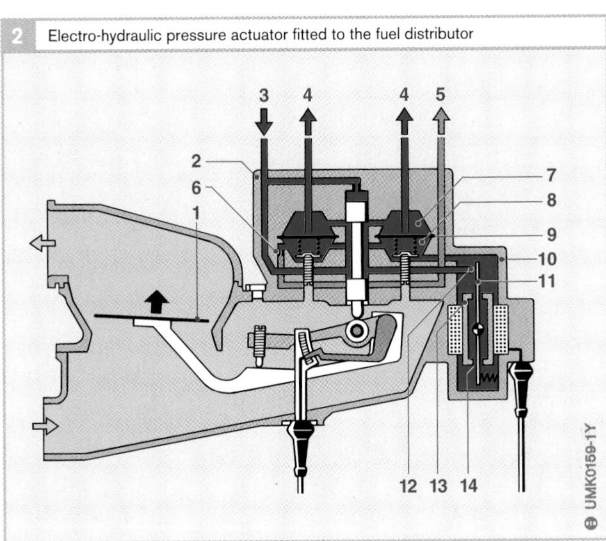

Fig. 2
1 Sensor plate
2 Fuel distributor
3 Fuel inlet (primary pressure)
4 Fuel to the injection valves
5 Fuel return to the pressure regulator
6 Fixed restriction
7 Upper chamber
8 Lower chamber
9 Diaphragm
10 Pressure actuator
11 Baffle plate
12 Nozzle
13 Magnetic pole
14 Air gap

3 Section through the electro-hydraulic pressure actuator

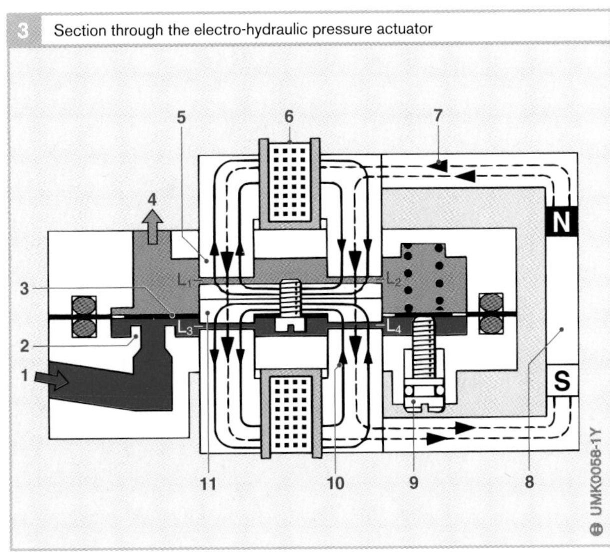

Fig. 3
1 Fuel inlet (primary pressure)
2 Nozzle
3 Baffle plate
4 Fuel outlet (primary pressure)
5 Magnetic pole
6 Electromagnet coil
7 Permanent-magnet flux
8 Permanent magnet (turned through 90 degrees from the focal plane)
9 Adjustment screw for basic moment of force
10 Electromagnetic flux
11 Armature (L_1 to L_4 are air gaps)

L-Jetronic

System overview

The L-Jetronic is an electronically controlled fuel-injection system which injects fuel intermittently into the intake ports. It does not require any form of drive. It combines the advantages of direct air-flow sensing and the special capabilities afforded by electronics.

As is the case with the K-Jetronic system, this system detects all changes resulting from the engine (wear, deposits in the combustion chamber and changes in valve settings).

The task of the gasoline injection system is to supply to each cylinder precisely the correct amount of fuel as is necessary for the operation of the engine at that particular moment. Due to the fact that the operating condition constantly changes, a speedy adaptation of the fuel delivery to the driving situation at any given moment is of prime importance. Electronically controlled gasoline injection meets these requirements. It enables a variety of operational data at any particular location of the vehicle to be registered and converted into electrical signals by sensors. These signals are then passed on to the control unit of the fuel-injection system which processes them and calculates the exact amount of fuel to be injected. This is influenced via the duration of injection.

Function

The electric fuel pump (Figure 1, Pos. 2) supplies the fuel to the engine and creates the pressure necessary for injection. Injection valves (5) inject the fuel into the individual intake ports and onto the intake valves. An electronic control unit (4) controls the injection valves. Fundamental function blocks are listed below.

1 Schematic diagram of an L-Jetronic system with lambda closed-loop control

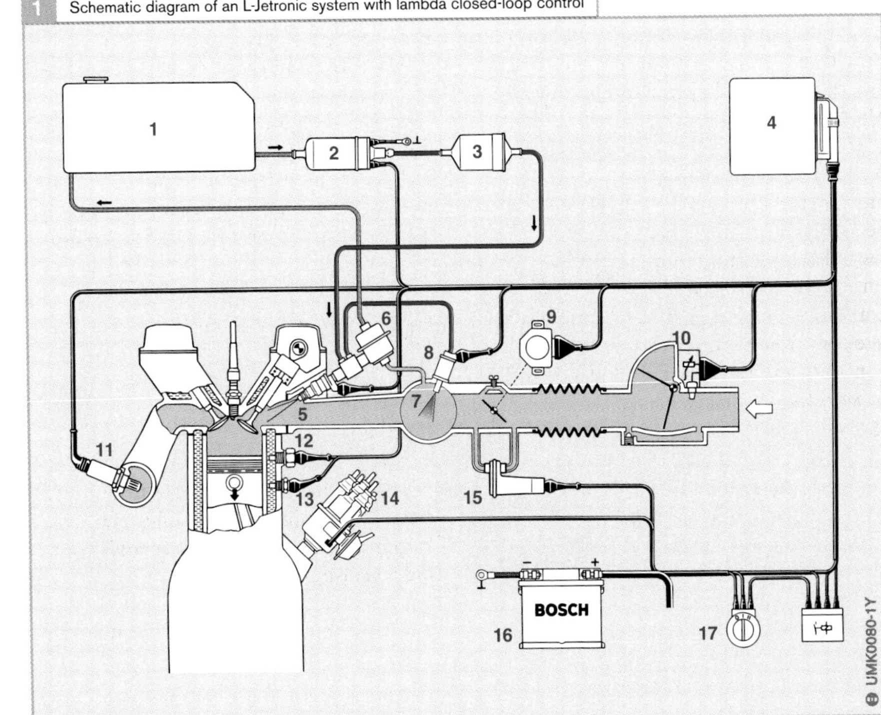

Fig. 1
1 Fuel tank
2 Electric fuel pump
3 Fuel filter
4 ECU
5 Fuel-injection valve (injector)
6 Fuel rail and fuel-pressure regulator
7 Intake manifold
8 Cold-start valve
9 Throttle-valve switch
10 Air-flow sensor
11 Lambda sensor
12 Thermo-time switch
13 Engine-temperature sensor
14 Ignition distributor
15 Auxiliary-air device
16 Battery
17 Ignition and starting switch

Fuel-supply system

The fuel system supplies fuel from the fuel tank to the injection valves, creates the pressure necessary for injection and maintains it at a constant level. At first, standard systems were used and later returnless fuel systems were used. Further details can be found in the "Fuel supply" chapter.

Operating-data sensing system

The sensors register the measured variables which characterize the operating mode of the engine.

The most important measured variable is the amount of air drawn in by the engine and registered by the air-flow sensor. Other sensors register the position of the throttle, the engine speed, the air temperature and the engine temperature (see "Sensors" chapter).

Fuel-metering system

The electronic control unit (ECU) evaluates the signals delivered by the sensors and generates the appropriate control pulses for the injection valves. The amount of fuel to be injected is defined by the opening time of the fuel injectors.

Lambda closed-loop control

The air-fuel ratio can be maintained precisely at $\lambda = 1$ by means of the lambda closed-loop control. In the control unit, the lambda-sensor signal is compared with an ideal value (setpoint). This controls a two-state controller. Intervention in the fuel metering process takes place by the opening time of the fuel injectors.

Advantages of the L-Jetronic system

Fuel consumption

In carburetor systems, due to segregation processes in the intake manifold, the individual cylinders of the engine do not all receive the same amount of air-fuel mixture. The fuel allocation is therefore not optimal. This would result in high fuel consumption and unequal stressing of the cylinders. In L-Jetronic systems, each cylinder has its own injection valve. The injection valves are controlled centrally; this ensures that each cylinder receives precisely the same amount of fuel, the optimum amount, at any particular moment and under any particular load.

Adaptation to operating conditions

The L-Jetronic system adapts to changing load conditions virtually immediately since the required quantity of fuel is computed by the control unit (ECU) within milliseconds and injected by the injection valves onto the engine intake valves.

Low-pollution exhaust gas

The concentration of pollutants in the exhaust gas is directly related to the air-fuel ratio. If the engine is to be operated with the least pollutant emission, then a fuel-management system is necessary which is capable of maintaining a given air-fuel ratio. L-Jetronic is capable of setting the air-fuel ratio at $\lambda = 1$.

Higher power output per litre

The L-Jetronic enables the intake passages to be designed aerodynamically in order to achieve optimum air distribution and cylinder charge and, thus, greater torque. Since the fuel is injected directly onto the intake valves, the engine receives only air through the intake manifold. This results in a power output per liter that meets the requirements and a realistic torque curve.

Operating-data sensing system

Sensors detect the operating mode of the engine and signal this condition electrically to the control unit. The sensors and ECU form the control system.

Measured variables

The ECU evaluates all measured variables together so that the engine is always supplied with exactly the amount of fuel required for the instantaneous operating mode. There are three different main groups of measured variables.

Main measured variables
The main measured variables are the engine speed and the amount of air drawn in by the engine. These determine the amount of air per stroke which then serves as a direct measure for the loading condition of the engine.

Measured variables for compensation
For operating conditions which deviate from normal operation (e.g. cold start, warm-up), the mixture must be adapted to the modified conditions. Detection is performed by sensors which transmit the engine temperature to the control unit. For compensating various load conditions, the load range (idle, part-load, full-load) is transmitted to the control unit via the throttle-valve switch.

2 Calculating engine speed with a breaker-triggered ignition system

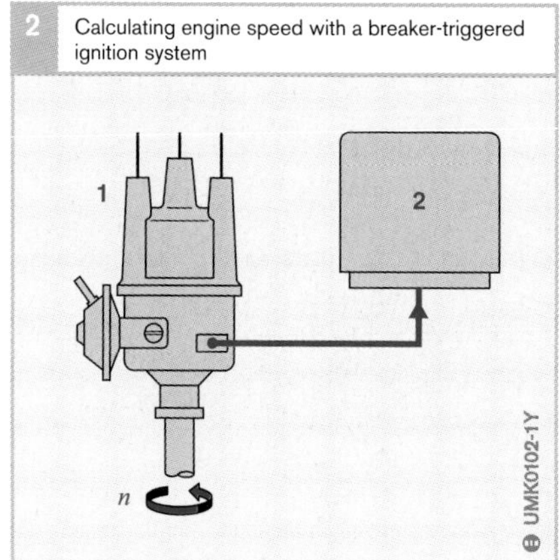

Fig. 2
1 Ignition distributor
2 ECU

n Engine speed

Measured variables for precision compensation
In order to achieve optimum driving behavior, further operating ranges and influences can be considered. The sensors mentioned above detect the data for transition response when accelerating, for maximum engine-speed limitation and during overrun. The sensor signals have a particular relationship to each other in these operating ranges. The control unit recognizes these relationships and influences the control signals of the injection valves accordingly.

Calculating engine speed

Information on engine speed and the start of injection is passed on to the L-Jetronic ECU in breaker-triggered ignition systems by the contact-breaker points in the ignition distributor, and, in breakerless ignition systems, by terminal 1 of the ignition coil (Fig. 2).

Measuring the air flow

The amount of air drawn in by the engine is a measure of its loading condition. The air-flow measurement system allows for all changes which may take place in the engine during the service life of the vehicle, e.g. wear, combustion-chamber deposits and changes to the valve settings.

Since the quantity of air drawn in must first pass through the air-flow sensor before entering the engine, this means that, during acceleration, the signal leaves the sensor before the air is actually drawn into the cylinder. This permits correct mixture adaptation at any time during load changes.

The sensor flap in the air-flow sensor measures the entire air quantity inducted by the engine, thereby serving as the main controlled variable for determining the load signal and basic injection quantity.

Fuel metering

As the central unit of the system, the ECU evaluates the data delivered by the sensors on the operating mode of the engine. From this data, control pulses for the injection valves are generated, whereby the quantity to be injected is determined by the length of time the injection valves are opened.

Electronic control unit (ECU)

Configuration

The L-Jetronic ECU is in a splash-proof sheet-metal housing which is fitted where it is not affected by the heat radiated from the engine. The electronic components in the ECU are arranged on printed-circuit boards; the output-stage power components are mounted on the metal frame of the ECU thus assuring good heat dissipation. By using integrated circuits and hybrid modules, it has been possible to reduce the number of parts to a minimum. The reliability of the ECU was increased by combining functional groups into integrated circuits (e.g. pulse shaper, pulse divider and division control multivibrator, Fig. 3) and by combining components into hybrid modules.

A multiple plug is used to connect the ECU to the injection valves, the sensors and the vehicle electrical system. The input circuit in the ECU is designed so that the latter cannot be connected with the wrong polarity and cannot be short-circuited. Special Bosch testers are available for carrying out measurements on the ECU and on the sensors. The testers can be connected between the wiring harness and the ECU with multiple plugs.

Operating-data processing

Engine speed and inducted air quantity determine the basic duration of injection. The timing frequency of the injection pulses is determined on the basis of the engine speed.

The pulses delivered by the ignition system for this purpose are processed by the ECU. First of all, they pass through a pulse-shaping circuit which generates square-wave pulses from the signal "delivered" in the form of damped oscillations, and then feeds these to a frequency divider. The frequency divider divides the pulse frequency given by the ignition sequence in such a manner that two pulses occur for each working cycle regardless of the number of cylinders. The start of the pulse is, at the same time, the start of injection for the injection valves. For each turn of the crankshaft, each injection valve injects once, regardless of the position of the intake valves. When the intake valve is

3 Block diagram of the ECU

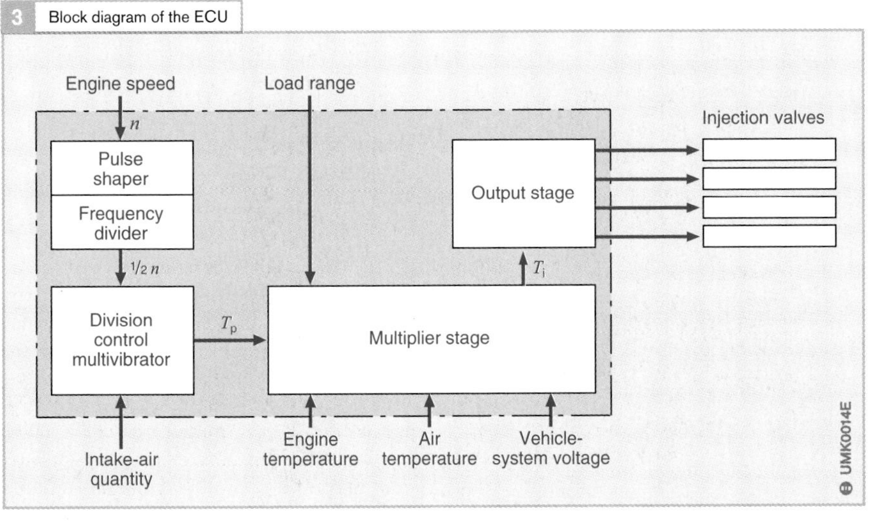

Fig. 3
T_i Injection pulses, corrected
T_p Basic injection duration
n Engine speed

closed, the fuel is stored and the next time it opens the fuel is drawn into the combustion chamber together with the air. The duration of injection depends on the amount of air measured by the air-flow sensor and the engine speed.

The ECU also evaluates the signal supplied by the potentiometer. Fig. 4 shows the interrelationships between intake air quantity, flap angle, potentiometer voltage and injected quantity.

Assuming a specific intake-air quantity Q_L flowing through the air-flow sensor (point Q), we thus obtain the theoretically required injection quantity Q_K (point D). In addition, a specific flap angle (point A) is established as a function of the air intake quantity. The potentiometer actuated by the air-flow sensor flap supplies a voltage signal U_S to the ECU (point B) which controls the injection valves, whereby point C represents the injected fuel quantity VE.

It can be seen that the fuel quantity injected in practice and the theoretically required injection quantity are identical (line C–D).

Generation of injection pulses (Fig. 5)

The generation of the basic injection duration is carried out in a special circuit group in the ECU, the division control multivibrator (DSM).

The DSM receives the information on speed n from the frequency divider and evaluates it together with the air-quantity signal U_S. For the purpose of intermittent fuel injection, the DSM converts the voltage U_S into square-wave control pulses. The duration T_p of this pulse determines the basic injection quantity, i.e. the quantity of fuel to be injected per intake stroke without considering any corrections. T_p is therefore regarded as the "basic injection duration". The greater the quantity of air drawn in with each intake stroke, the longer the basic injection duration. Two border cases are possible here: if the engine speed n increases at a constant air throughput Q_L, then the absolute pressure sinks downstream of the throttle valve and the cylinders draw in less air per stroke, i.e. the cylinder charge is reduced. As a result, less fuel is needed for combustion and the duration of the pulse T_p is correspondingly shorter. If the engine output and thereby the amount of air drawn in per minute increase and providing the speed remains constant,

4 Interrelationships between the controlled variables

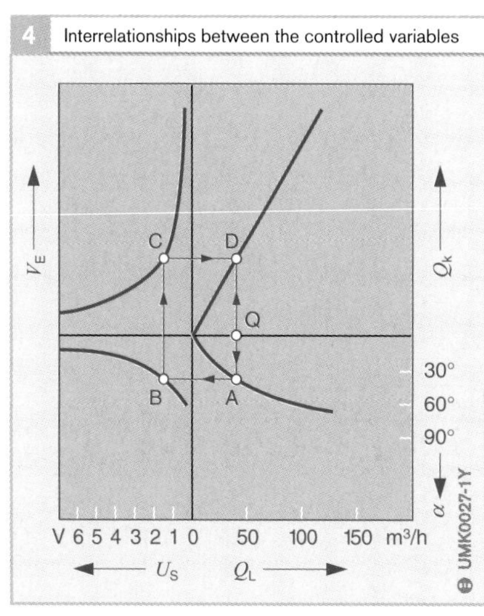

Fig. 4
A Flap angle dependent on the air quantity
B Size of the voltage signal dependent on the flap angle
C Injected fuel quantity dependent on the voltage signal
D Theoretically required injection quantity
Q Air quantity dependent on the flap angle

Q_K Required fuel quantity
Q_L Intake air quantity
U_S Voltage signal
V_E Injected fuel quantity

5 Signals and controlled variables at the ECU

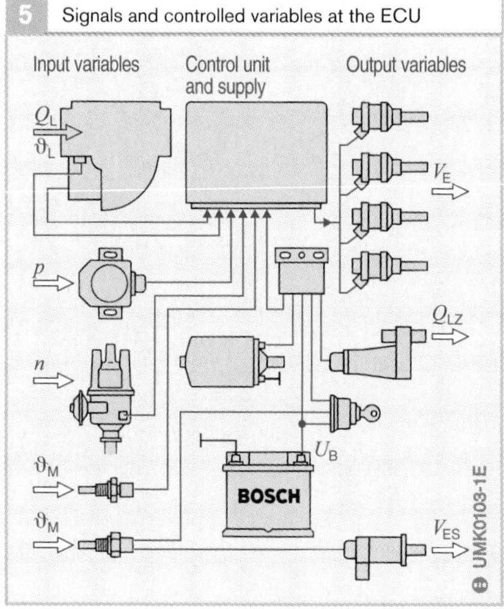

Fig. 5
Q_L Intake air quantity
ϑ_L Air temperature
n Engine speed
P Engine load range
ϑ_M Engine temperature
V_E Injected fuel quantity
Q_{LZ} Auxiliary air
V_{ES} Excess fuel for starting
U_B Vehicle-system voltage

then the cylinder charge will improve and more fuel will be required: the pulse duration T_p of the DSM is longer (Figs. 5 and 6).

During normal driving, engine speed and output usually change at the same time, whereby the DSM continually calculates the basic injection duration T_p. At a high speed, the engine output is normally high (full load) and this results ultimately in a longer pulse duration T_p and, therefore, more fuel per injection cycle. The basic injection duration is extended by the signals from the sensors depending on the operating mode of the engine.

Adaptation of the basic injection duration to the various operating conditions is carried out by the multiplying stage in the ECU (Fig. 6).

This stage is controlled by the DSM with the pulses of duration T_p. In addition, the multiplying stage gathers information on various operating modes of the engine, such as cold start, warm-up, full-load operation etc. From this information, the correction factor k is calculated. This is multiplied by the basic injection duration T_p calculated by the division control multivibrator. The resulting time is designated T_m. T_m is added to the basic injection duration T_p, i.e. the injection duration is extended and the air-fuel mixture becomes richer. T_m is therefore a measure of fuel enrichment, expressed by a factor which can be designated "enrichment

6 Schematic pulse-timing diagram of the L-Jetronic for 4-cylinder engines

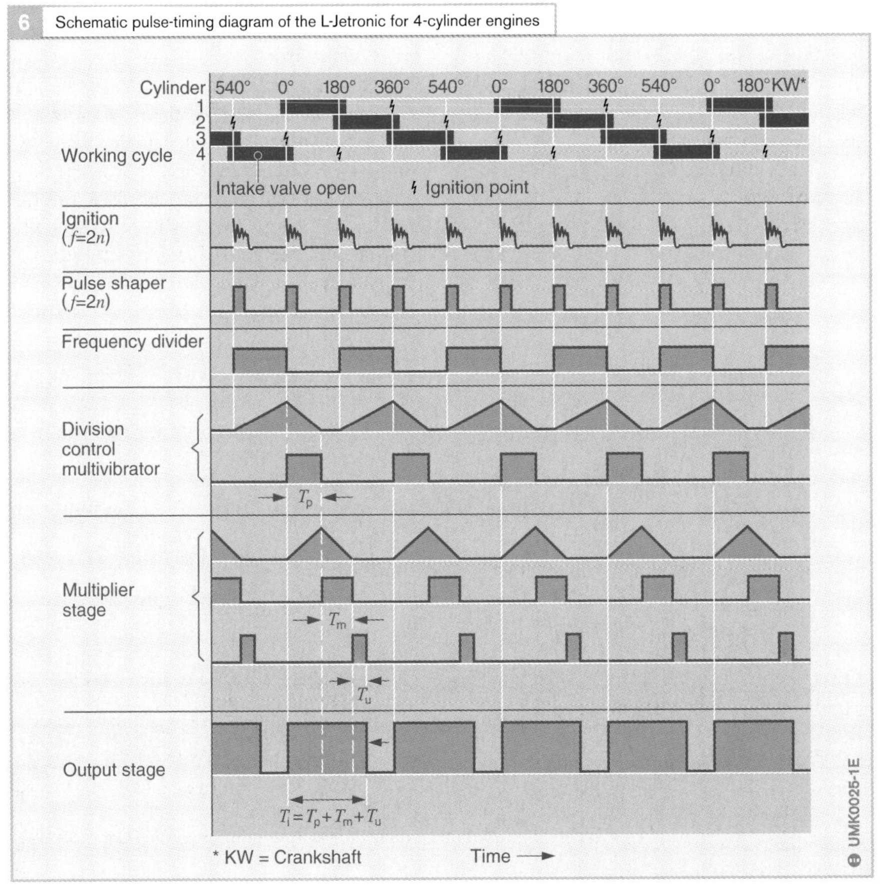

Fig. 6
f Ignition pulse, frequency or sparking rate
n Engine speed
T_p Basic duration of injection
T_m Pulse duration extension resulting from corrections
T_u Pulse duration extension resulting from voltage compensation
T_i Pulse control time. The actual injection duration per cycle differs from the pulse control time since both a response delay and a release delay change the injection duration

factor". When it is very cold, for example, the valves inject two to three times the amount of fuel at the beginning of the warm-up period (Figures 3 and 6).

Voltage correction

The pickup time of the fuel-injection valves depends very much on the battery voltage. Without electronic voltage correction, the response delay which results from a low-voltage battery would cause the injection duration to be too short and, as a result, insufficient fuel would be injected. The lower the battery voltage, the less fuel the engine would receive. For this reason, a low battery voltage, i.e. after starting with a heavily discharged battery, must be compensated for with an appropriate extension T_u of the precalculated pulse time in order that the engine receives the correct fuel quantity. This is known as "voltage compensation". For voltage compensation, the effective battery voltage is taken by the control unit as the controlled variable. An electronic compensation stage extends the valve control pulses by the amount T_u which is the voltage-dependent pickup delay of the injection valves. The total duration of the fuel-injection pulses T_i is thus the sum of T_p, T_m and T_u (Fig. 6).

Amplification of the injection pulses

The fuel-injection pulses generated by the multiplying stage are amplified in a following output stage. The injection valves are controlled with these amplified pulses. All the fuel-injection valves in the engine open and close at the same time. With each valve, a series resistor is wired into the circuit and functions as a current limiter.

The output stage of the L-Jetronic supplies 3 or 4 valves simultaneously with current. Control units for 6 and 8-cylinder engines have two output stages with 3 and 4 injection valves respectively. Both output stages operate in unison. The injection cycle of the L-Jetronic is selected so that for each revolution of the camshaft (= 1 working cycle) half the amount of fuel required by each working cylinder is injected twice.

In addition to controlling the fuel-injection valves through the series resistors, some control units have a regulated output stage. In these control units, the fuel-injection valves are operated without series resistors. Control of the fuel-injection valve takes place then as follows: as soon as the valve armatures have picked up at the beginning of the pulse, the valve current is regulated for the rest of the pulse duration to a considerably reduced holding current. Since these valves are switched on at the start of the pulse with a very high current, short response times are the result. By means of the reduction in current strength after switching on, the output stage is not subjected to such heavy loading. In this way, up to 12 fuel-injection valves can be switched with only one output stage.

Mixture formation

Mixture formation is carried out in the intake ports and in the engine cylinder.

The fuel-injection valve injects its fuel directly onto the engine intake valve and, when this opens, the cloud of fuel is entrained along with the air which is drawn in by the engine and an ignitable mixture is formed by the swirling action which takes place during the intake cycle.

Adaptation to operating modes

In addition to the basic functions described up to now, the mixture has to be adapted during particular operating modes. These adaptations (corrections) are necessary in order to optimize the power delivered by the engine, to improve the exhaust-gas composition and to improve the starting behavior and driveability. With additional sensors for the engine temperature and the throttle-valve position (load signal), the L-Jetronic ECU can perform these adaptation tasks. The characteristic curve of the air-flow sensor determines the fuel-requirement curve, specific to the particular engine, for all operating ranges.

Cold-start enrichment

When the engine is started, additional fuel is injected for a limited period depending on the temperature of the engine. This makes the mixture "richer". This is carried out in order to compensate for fuel condensation losses in the inducted mixture and in order to facilitate starting the cold engine.

There are two methods of cold-start enrichment:

Start control via control unit and fuel injectors
By extending the period during which the fuel-injection valves inject, more fuel can be supplied during the starting phase. The electronic control unit controls the start procedure by processing the signals from the ignition and starting switch and the engine-temperature sensor.

Control via thermo-time switch and cold-start valve
The cold-start valve is a solenoid-operated valve. The solenoid winding is located in the valve. In neutral position, a helical spring presses the movable solenoid armature against a seal, thereby shutting off the valve.

When a current is passed through the solenoid, the armature, which now rises from the valve seat, allows fuel to flow along the sides of the armature to a nozzle where it is swirled. The swirl nozzle atomizes the fuel very finely and as a result enriches the air in the intake manifold downstream of the throttle valve with fuel.

The thermo-time switch is an electrically heated bimetal switch which opens or closes a contact depending on its temperature. It is controlled via the ignition and starting switch. The thermo-time switch is attached in a position representative of the engine temperature. During a cold start, it limits the "on" period of the cold-start valve. The "on" period is determined by the thermo-time switch which is heated by the heat of the engine as well as by its own built-in electric heater. The electrical heating is necessary in order to ensure that the "on" period of the cold-start valve is limited under all conditions, and engine over-enrichment is prevented. During an actual cold start, the heat generated by the built-in heating winding is mainly responsible for the "on" period (switch-off, for instance, at –20 °C after approx. 7.5 s). With a warm engine, the thermo-time switch has already been heated so far by engine heat that it remains open and prevents the cold-start valve from going into action.

In the case of repeated start attempts, or when starting takes too long, the cold-start valve ceases to inject.

Post-start and warm-up enrichment

The warm-up phase follows the cold-start phase of the engine. During this phase, the engine needs substantially more fuel since some of the fuel condenses on the still cold cylinder walls. In addition, without supplementary fuel enrichment during the warm-up period, a major drop in engine speed would be noticed after the additional fuel from the cold-start valve has been cut off.

When the post-start enrichment has finished, the engine needs only a slight mixture enrichment, this being controlled by the engine temperature. The diagram (Figure 7) shows a typical enrichment curve with reference to time with a starting temperature of 22 °C. In order to trigger this control process, the electronic control unit must receive information on the engine temperature. This comes from the temperature sensor.

Part-load adaptation

By far the major part of the time, the engine will be operating in the part-load range. The fuel-requirement curve for this range is programmed in the ECU and determines the amount of fuel supplied. The curve is such that the fuel consumption of the engine is low in the part-load range.

Acceleration enrichment

During acceleration, the L-Jetronic meters additional fuel to the engine. If the throttle is opened abruptly, the air-fuel mixture is momentarily leaned-off, and a short period of mixture enrichment is needed to ensure good transitional response. With this abrupt opening of the throttle valve, the amount of air which enters the combustion chamber, plus the amount of air which is needed to bring the manifold pressure up to the new level, flow through the air-flow sensor. This causes the sensor plate to "overswing" past the wide-open-throttle point. This "overswing" results in more fuel being metered to the engine (acceleration enrichment) and ensures good acceleration response.

7 Warm-up enrichment curve

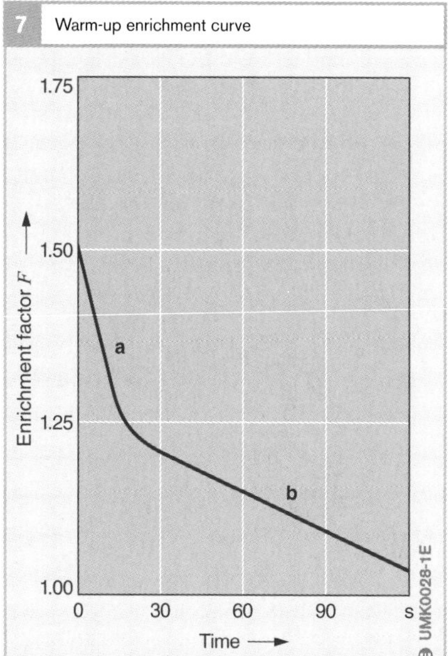

Fig. 7
Enrichment factor F as a function of time

a Proportion mainly dependent on time
b Proportion mainly dependent on engine temperature

Since this acceleration enrichment is not adequate during the warm-up phase, the control unit also evaluates a signal representing the speed at which the sensor flap deflects during this operating mode.

Full-load enrichment

The engine delivers its maximum torque at full load, when the air-fuel mixture must be enriched compared to that at part-load. This enrichment is programmed in the electronic control unit, specific to the particular engine. The information on the load condition is supplied to the control unit by the throttle-valve switch.

Throttle-valve switch

The throttle-valve switch (Figure 8) communicates the "idle" and "full load" throttle positions to the control unit. It is mounted on the throttle body and actuated by the throttle-valve shaft. A contoured switching guide closes the "idle" contact at one end of switch travel and the "full-load" contact at the other.

Controlling the idle speed

The air-flow sensor contains an adjustable bypass via which a small quantity of air can bypass the sensor flap. The idle-mixture-adjusting screw permits a basic setting of the air-fuel ratio by varying the bypass cross-section (Figure 9).

In order to achieve smoother running even at idle, the idle-speed control increases the idle speed. This also leads to a more rapid warm-up of the engine. Depending upon engine temperature, an electrically heated auxiliary-air device in the form of a bypass around the throttle plate allows the engine to draw in more air. This auxiliary air is measured by the air-flow sensor, and leads to the L-Jetronic providing the engine with more fuel.

Precise adaptation is by means of the electrical heating facility. The engine temperature then determines how much auxiliary air is fed in initially through the bypass, and the electrical heating is mainly responsible for subsequently reducing the auxiliary air as a function of time.

Auxiliary-air device

The auxiliary-air device incorporates a perforated plate which controls the cross-section of the bypass passage. Initially, the bypass cross-section opened by the perforated plate is determined by the engine temperature, so that during a cold start the bypass opening is adequate for the auxiliary air required. The opening closes steadily along with increasing engine temperature until, finally, it is closed completely. The auxiliary-air device is fitted in the best possible position on the engine for it to assume engine temperature. It does not function when the engine is warm.

Adaptation to the air temperature

The quantity of fuel injected is adapted to the air temperature. The quantity of air necessary for combustion depends upon the temperature of the air drawn in. Cold air is denser. This means that with the same throttle-valve position the volumetric efficiency of the cylinders drops as the temperature increases.

To register this effect, a temperature sensor is fitted in the intake duct of the air-flow sensor. This sensor measures the temperature of the air drawn in and passes this information on to the control unit which then controls the amount of fuel metered to the cylinders accordingly.

8 Idle/full-load correction

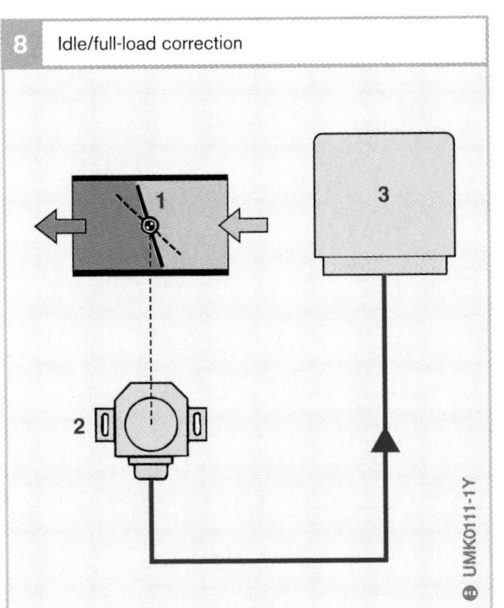

9 Idle-speed control

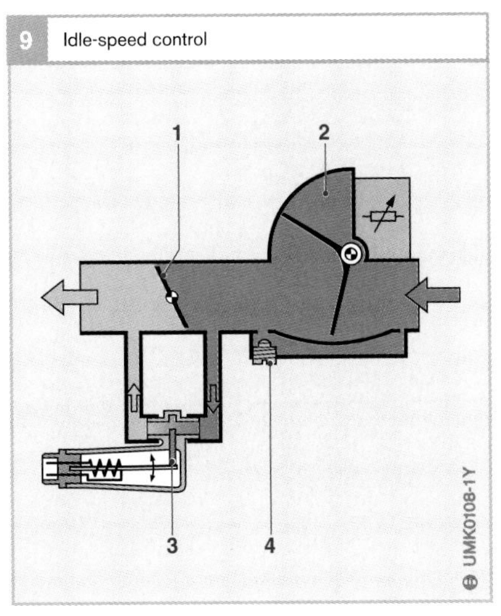

Fig. 8
1 Throttle valve
2 Throttle-valve switch
3 ECU

Fig. 9
1 Throttle valve
2 Air-flow sensor
3 Auxiliary-air device
4 Idle-mixture-adjusting screw

L3-Jetronic

Specific systems for specific markets have in the meantime been developed on the basis of the L-Jetronic. The most recent stage of development is the L3-Jetronic which differs from its predecessors in respect of the following details:

- The control unit, which is suitable for installation in the engine compartment, is attached to the air-flow sensor and thus no longer requires space in the passenger compartment.
- The combined unit of control unit and air-flow sensor with internal connections simplifies the cable harness and reduces installation expense.
- The use of digital techniques permits new functions with improved adaptation capabilities to be implemented as compared with the previous analog techniques used

The L3-Jetronic system is available both with (Figure 10) and without lambda closed-loop control. Both versions have what is called a "limp-home" function which enables the driver to drive the vehicle to the nearest workshop if the microcomputer fails. In addition the input signals are checked for plausibility, i.e. an implausible input signal (e.g. engine temperature lower than –40°C) is ignored and a default value stored in the control unit is used in its place.

Fuel supply

On this system, the fuel is supplied to the injection valves in the same way as on the L-Jetronic system.

Operating-data sensing system

The data is sensed via the same components as with the L-Jetronic. The air-flow sensor is an exception.

Air-flow sensor

The control unit and the air-flow sensor are integrated into a measuring and control unit

10 Schematic diagram of an L3-Jetronic system with lambda closed-loop control

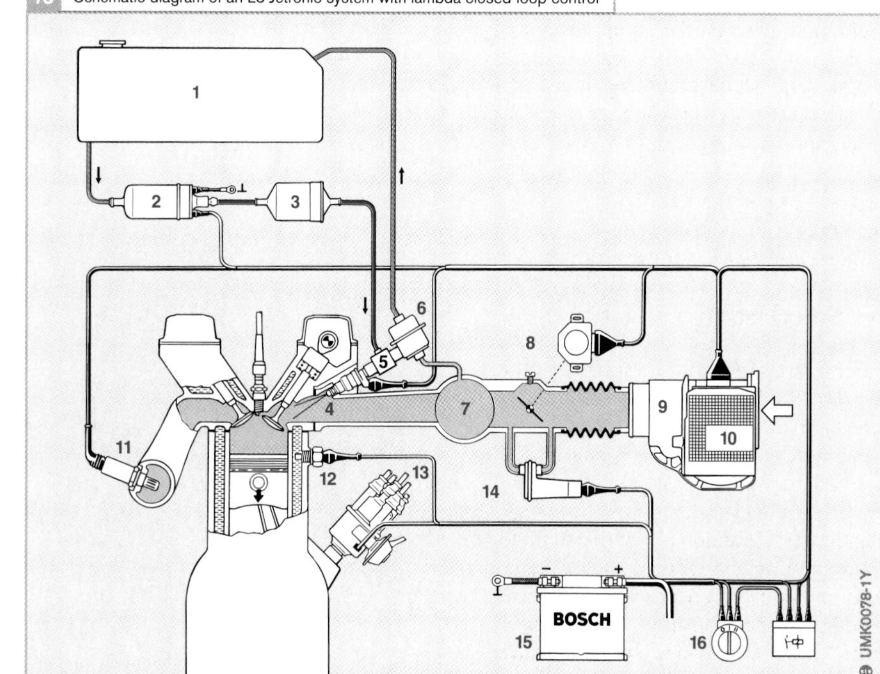

Fig. 10
1 Fuel tank
2 Electric fuel pump
3 Fuel filter
4 Fuel-injection valve (injector)
5 Fuel rail
6 Fuel-pressure regulator
7 Intake manifold
8 Throttle-valve switch
9 Air-flow sensor
10 ECU
11 Lambda sensor
12 Engine-temperature sensor
13 Ignition distributor
14 Auxiliary-air device
15 Battery
16 Ignition and starting switch

(Figure 11). The dimensions of the potentiometer chamber in the air-flow sensor and of the control unit have been reduced to such an extent that the overall height of the entire unit does not exceed that of the previous air-flow sensor alone. Other features of the new air-flow sensor (Figure 12) include the reduced weight due to the aluminum used in place of the zinc material for the housing, the extended measuring range and the improved damping behavior in the event of abrupt changes in the intake air quantity.

Fuel metering

Solenoid-operated injection valves injected the fuel onto the intake valves of the engine. One solenoid valve is assigned to each cylinder and is operated once per crankshaft revolution. In order to reduce the circuit complexity, all valves are connected electrically in parallel. The differential pressure between the fuel pressure and intake-manifold pressure is maintained constant at 2.5 or 3 bar so that the quantity of fuel injected depends solely upon the opening time of the injection valves. For this purpose, the control unit supplies control pulses, the duration of which depends upon the inducted air quantity, the engine speed and other actuating variables which are detected by sensors and processed in the control unit.

Electronic control unit (ECU)

By contrast with the L-Jetronic system, the digital control unit of this system adapts the air-fuel ratio by means of a load/engine-speed map. On the basis of the input signals from the sensors, the control unit computes the injection duration as a measure of the amount of fuel to be injected. The microcomputer system of the control unit permits the required functions to be influenced. The control unit for attachment to the air-flow sensor must be very compact and must have very few plug connections in addition to being resistant to heat, vibration and moisture. These conditions are met by the use of a special-purpose hybrid circuit and a small PC board in the control unit. In addition to accommodating the microcomputer, the hybrid circuit also accommodates 5 other integrated circuits, 88 film resistors and 23 capacitors. The connections from the ICs to the thick-film board comprise thin gold wires which are a mere 33 thousandths of a millimeter in thickness.

Adaptation to operating conditions

The L3-Jetronic performs corrective interventions in mixture formation with components such as the throttle-valve switch, auxiliary-air device, engine-temperature sensor and the lambda closed-loop control. The function is explained in the chapter on the L-Jetronic.

11 Measuring and control unit

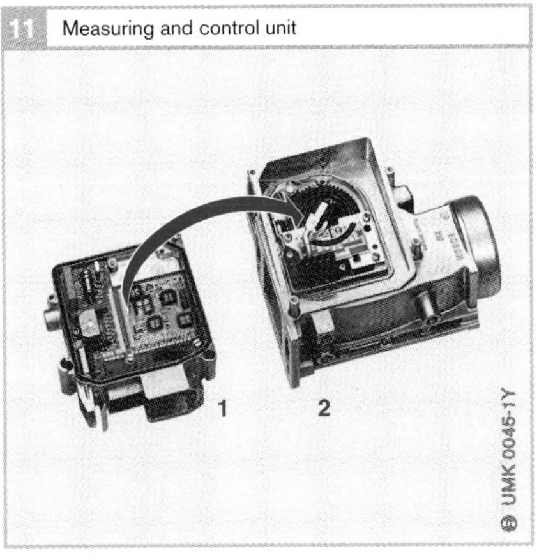

Fig. 11
1 ECU
2 Air-flow sensor with potentiometer

12 Air-flow sensor of the L3-Jetronic

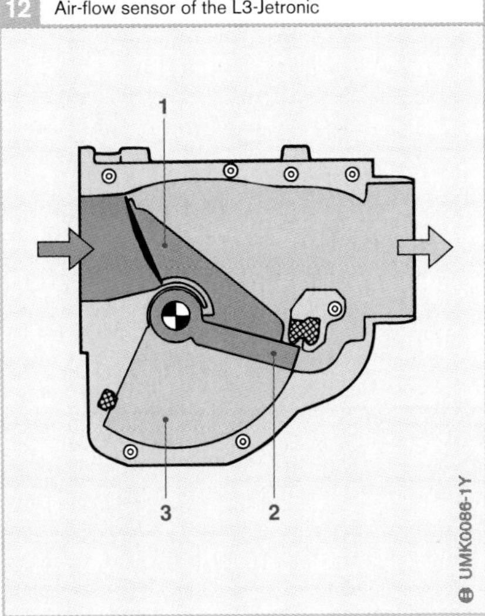

Fig. 12
1 Sensor flap
2 Compensation flap
3 Damping volume

LH-Jetronic

The LH-Jetronic (Fig. 13) is closely related to the L-Jetronic. The difference lies in the hot-wire air-mass meter which measures the air mass inducted by the engine. The result of measurement is thus independent of the air density which is itself dependent upon temperature and pressure.

Fuel supply

The fuel is supplied in the same way as with the L-Jetronic.

Operating-data sensing system

The data is transferred in the same way as with the L-Jetronic. The air-mass meter is an exception.

Air-mass meters

The hot-wire and hot-film air-mass meters are "thermal" load sensors. They are installed between the air filter and the throttle valve and register the air-mass flow [kg/h] drawn in by the engine. Both sensors operate according to the same principle.

Hot-wire air-mass meter

With the hot-wire air-mass meter (Figures 14 and 15), the electrically heated element is in the form of a 70 µm thick platinum wire. The intake-air temperature is registered by a temperature sensor. The hot wire and the intake-air temperature sensor are part of a bridge circuit in which they function as temperature-dependent resistances. A voltage signal which is proportional to the air-mass flow is transmitted to the ECU.

13 Schematic diagram of an LH-Jetronic system

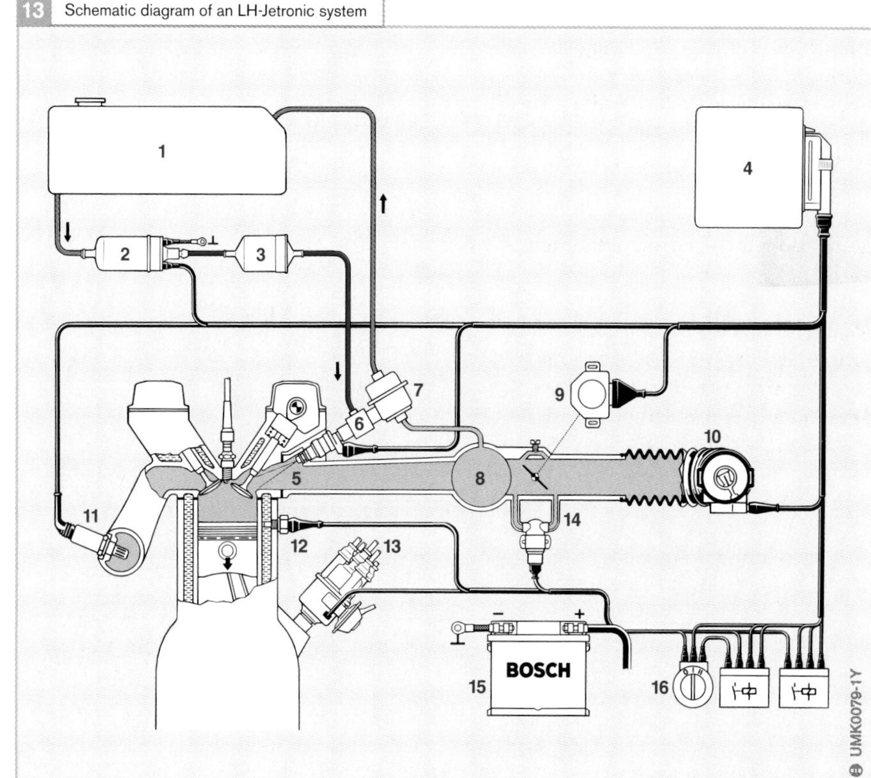

Fig. 13
1 Fuel tank
2 Electric fuel pump
3 Fuel filter
4 ECU
5 Fuel-injection valve (injector)
6 Fuel rail
7 Fuel-pressure regulator
8 Intake manifold
9 Throttle-valve switch
10 Hot-wire air-mass meter
11 Lambda sensor
12 Engine-temperature sensor
13 Ignition distributor
14 Rotary idle actuator
15 Battery
16 Ignition and starting switch

14 Hot-wire air-mass meter

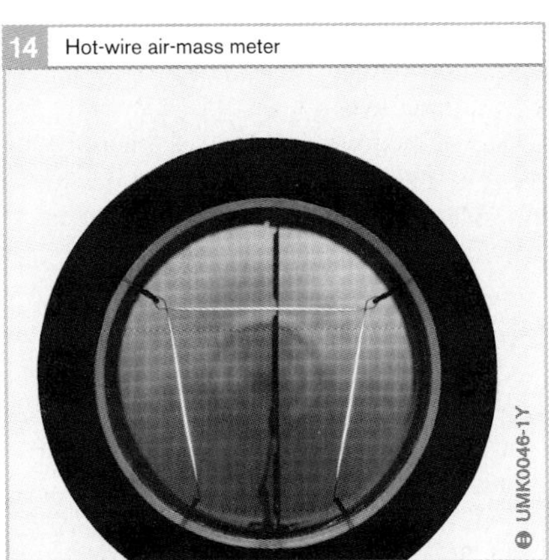

Fig. 14
The 70 μm thin platinum wire is suspended inside the measuring venturi.

Hot-film air-mass meter
With the hot-film air-mass meter (Figure 16), the electrically heated element is in the form of a platinum film resistor (heater element). The temperature of the heater element is registered by a temperature-dependent resistor (throughflow sensor). The voltage across the heater element is a measure for the air-mass flow. It is converted by the hot-film air-mass meter's electronic circuitry into a voltage which is suitable for the ECU.

Fuel metering
Fuel metering is controlled in the same way as with the L-Jetronic.

Adaptation to operating conditions
Various operating conditions are achieved using the same evaluation options as the L-Jetronic and used for interventions in mixture formation.

15 Hot-wire air-mass meter

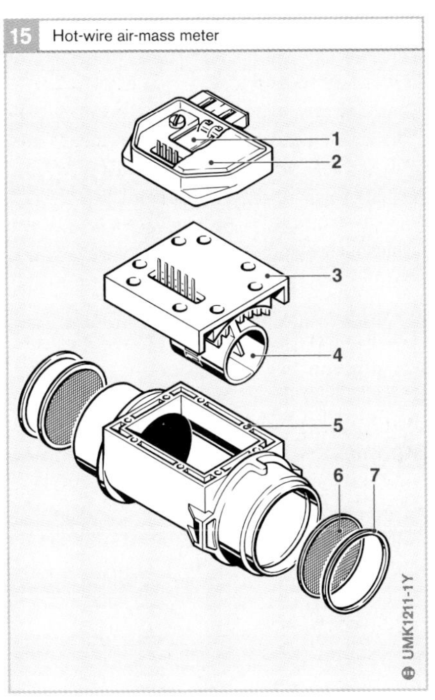

Fig. 15
1 Hybrid circuit
2 Cover
3 Metal insert
4 Venturi with hot wire
5 Housing
6 Screen
7 Retaining ring

16 Hot-film air-mass meter

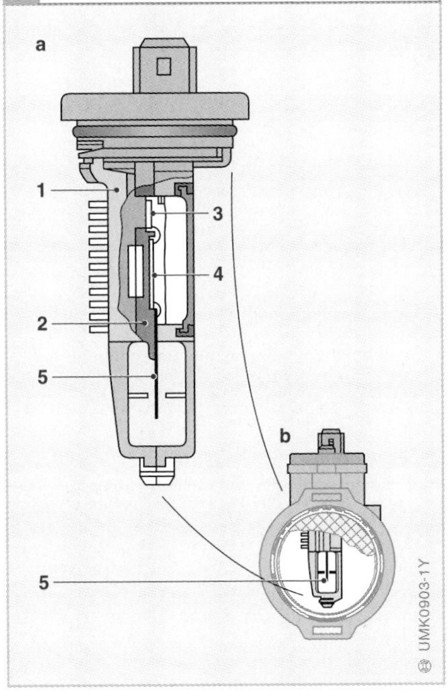

Fig. 16
a Hot-film sensor
b Venturi with built-in hot-film sensor

1 Heat sink
2 Intermediate module
3 Power module
4 Hybrid circuit
5 Sensor element (heater element)

Mono-Jetronic

System overview

Mono-Jetronic is an electronically controlled, low-pressure, single-point injection (SPI) system for 4-cylinder engines. While port injection systems such as KE and L-Jetronic employ a separate injector for each cylinder, Mono-Jetronic features a single, centrally-located, solenoid-controlled injection valve for the entire engine. The heart of the Mono-Jetronic is the central injection unit. It uses a single solenoid-operated injector for intermittent fuel injection above the throttle valve (Figure 1).

The intake manifold distributes the fuel to the individual cylinders. A variety of different sensors are used to monitor engine operation and furnish the essential control parameters for optimum mixture adaptation.

These include:

- Throttle-valve angle
- Engine speed
- Engine and intake-air temperature
- Throttle-valve positions (idle/full-throttle)
- Residual oxygen content of exhaust gas, and (depending on the vehicle's equipment level)
- Automatic transmission
- Air-conditioner settings and
- a/c compressor clutch status (engaged-disengaged)

Input circuits in the ECU convert the sensor data for transmission to the microprocessor, which analyzes the operating data to determine current engine operating conditions; this information, in turn, provides the basis for calculating control signals to the various final-control elements (actuators). Output amplifiers process the signals for transmis-

1 Mono-Jetronic schematic diagram

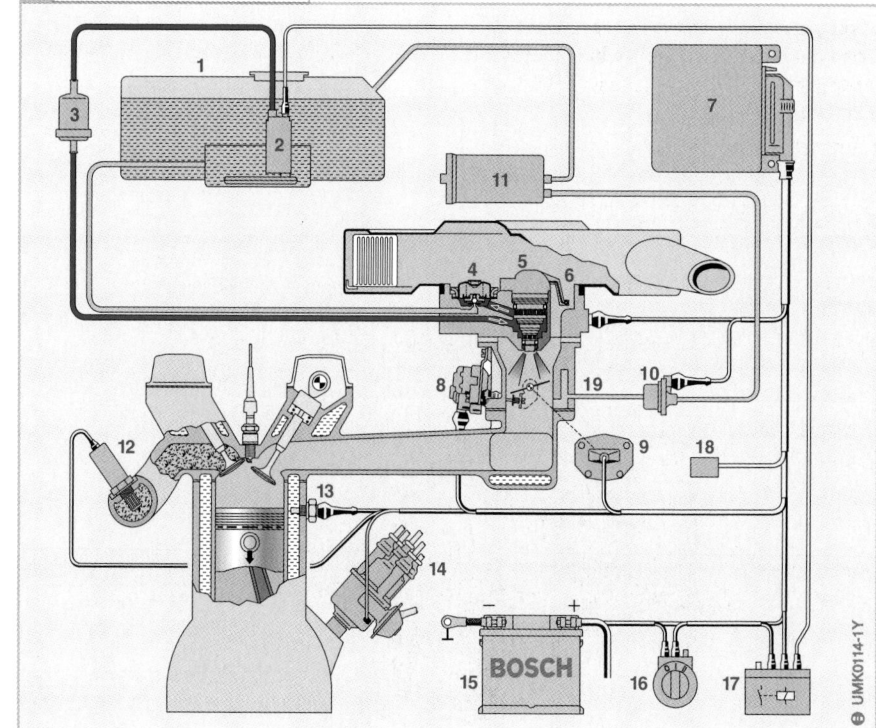

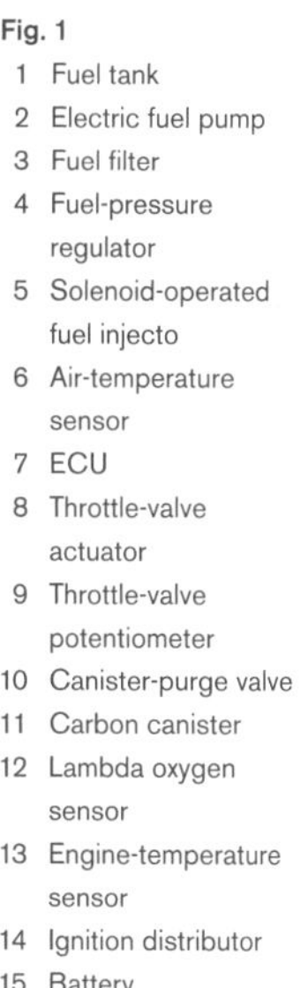
Fig. 1
1 Fuel tank
2 Electric fuel pump
3 Fuel filter
4 Fuel-pressure regulator
5 Solenoid-operated fuel injecto
6 Air-temperature sensor
7 ECU
8 Throttle-valve actuator
9 Throttle-valve potentiometer
10 Canister-purge valve
11 Carbon canister
12 Lambda oxygen sensor
13 Engine-temperature sensor
14 Ignition distributor
15 Battery
16 Ignition-start switch
17 Relay
18 Diagnosis connection
19 Central injection unit

sion to the injector, throttle-valve actuator and canister-purge valve (evaporative-emissions control system).

Versions

The following text and illustrations describe a typical Mono-Jetronic installation (Fig. 1). Other versions are available to satisfy any specific individual requirements that the manufacturers define for fuel-injection systems.

The Mono-Jetronic system discharges the following individual functions (Fig. 2):

- Fuel supply
- Acquisition of operating data and
- Processing of operating data

Basic function

Mono-Jetronic's essential function is to control the fuel-injection process.

Supplementary functions

Mono-Jetronic also incorporates a number of supplementary closed-loop and open-loop control functions with which it monitors operation of emissions-relevant components. These include

- Idle-speed control
- Lambda closed-loop control and
- Open-loop control of the evaporative-emissions control system

Central injection unit

The Mono-Jetronic central injection unit is bolted directly to the intake manifold. It supplies the engine with finely atomized fuel, and is the heart of the Mono-Jetronic system. Its design is dictated by the fact that contrary to multipoint fuel injection (e.g., L-Jetronic), fuel-injection takes place at a central point, and the intake-air quantity is measured indirectly through the two factors throttle-valve angle α and engine speed n, as already described (Figure 2).

Lower section

The lower section of the central injection unit comprises the throttle valve together with the throttle-valve potentiometer for measuring the throttle-valve angle. The throttle-valve actuator for the idle-speed control is mounted on a special bracket on the lower section.

2 Central-injection unit (part sectional drawing)

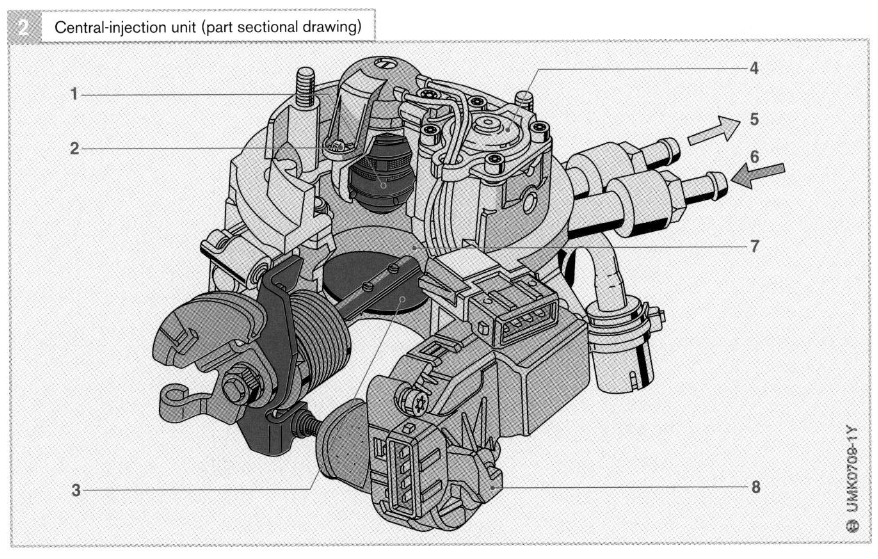

Fig. 2
1 Fuel injector
2 Air-temperature sensor
3 Throttle valve
4 Fuel-pressure regulator
5 Fuel return
6 Fuel inlet
7 Throttle-valve potentiometer (on throttle-valve shaft extension, not shown)
8 Throttle-valve actuator

Upper section
The complete fuel system for the central injection unit is in the upper section, consisting of the fuel injector, the fuel-pressure regulator, and the fuel channels to the fuel injector which are incorporated in the injector bracket. Two passages descend through the housing to supply the injector with fuel. The fuel flows to the injector via the lower passage.

The upper passage is connected to the lower chamber of the pressure regulator. From the fuel-pressure regulator, excess fuel enters the fuel-return line via the plate valve. This fuel-channel arrangement was selected so that even when the formation of fuel-vapor bubbles increases (as can occur for instance when the fuel-injection assembly heats up after the engine is switched off), enough fuel collects at the injector metering area to ensure trouble-free starting.

The open cross-section area between the return channel and the inlet channel is limited to a defined dimension by a shoulder on the fuel-injector strainer. This ensures that excessive fuel that has not been injected is divided into two partial streams, one of which flows through the injector and the other around it.

This form of intensive flushing cools the fuel injector, and the fuel-channel arrangement with fuel circulating around the injector as well as through it, results in the Mono-Jetronic's excellent hot-starting characteristics.

The temperature sensor for measuring the intake-air temperature is fitted in the central injection unit's cover cap.

Fuel supply
The fuel system supplies fuel from the tank to the solenoid-operated injector.

Fuel delivery
The electric fuel pump continuously pumps fuel from the tank and through the fuel filter to the central injection unit. For the Mono-Jetronic, an in-tank unit is used which incorporates

- The electric fuel pump
- A fuel filter on the intake side
- A fuel-level gauge
- A fuel-swirl pot which serves as a fuel reservoir and
- The electrical and hydraulic connections to the outside

The electric fuel pump is the most common type used in Mono-Jetronic applications. It provides ideal performance with this system's low primary pressure. It is a two-stage flow-type pump: a side-channel pump being used as the preliminary stage (pre-stage) and a peripheral pump as the main stage. Both stages are integrated in a single impeller wheel.

Fuel-pressure control
It is the task of the fuel-pressure regulator to maintain the differential between line pressure and the local pressure at the injector nozzle constant at 100 kPa. In the Mono-Jetronic, the pressure regulator is an integral part of the central injection unit's hydraulic circuit.

A rubber-fabric diaphragm divides the fuel-pressure regulator into the lower (fuel) chamber (Figure 3, Pos. 6) and the upper (spring) chamber (5). The pressure from the helical spring (4), is applied to

3 Fuel-pressure regulator

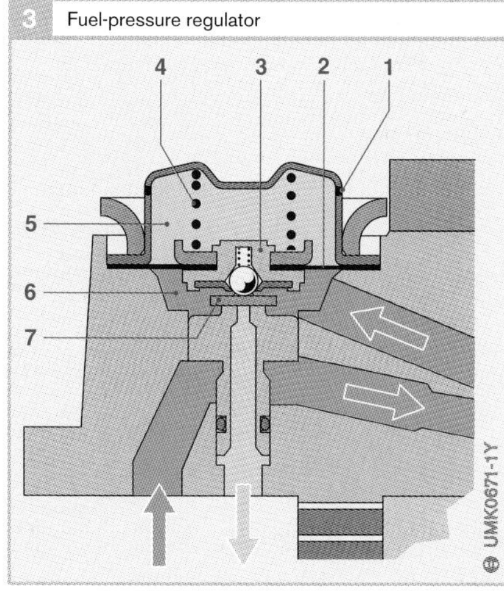

Fig. 3
1 Venting ports
2 Diaphragm
3 Valve holder
4 Compression spring
5 Upper chamber
6 Lower chamber
7 Valve plate

the diaphragm. A movable valve plate which is connected to the diaphragm through the valve holder is pressed onto the valve seat by spring pressure (flat-seat valve).

When the force generated by the fuel pressure against the diaphragm's area exceeds the opposing spring force, the valve plate is lifted slightly from its seat and the excess fuel can flow back to the tank through the open valve. In this state of equilibrium, the differential pressure between the upper and lower chambers of the pressure regulator is 100 kPa.

Vents maintain the spring chamber's internal pressure at levels corresponding to those at the injector nozzle. The valve-plate lift varies depending upon the delivery quantity and the actual fuel quantity required.

The spring characteristics and the diaphragm area have been selected to provide constant, narrow-tolerance pressure over a broad range of fuel-delivery rates. When the engine is switched off, fuel delivery also stops. The check valve in the electric fuel pump then closes, as does the pressure-regulator valve, maintaining the pressure in the supply line and in the hydraulic section for a certain period. Because this design configuration effectively inhibits the formation of vapor bubbles which can result from fuel-line heat build-up during pauses in operation, it helps ensure trouble-free warm starts.

Acquisition of operating data

Sensors monitor all essential operating data to furnish instantaneous information on current engine operating conditions. This information is transmitted to the ECU in the form of electric signals, which are converted to digital form and processed for use in controlling the various final-control elements.

Air charge

The system derives the data required for maintaining the required air-fuel mixture ratio by monitoring the intake-air charge for each cycle. Once this air mass (referred to as air charge in the following) has been measured, it is possible to adapt the injected fuel quantity by varying injection duration.

With the Mono-Jetronic, the air charge is determined indirectly, using coordinates defined by the throttle-valve angle α and engine speed n.

Air charge as a function of α and n is determined for a given engine on the engine dynamometer. Figure 4 shows a typical set of curves for an engine program; it illustrates the relative air-charge factors for various throttle-valve apertures α and engine speeds n. If the engine response data are already available, and assuming constant air density, air charge can be defined precisely using α and n exclusively (α/n system).

4 Engine map

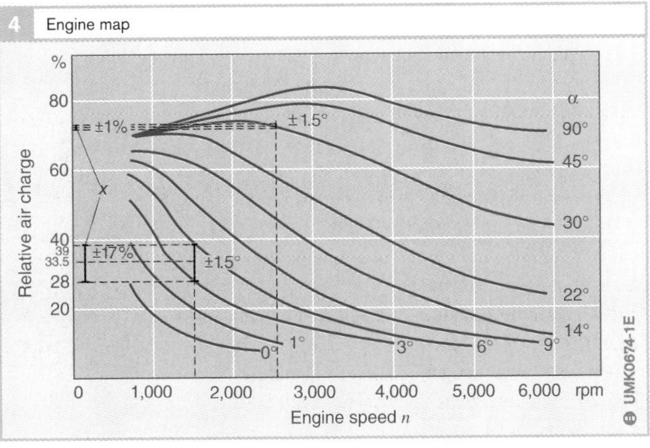

Fig. 4
Relative air charge as a function of engine speed n and throttle-valve angle α

x Relative change in air-charge

By actuating the throttle valve from the accelerator pedal, the driver controls the intake air flow of the gasoline engine, thereby specifying the desired operating point. Here, the throttle-valve potentiometer in the fuel-injection system component records the throttle-valve angle α. Besides the throttle valve's position α, the engine speed n and the air density are additional influencing variables for the air mass taken in by the engine. The engine-speed information is provided by the ignition system.

Because the differential between internal fuel pressure within the injector and air pressure at the nozzle is maintained at a constant level, injected fuel quantities are determined solely by the length of time (injection duration) the injector remains open for each triggering pulse.

This injection duration must be proportional to the monitored air charge to maintain a specific air-fuel ratio. In other words: The injection duration is a direct function of α and n. In the Mono-Jetronic, this relationship is governed by a Lambda program map with input variables α and n. The system is programmed to compensate for fluctuations in air density, which is a function of the intake-air temperature and the air pressure. The intake-air temperature, measured as the air enters the central injection unit, provides the ECU with the basic data required to determine the corresponding correction factor.

The Mono-Jetronic is always equipped with the Lambda closed-loop control, designed to maintain the air-fuel ratio at $\lambda = 1$. In addition, the Lambda closed-loop control is used to implement adaptive mixture adaptation, in other words, the system learns to adapt itself to the changing operating conditions. Correction factors for variations in atmospheric pressure (especially those associated with altitude changes) are supplemented by correction factors designed to compensate for differences in production tolerances and ongoing wear. When the engine is switched off, the system stores the correction factors so that they are effective immediately once the engine is started again.

This system of indirect determination of intake-air mass – with control based on the α/n control parameters – operates with adaptive mixture control and superimposed Lambda closed-loop control to accurately maintain a constant mixture, without any need for direct measurement of air mass.

Throttle-valve angle

The ECU employs the throttle-valve angle α to calculate the throttle-valve's position and angular velocity. Throttle-valve position is an important input parameter for determining intake-air volume, the basic factor in calculating injection duration. When the throttle is closed, the idle switch provides the throttle-valve actuator with a supplementary position signal.

Data on the throttle valve's angular velocity are used mainly for the transition compensation. The resolution of the α signal is determined by the air-charge measurement. To ensure good drive-ability and low emissions, the resolution of the air-charge measurement and injection duration must take place in the smallest-possible digital steps (quantization), to maintain the air-fuel ratio within a tolerance range of 2 %.

The program range that displays the largest intake-charge variations relative to α is defined by minimal apertures α and low engine speeds n, i.e., at idle and low part load. Within this range – as illustrated in Figure 4 – a change of ±1.5 ° in throttle-valve angle shifts the air-charge/lambda factor by ±17 %, whereas outside this range, with higher throttle-valve angles, a similar increment has an almost negligible effect. This means that a high angular resolution is necessary at idle and low part load.

The required high level of signal resolution is achieved by distributing the throttle-valve angle in the throttle-valve potentiometer for the range between idle and full-throttle between two resistance tracks. Track 1 covers the angular range from 0°...24°, and track 2 the range from 18°...90°. The angle signals (α) from each track are converted separately in the ECU, each in its own analog/digital converter circuit.

Engine speed
The engine-speed information required for α/n control is obtained by monitoring the periodicity of the ignition signal. These signals from the ignition system are then processed in the ECU. The ignition signals are either TD pulses already processed by the ignition trigger box, or the voltage signal available at Terminal 1 (U_S) on the low-voltage side of the ignition coil. At the same time, these signals are also used for triggering the injection pulses, whereby each ignition pulse triggers an injection pulse.

Engine temperature
Engine temperature has a considerable influence on fuel consumption. A temperature sensor in the engine coolant circuit measures the engine temperature and provides the ECU with an electrical signal.

Intake-air temperature
Intake-air density varies according to temperature. To compensate for this influence, a temperature sensor on the intake side of the throttle body monitors the temperature of the intake air for transmission to the ECU.

Operating modes
Idle and full-throttle must be registered accurately so that the injected fuel quantity can be optimized for these operating modes, and in order that full-load enrichment and overrun fuel cutoff can function correctly.

The idle operating mode is registered from the actuated idle contact of a switch in the throttle-valve actuator. The ECU recognizes full-throttle operation based on the electrical signal from the throttle-valve potentiometer.

Battery voltage
The solenoid fuel injector's pickup and release times vary according to battery voltage. If system voltage fluctuates during operation, the ECU adjusts the injection timing to compensate for delays in injector response times.

In addition, the ECU responds to the low system voltages encountered during starting at low temperatures by extending injection duration. The extended duration compensates for voltage-induced variations in the pumping characteristic of the electric fuel pump, which does not achieve maximum system pressure under these conditions.

The ECU receives the battery voltage in the form of a continuous signal transmitted to the microprocessor via the A/D converter.

Control signals from the air-conditioner and/or automatic transmission
When the air conditioner is switched on, or the automatic transmission is placed in gear, the resulting engine load would normally cause the idle speed to drop. To compensate, the air-conditioner modes "air-conditioner ON" and "compressor ON," as well as the "Drive" position on the automatic gearbox, are registered by the ECU as switching signals. The ECU then modifies the idle-speed control signal to compensate. It may be necessary to increase the idle speed to ensure that the air conditioner continues to operate effectively, and reductions in idle speed are often required when "Drive" is selected on automatic-transmission vehicles.

Mixture composition
Due to the exhaust-gas aftertreatment using a three-way catalytic converter, the correct air-fuel mixture composition must be precisely maintained. A Lambda oxygen sensor in the exhaust-gas flow provides the ECU with an electric signal indicating the current mixture composition. The ECU then uses this signal for the closed-loop control of the mixture composition to obtain a stoichiometric ratio. The Lambda oxygen sensor is

installed in the engine's exhaust system at a position which ensures that it is kept at the temperature required for correct functioning across the complete engine operating range.

Processing of operating data

The ECU processes the engine operating data received from the sensors. From this data, it uses the programmed ECU functions to generate the triggering signals for the fuel injector, the throttle-valve actuator, and the canister-purge valve.

Electronic control unit (ECU)

The ECU is housed in a fiberglass-reinforced, polyamide plastic casing. To insulate it from the engine's heat, it is located either in the passenger compartment or the ventilation plenum chamber between the engine and passenger compartments.

All of the ECU's electronic components are installed on a single printed-circuit board. The output amplifiers and the voltage regulator responsible for maintaining 5V supply to the electronic components are installed on heat sinks for better thermal dissipation. A 25-pin plug connects the control unit to the battery, the sensors and the actuators (final-control elements).

Analog-digital converter (A/D)

The signals from the throttle-valve potentiometer are continuous analog signals; as are the Lambda-sensor signal; the engine-temperature signal; the intake-air signal; the battery voltage; and a voltage-reference signal generated in the ECU. These analog signals are converted to data words by the analog-digital converter (A/D) and transmitted to the microprocessor via the data bus. An analog-digital input is used so that, depending upon the input voltage, various data records in the read memory can be addressed (data coding).

The engine-speed signal provided by the ignition is conditioned in an integrated circuit (IC) and transmitted to the microprocessor. The engine-speed signal is also used to control the fuel-pump relay via an output stage.

Microprocessor

The microprocessor is the heart of the ECU (Fig. 5). It is connected through the data and address bus with the programmable read-only memory (EPROM) and the random-access memory (RAM). The read memory contains the program code and the data for the definition of operating parameters.
In particular, the random-access memory serves to store the adaptation values (adaptation: adapting to changing conditions through self-learning). This memory module remains permanently connected to the vehicle's battery to maintain the adaptation data contained in the random-access memory when the ignition is switched off.

A 6 MHz quartz oscillator provides the stable basic clock rate needed for the arithmetic operations. A signal interface adapts the amplitude and shape of the control signals before transmitting them to the microprocessor for processing. These signals include the idle-switch setting, diagnosis activation, the position of the selector lever on automatic-transmission vehicles (Neutral, Drive), and a signal for "Air-conditioner ON" and "Compressor ON/OFF" in vehicles with air-conditioning.

Driver stages

A number of different driver stages generate the control signals for the fuel injector, the throttle-valve actuator, the canister-purge valve, and the fuel-pump relay. A fault lamp is installed in some vehicles to warn the driver in case of sensor or actuator faults. The fault lamp's output also serves as an interface for diagnosis activation and read-outs.

Lambda program map

The Lambda map is used for precise adaptation of the air-fuel ratio at all static operating points once the engine is warm. This map is stored electronically in the digital circuit section of the ECU (Figure 6); the reference data are determined empirically through tests on the engine dynamometer. For a Lambda closed-loop-controlled engine-management concept such as the Mono-Jetronic, this testing is employed to determine optimum injection timing and duration for a specific engine under all operating conditions (idle, part or full-throttle). The resulting program consistently maintains an ideal (stoichiometric) air-fuel mixture throughout the operating range.

The Mono-Jetronic Lambda program map consists of 225 control coordinates; these are assigned to 15 reference coordinates for the parameters throttle-valve angle α and n for engine speed. Because the α/n curves are extremely non-linear, necessitating high resolution accuracy at idle and in

6 Lambda map

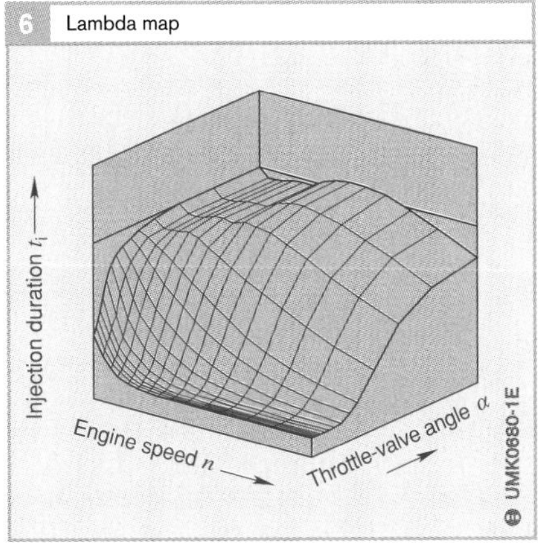

Fig. 6
Injection duration as a function of engine speed n and throttle-valve angle α

5 Schematic of the Mono-Jetronic ECU

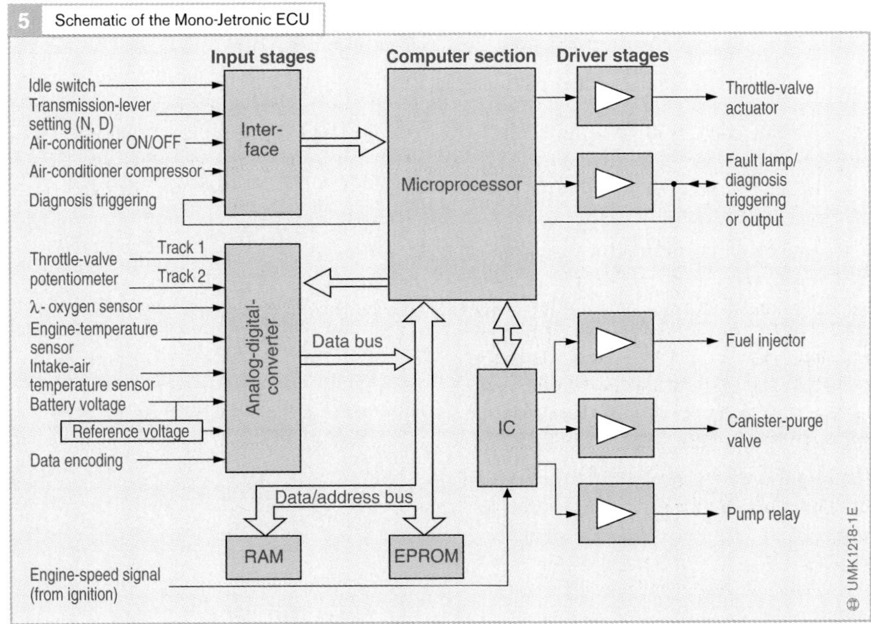

the lower part-load range, the data points are situated very closely together in this area of the map. Control coordinates located between these reference coordinates are determined in the ECU using linear interpolation.

Because the Lambda map is designed for the engine's normal operating and temperature range, it becomes necessary to correct the basic injection timing when engine temperatures deviate or when special operating conditions are encountered.

If the ECU registers deviations from $\lambda = 1$ in the signals from the Lambda sensor, and as a result is forced to correct the basic injection duration for an extended period of time, it generates mixture correction values and stores them in an internal adaptation process. From then on, these values are effective for the complete map and are continually updated. This layout ensures consistent compensation for individual tolerances and for permanent changes in the response characteristics of engine and injection components.

Fuel injection

The fuel-injection system must be able to accurately meter to the engine the minimum amount of fuel required (at idle or zero-load), as well as the maximum (at full throttle). The control coordinates for these conditions must be situated within the linear range on the injector curves (Fig. 7).

One of the Mono-Jetronic's most important assignments is the uniform distribution of the air-fuel mixture to all cylinders. Apart from intake-manifold design, the distribution depends mainly upon the fuel injector's location and position, and the quality of its air-fuel mixture preparation. The ideal fuel-injector position within the Mono-Jetronic housing was determined in the research and development phase. Special adaptations for operation in individual engines are not required.

The central injection unit's housing is centered in the intake-air flow by a special bracket, and has been designed for maximum aerodynamic efficiency. The housing contains the injector, which is installed directly above the throttle valve for intensive mixing of the injected fuel with the intake air. To this end, the injector finely atomizes the fuel and injects it in a cone-shaped jet between the throttle valve and the throttle-valve housing into the area with the highest air-flow speed.

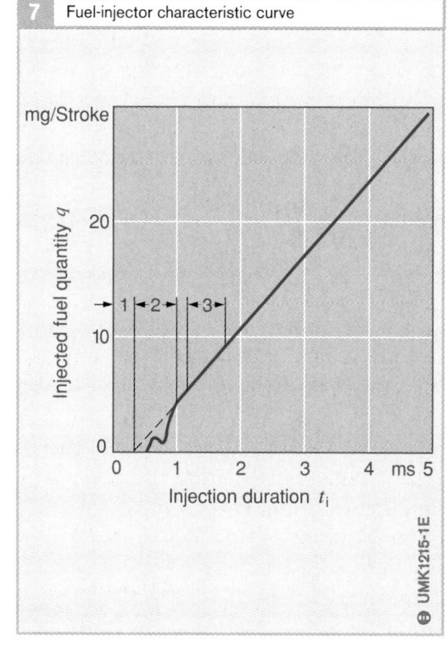

Fig. 7
At an engine speed of n = 900 rpm (corresponds to a 33 ms injection pulse train)

1 Voltage-dependent fuel-injector delay time
2 Non-linear characteristic section
3 Injection-duration section at idle or no-load operation

8 Fuel injector

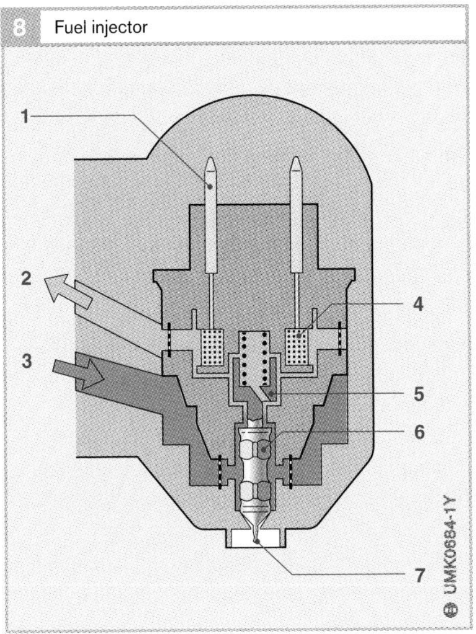

Injector
The fuel injector (Fig. 8) comprises the valve housing and the valve group. The valve housing contains the solenoid winding (4) and the electrical connections (1). The valve group includes the valve body which holds the valve needle (6) and its solenoid armature (5). When no voltage is applied to the solenoid winding, a helical spring assisted by the primary system pressure forces the valve needle onto its seat. With voltage applied and the solenoid energized, the needle lifts about 0.06 mm (depending upon valve design) from its seat so that fuel can exit through an annular gap. The pintle (7) on the front end of the valve needle projects from the valve-needle bore, and its shape ensures excellent fuel atomization.

The size of the gap between the pintle and the valve body determines the valve's static fuel quantity. In other words, the maximum fuel flow with the valve permanently open. The dynamic fuel quantity injected during intermittent operation depends also on the valve spring, the valve-needle mass, the magnetic circuit and the ECU driver stage. Because the fuel pressure is constant, the amount of fuel actually injected by the fuel injector depends solely upon the valve's opening time (injection duration).

Due to the frequency of the injection pulse train (every ignition pulse triggers an injection pulse) the fuel injector must feature extremely short switching times. Its pick-up and release times are kept to below one millisecond by the low mass of armature and needle, as well as by the optimized magnetic circuit. Precise metering of even the smallest amount of fuel is therefore guaranteed.

Mixture adaptation

Starting phase
When the engine is cold, effective fuel vaporization is inhibited by the following factors:

- Cold intake air
- Cold manifold walls
- High manifold pressure
- Low air-flow velocity in the intake manifold and
- Cold combustion chambers and cylinder walls

These factors mean that part of the fuel metered to the engine condenses on the cold manifold walls and covers them with a film of fuel. To ensure that the fuel leaving the injector reaches the engine for combustion, this condensation process must be terminated as soon as possible. To do so, more fuel is metered to the engine when starting than would otherwise be needed for the combustion of the intake air. Being as the amount of fuel condensation mainly depends upon the intake-manifold temperature, the injection times for starting are specified by the ECU according to engine temperature (Fig. 9a, next page).

The fuel condensation rate not only depends upon the manifold-wall temperature but also on the air velocity in the intake manifold. The higher the air-flow velocity, the less fuel is deposited on the manifold walls. The injection duration is reduced as engine speed increases (Fig. 10a).

Fig. 8
1 Electrical connection
2 Fuel return
3 Fuel inlet
4 Solenoid winding
5 Solenoid armature
6 Valve needle
7 Pintle

In order to achieve very short starting times, the build-up of the fuel film on the manifold walls must take place as quickly as possible, in other words a large quantity of fuel must be metered to the engine in a short period. At the same time, precautions must be taken to prevent over-enrichment of the engine. These demands are in opposition to each other, but are complied with by reducing the injection duration the longer engine cranking continues (Fig. 10b). The engine is considered to have started as soon as the end-of-starting speed, which is dependent upon engine temperature, is exceeded (Fig. 9b).

9 Corrections as a function of engine temperature. During the cranking phase, the post-start phase, and the warm-up phase

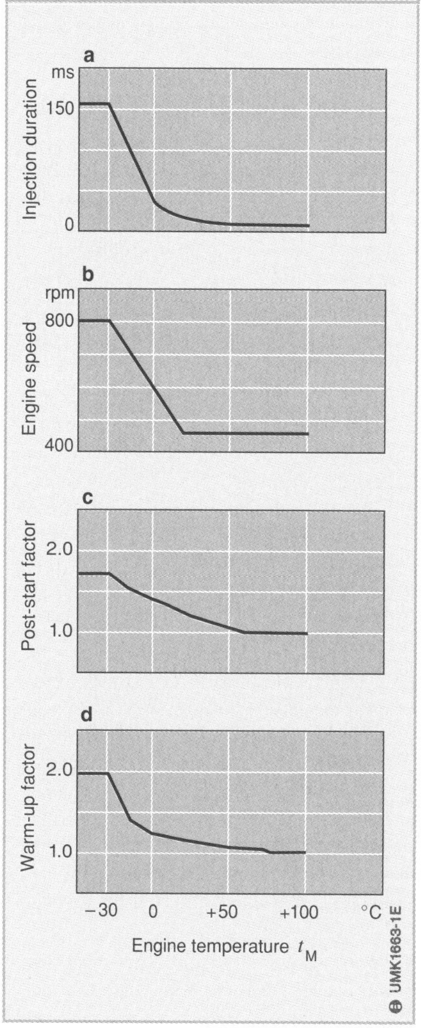

Fig. 9
a Injection duration at start of cranking
b Engine speed at end of cranking
c Post-start factor
d Warm-up factor

Post-start and warm-up phase

As soon as the engine has started, and depending upon throttle-valve position and engine speed, the fuel injector is triggered with the injection durations stored in the Lambda map. From this point on, and until engine operating temperature has been reached, fuel enrichment continues to be necessary due to the fuel condensation on the combustion chamber and manifold walls, which are still cold.

Immediately following a successful start, the enrichment must be increased briefly, but this increase is then followed by normal enrichment which depends solely on the engine temperature. Two functions determine the engine's fuel requirement in the phase between end-of-start and reaching operating temperature:

10 Reduction of injected-fuel quantity during starting

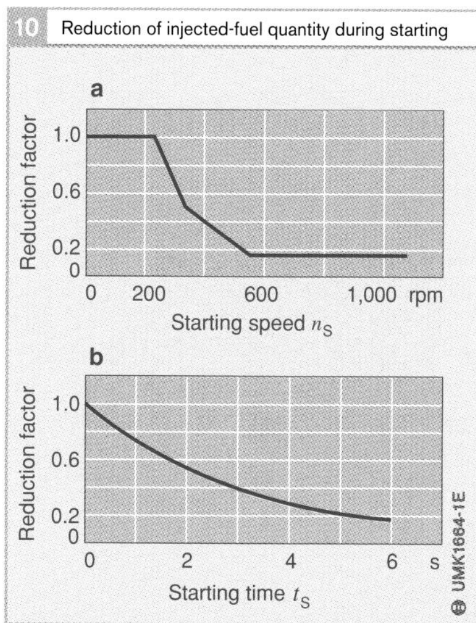

Fig. 10
a Reduction as a function of speed
b Reduction as a function of time

- Post-start enrichment is stored as an engine-temperature-dependent correction factor (Fig. 9c). This post-start factor is used to correct the injection durations calculated from the Lambda map. The reduction of the post-start factor to the value 1 is time-dependent.
- The warm-up enrichment is also stored as an engine-temperature-dependent correction factor. Reduction of this factor to value 1 depends solely on engine temperature (Fig. 9d).

Both functions operate simultaneously; the injection durations calculated from the Lambda map are adapted with the post-start factor as well as with the warm-up factor.

Mixture adaptation as a function of intake-air temperature
The air mass required for the combustion is dependent upon the intake air's temperature. Taking a constant throttle-valve setting, this means that cylinder charge reduces along with increasing air temperature. The Mono-Jetronic central injection unit therefore is equipped with a temperature sensor which reports the intake-air temperature to the ECU which then corrects the injection duration, e.g., the injected fuel quantity, by means of an intake-air-dependent enrichment factor (Fig. 11).

11 Enrichment factor as a function of intake-air temperature

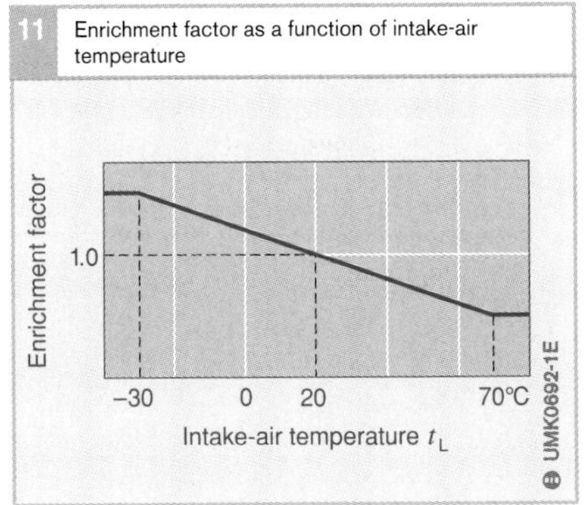

Transition compensation
The dynamic mixture correction which is necessary to compensate for the load changes due to throttle-valve movements is provided by the transition compensation. This facility is required in order to achieve best-possible driveability and exhaust-gas behavior, and in the case of a single-point fuel-injection system its functional complexity is somewhat higher than with a multipoint system. This is due to the fact that mixture distribution is via the intake manifold in a single-point system, which means that for transition compensation three different factors must be taken into consideration with regard to the transportation of the fuel:

- Fuel vapor in the central injection unit or in the intake manifold, or which forms on the manifold walls due to the evaporation of the fuel film. This vapor is transported very quickly at the same speed as the intake air.
- Fuel droplets, which are transported at different speeds, but nevertheless at about the same speed as the intake-air. Some of these droplets are flung against the intake-manifold walls though, where they contribute to the evaporation of the fuel film.
- Liquid fuel, transported to the combustion chamber at reduced speed, stemming from the fuel film on the intake-manifold walls. There is a time lag in the availability of this portion of the fuel for combustion.

At low intake-manifold pressures, that is at idle or lower part load, the fuel in the intake manifold is almost completely in vaporous form and there is practically no fuel film on the manifold walls. When the manifold pressure increases though, i.e. when the throttle valve is opened (or when speed drops), the proportion of fuel in the wall film increases. The result is that when the throttle valve is operated during a transition, the balance between the increase and decrease in the amount of fuel in the wall film is disturbed. This means that when the throttle valve is opened, some form of compensation is necessary in order to prevent the mixture leaning-off due to the increase in the amount of fuel deposited on the manifold walls. This compensation is provided by the acceleration-enrichment facility. Correspondingly, when the throttle valve is closed, the wall film reduces by releasing some of its fuel, and without some form of compensation this would lead to mixture enrichment at the cylinders during the transitional phase. Here, the deceleration lean-off facility comes into effect.

In addition to the tendency of the fuel to evaporate as a result of the intake-manifold pressure, the temperature also plays an important role. Therefore, when the intake manifold is still cold, or with low intake-air temperatures, there is a further increase in the amount of fuel held in the wall film.

In the Mono-Jetronic, complex electronic functions are applied to compensate for these dynamic mixture-transportation effects, and these functions ensure that the air-fuel ratio remains as near as possible to λ = 1 during transition modes.The acceleration enrichment and deceleration lean adjustment functions are based on throttle-valve angle, engine speed, intake-air temperature, engine temperature and the throttle-valve's angular velocity.

Acceleration enrichment and trailing-throttle lean adjustment are triggered when the throttle valve's angular velocity exceeds the respective trigger threshold (Figure 12). For acceleration enrichment, the trigger threshold is in the form of a characteristic curve based on the throttle valve's angular velocity, and for deceleration lean adjustment it is constant.

Also based on the throttle valve's rate of travel are the two dynamic-response correction factors for acceleration enrichment and trailing-throttle lean adjustment. Both of these dynamic mixture-correction factors are stored as characteristic curves.

The program's compensation mode thus functions as a comprehensive transition factor for adapting both injection timing and duration. Because changes in load factor can be quite rapid relative to injection periodicity, the system is also capable of generating a supplementary injection pulse to provide additional compensation.

Lambda closed-loop control

The Lambda closed-loop control circuit maintains an air-fuel mixture of precisely $\lambda = 1$. The ECU receives continuous signals from the Lambda oxygen sensor located in the exhaust-gas stream.

Using this signal, the ECU monitors the instantaneous level of the residual oxygen in the exhaust gas. The ECU employs these signals to monitor the current mixture ratio and to adjust injection duration as required.

12 Trigger thresholds for transition compensation

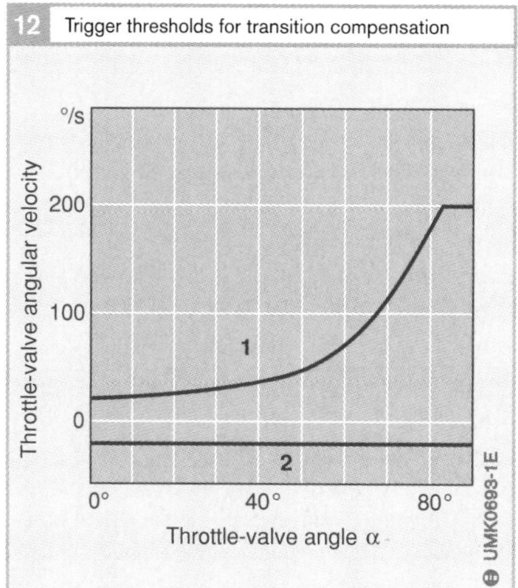

Fig. 12
1 Acceleration enrichment
2 Deceleration lean-off
No triggering

The Lambda closed-loop control is superimposed on the basic mixture-control system. It ensures that the system is always optimally matched to the 3-way catalytic converter. Deviations from the stoichiometric air-fuel ratio are detected and corrected with the aid of the Lambda oxygen sensor. The Lambda sensor (two-point sensor) is installed so that it extends into the exhaust gas and acts as a probe which delivers information indicating whether the mixture is richer or leaner than $\lambda = 1$. If the mixture deviates from $\lambda = 1$, the sensor output voltage changes abruptly and this change is evaluated by the ECU control circuit. A high (approx. 800 mV) sensor output voltage indicates a mixture which is richer than $\lambda = 1$, and a low output signal (approx. 200 mV) a mixture which is leaner. The Lambda sensor's signal periodicity is illustrated in Figure 13. The Lambda control stage receives a signal for every transition from rich to lean mixture and from lean to rich.

The Lambda correction factor is applied to adjust injection duration. When a sensor voltage jump occurs, in order to generate a correction factor as soon as possible, the air-fuel mixture is changed immediately by a given amount. The manipulated variable then follows a programmed adaptation function until the next Lambda-sensor voltage jump takes place. The result is that within a range very close to $\lambda = 1$, the air-fuel mixture permanently changes its composition either in the rich or in the lean direction.

If it were possible to adapt the Lambda map to the ideal value $\lambda = 1$, the manipulated variable for the Lambda control stage (the Lambda correction factor) would control permanently to the neutral value $\lambda = 1$. Unavoidable tolerances make this impossible though, so the Lambda closed-loop control follows the deviations from the ideal value and controls each map point to $\lambda = 1$.

The Lambda oxygen sensor needs a temperature above about 350 °C before it delivers a signal which can be evaluated. There is no closed-loop control below this temperature.

Mixture adaptation

The mixture-adaptation function provides separate, individual mixture adjustment for specific engine-defined operating environments. It also furnishes the mixture-control circuits with reliable compensation for variations in air density. The mixture adaptation program is designed to compensate for the effects of production tolerances and wear on engine and injection-system components.

13 Voltage characteristic of the Lambda-sensor signal

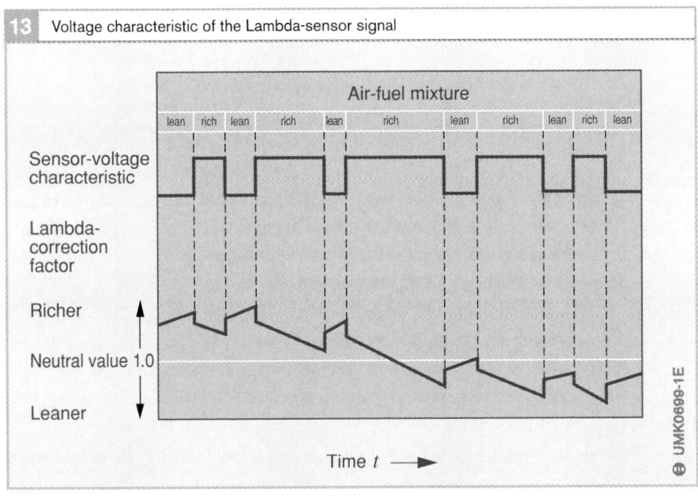

The system must compensate for three variables:

- Influences due to air-density changes when driving at high altitudes (air-flow multiplicative influence).
- Influences related to vacuum leakage in the intake tract. Leakage downstream from the throttle valve can also vary as deposits reseal the affected areas (air-flow additive influence).
- Influences due to individual deviations in the injector response delay (injection-duration additive influence).

As there are certain map sectors where these influences have a very marked effect, the map is subdivided into three mixture-adaptation sectors:

- Air-density changes which have the same effect over the complete map. The mixture-adaptation sector for the adaptation variable which takes the air density into account (air-flow multiplicative value) therefore applies to the complete map.
- Changes in the leakage-air rate have a pronounced effect at low air-flow rates (e.g. in the vicinity of idle). An additional adaptation value is therefore calculated in a second sector (airflow-additive value).
- Changes in the injected fuel quantity for every injection pulse are particularly effective when the injection duration is very short. A further adaptation value is therefore calculated in a third sector (injection-duration additive value).

The mixture-adaptation variable is calculated as follows: The Lambda control-stage manipulated variable is a defined quantity. When a mixture fault is detected, this variable is changed until the mixture has been corrected to $\lambda = 1$. Here the mixture correction for the Lambda controller is defined by the deviation from the Lambda control value. For mixture adaptation, these Lambda controller parameters are evaluated using a weighting factor before being added to the adaptation variables for the individual ranges. The adaptation variable thus varies in fixed increments, with the size of the increments being proportional to the current Lambda mixture-correction factor. Each increment provides compensation for another segment of the mixture correction (Figure 14).

Depending upon the engine load and speed, the incremental periodicity is between 1 s and a few 100 ms. Adaptation variables are updated so quickly that any effects of tolerance and drift on driveability and exhaust-gas composition are compensated for completely.

14 Cyclic change between mixture adaptation and adaptation of the cylinder-charge factor

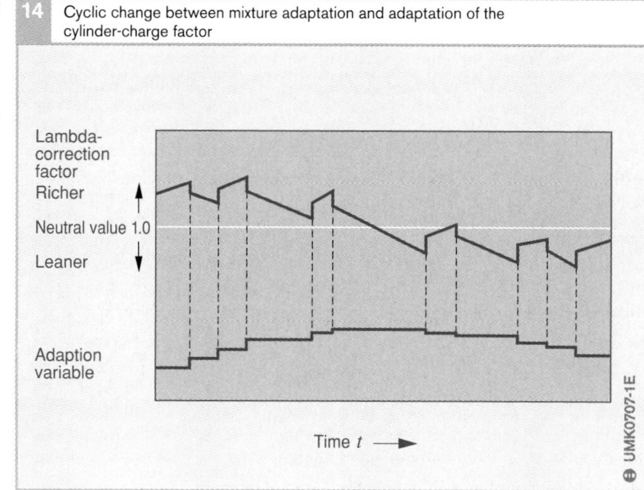

Idle-speed control

The idle-speed (closed-loop) control is used to reduce and stabilize the idle speed, and maintains a consistent idle speed throughout the whole of the vehicle's service life. In this type of idle-speed control, the throttle-valve actuator, which opens the throttle valve by means of a lever, is so controlled that the stipulated idle speed is maintained exactly under all operating conditions. This applies no matter whether the vehicle electrical system is heavily loaded, or the air-conditioner is switched on, or the automatic gearbox is at "Drive", or the power-assisted steering is at full lock etc. Engine temperature also has no effect, nor do high altitudes, where larger throttle-valve angles are required to compensate for lower barometric pressure.

For the idle-speed control facility there are two engine-temperature-dependent curves stored in the ECU (Fig. 15a):

- Curve 1 for automatic-transmission vehicles when "Drive" is engaged.
- Curve 2 for manually-shifted transmissions, or automatic gearboxes if "Drive" is not engaged (Neutral).

The idle speed is usually reduced on automatic vehicles in order to reduce their tendency to creep when "Drive" is engaged. When the air-conditioner is switched on, the idle speed is often increased to maintain a defined minimum idle and ensure adequate cooling (Curve 3). In order to prevent engine-speed fluctuations when the air-conditioner compressor engages and disengages, the high idle speed is maintained even when the compressor is not engaged.

From the difference between the actual engine speed and the set speed (n_{set}), the engine-speed control stage calculates the appropriate throttle-valve setting.

With the idle switch closed, the control signal to the throttle-valve actuator comes from a position controller. This generates the control signal from the difference between the actual throttle-valve setting as registered by the throttle-valve potentiometer, and the desired throttle-valve setting as calculated by the engine-speed control stage.

In order to avoid sudden drops in engine speed during transitions, for instance, in the transition from overrun to idle, the throttle valve must not be closed too far. This is ensured by applying pilot control characteristics which electronically limit the minimum correcting range of the throttle-valve actuator. This necessitates the ECU containing a temperature-dependent throttle-valve pilot-control characteristic for "Drive" and for "Neutral" (Fig. 15b).

15 Idle-speed control

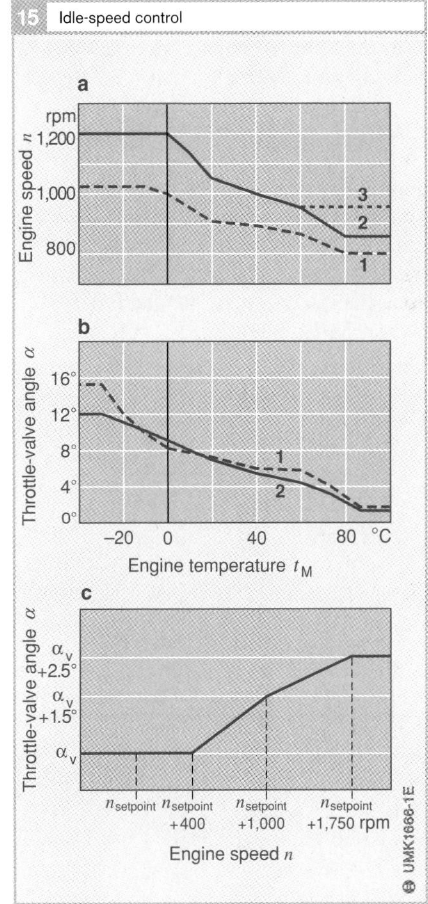

Fig. 15
a Desired engine speed
b Throttle-valve pilot control
c Vacuum limiting

1 Drive
2 Neutral
3 Air conditioner ON

V Throttle-valve pilot control

In addition, a number of different pilot-control corrections come into effect when the air-conditioner is switched on, dependent on whether the compressor is engaged or not. So that the pilot control is always at optimum value, pilot-control corrections are also adapted to cover all possible combinations resulting from the input signals for Transmission setting (Drive/Neutral), "Air-conditioner ready" (YES/NO), and "Compressor engaged" (YES/NO). It is the object of this adaptation to select the overall effective pilot-control value so that at idle this is at a specified number of degrees in advance of the actual throttle-valve setting.

In order that the correction of the pilot-control values becomes effective before the first idle phase when driving at high-altitudes, an additional air-density-dependent pilot-control correction is applied. The possibility of also being able to operate the throttle-valve outside the idle range (if the driver is not pressing the accelerator pedal) is taken advantage of and applied as a vacuum-limiting function. During overrun, this function opens the throttle valve in accordance with an engine-speed-dependent characteristic curve (Fig. 15c), just far enough to avoid operating points which have a very low cylinder charge (incomplete combustion).

16 Throttle-valve actuator

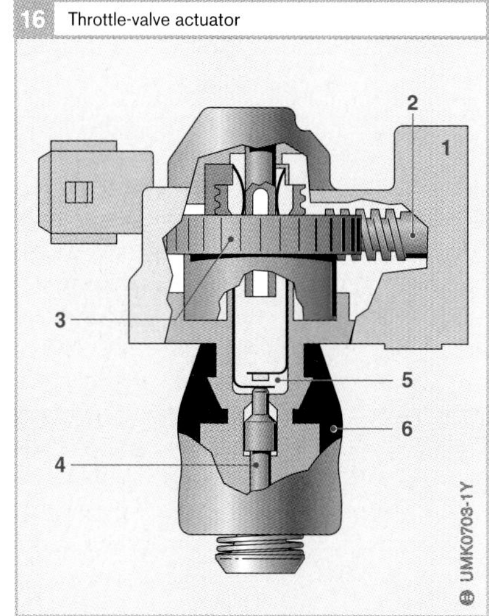

Fig. 16
1 Housing with electric motor
2 Worm
3 Wormwheel
4 Setting shaft
5 Idle contact
6 Rubber bellows

Throttle-valve actuator
Through its control shaft (4), the throttle-valve actuator (Figure 16) can adjust the throttle-valve lever and thereby influence the amount of air available to the engine. The actuator is powered by a DC motor (1) which drives the setting shaft through a worm (2) and roller gearset (3). Depending upon the motor's direction of rotation (which in turn depends upon the polarity applied to it), the setting shaft either extends and opens the throttle valve or retracts and reduces the throttle-valve angle.

The control shaft incorporates a switching contact which closes when the shaft abuts against the throttle-valve lever and provides the ECU with the idle signal.

Full-load enrichment
The driver presses the accelerator pedal to the floor to obtain maximum power from the engine. An IC engine develops maximum power with an air-fuel mixture which is about 10...15 % richer than the stoichiometric air-fuel ratio. The degree of enrichment is stored in the ECU in the form of a factor which is used to multiply the injection duration figures calculated from the Lambda map. Full-load enrichment comes into effect as soon as a specified throttle-valve angle is exceeded (this is a few degrees before the throttle-valve stop).

Engine-speed limiting
Extremely high rotational speeds can destroy the engine. The limiting circuit prevents the engine from exceeding a given maximum speed. This speed n0 can be defined for every engine, and as soon as it is exceeded the ECU suppresses the injection pulses. When the engine speed drops below n_0, the injection pulses are resumed again. This cycle is repeated in rapid succession within a tolerance band centered around the maximum permitted engine speed.

Overrun (trailing throttle)

When the driver takes his foot off of the accelerator pedal while driving the vehicle, the throttle valve closes completely, and the vehicle is driven by its own kinetic energy (overrun, trailing throttle). In order to reduce exhaust emissions and to improve driveability, a number of functions come into operation during overrun.

If the engine speed is above a given threshold (Figure 17, rpm threshold 2) when the throttle valve is closed, injector triggering stops, the engine receives no more fuel, and its speed falls. Once the next threshold is reached (rpm threshold 3), fuel injection is resumed. If the engine speed drops abruptly during overrun, as can occur when the clutch is pressed, fuel injection is resumed at a higher engine speed (rpm threshold 1), otherwise the engine speed could fall below idle or the engine could even stall (Fig. 17).

If the throttle valve is closed at high engine speeds, on the one hand the vehicle is decelerated abruptly due to the braking effect of the motored engine. On the other, the emission of hydrocarbons increases because the drop in manifold pressure causes the fuel film on the manifold walls to evaporate. This fuel though cannot combust completely because there is insufficient combustion air. To counteract this effect, during overrun the throttle-valve actuator opens the throttle valve slightly as a function of engine speed. If on the other hand the engine-speed drop is very abrupt during overrun, the throttle-valve opening angle is no longer a function of the fall in engine speed but instead the reduction in opening angle is slower and follows a time function.

During overrun the film of fuel deposited on the intake-manifold walls evaporates completely and the intake manifold dries-out. When the overrun mode is completed, fuel must be made available to build up this fuel film on the manifold walls again. This results in a slightly leaner air-fuel mixture in the transitional period until equilibrium returns. Fuel-film buildup is aided by an additional injection pulse, the length of which depends upon the overrun duration.

17 Fuel injection during overrun

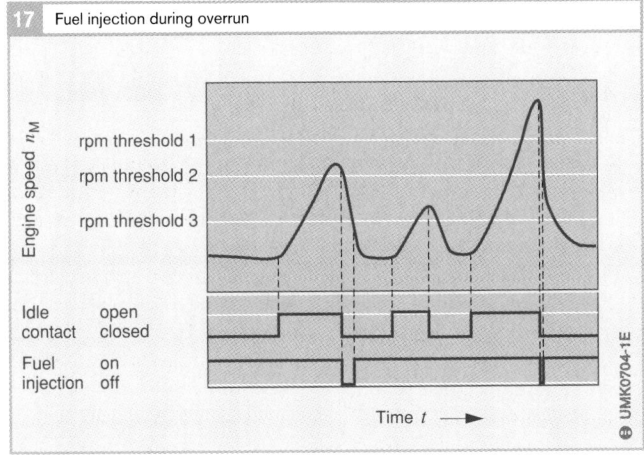

System voltage compensation

Fuel-injector voltage compensation

A feature of the solenoid-operated fuel injector is that due to self-induction it tends to open more slowly at the beginning of the injection pulse and to close more slowly at the end. Opening and closing times are in the order of 0.8 ms. The opening time depends heavily on the battery voltage, whereas the closing time depends only slightly on this factor. Without electronic voltage correction, the resulting delay in injector pickup would mean a far too short injection duration and therefore insufficient fuel would be injected. Therefore, a reduction in battery voltage must be counteracted by a voltage-dependent increase in injection duration, the so-called additive injector correction factor (Fig. 18a). The ECU registers the actual battery voltage and increases the injector-triggering pulse by the voltage-dependent injector pickup delay.

Fuel-pump voltage compensation

The speed of the fuel-pump motor is very sensitive to variations in battery voltage. For this reason, if the battery voltage is low (for instance during a cold start), the fuel pump, which functions according to the hydrodynamic principle, is unable to bring the primary pressure up to its specified level. This would result in insufficient fuel being injected. To compensate for this effect, particularly at very low battery voltages, a voltage-correction function is applied to correct the injection duration (Fig. 18b).

If a positive-displacement electric fuel pump is used, this voltage correction function is unnecessary. The ECU is provided with a coding input, therefore, which enables the voltage-correction function to be activated depending upon pump type.

Controlling the regeneration-gas flow

When fresh air is drawn through the carbon in the carbon canister (purging), it transports the fuel trapped in the carbon to the engine for combustion. The canister-purge valve between the carbon canister and the central injection unit controls the regeneration-gas flow. The control ensures that under all operating conditions as much of the trapped fuel as possible is transported away to the engine for burning. In other words, the regeneration-gas flow is kept as high as possible, without driveability being impaired.

In order to ensure the correct functioning of the mixture adaptation it is imperative that a cyclic change is made between normal operation, which makes the mixture adaptation possible, and regeneration operation. Furthermore, it is necessary during the regeneration phase to ascertain the fuel content in the regeneration gas and to use this value for adaptation. The same as with the mixture

18 Correction of injection duration as a function of battery voltage

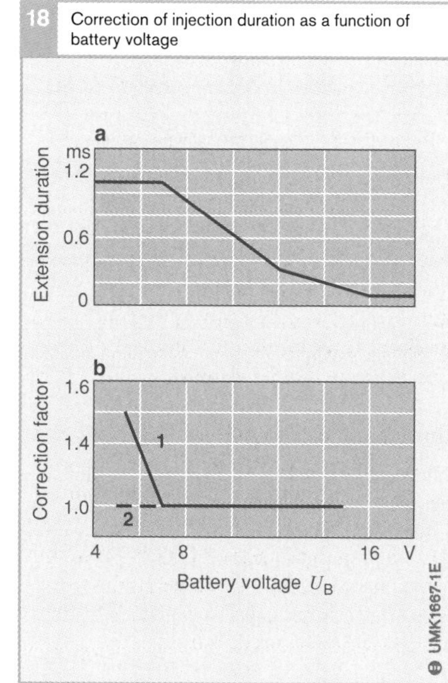

Fig. 18
a Voltage compensation, fuel injector
b Voltage compensation, electric fuel pump:

1 Flow-type pump
2 Positive-displacement pump

adaptation, this takes place by evaluating the Lambda control stage's deviation from the $\lambda = 1$ value. Once the fuel content is known, the injection duration can be increased or decreased accordingly when the cycle changes so that during transition the mixture is maintained within tight limits around $\lambda = 1$.

To specify the fuel content in the regeneration-gas stream as a function of the engine operating mode, as well as to adapt the proportion of fuel in the regeneration-gas stream, the relationship between the regeneration-gas stream and the amount of air drawn in by the engine via the throttle valve must be determined. Both partial streams are practically proportional to the open cross-section areas through which they flow.

Whereas it is relatively simple to calculate the cross-section area in the throttle housing from the throttle-valve angle, the open cross-section area of the canister-purge valve changes according to the applied differential pressure.

The differential pressure applied at the canister-purge valve depends upon the engine's operating point and can be derived from the injection durations stored in the Lambda map.

The ratio of regeneration-gas stream to air stream drawn in by the engine can be calculated for each engine operating point as defined by the throttle-valve angle and the engine speed. The regeneration-gas stream can be further reduced by cycling the canister-purge valve and in this manner adjusted precisely to the required ratio necessary to ensure acceptable driveability.

Limp-home and diagnosis

All sensor signals are continuously checked for their plausibility by monitoring functions incorporated in the ECU. If one of the sensor signals deviates from its defined, plausible range, this means that either the sensor itself is defective or there is a fault in one of its electrical connections.

If one of the sensor signals fails or is no longer plausible, it must be replaced with a substitute signal in order that the vehicle does not break down completely but can still safely reach the next specialist workshops under its own power even though with some restrictions in driveability and comfort.

For instance, if temperature signals fail, signals are substituted which normally prevail when the engine is at its operating temperature: intake air 20 °C and engine coolant 100 °C. If a fault occurs in the Lambda closed-loop control, the complete Lambda control facility is closed down, that is, the injection durations from the Lambda map are only corrected with the mixture-adaptation value (if available).

If the signals from the throttle-valve potentiometer are not plausible, this means that one of the main controlled variables is missing. In other words, access is no longer possible to the injection durations stored in the Lambda map. In this case. the fuel injector is triggered with defined constant-length pulses. Two different injection durations are available, and switching between them depends upon engine speed.

In addition to the sensors, the throttle-valve actuator (idle-speed control) is also monitored continuously.

As soon as a sensor failure or throttle-valve actuator malfunction is registered, a corresponding entry is made in the fault memory. This entry remains accessible for a number of operating cycles (that is, even when the engine is switched off and the vehicle is left standing overnight a number of times) so that the workshop is able to localize the fault. This also applies to sporadic faults such as loose contacts etc.

Fuel supply

The fuel injectors in a gasoline-injection system inject the fuel either into the intake manifold (manifold injection system) or directly into the combustion chamber (gasoline direct injection). To accomplish this, fuel must be conveyed to the fuel injectors at a specific pressure.

This chapter describes the components handling the delivery of the fuel from the fuel tank to the injector valves or, in the case of gasoline direct injection, from the fuel tank to the high-pressure pump.

Overview

The fuel supply system basically comprise the following components (Fig. 1, at right):
- Fuel tank (1)
- Electric fuel pump (2)
- Fuel filter (3)
- Fuel-pressure regulator (5)
- Fuel lines (4 and 8)

The electric fuel pump continuously conveys the fuel from the fuel tank via the fuel filter to the engine. The fuel-pressure regulator maintains a specific pressure within the fuel supply circuit. The pressure level depends on the fuel-injection system.

In the manifold injection system, the fuel delivered by the electric fuel pump flows to the fuel injectors (6) via the fuel rail section (7 or 9). On engines with gasoline direct injection, the fuel is conveyed to the high-pressure circuit via the high-pressure pump.

In order to ensure the availability of the required fuel pressure in all operating conditions, the fuel quantity delivered by the electric fuel pump is greater than that required by the engine. Excess fuel is returned to the fuel tank (exception: demand-controlled systems).

In older systems, the electric fuel pump is located in the fuel line outside of the fuel tank (in-line pump). On more recent systems, the electric fuel pump is located inside the fuel tank (in-tank pump). The pump may also be integrated in an in-tank unit together with additional components (e.g. preliminary filter, fuel level sensor).

To ensure the pressure buildup required for engine starting, the electric fuel pump starts up the instant the ignition or starting switch is actuated. In the event that the engine is not started, the fuel pump shuts off again within one second.

The pressure built up by the electric fuel pump basically prevents the formation of vapor bubbles in the fuel. An integrated non-return valve isolates the fuel system from the fuel tank by preventing the return of fuel to the fuel tank. The non-return valve maintains primary pressure for a period of time after the electric fuel pump has been switched off. This prevents the formation of vapor bubbles in the fuel system if there are high fuel temperatures after engine shutdown.

Fuel supply on manifold-injection systems

Standard system

An electrically powered fuel-supply pump (EKP) delivers the fuel and generates the injection pressure. The fuel is drawn from the fuel tank, and passes through a paper filter element (fuel filter) into a high-pressure line. From there, it runs to the engine-mounted fuel rail with its fuel injectors (Fig. 1a). The pressure regulator is attached to the fuel rail. Independently of engine load (intake-manifold pressure), it maintains a constant differential pressure at the metering orifice.

The fuel volume not used by the internal combustion engine flows through the fuel rail, and then returns to the fuel tank via a return line connected to the pressure regulator. On its way to the fuel tank, the returning fuel is warmed up. The result is an increase in the fuel temperature in the fuel tank. Fuel vapors are formed in the tank as a factor of fuel temperature. Ensuring adherence to environmental-protection regulations, the vapors are routed through a tank ventilation system for intermediate storage in an activated-charcoal canister until they can be returned through the intake manifold for combustion in the engine (evaporative-emissions control system).

Returnless system

Compared with a standard system, the returnless fuel supply system (Fig. 1b) reduces the tendency of the fuel in the fuel tank to heat up. This makes it easier to ensure compliance with legal requirements governing evaporative emissions in motor vehicles.

The pressure regulator is located inside the fuel tank or in its immediate vicinity. A return line from the engine to the fuel tank is therefore redundant. Only the fuel quantity used by the injectors flows into the non-flushed fuel rail. The excess-flow volume delivered by the electric fuel pump returns directly to the fuel tank without taking the detour through the engine compartment. If operating conditions are constant, this system can reduce in-tank fuel temperatures by up to 10 K, decreasing the vaporized fuel volume to roughly one-third.

Demand-controlled system

A further reduction of in-tank fuel temperature can be achieved by using a demand-controlled system. This also has the side effect of reducing fuel consumption. In this fuel circuit, the fuel-supply pump delivers only the fuel quantity that is required by the engine and to set up the necessary fuel pressure (Fig. 2). A mechanical pressure regulator is redundant. Pressure regulation is performed by a closed control loop in the engine control unit. It also includes pressure

1 Fuel supply on manifold injection systems

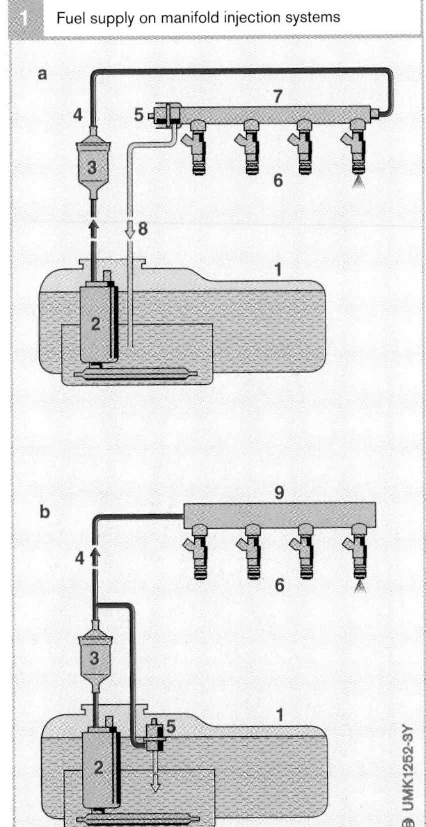

Fig. 1
a Standard system
b Returnless system

1 Fuel tank
2 Electric fuel pump
3 Fuel filter
4 Fuel-injection tubing
5 Pressure regulator
6 Fuel injectors
7 Fuel rail (continuous flow)
8 Return line
9 Fuel rail (no return flow)

measurement via a low-pressure sensor. To adjust the delivery volume of the fuel-supply pump, its operating voltage is altered by means of a clock module that is triggered by the engine control unit. The system is completed by a pressure-relief valve that prevents excessive fuel pressure caused by fuel heating at overrun fuel cutoff and engine shutdown.

Besides further reducing fuel temperature and dropping consumption by up to 0.1 *l*/100 km, the variably adjustable pressure can be used to increase fuel pressure in hot-start conditions, or to extend the metering range of the fuel injectors in turbocharge applications. In addition, there are much better options for fuel-system diagnostics compared with earlier systems. Another advantage is that fuel pressure is included in calculating injection time in the engine ECU, in particular while the engine is building up pressure after starting.

2 Demand-controlled fuel supply on manifold injection systems

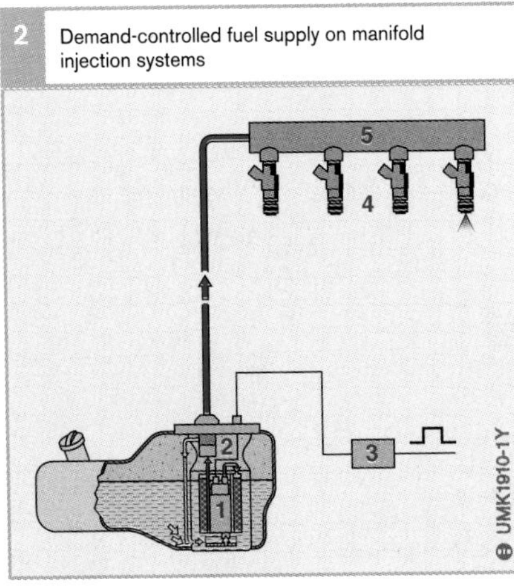

Fig. 2
Demand-controlled system

1 Electric fuel pump with fuel filter (fuel filter may also be mounted outside of tank)
2 Pressure-relief valve and pressure sensor
3 Clock module
4 Fuel injectors
5 Fuel rail (no return flow)

Fuel supply for gasoline direct injection

Compared with fuel injection in the intake manifold, the fuel must overcome higher pressures to inject fuel directly into the combustion chamber in stratified-charge mode. In addition, only a reduced time window is available for fuel injection. For this reason, fuel systems for gasoline direct injection require higher fuel pressure levels.

The fuel system is divided into:

- A low-pressure system and
- A high-pressure system

Low-pressure system

Low-pressure systems for gasoline direct injection essentially use the fuel systems and components that are familiar in manifold injection systems. Due to the fact that currently used high-pressure pumps require elevated inlet pressure in order to prevent vapor-bubble formation during hot starts and high-temperature operation, varying the low pressure may be of advantage. This is ideally accomplished by using a demand-controlled low-pressure system. However, other systems are used. They may be returnless systems with selectable inlet pressure (controlled by means of a shutoff valve), or systems featuring a constant, high inlet pressure.

High-pressure system

The high-pressure system is composed of:

- High-pressure pump
- Fuel rail
- Pressure sensor and, depending on the system
- Pressure-control valve or
- Pressure limiter

A distinction is made between permanent-delivery and demand-controlled systems.

Depending on the operating point, a primary pressure of between 5 and 12 MPa is set via a high-pressure regulator on the engine ECU. The high-pressure injectors injecting the fuel directly into the engine's combustion chamber are mounted on the fuel rail.

Continuous-delivery system

A pump driven by the engine camshaft, e.g. the three-barrel radial-piston pump HDP1, forces fuel into the rail against the primary pressure. The pump's delivery quantity is not adjustable. The excess fuel not required for fuel injection, or to maintain the pressure, is depressurized by the pressure-control valve, and returned to the low-pressure system (Fig. 3). The pressure-control valve is triggered by the engine control unit to obtain the pressure required at a given operating point.The pressure-control valve doubles up as a mechanical pressure limiter.

In permanent-delivery systems, most of the operating points cause more fuel to be brought up to primary pressure than is needed by the – engine. This takes place at the expense of additional power, and the excess fuel quantity depressurized via the pressure-control valve helps to raise the fuel temperature. The above demonstrates a clear advantage for demand-controlled systems.

Demand-controlled system

A fuel pump featuring adjustable delivery quantity forces only the fuel quantity (injected into the rail against primary pressure) that is needed for injection and to maintain pressure (Fig. 4). A typical fuel pump is the single-cylinder radial-piston pump HDP2. The pump is driven by the engine camshaft. The regulating function on the engine ECU triggers the pump in order to obtain the primary pressure required in the rail at a given operating point. A mechanical pressure limiter is required in the rail. Compared with the permanent-delivery system based on the single-cylinder radial-piston pump, the demand-controlled system requires a greater rail volume to compensate for pressure dips caused by fuel-injection cycles, and to maintain constant pressure levels.

3 Continuous-delivery system

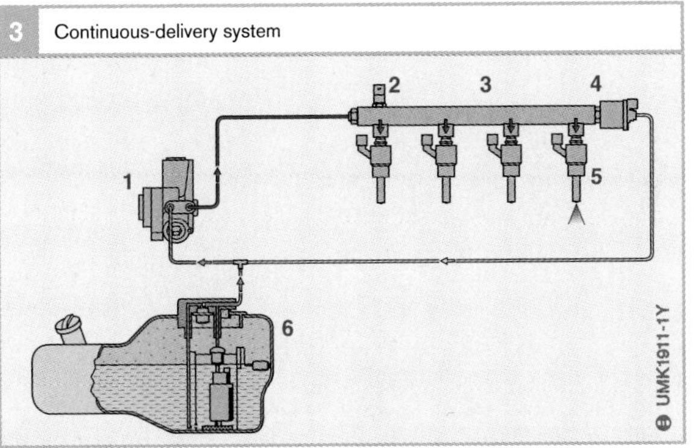

Fig. 3
1 High-pressure pump HDP1
2 High-pressure sensor
3 Fuel rail
4 Pressure-control valve
5 High-pressure fuel injectors
6 Fuel tank with pump module, incl. presupply pump

4 Demand-controlled system

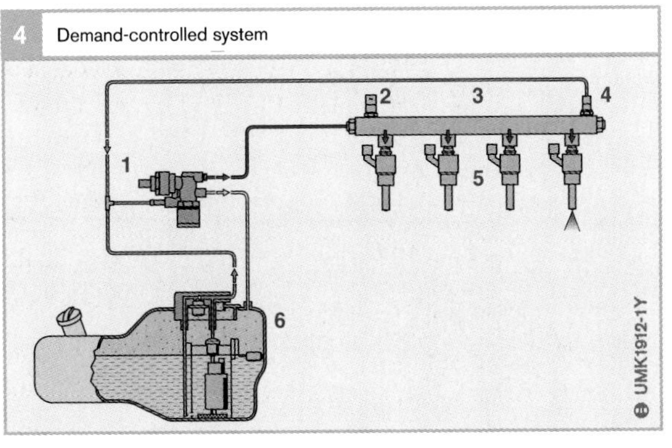

Fig. 4
1 High-pressure pump HDP2
2 High-pressure sensor
3 Fuel rail
4 Pressure limiter
5 High-pressure fuel injectors
6 Fuel tank with pump module, incl. presupply pump

Integration in the vehicle: In-tank unit

In the early years of electronically controlled gasoline injection, the electric fuel pump was always installed in the fuel line ("in-line") outside the fuel tank. Today, on the other hand, the majority of electric fuel pumps are of the "in-tank" type and, as the name implies, are part of an "in-tank unit", the so-called fuel-supply module. This contains an increasing number of other components, for instance:

- A preliminary filter
- A fuel-level sensor
- Electric and hydraulic connections and
- A special fuel reservoir for maintaining the fuel supply when cornering or in sharp bends

Usually, a jet pump or a separate stage in the electric fuel pump keep this reservoir full.

On RLFS systems, the fuel-pressure regulator (4), is usually integrated in the in-tank unit where it is responsible for the fuel return. The pressure-side fine fuel filter can also be located in the in-tank unit.

In future, the fuel-supply module will incorporate further functions, for instance diagnosis devices for detecting tank leaks, or the timing module for triggering the electric fuel pump.

In-tank unit: The complete unit for a returnless fuel system (RLFS)

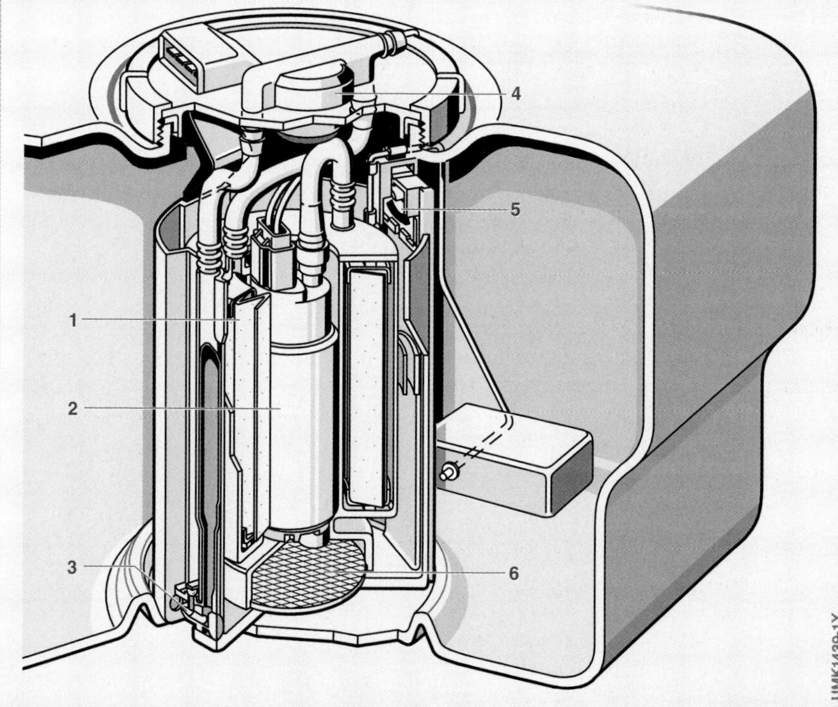

1 Fuel filter
2 Electric fuel pump
3 Jet pump (closed-loop controlled)
4 Fuel-pressure regulator
5 Fuel-level sensor
6 Preliminary filter

Evaporative-emissions control system

In order to comply with the legal limits for evaporative hydrocarbon emissions, vehicles are being equipped with evaporative-emissions control systems. This system prevents fuel vapor escaping to the atmosphere from the fuel tank.

Fuel-vapor generation

More fuel vapor escapes from the fuel tank under the following circumstances:

- When the fuel in the fuel tank warms up, due either to high surrounding temperatures, or to the return to tank of excess fuel which has heated up in the engine compartment and
- When the surrounding pressure drops, for instance when driving up a hill in the mountains

Design and operating concept

The evaporative-emissions control system (Fig. 1) comprises the carbon canister (3), into which is led the venting line (2) from the fuel tank (1), together with the so-called canister-purge valve (5) which is connected to both the carbon canister and the intake manifold (8).

The activated carbon in the carbon canister absorbs the fuel contained in the fuel vapor and thus permits only air to escape into the atmosphere. As soon as the canister-purge valve opens the line (6) between the carbon canister and the intake manifold, the vacuum in the manifold causes fresh air to be drawn through the activated carbon. The absorbed fuel is then entrained with the fresh air (purging or regeneration of the activated carbon) and burnt in the normal combustion process. The system control reduces the injected fuel quantity by the amount returned through canister-purge valve. Regeneration is a closed-loop control process, whereby the fuel concentration in the canister-purge gas flow is continuously calculated based on the changes it causes in the excess-air factor λ.

The canister-purge gas quantity is controlled as a function of the working point and can be very finely metered using the canister-purge valve. In order to ensure that the carbon canister is always able to absorb fuel vapor, the activated carbon must be regenerated at regular intervals.

Gasoline direction injection: Special features

During stratified-charge operation on gasoline direct-injection engines, the possibility of regenerating the carbon canister's contents is limited due to the low level of vacuum in the intake manifold (caused by practically 100 % "unthrottled" operation) and the incomplete combustion of the homogeneously distributed canister-purge gas. This results in reduced canister-purge gas flow compared to homogeneous operation.
For instance, if the canister-purge gas flow is inadequate for coping with high levels of gasoline evaporation, the engine must be operated in the homogeneous mode until the high concentrations of gasoline in the canister-purge gas flow have dropped far enough.

1 Evaporative-emissions control system

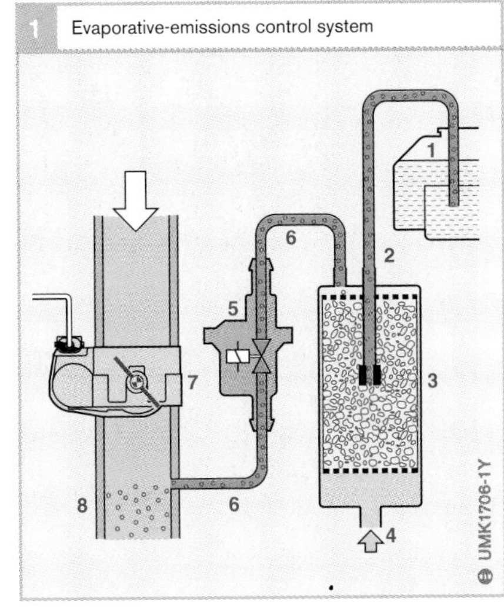

Fig. 1
1 Fuel tank
2 Fuel-tank venting line
3 Carbon canister
4 Fresh air
5 Canister-purge valve
6 Line to the intake manifold
7 Throttle valve
8 Intake manifold

Electric fuel pump

Assignment

The electric fuel pump (EKP) must at all times deliver enough fuel to the engine at a high enough pressure to permit efficient fuel injection. The most important performance demands made on the pump are:

- Delivery quantity between 60 and 200 l/h at rated voltage
- Pressure in the fuel system between 300 and 450 kPa (3...4.5 bar)
- System-pressure buildup even down to as low as between 50 and 60 % of rated voltage

Apart from this, the EKP is increasingly being used as the pre-supply pump for the modern direct-injection systems used on diesel and gasoline engines.

On gasoline direct-injection systems for instance, pressures of up to 700 kPa are sometimes required during hot-delivery operations.

1 Electric fuel pump: Design and construction using a turbine pump as an example

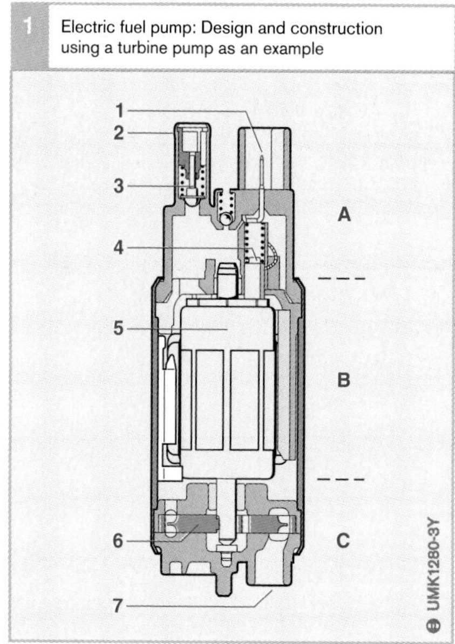

Fig. 1
1 Electric connections
2 Hydraulic connections (fuel outlet)
3 Non-return valve
4 Carbon brushes
5 Permanent-magnet motor armature
6 Turbine-pump impeller ring
7 Hydraulic connection (fuel inlet)

Design and construction

The electric fuel pump is comprised of:

- End plate (Fig. 1, A), incorporating spark-suppression elements if required
- Electric motor (B) and
- Pump element (C), designed as either positive-displacement or turbine pump (for description, see Section "Types" below)

Types

Positive-displacement pumps

In this type of pump, the fuel is drawn in, compressed in a closed chamber by rotation of the pump element, and transported to the high-pressure side. For the EKP, internal-gear pumps or roller-cell pumps (Figs. 2a, 2b) are used. When high system pressures are needed (400 kPa and above), positive-displacement pumps are particularly suitable. These feature a good low-voltage characteristic, that is, they have a relatively flat delivery-rate characteristic as a function of the operating voltage. Efficiency can be as high as 25 %.

Pressure pulsations, which are unavoidable, can cause audible noise depending upon the particular design details and installation conditions. The fact that the delivery rate can drop when the fuel is hot is another disadvantage which can occur in exceptional cases. This is due to vapor bubbles being pumped instead of fuel, and for this reason conventional positive-displacement pumps are equipped with peripheral preliminary stages for degassing purposes.

Whereas in electronic gasoline-injection systems the positive-displacement pump has to a great extent been superseded by the turbine pump for the classical fuel-pump requirements, it has captured a new field of application as the presupply pump on direct-injection systems wihich operate with far higher fuel-pressures.

Turbine pumps

This type of pump comprises an impeller ring with numerous blades inserted in slots around its periphery (Fig. 2c, Pos. 6). The impeller ring with blades rotates in a chamber formed from two fixed housing sections, each of which has a passage (7) adjacent to the blades which starts at the level of the intake port (A) and terminates where the fuel is forced out of the pump at system pressure through the fuel outlet (B). The "Stopper" between start and end of the passage prevents internal leakage.

At a given angle and distance from the intake opening a small degassing bore has been provided which provides for the exit of any gas bubbles which may be in the fuel. This, although improving the hot-delivery characteristics, is at the cost of very slight internal leakage. The degassing bore is not needed with diesel applications.

Pressure builds up along the passage (7) as a result of the exchange of pulses between the ring blades and the liquid particles. This leads to spiral-shaped rotation of the liquid volume trapped in the impeller ring and in the passages. In the case of the peripheral pump (Fig. 2c), the ring blades around the periphery of the ring are surrounded completely by the passage (hence the word "peripheral"). On the side-channel pump, the two channels are located on each side of the impeller ring adjacent to the blades.

Turbine pumps feature a low noise level since pressure buildup takes place continuously and is practically pulsation-free. Efficiency is between 10 % and about 20 %. Construction though is far simpler than that of the positive-displacement pumps.

Single-stage pumps can generate system pressures of up to 450 kPa. In future, turbine pumps will also be suitable for the higher system pressures that will be needed for brief periods on highly supercharged engines and gasoline direct-injection engines.

2 Principle of functioning of electric fuel pumps

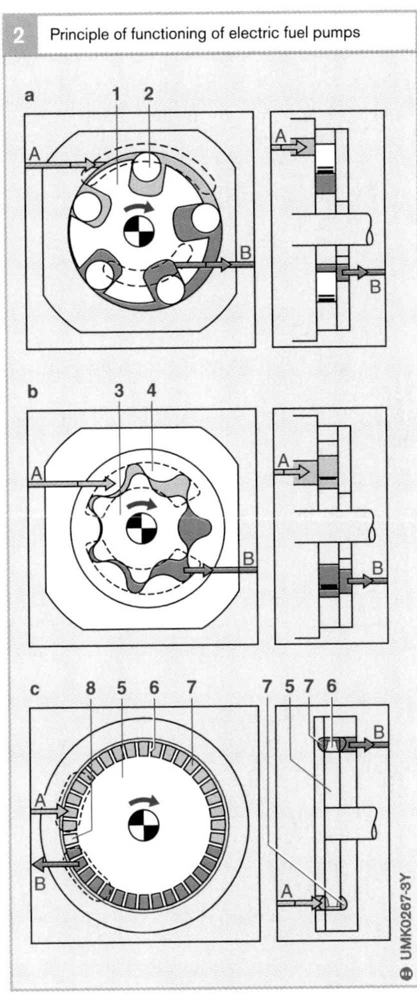

Fig. 2
a Roller-cell pump (RZP)
b Inner-gear pump (IZP)
c Peripheral pump (PP)

A Intake port
B Outlet

1 Slotted rotor (eccentric)
2 Roller
3 Inner drive wheel
4 Rotor
5 Impeller ring
6 Impeller-ring blades
7 Passage (peripheral)
8 "Stopper"

For costs reasons, and due to their being quieter, turbine pumps are used almost exclusively on newly designed gasoline-engine automobiles.

Fuel filter

The injection systems for automobile spark-ignition (SI) engines operate with extreme precision. In order not to damage their precision parts, it is imperative that the fuel is efficiently cleaned. Filters in the fuel circuit remove the solid particles which could cause wear. Such filters are either replaceable in-line filters, or are integrated in the fuel tank as "lifetime" in-tank filters. Apart from the filter's purely straining or filtering effect, a number of different processes are applied in order to remove the contaminants from the fuel. These include impact, diffusion, and blocking effects.

1 Section through a fuel filter

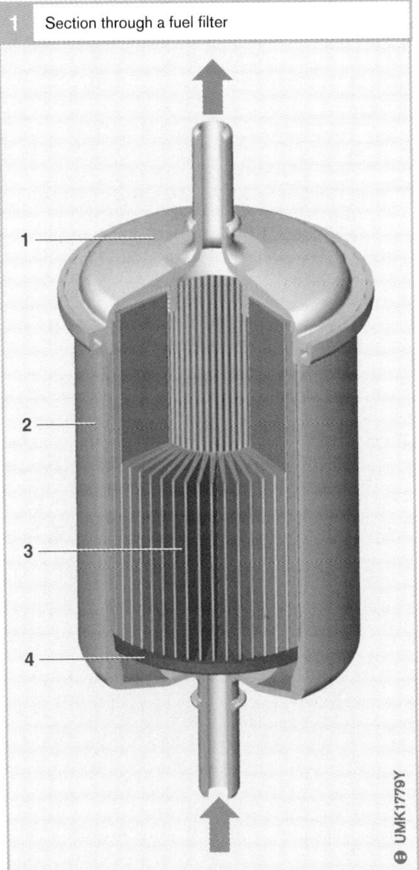

Fig. 1
1 Filter cover
2 Filter housing
3 Filter element
4 Support plate

The filtration efficiency of the individual effects is a function of the size and the flow speed of the contaminant particles, and the filter medium is matched to these factors.

Pleated paper, which is sometimes specially impregnated, has come to the forefront as the filter medium (Fig. 1, Pos. 3). The filter medium is arranged in the fuel circuit so that the velocity of the fuel flow through all sections of its surface is as uniform as possible.

Whereas on manifold-injection systems the filter element has a mean pore size of 10 µm, far finer filtering is needed for gasoline direct-injection systems where up to 85 % of the particles larger than 5 µm must be reliably filtered out of the fuel.

In addition, for gasoline direct injection, when a new filter is fitted the traces of contaminant remaining in the filter after manufacture are an important factor: Metal, mineral, plastic, and glass-fiber particles must not exceed 200 µm.

Depending upon the filter volume, the useful life (guaranteed mileage) of the conventional in-line filter is somewhere between 37,500 and 55,000 miles (60,000...90,000 km). Guaranteed mileages of 100,000 miles (160,000 km) apply for in-tank filters. There are in-tank and in-line filters available for use with gasoline direct-injection systems which feature service lives in excess of 150,000 miles (250,000 km).

Filter housings (2) are either steel, aluminum, or plastic (100 % free from metal). Connections of the threaded, hose, or quick-connect type are used.

Filter efficiency depends on the throughflow direction. When replacing in-line filters, it is imperative that the flow direction given by the arrow is observed.

Fuel rail

Manifold injection

The fuel rail has the following assignments:

- Mounting and location of the injectors
- Storage of the fuel volume
- Ensuring that fuel is distributed evenly to all injectors

In addition to the injectors, the fuel rail usually accomodates the fuel-pressure regulator and possibly even a pressure damper. Local fuel-pressure fluctuations caused by resonance when the injectors open and closed, is prevented by careful selection of the fuel-rail dimensions. As a result, irregularities in injected fuel quantity which can arise as a function of load and engine speed are avoided.

Depending upon the particular requirements of the vehicle in question, plastic or stainless-steel fuel rails are used. The fuel rail can incorporate a diagnosis valve for workshop testing purposes.

Gasoline direct injection

On gasoline DI systems, the rail is located downstream of the high-pressure pump, and is an integral part of the high-pressure stage.

Fuel-pressure regulator

Manifold injection

The amount of fuel injected by the injector (injected fuel quantity) depends upon the injection period and the difference between the fuel pressure in the fuel rail and the counterpressure in the manifold. On fuel systems with return, the influence of pressure is compensated for by a pressure regulator which maintains the difference between fuel pressure and manifold pressure at a constant level. This pressure regulator permits just enough fuel to return to the tank so that the pressure drop across the injectors remains constant. In order to ensure that the fuel rail is efficiently flushed, the fuel-pressure regulator is normally located at the end of the rail which leads the fuel tank.

On returnless fuel systems (RLFS), the pressure regulator is part of the in-tank unit installed in the fuel tank. The fuel-rail pressure is maintained at a constant level with reference to the surrounding pressure. This means that the difference between fuel-rail pressure and manifold pressure is not constant and must be taken into account when the injection duration is calculated.

1 Fuel-pressure regulator DR2

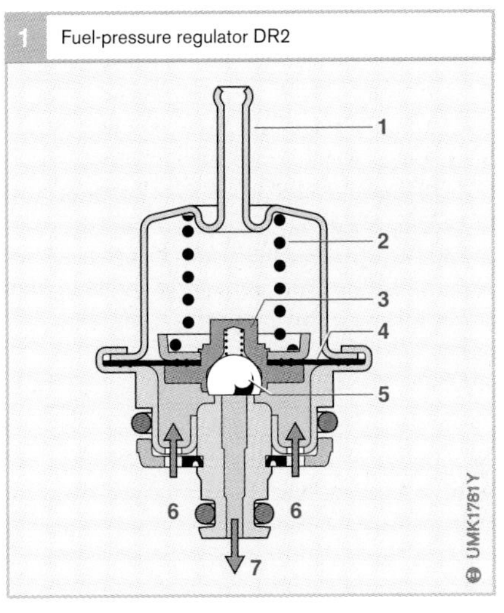

Fig. 1
1 Intake-manifold connection
2 Spring
3 Valve holder
4 Diaphragm
5 Valve
6 Fuel inlet
7 Fuel return

The fuel-pressure regulator (Fig. 1) is of the diaphragm-controlled overflow type. A rubber-fabric diaphragm (4) divides the pressure regulator into a fuel chamber and a spring chamber. Through a valve holder (3) integrated in the diaphragm, the spring (2) forces a movable valve plate against the valve seat so that the valve closes. As soon as the pressure applied to the diaphragm by the fuel exceeds the spring force, the valve opens again and permits just enough fuel to flow back to the fuel tank that equilibrium of forces is achieved again at the diaphragm.

On multipoint fuel-injection systems, in order that the manifold vacuum can be applied to the spring chamber, this is connected pneumatically to the intake manifold at a point downstream of the throttle plate. There is therefore the same pressure ratio at the diaphragm as at the injectors. This means that the pressure drop across the injectors is solely a function of spring force and diaphragm surface area, and therefore remains constant.

Gasoline direct injection

On gasoline direct-injection systems, it is necessary to regulate the pressures in the high-pressure and the low-pressure stage, whereby the same fuel-pressure regulators are used for the low-pressure stage as for manifold injection.

Fuel-pressure damper

The repeated opening and closing of the injectors, together with the periodic supply of fuel when electric positive-displacement fuel pumps are used, leads to fuel-pressure oscillations. These can cause pressure resonances which adversely affect fuel-metering accuracy. It is even possible that under certain circumstances, noise can be caused by these oscillations being transferred to the fuel tank and the vehicle bodywork through the mounting elements of the fuel rail, fuel lines, and fuel pump.

These problems are alleviated by the use of special-design mounting elements and fuel-pressure dampers.The fuel-pressure damper is similar in design to the fuel-pressure regulator. Here too, a spring-loaded diaphragm separates the fuel chamber from the air chamber. The spring force is selected such that the diaphragm lifts from its seat as soon as the fuel pressure reaches its working range. This means that the fuel chamber is variable and not only absorbs fuel when pressure peaks occur, but also releases fuel when the pressure drops. In order to always operate in the most favorable range when the absolute fuel pressure fluctuates due to conditions at the manifold, the spring chamber can be provided with an intake-manifold connection.

Similar to the fuel-pressure regulator, the fuel-pressure damper can also be attached to the fuel rail or installed in the fuel line. In the case of gasoline direct injection, it can also be attached to the high-pressure pump.

Fuel tank

As its name implies, the fuel tank is used as the reservoir for the fuel. It must be non-corroding and must remain free of leaks at up to twice working pressure, or up to at least 0.03 MPa (0.3 bar) gauge pressure. Openings or safety valves must be provided for excess pressure to escape automatically. During cornering, on inclines, and in case of shock or impact, no fuel may leak out through the filler cap or pressure-compensation devices. The fuel tank must be situated far enough from the engine to avoid ignition of escaping fuel in case of an accident.

Fuel lines

The fuel lines serve to carry the fuel from the fuel tank to the fuel-injection system. Seamless, flexible metal conduit or fuel-resistant hardly combustible material can be used for the fuel lines. These must be routed so that mechanical damage is avoided, and fuel which has evaporated or dripped as a result of malfunctions cannot accumulate or ignite. All fuel-carrying components must be protected against heat that could interfere with correct performance. Gravity feed must not be used in the fuel-supply circuit.

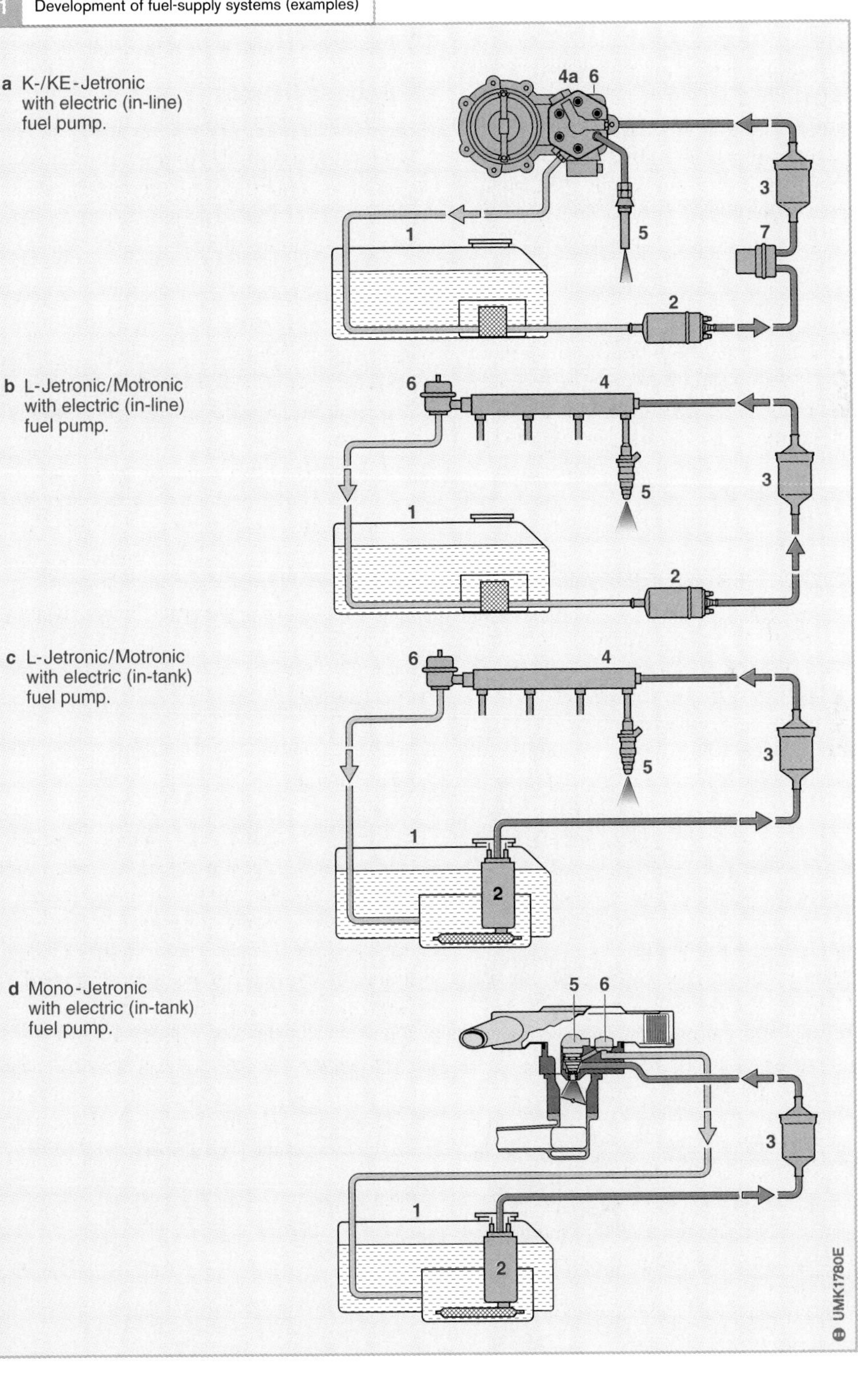

Fig. 1

1 Fuel tank
2 Electric fuel pump (EKP)
3 Fuel filter
4 Fuel rail
4a Fuel distributor (K-/KE-Jetronic)
5 Injector
6 Pressure regulator
7 Fuel accumulator (K-/KE-Jetronic)

Manifold fuel injection

Manifold-injection engines generate the A/F mixture in the intake manifold and not in the combustion chamber. Since they were introduced to the market, these engines and their control systems have been vastly improved. Their superior fuel-metering characteristics have enabled them to completely supersede the carburetor engine which also operates with external A/F-mixture formation.

Overview

Regarding smooth running and exhaust-gas behaviour very high demands are made on modern-day vehicles which correspond to the latest state-of-the-art. This leads to strict requirements with respect to the composition of the A/F mixture. Apart from the precision metering of the injected fuel mass as a function of the air drawn in by the engine, it is also imperative that injection of the fuel takes place at exactly the right instant in time.

As a direct result of increasingly severe emission-control legislation, these technical stipulations are being increasingly tightened so that fuel-injection system development is forced to keep pace.

In the manifold-injection field, the electronically controlled multipoint fuel-injection system represents the state-of-the-art. This system injects the fuel intermittently, and individually, for each cylinder directly onto its intake valve(s) (Fig. 1).

Mechanically controlled continuous-injection multipoint systems no longer have any significance for new developments in this field, nor do the single-point (TBI) systems which inject intermittently through a single injector into the intake manifold upstream of the throttle valve.

1 Manifold injection

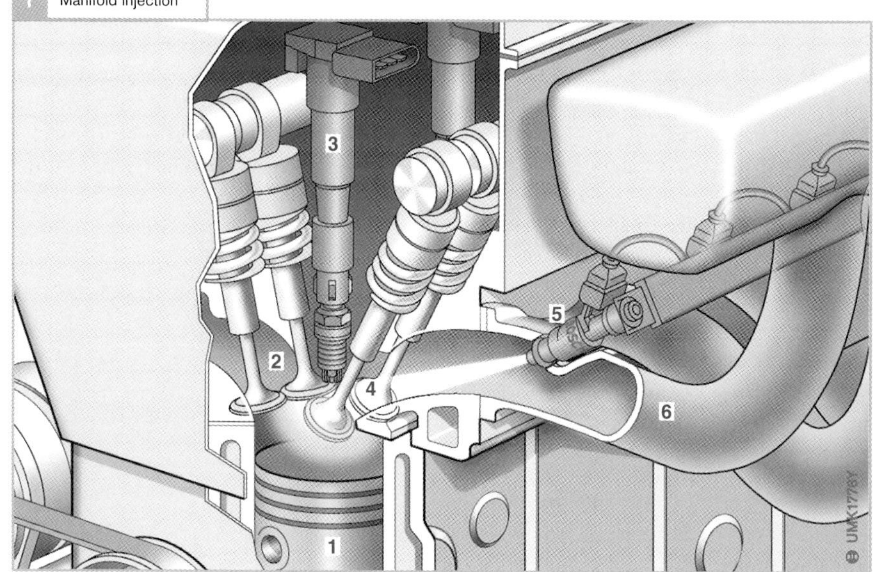

Fig. 1
1 Cylinder with piston
2 Exhaust valves
3 Ignition coil with spark plug
4 Intake valves
5 Injector
6 Intake manifold

Operating concept

Gasoline injection systems of the manifold-injection type are characterized by the fact that they generate the A/F mixture outside the combustion chamber, in other words, in the intake manifold (Fig. 1), see "External A/F-mixture formation". The injector (5) sprays the fuel directly onto the intake valves (4) where together with the intake air it forms the A/F mixture which is then drawn into the cylinder (1) past the open intake valves during the subsequent induction stroke. One, two, or even three, intake valves can be used per cylinder.

The intake valves are designed so that the engine's fuel requirements are covered irrespective of operating conditions – at full load and at high engine revs.

A/F-mixture formation

Fuel injection

The electric fuel pump delivers the fuel to the injectors where it is then available for injection at system pressure. Each cylinder is allocated its own injector which injects intermittently into the intake manifold directly onto the intake valve (6). Here the finely atomized fuel evaporates to a great extent, and together with the intake air entering via the throttle plate generates the A/F mixture. In order that enough time is available for the generation of the A/F mixture, the fuel is best sprayed onto the closed intake valve and "stored" there.

Some of the fuel is deposited as a film on the manifold walls in the vicinity of the intake valves. The thickness of the film is a function of the manifold pressure and, therefore, of engine load. For good dynamic engine response, the fuel mass in the wall film must be kept to a minimum. This is achieved by appropriate manifold design and fuel-spray geometry. Since the injector is situated directly opposite the intake valve, the wall-film effects with multipoint injection systems are far less serious than they were with the former TBI and carburetor systems.

Provided the A/F mixture is stoichiometric ($\lambda = 1$), the pollutants generated during the combustion process can to a great extent be removed using the three-way catalytic converter. At the majority of their operating points, manifold-injection engines are therefore operated with this A/F mixture composition.

Measuring the air mass

In order that the A/F mixture can be precisely adjusted, it is imperative that the mass of the air which is used for combustion can be measured exactly. The air-mass meter is situated upstream of the throttle valve. It measures the air-mass flow entering the intake manifold and sends a corresponding electric signal to the engine ECU. As an alternative, there are also systems on the market which use a pressure sensor to measure the intake-manifold pressure. Together with the throttle-valve setting and the engine speed, this data is then used to calculate the intake-air mass. The ECU then applies the data on intake air mass and the engine's instantaneous operating mode to calculate the required fuel mass.

Injection duration

A given length of time is needed for the injection of the calculated fuel mass. This is termed the injection duration, and is a function of the injector's opening cross section and the difference between the intake-manifold pressure and the pressure prevailing in the fuel-supply system.

Electromagnetic fuel injectors

Assignment

The electromagnetic (solenoid-controlled) fuel injectors spray the fuel into the intake manifold at system pressure. They permit the precise metering of the quantity of fuel required by the engine. They are triggered via ECU driver stages with the signal calculated by the engine-management system.

Design and operating concept

Essentially. the electromagnetic injectors (Fig. 1) are comprised of the following components:

- The injector housing (9) with electrical (8) and hydraulic (1) connections
- The coil for the electromagnet (4)
- The movable valve needle (6) with solenoid armature and sealing ball
- The valve seat (10) with the injection-orifice plate (7) and the
- Spring (5)

In order to ensure trouble-free operation, stainless steel is used for the parts of the injector which come into contact with fuel. The injector is protected against contamination by a filter strainer (3) at the fuel input.

Connections

On the injectors presently in use, fuel supply to the injector is in the axial direction, that is, from top to bottom ("top feed"). The fuel line is fastened to the injector using a special clamp. Retaining clips ensure reliable alignment and fastening. The seal ring (2) on the hydraulic connection (1) seals off the injector at the fuel rail.

The injector is connected electrically to the engine ECU.

Injector operation

With no voltage across the solenoid (solenoid de-energised), the valve needle and sealing ball are pressed against the cone-shaped valve seat by the spring and the force exerted by the fuel pressure. The fuel-supply system is thus sealed off from the manifold. As soon as the solenoid is energised (excitation current), this generates a magnetic field which pulls in the valve-needle armature. The sealing ball lifts off the valve seat and the fuel is injected. As soon as the excitation current is switched off, the valve needle closes again due to spring force.

Fuel outlet

The fuel is atomized by means of an injection-orifice plate in which there are a number of holes. These holes (spray orifices) are stamped out of the plate and ensure that the injected fuel quantity remains highly reproducible. The injection-orifice plate is insensitive to fuel deposits. The spray pattern of the fuel leaving the injector is a function of the number of orifices and their configuration.

The highly efficient injector sealing at the valve seat is due to the cone/ball sealing principle.

The injector is inserted into the opening provided for it in the intake manifold. The bottom seal ring provides the seal between the injector and the intake manifold. Essentially, the injected fuel quantity per unit of time is determined by

- The fuel-supply system pressure
- The counterpressure in the intake manifold
- The geometry of the fuel-exit area

Types of construction

In the course of time, the injectors have been further and further developed to match them to the ever-increasing demands regarding engineering, quality, reliability, and weight. This has led to a variety of different injector designs.

EV6 injector

The EV6 injector is the standard injector for today's modern fuel-injection systems (Figs. 1 and 2a). It is characterized by its small external dimensions and its low weight. This injector therefore already provides one of the prerequisites for the design of compact intake modules.

In addition, the EV6 is also outstanding with regard to its hot-fuel behaviour, that is, there is very little tendency for vapor-bubble formation when using hot fuel. This facilitates the use of RLFS fuel-supply systems in which the fuel temperature in the injector is higher than with systems featuring fuel return. Thanks to wear-resisting surfaces, the fuel quantities injected by the EV6 remain highly reproducible over long periods of time, and the injector features a long useful life.

Thanks to their highly efficient sealing, these injectors fulfill all future requirements regarding zero evaporation. That is, no fuel vapor escapes from them.

The EV6 variant with "air shrouding" was developed especially to comply with requirements for even better fuel atomization. Finely vaporized fuel can be generated using other methods: In future, in addition to 4-hole injection-orifice plates, multi-orifice plates with between 10 and 12 holes will be used. Injectors equipped with these multi-orifice plates generate a very fine fuel fog.

There are a wide variety of injectors available for different areas of application. These feature different lengths, flow classes, and electrical characteristics. The EV6 is also suitable for use with fuels having an ethanol content of as much as 85 %.

EV14 injector

Further injector development has led to the EV14 (Fig. 2b) which is based on the EV6. It is even more compact, a fact which facilitates its integration in the fuel rail.

The EV14 is available in 3 different lengths (compact, standard, long). This makes it possible to adapt individually to the engine's intake-manifold geometry.

1 Design of the EV6 electromagnetic injector

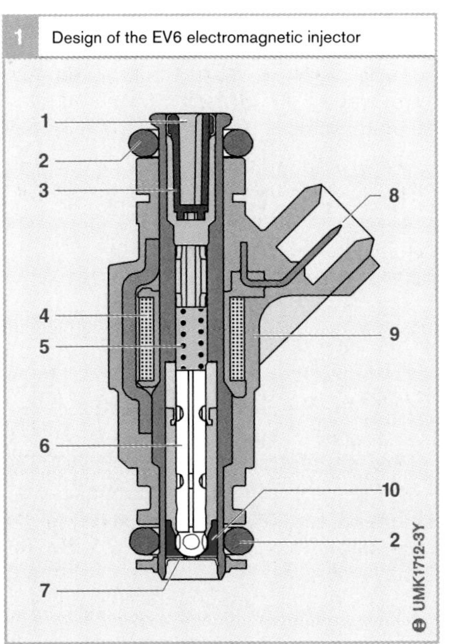

2 Injector versions

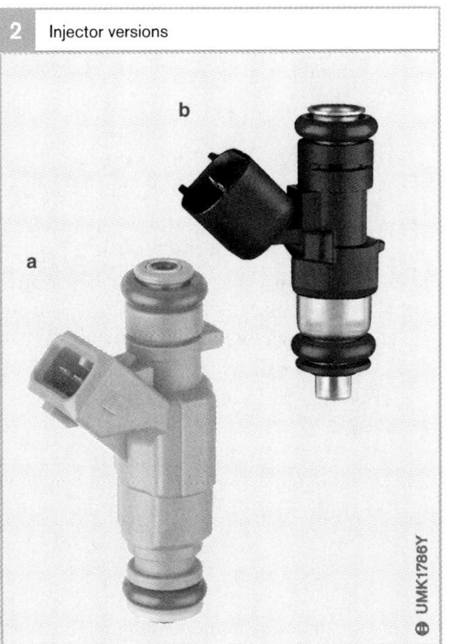

Fig. 1
1 Hydraulic connection
2 Seal rings (O-rings)
3 Filter strainer
4 Coil
5 Spring
6 Valve needle with armature and sealing ball
7 Injection-orifice plate
8 Electrical connection
9 Injector housing
10 Valve seat

Fig. 2
a EV6 Standard
b EV14 Compact

Spray formation

The injector's spray formation, that is, its fuel-spray shape, spray-dispersal angle, and fuel-droplet size, have a considerable influence upon the generation of the A/F mixture. Different versions of spray formation are required in order to comply with the demands of individual intake-manifold and cylinder-head geometries. Fig. 3 shows the most important fuel-spray shapes.

"Pencil" spray

A thin, concentrated, and highly-pulsed fuel spray results from using a single-hole injection-orifice plate. This form of spray practically eliminates the wetting of the manifold wall. Such injectors are most suitable for use with narrow intake manifolds, and in installations in which the fuel has to travel long distances between the point of injection and the intake valve.

The pencil-spray injector is only used in isolated cases due to its low level of fuel atomization.

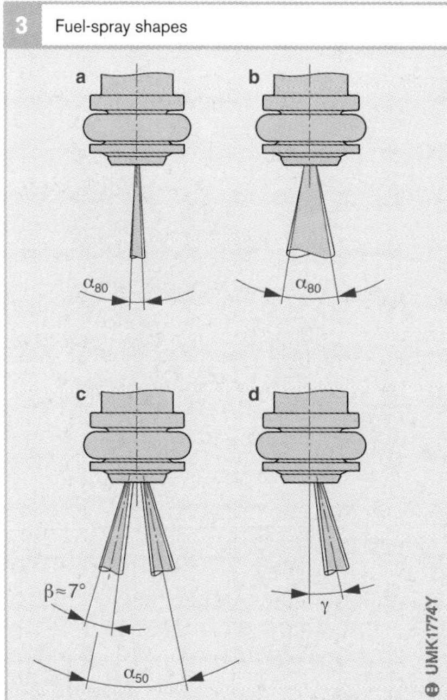

Fig. 3
a Pencil spray
b Tapered spray
c Dual spray
d Spray offset angle

α_{80}: 80 % of the injected fuel is within the angle defined by α
α_{50}: 50 % of the injected fuel is within the angle defined by α
β: 70 % of the injected fuel in a single spray is within the angle defined by β
γ: Spray offset angle

Tapered spray

A number of individual jets of fuel leave the injection-orifice plate. The tapered spray cone results from the combination of these fuel jets.

Although engines with only 1 intake valve per cylinder typically use tapered-spray injectors, they are also suitable for engines with 2 intake valves per cylinder.

Dual spray

The dual-spray formation principle is often applied on engines with 2 intake valves per cylinder. Engines with 3 intake valves per cylinder must be equipped with dual-spray injectors.

The holes in the injection-orifice plate are so arranged that two fuel sprays leave the injector and impact against the respective intake valve or against the web between the intake valves. Each of these sprays can be formed from a number of individual sprays (2 tapered sprays).

The spray offset angle

Referred to the injector's principle axis, the fuel spray in this case (single spray and dual spray) is at an angle, the spray offset angle (γ). Injectors with this spray shape are mostly used when installation conditions are difficult.

Types of fuel injection

In addition to the duration of injection, a further parameter which is important for optimisation of the fuel-consumption and exhaust-gas figures is the instant of injection referred to the crankshaft angle. Here, the possible variations are dependent upon the type of injection actually used (Fig. 1).

The new injection systems provide for either sequential fuel injection or cylinder-individual fuel injection (SEFI and CIFI respectively).

Simultaneous fuel injection

All injectors open and close together in this form of fuel injection. This means that the time which is available for fuel evaporation is different for each cylinder. In order to nevertheless obtain efficient A/F-mixture formation, the fuel quantity needed for the combustion is injected in two portions. Half in one revolution of the crankshaft and the remainder in the next. In this form of injection, the fuel for some of the cylinders is not stored in front of the particular intake valve but rather, since the valve has opened, the fuel is injected into the open intake port. The start of injection cannot be varied.

Group injection

Here, the injectors are combined to form two groups. For one revolution of the crankshaft, one injector group injects the total fuel quantity required for its cylinders, and for the next revolution the second group injects.

This configuration enables the start of injection to be selected as a function of engine-operating point. Apart from this, the undesirable injection into open inlet ports is avoided. Here too, the time available for the evaporation of fuel is different for each cylinder.

Sequential fuel injection (SEFI)

The fuel is injected individually for each cylinder, the injectors being actuated one after the other in the same order as the firing sequence. Referred to piston TDC, the duration of injection and the start of injection are identical for all cylinders, and the fuel is stored in front of each cylinder.

Start of injection is freely programmable and can be adapted to the engine's operating state.

Cylinder-individual fuel injection (CIFI)

This form of injection provides for the greatest degree of design freedom. Compared to sequential fuel injection, CIFI has the advantage that the duration of injection can be individually varied for each cylinder. This permits compensation of irregularites, for instance with respect to cylinder charge.

1 Manifold fuel injection: Types of fuel injection

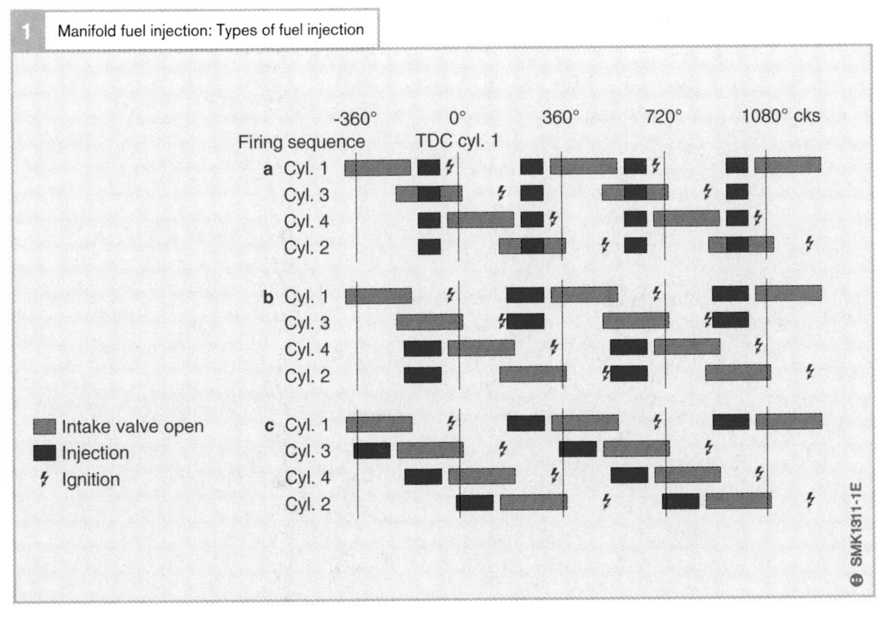

Fig. 1
a Simultaneous fuel injection
b Group fuel injection
c Sequential fuel injection (SEFI) and cylinder-individual fuel injection (CIFI)

Gasoline direct injection

Gasoline direct-injection engines generate the A/F mixture in the combustion chamber. During the induction stroke, only the combustion air flows past the open intake valve and into the cylinder. The fuel is injected directly into the cylinders by special injectors.

Overview

The demand for higher-power engines, coupled with the requirement for reduced fuel consumption, were behind the "re-discovery" of gasoline direct injection. As far back as 1937, an engine with mechanical gasoline direct injection took to the air in an airplane. In 1952, the "Gutbrod" was the first passenger car with a series-production mechanical gasoline direct-injection engine, and in 1954 the "Mercedes 300 SL" followed.

At that time, designing and building a direct-injection engine was a very complicated business. Moreover, this technology made extreme demands on the materials used. The engine's service life was a further problem.

These facts all contributed to it taking so long for gasoline direct injection to achieve its breakthrough.

1 Gasoline direct injection: Components

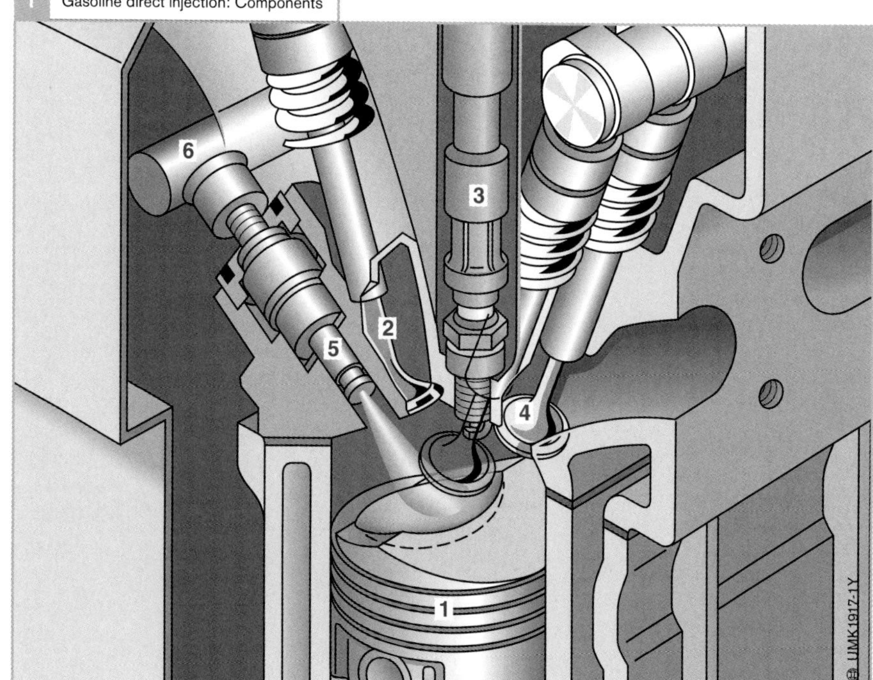

Fig. 1
1 Piston
2 Intake valve
3 Ignition coil with plugged-in spark plug
4 Exhaust valve
5 High-pressure injector
6 Fuel rail

Operating concept

Gasoline direct-injection systems are characterized by injecting the fuel directly into the combustion chamber at high pressure. Similar to the diesel engine, A/F-mixture formation takes place inside the cylinder (internal A/F-mixture formation).

Generation of high-pressure

The electric fuel pump delivers fuel to the high-pressure pump at a primary pressure of 0.3...0.5 MPa (3...5 bar). Depending on the engine operating point (required torque and engine speed), the high-pressure pump then generates the system pressure which forces the fuel, which is now at high pressure, into the rail (Fig. 1, Pos. 6) where it is stored until required for injection.

The fuel pressure is measured by the high-pressure sensor and adjusted to values between 5...12 MPa by the pressure-control valve.

The high-pressure injectors (5) are installed in the rail (also referred to as the "Common rail") and, when triggered by the engine ECU, inject the fuel directly into the combustion chambers.

A/F-mixture formation

The injected fuel is finely atomized due to the very high injection pressure. Together with the drawn-in air, it forms the A/F mixture in the combustion chamber. Depending upon the engine's operating mode, the fuel is injected in such a manner that an A/F mixture with $\lambda \leq 1$ is evenly distributed throughout the complete combustion chamber (homogeneous operation), or a stratified-charge A/F-mixture cloud ($\lambda \leq 1$) is formed around the spark plug (lean-burn operation or stratified-charge operation). During stratified-charge operation, the remainder of the cylinder is filled with either freshly drawn-in air, with inert gas returned to the cylinder by EGR, or with a very lean A/F mixture. The overall A/F mixture then has ltotal $\lambda_{total} > 1$.

The various methods of running the engine as listed above are referred to as the engine's operating modes. On the one hand, the selection of the operating mode to be applied is a function of engine speed and desired torque, and on the other it depends upon functional requirements such as the regeneration of the accumulator-type catalytic converter.

Torque

During stratified-charge operation, the injected fuel mass is decisive for the generated torque. The excess air permits "unthrottled" operation, also at part load, with the throttle opened wide. This measure reduces the pumping (exhaust and refill) work, and therefore also serves to lower the fuel consumption.

In homogeneous and lean-burn operation at $\lambda > 1$ and homogeneous A/F-mixture distribution, "unthrottling" also results in fuel savings, although not to the same extent as in stratified-charge operation.

In homogeneous operation at $\lambda \leq 1$, the gasoline direct-injection engine for the most part behaves the same as a manifold-injection engine.

Exhaust treatment

The catalytic converter is responsible for removing the pollutants from the exhaust gas. In order to operate with maximum efficiency, the 3-way catalytic converter needs a stoichiometric A/F mixture. Due to excess air, lean-burn operation results in increased levels of NO_x emissions which are stored temporarily in an accumulator-type NO_x catalytic converter. These are then reduced to nitrogen, carbon dioxide and water by running the engine briefly with excess air.

Rail

The function of the rail (fuel rail) is to store the fuel delivered by the high-pressure pump, and to distribute it to the high-pressure fuel injectors. The rail volume is sufficiently large to compensate for pressure pulses in the fuel system.

The rail is made of aluminum. Design variants (volume, dimensions, weight, etc.) are engine- and system-specific.

The rail features connections for the other components in the fuel-injection system (high-pressure pump, pressure-control valve, high-pressure sensor, high-pressure fuel injectors). The rail design guarantees that it does not leak and that it is well sealed to its various interfaces.

High-pressure pump

Function

The high-pressure pump is required to pressurize a sufficient quantity of the fuel delivered by the electric fuel pump at an admission pressure of 0.3...0.5 MPa to the high-pressure fuel-injection system (gasoline direct injection) compressed to a pressure of 5...12 MPa.

When the engine is started, the fuel is initially injected at inlet pressure. Pressure buildup occurs after the engine has run up to speed.

When the high-pressure pump is in operation, the fuel conveyed is the only cooling and lubricating medium.

Common engine configurations utilize three-barrel radial-piston pumps as well as demand-controlled single-barrel pumps.

1 Three-barrel pump HDP1

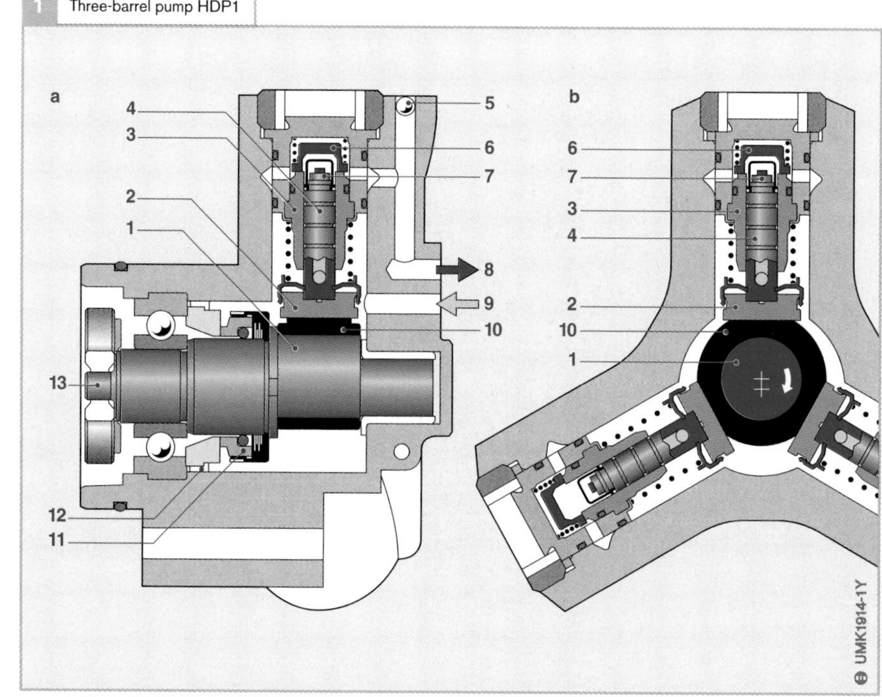

Fig. 1
a Longitudinal section
b Cross-section

1 Eccentric element
2 Slipper
3 Pump barrel
4 Pump plunger (hollow piston, fuel inlet)
5 Sealing ball
6 Outlet valve
7 Inlet valve
8 High-pressure connection to the rail
9 Fuel inlet (low-pressure)
10 Cam ring
11 Axial seal (sleeve seal)
12 Static seal
13 Input shaft

Three-barrel pump HDP1

The pump is a radial-piston pump with three pistons placed at circumferential offsets of 120°. Fig. 1 shows the implemented solution in both longitudinal and cross-sections.

Driven by the engine camshaft, the rotation of the input shaft with its eccentric element provides the lifting motion for the pump plunger inside the pump barrel. With the downward movement of the plunger, fuel at an inlet pressure of 0.3...0.5 MPa flows from the fuel line through the hollow pump plunger, and passes through the intake valve to the delivery chamber. As the plunger moves up, the fuel quantity is compressed. After exceeding the rail pressure, it flows through the outlet valve to the high-pressure fitting.

The selected plunger layout results in an alternating delivery overlap, which in turn results in low fuel stream pulsing, i.e. negligible pressure pulsation in the rail. The delivery quantity is proportional to engine speed.

To ensure that the primary pressure can be varied at sufficient speed as a factor of the engine's fuel demands even at maximum injected-fuel quantities, the maximum inject-fuel quantity of the HDP is designed larger by a specific variable. When operating at constant rail pressure, or at part-load, the pressure-control valve depressurizes the excess delivered fuel quantity to inlet pressure level, and the fuel is returned via the suction side of the HDP.

Single-barrel pump HDP2

The single-barrel radial-piston pump HDP2 is a plug-in pump. It is driven directly by the engine camshaft (Fig. 2). The transfer of the up-and-down motion to the pump plunger is accomplished by a bucket tappet located in the cylinder head.

2 Single-barrel pump HDP2

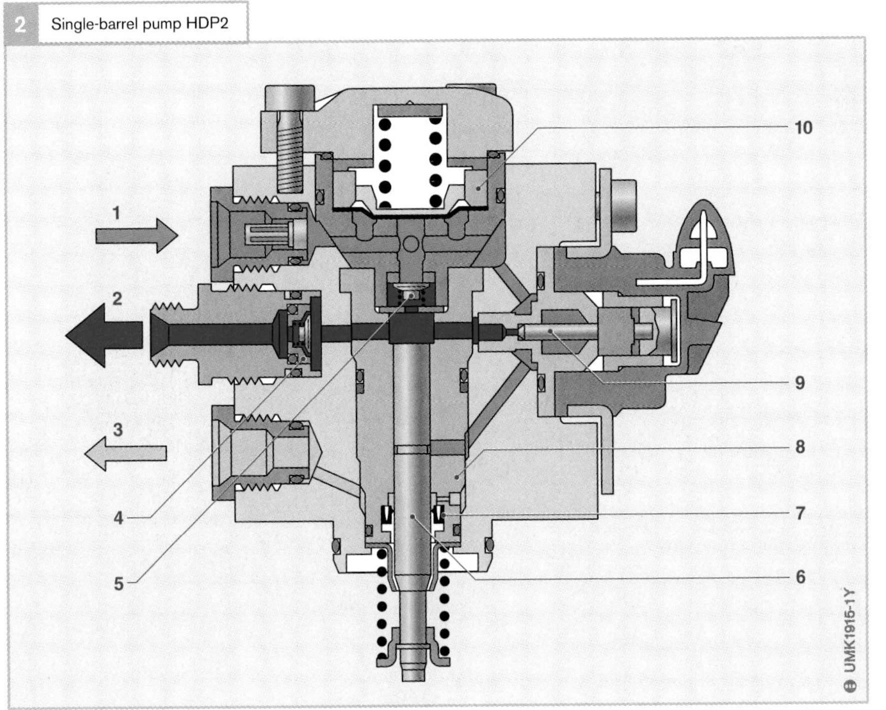

Fig. 2
1 Fuel inlet (low-pressure)
2 High-pressure connection to rail
3 Leakage return
4 Outlet valve
5 Inlet valve
6 Pump plunger
7 Piston seal
8 Pump barrel
9 Fuel-quantity control valve
10 Fuel-pressure attenuator

Dark blue: High-pressure range
Medium blue: Low-pressure range
Light blue: Pressure-less range (return)

The diagram shows the basic design of the component. An electrically switched fuel-quantity control valve with a return to the fuel inlet has been added to the original inlet and outlet valves in the delivery chamber. In the normally open position, high-pressure delivery does not take place because the entire delivery quantity flows back to the return line. During active operation, the valve closes when the pump plunger is at bottom dead center. It opens again to complete fuel delivery when the specified rail pressure has been reached. The fuel that continues to be delivered up to top dead center flows back to the fuel inlet.

This control process ensures exact delivery of the fuel quantity the engine requires. In turn, this reduces the power input of the pump and also fuel consumption.

To dampen fuel pressure pulses resulting from the delivery characteristics of the single-piston pump, a fuel-pressure attenuator is located in the fuel inlet directly upstream of the inlet valve. It is designed as a spring-type diaphragm accumulator used in manifold-injection systems.

Another key functional element is the plunger seal located in the pump barrel. It acts as a "separator" between the fuel and engine lube-oil sections within the pump. To increase operational reliability, the seal is pressure-relieved via a connection to the leakage return line that drains to the fuel tank.

Pressure-control valve

Function

The pressure-control valve is located between the fuel rail and the low-pressure side of the high-pressure pump HDP1. It provides the desired internal fuel-rail pressure by restricting the delivery flow of the HDP1. The excess fuel quantity is returned to the low-pressure system.

Design and mode of operation

A pulse-width modulated signal energizes the coil (Fig. 1, Item 3). Dependent on the pulse-duty factor, the valve ball (7) lifts off the valve seat (8) at a higher or lesser clearance. This changes the valve's effective flow cross-section.

In order to ensure the required fuel-rail pressure even if the electric control signal fails, the pressure-control valve is normally closed. An integrated mechanical pressure-limiting function protects the components against excessively high rail pressures.

1 Cross-section of pressure-control valve

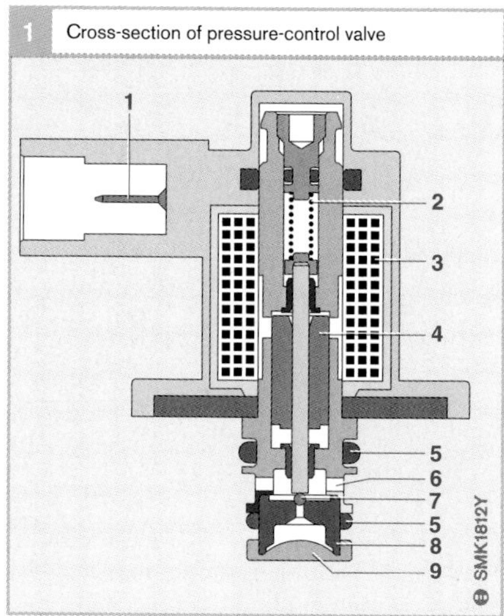

Fig. 1
1 Electrical connection
2 Compression spring
3 Coil
4 Solenoid armature
5 O-rings
6 Outlet passage
7 Valve ball
8 Valve seat
9 Fuel inlet with inlet strainer

Exploiting GDI to reduce fuel consumption

Emissions of carbon dioxide (CO_2) are one of the factors responsible for the earth's greenhouse effect (global warming). Part of the overall emissions is contributed by motor traffic. Given the rising number of miles traveled worldwide, the simultaneous reduction of CO_2 emissions can be achieved only by reducing fuel consumption per mile driven.

The European Automobile Manufacturers Association (ACEA, **A**ssociation des **C**onstructeurs **E**uropéens d'**A**utomobiles) has made a commitment to ensure a 25-percent reduction in automotive CO_2 emissions by the year 2008 (Fig. 1) based on 1995 emission levels.

There are several ways to reduce fuel consumption of vehicles powered by gasoline engines (Fig. 2). Gasoline direct injection (GDI) provides the greatest inherent potential. For the short term, it will achieve fuel savings of 10 to 15 percent, whereas the theoretical potential is as high as 25 percent. Provided that other measures are combined with gasoline direct injection, the resulting overall reduction – assuming a basic value of 12 percent for GDI – amounts to 25 percent (Fig. 3). This is the value demanded by the ACEA.

2 Fuel-saving measures

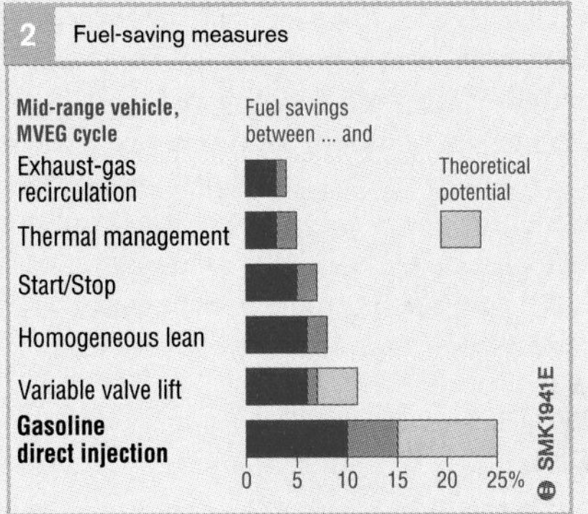

3 Fuel-saving measures in combination with gasoline direct injection (GDI)

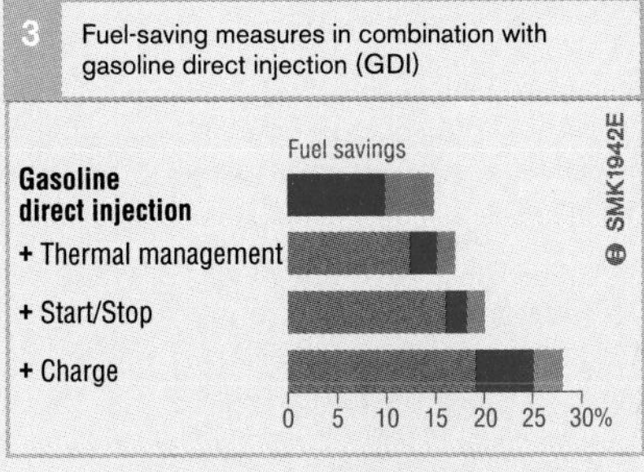

1 Anticipated fuel savings in passenger cars

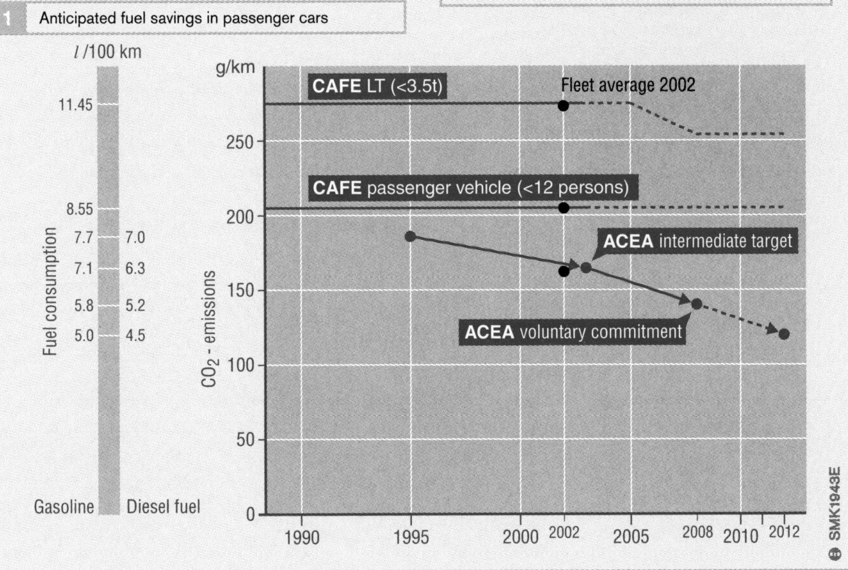

High-pressure injector

Assignment

The high-pressure injector represents the interface between the rail and the combustion chamber. Its job is to meter the fuel, and by means of the fuel's atomisation achieve controlled mixing of the fuel and air in a specific area of the combustion chamber. Depending upon the required operating mode, the fuel is either concentrated in the vicinity of the spark plug (stratified charge), or evenly distributed throughout the combustion chamber (homogeneous distribution).

1 High-pressure injector (HDEV): Design

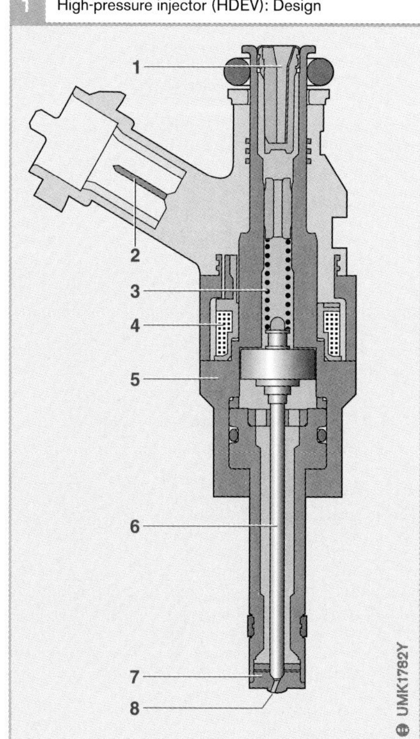

Fig. 1
1 Fuel inlet with fine strainer
2 Electrical connections
3 Spring
4 Solenoid
5 Injector housing
6 Nozzle needle with solenoid armature
7 Valve seat
8 Injector outlet passage

Design and operating concept

The high-pressure injector (Fig. 1) comprises the following components:

- Injector housing (5)
- Valve seat (7)
- Nozzle needle with solenoid armature (6)
- Spring (3) and
- Solenoid (4)

A magnetic field is generated when the solenoid coil is energized (current flows). This lifts the valve needle from the valve seat against the force of the spring and opens the injector outlet passage (8). Fuel is then injected into the combustion chamber due to the difference between rail pressure and combustion-chamber pressure.

When the energising current is switched off, the spring forces the needle back down against its seat and injection stops.

The injector opens very quickly, guarantees a constant opened cross-section during the time it is open, and closes against the rail pressure. Taking a given opened cross-section, the injected fuel quantity is therefore dependent upon the rail pressure, the counter-pressure in the combustion chamber, and the length of time the injector remains open. Excellent fuel atomisation is achieved thanks to the special nozzle geometry at the injector tip.

Compared to manifold injection, gasoline direct injection can boast faster injection, improved precision of spray alignment, and better formation of the fuel spray.

Technical requirements

Compared with manifold injection, gasoline direct injection differs mainly in its higher fuel pressure and the far shorter time which is available for directly injecting the fuel into the combustion chamber.

Fig. 2 underlines the technical demands made on the injector. With manifold injection, two revolutions of the crankshaft are available for injecting the fuel into the manifold. At an engine speed of 6,000 rpm, this corresponds to 20 ms.

In the case of gasoline direct injection though, considerably less time must suffice. During homogeneous operation, the fuel must be injected in the induction stroke. In other words, only half a crankshaft rotation is available for the injection process. Referred to the same engine speed as with manifold injection (6,000 rpm), this corresponds to an injection duration of only 5 ms.

For gasoline direct injection, the fuel requirement at idle referred to that at WOT is far lower than with manifold injection (factor 1:12). At idle, this results in an injection duration of approx. 0.4 ms.

Triggering the HDEV high-pressure injector

The injector must be triggered with a highly complex current characteristic in order to comply with the requirements for defined, reproducible fuel-injection processes (Fig. 3). The initial triggering signal from the microcontroller in the engine ECU is simply a digital signal (a). A special triggering module uses this signal to generate the actual triggering signal (b) with which the HDEV driver stage triggers the injector.

A booster capacitor is used to generate the 50...90 V trigger voltage which is high enough to provide a high current at the start of the switch-on process so that the valve needle can lift off of the valve seat very quickly (c). Once the valve needle has lifted (maximum needle lift), only a very low triggering current suffices to maintain the needle at a constant opened position. With the needle's opened position constant, the injected fuel quantity is proportional to the injection duration (d).

The calculations for the duration of injection take into account the premagnetisation time before the valve needle actually lifts.

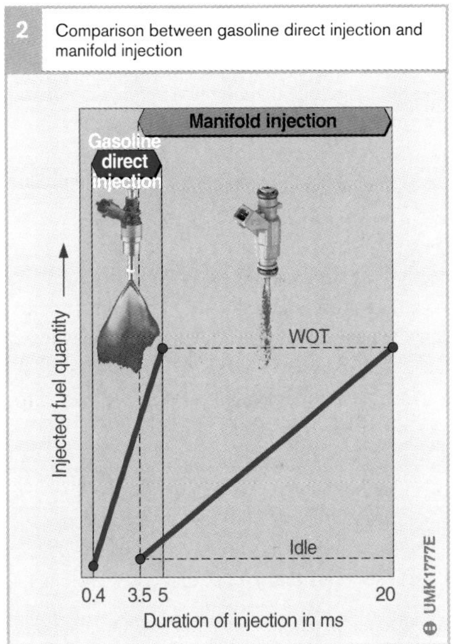

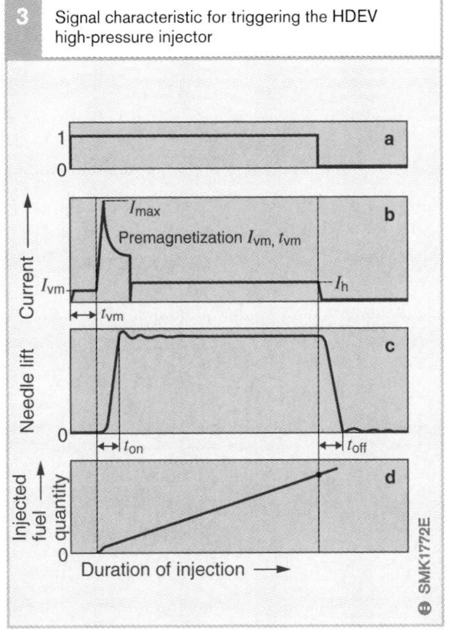

Fig. 2
Injected fuel quantity as a function of the duration of injection

Fig. 3
- a Triggering signal
- b Injector current characteristic
- c Needle lift
- d Injected fuel quantity

Combustion process

The combustion process is defined as the way in which A/F-mixture formation and energy conversion take place in the combustion chamber.

Depending upon the combustion process concerned, flows of air are generated in the combustion chamber. In order to obtain the required charge stratification, the injector injects the fuel into the air flow in such a manner that it evaporates in a defined area. The air flow then transports the A/F-mixture cloud in the direction of the spark plug so that it arrives there at the moment of ignition.

Two basically different combustion processes are possible.

1 Air-flow conditions for the various combustion processes

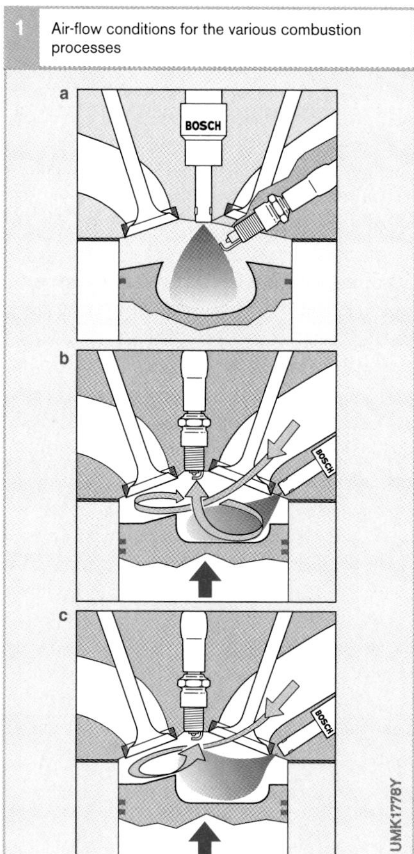

Fig. 1
a Spray-guided
b Wall-guided swirl air flow
c Wall-guided tumble air flow

Spray-guided combustion process

The spray-guided process is characterised by the fuel being injected in the spark plug's immediate vicinity where it also evaporates (Fig. 1a). In order to be able to ignite the A/F mixture at the correct moment in time (ignition point), it is imperative that spark plug and injector are exactly positioned, and that the spray direction is precisely aligned.

With this process, the spark plug is subjected to considerable thermal stressing since under certain circumstances the hot spark plug can be directly impacted by the relatively cold jet of injected fuel.

Wall-guided combustion process

In the case of the wall-guided process, one differentiates between two possible flows of air which are the result of specific intake-port and piston design. The injector injects into this air flow which transports the resulting A/F mixture to the spark plug in the form of a closed A/F-mixture cloud.

Swirl air flow

The air drawn by the piston through the open intake valve and into the cylinder generates a turbulent flow (rotational air movement) along the cylinder wall (Fig. 1b). This process is also designated "swirl combustion process".

Tumble air flow

This process produces a cylindrical air flow, or tumbling air flow, which in its movement from top to bottom is deflected by a pronounced piston recess so that it then moves upwards in the direction of the spark plug (Fig. 1c).

A/F-mixture formation

Assignment

It is the job of the A/F-mixture formation to provide a combustible A/F mixture which is to be as homogeneous as possible.

Technical requirements

In the "homogeneous" mode of operation (homogeneous $\lambda \leq 1$ and homogeneous lean-burn), this A/F mixture is distributed homogeneously throughout the whole of the combustion chamber. During stratified-charge operation on the other hand, the A/F mixture is only homogeneous within a restricted area, while the remaining areas of the combustion chamber are filled with inert gas or fresh air.

All fuel must have evaporated before a gas mixture or gas-vapor mixture can be termed homogeneous. A number of factors influence this process:

- Combustion-chamber temperature
- Fuel-droplet size and
- The time which is available for fuel evaporation

Influencing factors

Temperature influence

Depending upon temperature, pressure, and combustion-chamber geometry, an A/F mixture (air/gasoline) is combustible at $\lambda = 0.6...1.6$. Since gasoline cannot evaporate completely at low temperatures, this means that under these conditions more fuel must be injected in order to obtain a combustible A/F mixture.

A/F-mixture formation in the homogeneous operation mode

The fuel is injected as soon as possible so that the maximum length of time is available for formation of the A/F mixture. This is why the fuel is injected in the induction stroke during homogeneous operation. The intake air can then assist in achieving rapid evaporation of the fuel and efficient homogenisation of the A/F mixture.

A/F-mixture formation in the stratified-charge mode

The configuration of the combustible A/F-mixture cloud which is in the vicinity of the spark plug at the instant of ignition is decisive for the stratified-charge mode. This is why the fuel is injected during the compression stroke so that a cloud of A/F mixture is generated which can be transported to the vicinity of the spark plug by the air flows in the combustion chamber, and by the piston as it moves upwards. The ignition point is a function of the engine speed and the required torque.

Penetration depth

The fuel-droplet size in the injected fuel is a function of injection pressure and combustion-chamber pressure. Higher injection pressures result in smaller droplets which then vaporize quicker. Taking a constant combustion-chamber pressure, the so-called penetration depth increases along with increasing injection pressure. The penetration depth is defined as the distance travelled by the individual fuel droplet before it vaporizes completely.

The cylinder wall or the piston will be wetted with fuel if the distance needed for full vaporization exceeds the distance from the injector to the combustion-chamber wall. If the fuel on the cylinder wall and piston has not vaporized before the ignition point, either no combustion takes place at all, or it is incomplete.

Operating modes

There are six operating modes in use with gasoline direct injection (Fig. 1):

- Stratified-charge mode
- Homogeneous mode
- Homogeneous and lean-burn
- Homogeneous and stratified-charge
- Homogeneous/anti-knock
- Stratified-charge/cat-heating

These operating modes permit the best-possible adaptation for each and every engine operating state. During actual driving, the driver does not notice the change-overs between operating modes since these take place without torque surge.

The lines in the diagram (Fig. 1) show which operating modes are passed through when accelerating strongly (pronounced changes in torque with at first unchanged engine speed), and when accelerating gently (slight changes in torque with increasing engine speed).

Stratified-charge mode

In the lower torque range at speeds up to approx. 3,000 rpm, the engine is operated in the stratified-charge mode. Here, the injector injects the fuel during the compression stroke shortly before the ignition point. During the brief period before the ignition point the air flow in the combustion chamber transports the A/F mixture to the spark plug. Due to the late injection point, there is not sufficient time to distribute the A/F mixture throughout the complete combustion chamber.

Referred to the combustion chamber as a whole, the A/F mixture is very lean in the stratified-charge mode. The untreated NO_x emissions are very high when the excess-air level is high. In this operating mode, the best remedy is to use a high EGR rate, whereby the recirculated exhaust gas reduces the combustion temperature and, as a result, lowers the temperature-dependent NO_x emissions.

The parameters "Engine speed" and "Torque" define the limits for stratified-charge operation. In the case of excessive torque, soot is generated due to zones of local rich-mixture. If engine speed is too high, charge stratification and efficient transport of the A/F mixture to the spark plug can no longer be maintained due to excessive turbulence.

1 Operating-mode characteristic curves for gasoline direct injection

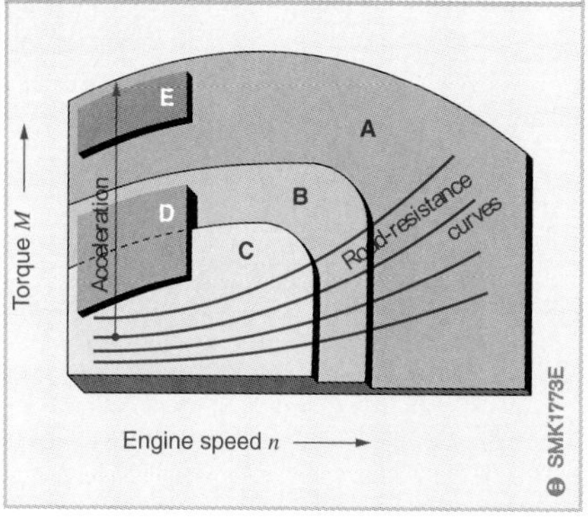

Fig. 1
A Homogeneous operation with $\lambda = 1$, this operating mode is possible in all operating ranges
B Lean-burn or homogeneous operation, $\lambda = 1$ with EGR; this operating mode is possible in area C and area D
C Stratified-charge operation with EGR

Operating modes with dual injection:
C Stratified-charge/cat-heating mode, same area as stratified-charge operation with EGR
D Homogeneous and stratified-charge
E Homogeneous/anti-knock

Homogeneous mode

For high torques and high engine speeds the engine is operated in the homogeneous mode $\lambda = 1$ (in exceptional cases with $\lambda < 1$) instead of in the stratified-charge mode. Injection starts in the induction stroke, so that there is sufficient time for the A/F mixture to be distributed throughout the whole of the combustion chamber. The injected fuel mass is such that the A/F mixture ratio is stoichiometric or, in exceptional cases, has slightly excessive fuel ($\lambda \leq 1$).

Since the whole of the combustion chamber is utilised, the homogeneous mode is required when high levels of torque are demanded. In this operating mode, emissions of untreated exhaust gas are also low due to the stoichiometric A/F mixture.

In homogeneous operation, combustion to a great extent corresponds to the combustion for manifold injection.

Homogeneous and lean-burn mode

In the transitional range between stratified-charge and homogeneous mode, the engine can be run with a homogeneous lean A/F mixture (λ>1). Since the pumping losses are lower due to "non-throttling", in the homogeneous and lean-burn mode, fuel consumption is lower than in the homogeneous mode with $\lambda \leq 1$.

Homogeneous and stratified-charge mode

The complete combustion chamber is filled with a homogeneous lean A/F mixture. This mixture is generated by injecting a small quantity of fuel during the induction stroke.

Fuel is injected a second time (dual injection) during the compression stroke. This leads to a richer zone forming in the area of the spark plug. This stratified charge is easily ignitable and then ignites the rest of the homogeneous mixture in the remainder of the combustion chamber.

The homogeneous and stratified-charge mode is activated for a number of cycles during the transition between stratified-charge and homogeneous mode. This enables the engine management system to better adjust the torque during the transition. Due to the conversion to energy of the very lean A/F mixture $\lambda > 2$, the NO_x emissions are also reduced.

The distribution factor between each injection is approx. 75 %. That is, 75 % of the fuel is injected in the first injection which is responsible for the homogeneous basic A/F mixture.

Steady-state operation using dual injection at low engine speeds in the transitional range between stratified-charge and homogeneous mode reduces the soot emissions compared to stratified-charge operation, as well as lowering fuel consumption compared to homogeneous operation.

Homogeneous/anti-knock mode

In this operating mode, since the charge stratification hinders knock, the use of dual injection at WOT, together with ignition-angle shift in the retard direction as needed to avoid "knock", can be dispensed with. At the same time, the favorable ignition point also leads to higher torque.

Stratified-charge/cat-heating

Another form of dual injection makes it possible to quickly heat up the exhaust-gas system, although this must be optimized before this solution can be applied. Here, therefore, in stratified-charge operation with high levels of excess air, injection takes place once in the compression stroke (similar to the "stratified-charge mode"), and then again in the combustion (power) cycle whereby the fuel injected here combusts very late and thus heats up the exhaust side and the exhaust system to a very high temperature.

A further important application is for heating up of the NO_x catalytic converter to temperatures in excess of 650 °C as needed to initiate the desulphurization of the catalytic converter. Here, it is imperative that dual injection is used since with conventional heating methods the high temperature which is required here cannot always be reached in all operating modes.

Ignition-system overview

The spark-ignition engine is an internal-combustion machine that relies on an external source of ignition-energy to run. A spark ignites the mixture of air and fuel contained within the combustion chamber to initiate the combustion process. This spark is actually the result of an arc generated between the electrodes of a spark plug extending into the combustion chamber. The ignition system must generate adequate levels of high-voltage energy to generate this arc while also ensuring that flashover occurs at precisely the right instant.

Overview

Producing the ignition spark in the combustion chamber entails exceeding a specific voltage level: the ignition voltage. Depending upon factors such as the engine's instantaneous operating conditions and the state of the spark plugs, this voltage can extend upward as high as 30,000 volts (encountered on turbocharged engines). The combustion process is initiated when the spark plug's energy is transferred to the A/F mixture following flashover.

Applications

Inductive (coil) ignitions have become standard in passenger cars (Fig. 1). This system stores ignition energy in the coil's magnetic field, then transforms it into high-tension voltage for transfer to the mixture when ignition is triggered.

Precise ignition timing and equally exact deactivation of the coil's current are absolutely essential in ensuring satisfactory engine performance and effective emissions control. Complex process control is vital. This function has now migrated from the separate, dedicated ignition-system ECUs used until 1998 into the centralized engine-management system. At Bosch the designation for the latter is Motronic.

Ignition systems with capacitive energy storage are available for high-performance and competition power plants. These store the ignition energy in a capacitor's electrical field.

1 Cutaway view of a four-cylinder engine with direct gasoline injection and fully-electronic ignition (distributorless high-voltage distribution)

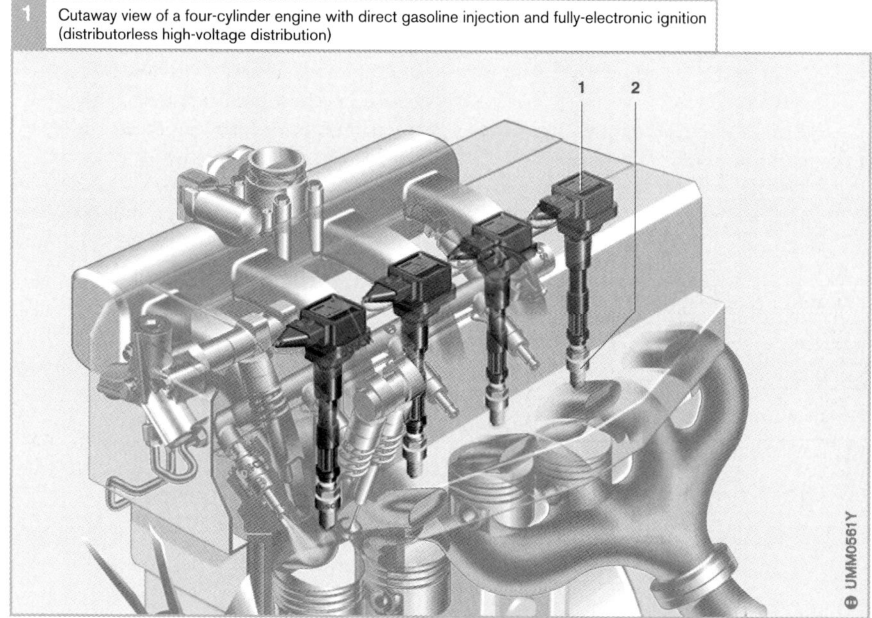

Fig. 1
1 Single-spark ignition coil
2 Spark plug

Ignition-system evolution (overview)

The low-tension magneto marked the advent of the ignition system at Bosch. Soon enough, however, this ignition concept based on an interruption arc would prove unequal to the demands posed by new automotive applications. A high-tension ignition system presented in 1902 did away with the mechanical interrupter contacts; combined with the spark plug it also allowed higher engine speeds for more power. The Bosch spark plug is thus already one hundred years old.

The era between the two world wars witnessed massive growth in motorization. In this period Bosch pursued development of battery-based ignitions that used an ignition coil to generate high-tension voltage and generate an arc at the spark plug. Over the years demand for enhanced power-plant performance and lower emissions expanded, and the ignition coil evolved to keep pace.
One important factor in this evolutionary process was the increased application of electronics (Fig. 2).

2 Development of inductive ignition systems

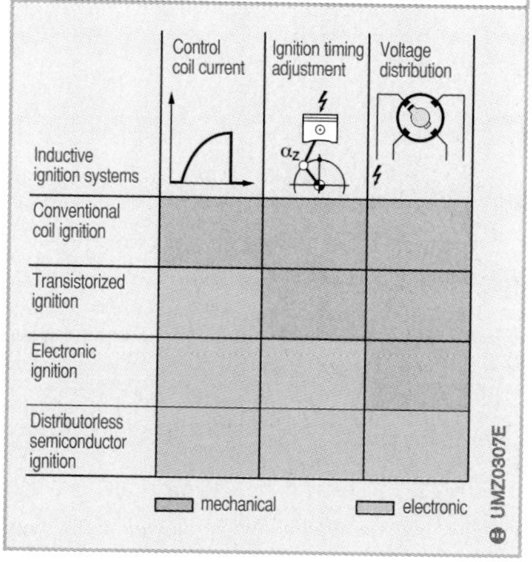

Inductive ignition systems | Control coil current | Ignition timing adjustment | Voltage distribution

Conventional coil ignition

Transistorized ignition

Electronic ignition

Distributorless semiconductor ignition

mechanical electronic

UMZ0307E

Conventional coil ignition

(1934 ... 1986)
Mechanical contact-breaker points control current flow through an ignition coil (energy storage and ignition triggering). A mechanical, centrifugal advance unit combines with an aneroid capsule to adjust ignition timing in response to variations in engine speed and load factor.

Distribution of high-tension voltage to the individual spark plugs is controlled by a rotor revolving within the ignition distributor (rotating high-tension distribution).

Transistorized ignition

(1965 ... 1993)
In this stage the mechanical contact-breaker points are replaced by wear-free power transistors in a transistorized switching unit; the spark is triggered by mechanical contacts or by proximity switching using an inductive sensor or Hall-effect trigger unit. This concept avoids the negative effects of contact-point wear experienced with the earlier system.

Electronic ignition

(1982 ... 1998)
While distribution of high-voltage energy is still based on a rotating mechanism, mechanical adjustment of ignition advance has already been consigned to redundancy. Instead, electronic components monitor the engine's speed and load factor, which are then employed as parameters for selecting an ignition advance angle stored in a semiconductor-based program map. This type of system relies on an ignition-system ECU featuring a microcontroller for operation.

Distributorless semiconductor ignition

(1983... 1998)
The mechanical distributor has been replaced by a 100 % electronic ignition control unit (stationary voltage distribution). Such fully-electronic ignition systems now operate without wear-sensitive moving parts.

The Volta pistol combined two basic elements of engine technology: it used a mixture of air and gas, and relied on an electrical spark. It is here that the story of electric ignition begins.

Early ignition evolution

Long before the first engines appeared at the end of the 19th century, inventors were engaged in efforts to evolve internal-combustion machines suitable for replacing the steam engines then used in stationary applications.

The first known attempt to create a thermal-energy machine to replace boiler, burner and steam with internal combustion was undertaken by Christiaan Huygens in the year 1673 (Fig. 1). The fuel was gunpowder (1), and ignition relied on a wick fuse (2). The basic concept is not complicated: following ignition, the combustion gases escape from the tube (3) through one-way valves (4). This creates a vacuum, causing the piston (5) to respond to atmospheric pressure by descending and performing work by lifting the weight G (7).

Because the device had to be reloaded after each ignition, it could not serve as a true engine by providing continuous power.

Over 100 years later, in 1777, Alessandro Volta determined that it was possible to ignite gas (swamp gas) that had collected within a well. He experimented further and found that it was possible to induce an explosion by mixing this gas with atmospheric air and applying a spark. Spark generation was provided by the electrophorous tube which he had invented in 1775. Volta exploited this effect in his Volta pistol.

In 1807 Isaak de Rivaz developed an atmospheric piston machine. Using the concept behind Volta's gas pistol as a starting point, he employed an electric spark to ignite a flammable mixture of air and gas. Rivaz installed a device based on his patent documents (Fig. 2) in a test vehicle, but soon abandoned his efforts in response to less than satisfactory results. The basic operating concept was similar to that used by Huygens in his powder-powered machine. Although the piston, impelled upward by an explosion and then forced back down by atmospheric pressure, succeeded in moving the vehicle several metres, it was necessary to inject a fresh combustion mixture into the cylinder and then ignite it for each successive cycle.

Required for application in motor vehicles were power plants with the ability to supply a continuous and uninterrupted flow of power. One of the major problems encountered in this period was how to ignite the flammable mixture once it was in the combustion chamber. Various systems evolved simultaneously as power-plant constructors searched for solutions.

1 Concept of Christiaan Huygens' powder machine from 1673

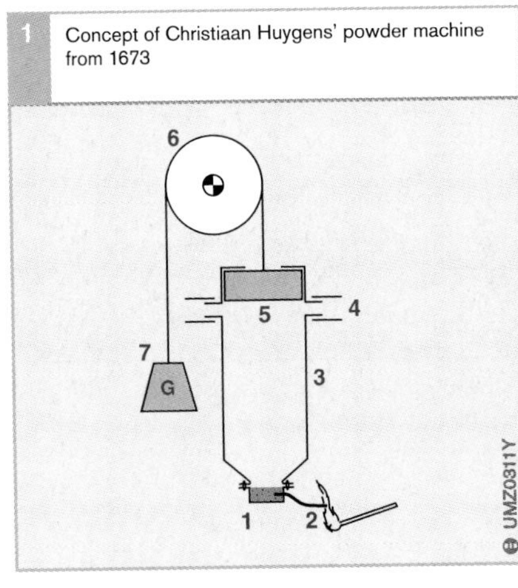

Fig. 1
1 Capsule with gunpowder
2 Fuse
3 Tube
4 1-way check valve
5 Piston
6 Idler pulley
7 Effective load G

2 Illustration showing vehicle designed by Isaak de Rivaz with atmospheric reciprocating piston, based on 1807 patent application

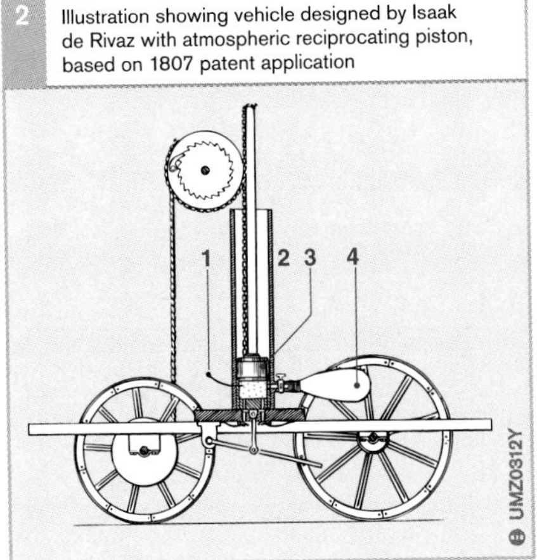

Fig. 2
1 Ignition-spark button
2 Cylinder
3 Piston
4 Bladder, filled with water

High-tension vibrator ignition

A concept for a battery-based ignition system had been available since 1860, when the Frenchman Etienne Lenoir constructed a "high-tension vibrator ignition" system for his stationary gas engine (Fig. 3). The ignition current was provided by a Ruhmkorff spark inductor (2) drawing energy from a Volta tube – serving as galvanic element – or some other source. The electrodes employed to generate the arc within the engine were two platinum wires (6) with porcelain insulators. Lenoir had thus invented the precursor of all spark plugs. Lenoir used a high-tension distributor on contact rails (5) to control current flow to the two spark plugs on the dual-action engine.

A basic characteristic associated with the Ruhmkorff spark inducer is the way it generates a magnetic field when the primary electrical circuit is closed. Self-induction slows the rise in primary current. But once primary current reaches a specific level, the armature (4) pulls in and the vibrator contacts (3) open to interrupt it. The magnetic field responds by collapsing rapidly, inducing high voltage in the secondary circuit and generating an arc in the spark path. The armature soon overcomes its intrinsic inertia to return to its original position and reclose the primary circuit. The cycle now starts again, with a potential maximum frequency of 40...50 cycles per second in then-contemporary designs. The vibrator system emitted a characteristic buzzing sound during operation.

Two basic considerations prevented this system from achieving widespread popularity in automotive applications. The first stems from the fact that the system actually generated an entire series of sparks during the ignition stroke, which prevented efficient combustion at higher engine speeds. It was, at best, suited for use in the very earliest vehicles, equipped as they were with engines that rarely exceeded 300 rpm. The second liability was simply that in those days no option was available for generating the required ignition energy with the vehicle underway.

During the course of 1886 Carl Benz applied himself to refining the high-voltage vibrator ignition's design. The result allowed the new engine to run at higher speeds than his first vehicular power plant (250 rpm and 0.8 PS). The spark plugs were designed and built by Benz personally. The electrical power source continued to pose problems, as the galvanic elements responsible for supplying current were ready for replacement after only 10 kilometers.

The ignition was – as Carl Benz once observed – "the crux of all our problems". "If there is no spark, then everything else has been in vain, and the most brilliant design is worthless."

It was not for nothing that, instead of wishing a pleasant journey, French motorists at the turn of the century saluted each other with "Good ignition!" ("Bon Allumage!")

3 Lenoir high-tension vibrator ignition

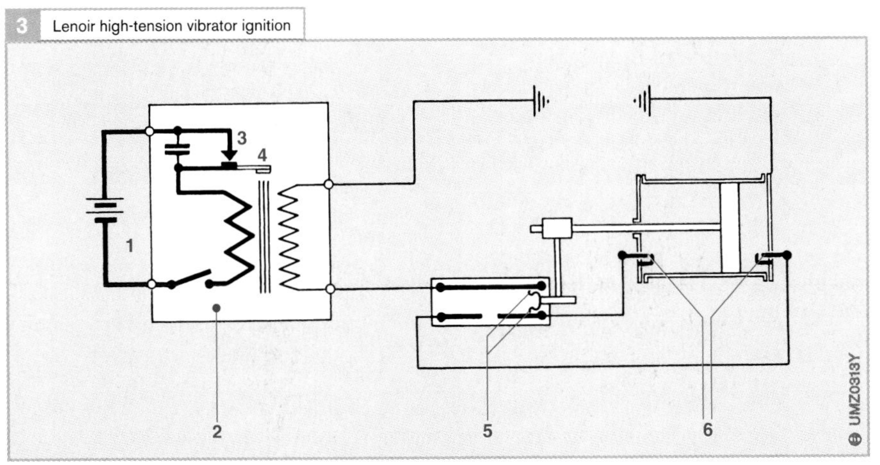

Fig. 3
1 Battery (galvanic element)
2 Ruhmkorff spark inductor
3 Oscillating contacts
4 Armature
5 Distributor with contact spring
6 Spark plug

Hot-tube ignition

Increases in engine operating speeds were essential if adequate power for automotive applications was to be extracted from power plants of reasonable displacement. Unfortunately, the control mechanisms employed for the flame ignition systems on stationary powerplants were too slow for use at high speeds.

In 1883 patent rights were granted on the continuous-operation, hot-tube ignition developed by Gottlieb Daimler. This ignition system (Fig. 4) consisted of a passage permanently connected to the cylinder's combustion chamber and terminating in a gas-tight glow tube (2). A small burner (3) kept the tube glowing continuously. During the compression stroke, mixture was forced into the glow tube, where it ignited. This mixture then initiated combustion of the remaining mixture in the combustion chamber.
The glow tube's heat had to be regulated to ensure that ignition would not occur before the end of the compression stroke. Depending upon the individual system, speeds as high as 700...900 rpm were possible.

For more than a decade, glow-tube ignition systems were the standard for automotive constructors. The concept fostered widespread acceptance of both the Daimler engine and the motor vehicle in general. The difficulty though, was the fact that the glow tube had to be maintained at precisely the right temperature. The flame could also be extinguished by exposure to rain or wind, while clumsiness could lead to fires. It was not for nothing that Wilhelm Maybach authored an 1897 memorandum based on the thesis that every car with glow-tube ignition was bound to catch fire sooner or later.

Magneto ignition had evolved into a viable option by 1897, when Daimler decided that the concept warranted his personal interest.

Low-tension, oscillating-magneto ignition

In the low-tension trip-lever magneto developed by Nikolaus August Otto in 1884, a magnetic inductor with an oscillating armature and rod-shaped permanent magnets provided low-voltage ignition current (see Fig. 5 for concept). Interrupting the current flow produced an arc at the contact points in the cylinder. The armature's spring-loaded return mechanism and the push rod controlling the ignition contact's trip lever, were coordinated to open the circuit at precisely the instant when armature current peaked. This produced a powerful ignition spark.

Because the 4-stroke engine produced by Otto in 1876 was powered by municipal gas, it was only suitable for stationary applications. While the advent of the low-tension armature magneto made it possible to run engines on gasoline, low operating speeds meant that application options remained restricted to low-revving stationary powerplants.

4 Hot-tube ignition on an 1885 Daimler engine

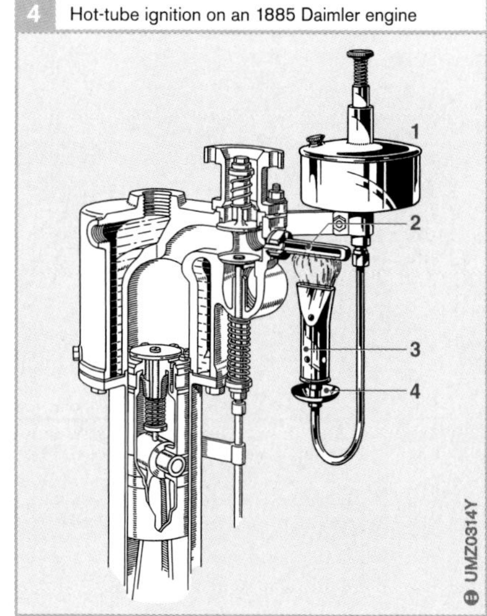

Fig. 4
1 Fuel container for burner
2 Hot tube
3 Burner
4 Preheater bowl

Magneto ignition

The search for a more suitable solution to the ignition issue in automotive power plants created the right conditions for participation by a specialist firm. Instead of constructing complete engines, this firm focused on supplying ignition systems for low-revving stationary power plants: the Workshop for Precision Mechanics and Electrotechnology founded by Robert Bosch in Stuttgart, in the year 1886.

Low-tension oscillating-magneto ignition from Bosch

In order to be able to sell them as equipment for stationary spark-ignition engines, Bosch continued development of the low-voltage magnetos featuring Otto's release mechanism (Fig. 5). The system's primary asset was its ability to operate without a battery. While the unit's substantial weight combined with its lethargic ignition mechanism to prevent application on automotive engines, the development project did provide Bosch with valuable experience.

Low-tension magneto ignition

Bosch developed the lethargic release mechanism into a faster and lighter interruption-arc magneto ignition suitable for high-speed automotive engines.

Instead of allowing the heavy, wire-bound armature to oscillate, this system used a sleeve suspended between the poles and the fixed armature (Fig. 6). The sleeve, which served to bundle the lines of force, was driven by bevel gears, which also provided the timing adjustment. A cam lobe with a gradual profile turned the snap lever. When spring force pushed the lever off the lobe, the ignition lever inside the cylinder abruptly separated from the ignition contact to produce a spark.

Because the arrangement met the era's speed demands, the sleeve-magneto concept and the positive drive were an immediate success.

The double-T armature became the Bosch armature, the symbol of the Robert Bosch GmbH.

5 Design of the Bosch low-tension ignition magneto with snap release and ignition flange from 1887

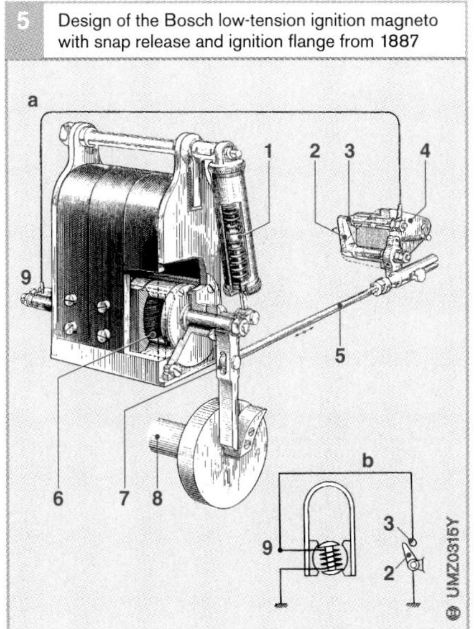

6 Design of the Bosch low-tension ignition magneto with oscillating shell, 1897 version

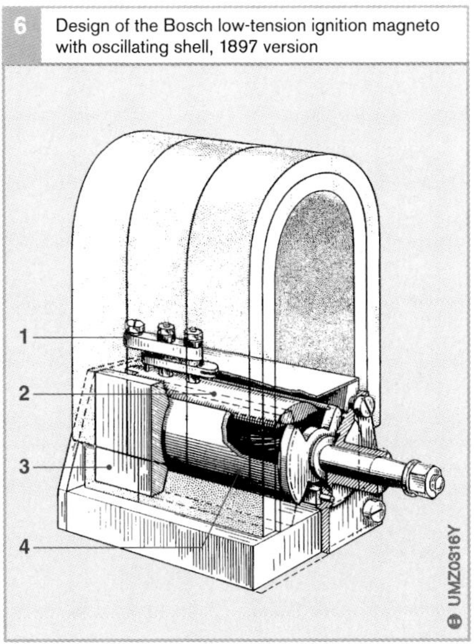

Fig. 5
a Design
b Illustration of concept (cutaway)

1 Spring locations
2 Ignition lever
3 Ignition pin
4 Ignition flange
5 Push rod
6 Double-T armature
7 Elbow lever
8 Control shaft
9 Clamp

Fig. 6
1 Clamp
2 Double-T armature (fixed position)
3 Pole shoes
4 Sleeve (oscillating)

Soon afterward it was presented to Gottlieb Daimler, Wilhelm Maybach and Carl Benz. Daimler had one of these ignition systems installed in a vehicle in 1898, and then proceeded to road test it by driving from Stuttgart to Tyrol, a trial the unit absolved with success. The Daimler engines in the first Zeppelin airships were also equipped with Bosch snap ignition, as the highly inflammable gas used to inflate the dirigibles meant that a hot-tube system could never be considered as a serious option.

This was still a low-tension magneto unit relying on a snap lever and opening mechanism to generate a spark at the ignition contacts in the combustion chamber. The contacts themselves evolved from purely mechanical to electromechanical actuation in this period.

High-tension magneto ignition

Higher engine speeds, compression ratios and combustion temperatures all combined to produce ignition demands that oscillating ignition could not satisfy. Until problems with batteries could be resolved, magneto ignition using spark plugs instead of snap contacts represented the only viable option. For this, a source of high-tension ignition current was essential.

Robert Bosch assigned Gottlob Honold to design a magneto-based ignition system in which the snap contacts would be replaced by permanent ignition electrodes. The project brief called for an ignition system suitable for sale to engine manufacturers as a single, complete package.

Honold's starting point was a low-voltage magneto with an oscillating sleeve, which he then proceeded to modify. The double-T armature had two windings; one consisted of a limited number of thick-wire loop, while the second comprised a larger number of loops thin-wire (Fig. 7 and 8). Rotating the sleeve generated low tension in the armature winding. Simultaneously contacts closed to short the coil element with the limited number of loops (Fig. 7, Pos. 3). This produced the strong current that was subsequently interrupted to induce a brief burst of high voltage in the coil element with the large number of loops. This voltage jumped across the

Presentation of the Bosch low-tension ignition magneto

Berlin's "First International Motorcar Exhibition" in 1899 provided the first opportunity to show the new interrupter ignition. The Bosch stand consisted of a table measuring roughly two meters square. Although Robert Bosch was personally on hand to extol the new ignition's virtues, the visitors seemed less than entirely convinced, and many simply shook their heads. Unfortunately it had not been possible to install the new ignition system in any of the display vehicles. As a result, it received only a bronze medal. It was only in the following year, 1900, that the Bosch ignition received recognition in the form of gold medals in Nuremburg and Vienna.

7 Electrical diagram from the patent application No. 156117 of 7 January, 1902, showing the Bosch high-tension ignition magneto

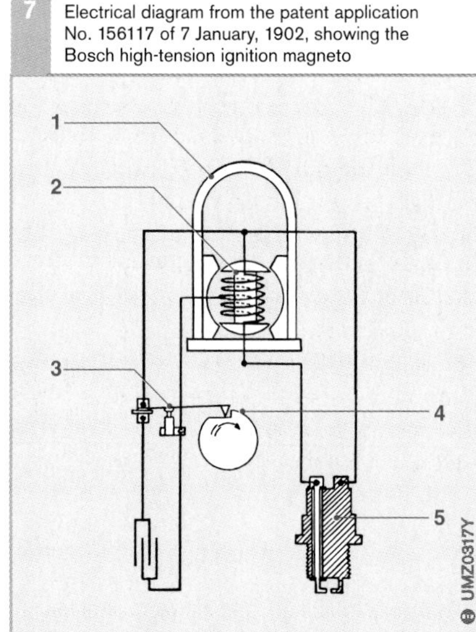

Fig. 7
1 Magnet
2 Double-T armature
3 Breaker contacts
4 Cam
5 Spark plug

spark plug's spark gap (5) to make it conductive. After this a second voltage was induced in the same windings. Although substantially lower than the first, it was sufficiently powerful to flow through the conductive path produced by the first burst of energy and produce an arc at the spark plug's electrodes, emulating the snap contact system.

The contact breakers were operated mechanically by a cam (4) to allow precise timing of the process in which they opened and closed the low-voltage winding's circuit. A capacitor was installed in parallel with the contact breakers to inhibit arcing across the points when they were open.

By December of 1901 Honold had already completed the first provisional test unit, and it was just at this point that Bosch returned from an extended trip abroad. Having been away, Bosch was not aware of the project's current status. Observing the trial unit in operation, he immediately appreciated the new design's significance. He expressed his delight with the words, "You really hit the bull's eye with that one."

The spark plugs also needed development work. The plugs used with the earlier battery-powered ignition systems would never be able to cope with the high-intensity arc produced by the new ignition. The electrodes burned away too quickly. The only solution, therefore, was for Bosch to start producing his own spark plugs. This means that the development of the spark plug also stems from this era. Extensive effort also had to be invested in developing reliable contact breakers that would form the centerpiece of the high-tension magneto system right from the start.

Yet another version of the magneto ignition was developed by Ernst Eisemann. This system's high-tension voltage was generated by a separate transformer fed by a low-tension magneto. Although early versions of this unit relied on a synchronous-motion contact unit to short the circuit several times in the course of each armature rotation, it was later realized that a single short circuit was adequate.

In 1902 Eisemann attempted to recruit customers for his development among the most important German firms. He received the cold shoulder: the unanimous opinion was that the spark was much too small for use in an engine. Eisemann was more fortunate in France, where the Count de la Valette – an engineer – acquired exclusive French distribution rights for the Eisemann high-tension magneto ignition on 15 September, 1902. Later Eisemann abandoned the separate coil in favour of the Bosch design featuring the familiar double-T armature with its two windings.

24 September, 1902
The first Bosch high-tension ignition magneto, including the spark plugs, is delivered to the Daimler company

8 Design of the first Bosch high-tension ignition magneto to enter series production, the Type HDh with rotating sleeve from 1902

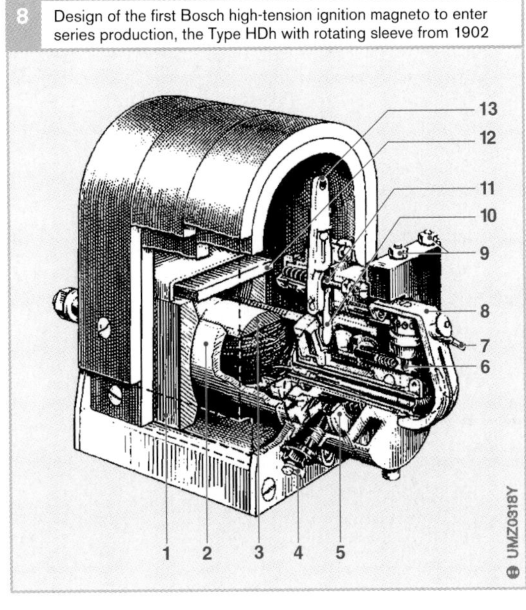

Fig. 8
1 Pole shoe
2 Sleeve (rotating)
3 Double-T armature
4 Current contact with rail providing link to spark-plug terminal
5 Distributor disc with collector ring
6 Connector to distributor disc (secondary)
7 to ignition switch
8 Connection to breaker (primary)
9 Terminals to spark plugs
10 Contact-breaker lever
11 Contact breakers
12 Capacitor
13 Ignition-timing adjustment

Magneto ignition applications

Bosch magneto ignition in motor racing

Bosch low-tension magneto ignition systems successfully absolved the acid test in the first car with the name Mercedes, which won three French races as well as achieving other victories in the course of 1901. One particularly significant event was the Irish Gorden Bennett race in 1903. With the Belgian driver Camille Jenatzy at the helm, the 60 bhp Mercedes posted an impressive triumph – a success to which the reliability and superior performance of Bosch magneto ignition made a major contribution. By the time the 1904 Gorden Bennett rolled around, the five fastest cars were all equipped with Bosch ignition.

Camille Jenatzy as Bosch Mephisto on a Bosch advertising poster from 1911

In June of 1902 a "light touring car" from Renault was the first to reach Vienna's Trabrennplatz at the culmination of the Paris to Vienna long-distance race. At the wheel was Marcel Renault, whose brother had already attracted considerable attention while at the same time laying the foundation for a major automotive marque with his "voiturette" in 1898. Renault's winning car was equipped with the new Bosch high-tension magneto ignition, an innovation still not available on standard vehicles at the time.

In 1906, victory at the French Grand Prix also went to a vehicle equipped with the Bosch high-tension magneto system. This system soon found favored status as the system of choice among automotive manufacturers, resulting in a massive sales increase.

Magneto ignition in aircraft

It was in May, 1927, that postal aviator Charles Lindbergh embarked upon his historic flight across the Atlantic. His single-engine "Spirit of St. Louis" made the non-stop trip from New York to Paris in 33.5 hours. Trouble-free ignition during the journey was furnished by a magneto manufactured by Scintilla in Solothurn, Switzerland, now a member of the Bosch group.

In April, 1928, aviation pioneers Hermann Köhl, Günther Freiherr von Hünefeld and James Fitzmaurice achieved the first non-stop airborne traversal of the Atlantic from East to West in a Junkers W33 featuring a fuselage of corrugated sheet metal. They took off from Ireland and landed 36 hours later in Greenly Island, Canada. They were unable to reach their original objective, New York, owing to violent weather. But: "the flight was successful with Bosch spark plugs and a Bosch magneto" (see illustration).

Battery ignition

Magneto ignition was still predominant in the automotive industry when the Robert Bosch AG presented its battery ignition in 1925. In those days magnetos were thought to provide the most reliable ignition. But vehicle manufacturers were demanding a less expensive system. After becoming established in series production in the USA, battery ignition started to take hold on both passenger cars and motorcycles in Europe.

First series production in the USA

By 1908 the American Charles F. Kettering had improved the battery ignition to the point where it was ready for series production at Cadillac in 1910. Despite problems, it became increasingly popular during the First World War. The public demand for mass motorization in affordable vehicles made cost reductions imperative, even when this meant becoming dependent on a vehicle battery. A generator was installed to charge the battery during vehicle operation.

Bosch introduces the battery ignition to Europe

In the initial years following World War I automobiles were restricted to a small segment of the population. But the slowly emerging mass market soon fostered demand for less expensive products, just as it had earlier in the USA. In the 1920s conditions were ripe for widespread application of ignition systems using power from a battery. Bosch had long been in possession of the expertise required to design such a system for series production. Before 1914 Bosch was already supplying ignition coils – the core of battery ignition – to the US market. In 1925 Bosch became one of the first manufacturers to respond to the new situation by offering a battery-ignition system – consisting of the TA coil and the VA distributor – for sale in Europe. Initially use was restricted to the Brennabor 4/25. But by 1931 46 of the 55 automotive models available in Germany were equipped with the system.

Design and operation of the battery ignition system

The new battery ignition systems generally consisted of two separate devices: the engine-driven ignition distributor and the stationary ignition coil (Fig. 9). The coil (7) – already familiar from Eisemann's high-tension magneto ignition – included an iron core along with the primary and secondary windings. The distributor (8) comprised the stationary breaker-point assembly (5) and the rotating actuator cam (4) along with a mechanism to distribute the secondary current. The capacitor (3) protected the points against premature wear by suppressing arcing.

The only moving parts in the system were the breaker cam and the distributor shaft. The system also contrasted with magneto-based assemblies by requiring only negligible levels of force to sustain operation.

Another difference relative to the magneto was that the battery ignition obtained its primary operating current from the vehicle's electrical system. The concepts for generating high voltage were broadly similar: the current flow creates a magnetic flux field in the primary winding which is then interrupted when the mechanical contacts open. The collapse of the flux field induces high voltage in the secondary winding.

9 Design of breaker-point ignition system

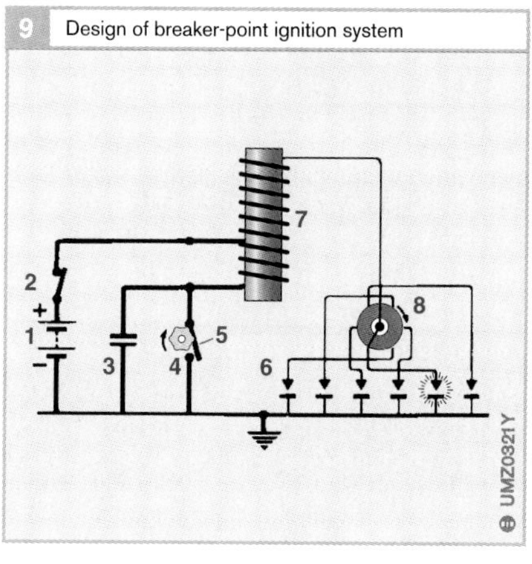

Fig. 9
1 Battery
2 Ignition switch
3 Ignition capacitor
4 Breaker cam
5 Contact-breaker points
6 Spark plugs
7 Ignition coil
8 Ignition distributor

Ignition performance demands for "modern times"

The performance demands placed on ignition systems for internal-combustion engines entered a period of unusually pronounced expansion. Engines started using leaner mixtures and higher compression ratios while operating speeds also increased. Other new entries in the demand catalog for ignition systems included:

- Noise reductions
- Flawless idling
- Extended maintenance intervals
- Durability in the face of harsh operating conditions
- Low weight
- Modest dimensions and
- Low cost

Higher compression ratios combined with more leaner carburetor mixtures meant that higher ignition voltages were needed to ensure reliable spark generation. Meanwhile, wider spark-plug electrode gaps were required for smooth idling, and this also led to additional demands for higher ignition voltages, which had to rise to more than twice their earlier level. This, in turn, had implications for the conductive elements in the high-tension circuit, which had to be designed to resist arcing.

Also required was a way to adjust ignition timing to accomodate the expanded engine-speed range. Ignition timing had to adjusted through a larger range to compensate for the increased lag between firing point and flame-front formation encountered at high rpm. In systems developed for multicylinder engines, the breaker points for opening the primary circuit, and the mechanism for distributing the high-tension current from the coil, were integrated in a single distributor unit, where they shared a common drive shaft. Ignition timing was regulated by shifting the position of the breaker-point lever relative to the shaft's cam. Originally, the driver was responsible for this adjustment which demanded experience and a certain level of dexterity.

Later, automatic centrifugal advance mechanisms, which had appeared as early as 1910 in magneto ignitions, were adopted for battery-ignition systems.

Since fuel economy progressively became a more important consideration, it became necessary for the timing adjustment to take into account the effects of engine loading on the combustion process. The answer was to install a diaphragm which adjusted the distributor by responding to the vacuum in the intake manifold upstream of the throttle plate. This supplemented the centrifugal adjustment with a second correction factor. This vacuum advance mechanism was introduced by Bosch early on, in 1936.

In developing the contact-breaker points Bosch was able to draw on experience already garnered while working with magnetos. All of the components in the battery ignition underwent improvement over the course of time. Eventually technological advances – especially in the realm of semiconductors – paved the way to new ignition systems. While the basic concept mirrored that of the original battery-ignition system, the designs were radically different.

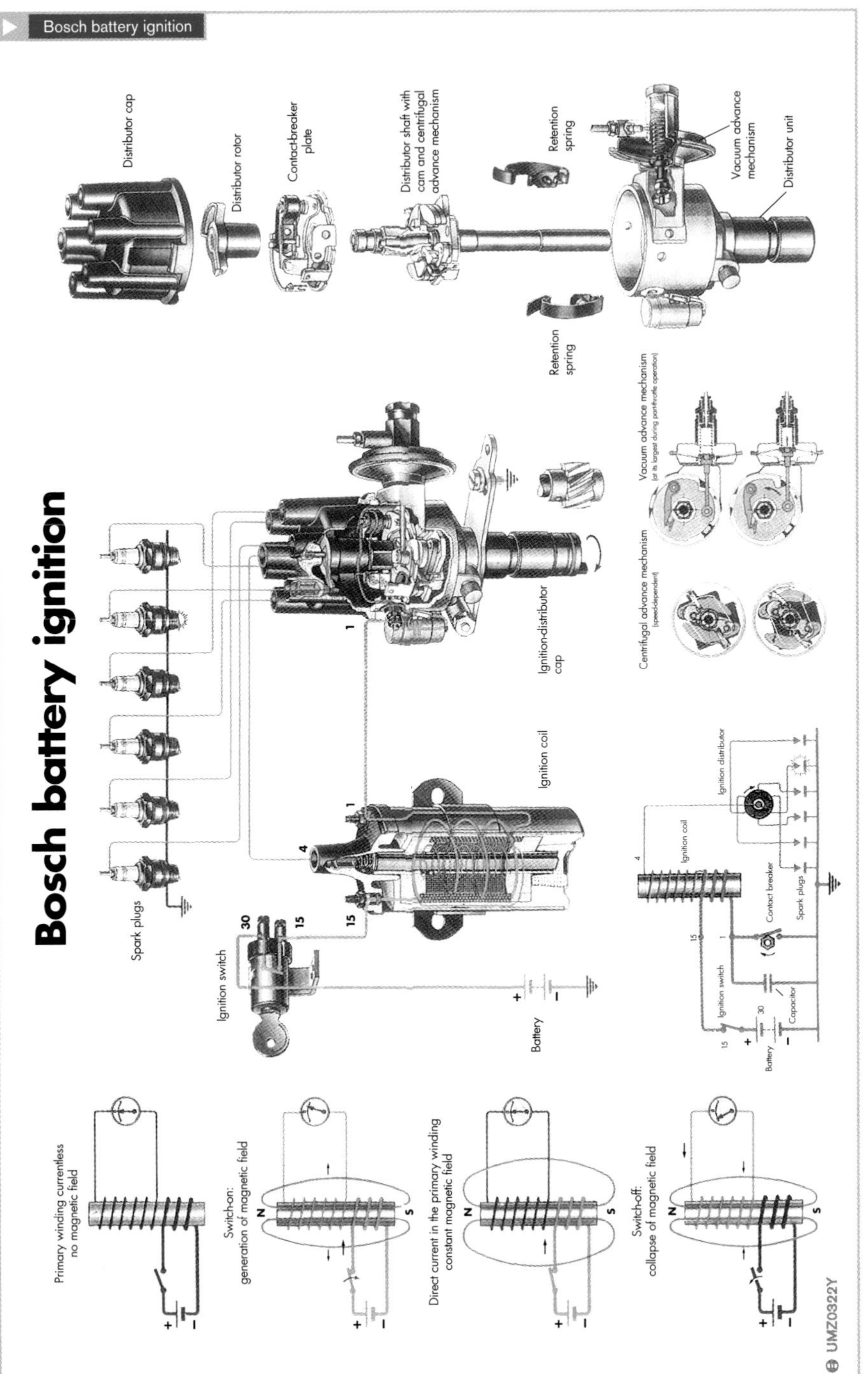

A training chart from 1969 showing Bosch battery ignition

Battery ignition systems over the years

The period between the appearance of Bosch battery ignition in 1925 and the final versions of this system many years later was marked by constant change and continuous evolution.

There were no substantive changes in the basic concept behind battery ignition in this time. Most of the modifications focused on the mechanisms employed to adjust ignition timing. These were reflected in the changes to system components. Ultimately the only components remaining from the original battery ignition were the coil and the spark plug. Finally, at the end of the 1990s, control of ignition functions was incorporated in the Motronic engine-management system. Thus ignition systems with separate ignition control units – as described in the following section – are now history.

Conventional coil ignition (CI)

Conventional coil-ignition systems are controlled by contact-breaker points. The contact breakers in the distributor open and close the circuit to control current flow within the ignition coil.

Design and operation

The components in the conventional coil-ignition system (Fig. 1) are the

- Ignition coil (3)
- Ignition distributor (4) with breaker points (6), ignition capacitor (5), centrifugal and vacuum advance mechanisms (7) and the
- Spark plugs (9)

During operation battery voltage flows through the ignition switch (2) on its way to the coil's Terminal 15. When the points close, current flows through the ignition coil's primary winding (asphalt coil, refer to section on ignition coils) and to ground. This flow produces a magnetic flux field in which ignition energy is stored. The rise in current flow is gradual owing to inductance and primary resistance in the primary winding. The time available for charging is determined by the dwell angle. The dwell angle, in turn, is defined by the contours of the distributor-cam lobes, which open and close the breaker points by pushing against the cam follower (Fig. 2b). At the end of the dwell period the cam lobe opens the contacts to interrupt current flow in the coil. The number of lobes on the cam corresponds to the number of cylinders in the engine.

Points must be replaced at regular intervals owing to wear on the cam follower as well as burning and pitting on the contact surfaces.

1 Conventional coil ignition system

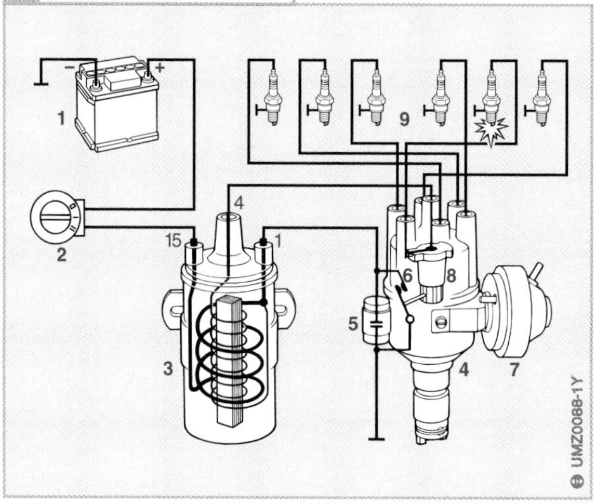

Fig. 1
1 Battery
2 Ignition/starter switch
3 Ignition coil
4 Distributor
5 Condenser
6 Contact-breaker points
7 Aneroid capsule
8 Rotor
9 Spark plug
1, 4, 15 Terminals

Current, dwell time and the number of secondary windings in the coil are the primary determinants of the ignition voltage induced in the coil's secondary circuit.

A capacitor in parallel with the points prevents arcing between the contact surfaces, which would allow current to continue flowing after they open.

The high-tension voltage induced in the ignition coil's secondary winding is conducted to the distributor's centre contact. As the rotor (Fig. 1, Pos. 8) turns it establishes an electrical path between this center contact and one of the peripheral electrodes. The current flows through each electrode in sequence, conducting high voltage to the cylinder that is currently approaching the end of its compression stroke to generate an arc at the spark plug. The distributor must remain synchronized with the crankshaft for its operation to remain in rhythm with the pistons in the individual cylinders. Synchronization is assured by a positive mechanical link between the distributor and either the camshaft or another shaft coupled to the crankshaft at a 2:1 step-down ratio.

Ignition advance adjustment

Because of the positive mechanical coupling between distributor shaft and crankshaft, it is possible to adjust the ignition timing to the specified angle by rotating the distributor housing.

Centrifugal advance adjustment

The centrifugal advance mechanism varies ignition timing in response to shifts in engine speed. Flyweights (4) are mounted in a support plate (Fig. 2a, Pos 1) that rotates with the distributor shaft. These flyweights spin outward as engine and shaft speed increase. They shift the base plate (5) along the contact path (3) to turn it opposite the distributor shaft's (6) direction of rotation. This shifts the relative positions of the point assembly and distributor cam by the adjustment angle α. Ignition timing is advanced by this increment.

Vacuum advance adjustment

The vacuum-advance mechanism adjusts ignition timing in response to variations in the engine's load factor. The index of load factor is manifold vacuum, which is relayed via hose to the two aneroid capsules (Fig, 2b).

Falling load factors are accompanied by higher vacuum levels in the advance unit which pull the diaphragm (11) and its advance/retard arm (16) to the right. In doing so, the arm turns the breaker-point assembly's base plate (8) in the opposite direction to that of the distributor shaft's rotation and thus increases the ignition advance.

Vacuum in the retard unit, for which the manifold vacuum connection is behind the throttle plate instead of in front of it, moves the annular diaphragm (15) and its advance/retard arm to the left to retard the timing. This spark retardation system is used to improve engine emissions under certain operating conditions (idle, trailing throttle, etc.). The vacuum advance is the priority system.

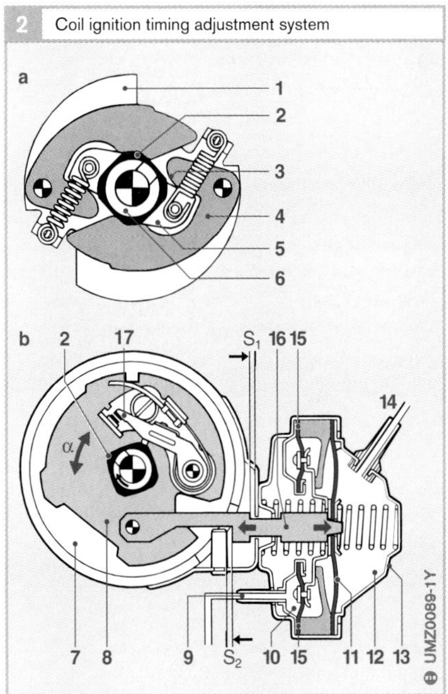

Fig. 2
a Centrifugal advance mechanism (illustrated in passive state)
b Vacuum advance and retard mechanism

1 Support plate
2 Distributor cam
3 Contact path
4 Flyweight
5 Base plate
6 Distributor shaft
7 Distributor
8 Breaker-point base plate
9 Manifold connection for retard unit
10 Retard unit
11 Diaphragm (ignition advance system)
12 Advance unit
13 Aneroid unit
14 Manifold connection for advance unit
15 Annular diaphragm (retard system)
16 Advance/retard arm
17 Contact-breaker points

s_1 Total timing advance
s_2 Total timing retardation
α Timing adjustment angle

Breaker-triggered transistorized ignition

Design and operating concept

The distributors used in transistorized breaker-triggered ignition systems are identical to those employed with coil ignition. The difference is in the control of the primary ignition circuit. Instead of being opened and closed by contact-breaker points, the circuit is now controlled by a transistor – installed along with supplementary electronics in the ignition trigger box. In this system only the control current for the transistorized ignition system is switched by the breaker points. Thus ultimate control of the system still resides with the points. Figure 3 compares the two designs.

When the breaker points (7) are closed, control current flows to the base B, making the path between the emitter E and the collector C on the transistor conductive. This charges the coil. When the breaker points open, no current flows to the base, and the transistor blocks the flow of primary current.

The ballast resistors (3) limit the primary current to the low-resistance, fast-charging coil used in this ignition system. During starting, compensation for the reduced battery voltage is furnished by bypassing one of these resistors at the starter's Terminal 50.

Advantages

Two major assets distinguish breaker-triggered transistorized ignition from conventional coil systems. Because there is only minimal current flow through the points, their service life is increased dramatically. Yet another advantage is the fact that the transistor can control higher primary currents than mechanical contact breakers. This higher primary current increases the amount of energy stored in the coil, leading to improvements in all high-voltage data, including voltage levels, spark duration and spark current.

Transistorized ignition with Hall-effect trigger

Design

In this transistorized ignition system the contact breakers that were still present in the breaker-triggered system are replaced by a Hall-effect sensor integrated within the distributor assembly. As the distributor shaft turns, the rotor's shutters (Fig. 4a, Pos 1) rotate through the gap (4) in the magnetic triggering unit. There is no direct mechanical contact. The two soft-magnetic conductive elements with the permanent magnets (2) generate a flux field. When the gap is vacated, the flux field penetrates the Hall IC (3). When the shutters enter the gap, most of the magnetic flux is dissipated around them instead of impacting on the IC. This process produces a digital voltage signal (Fig. 4b).

3 Comparison of conventional coil ignition and breaker-triggered transistorized ignition

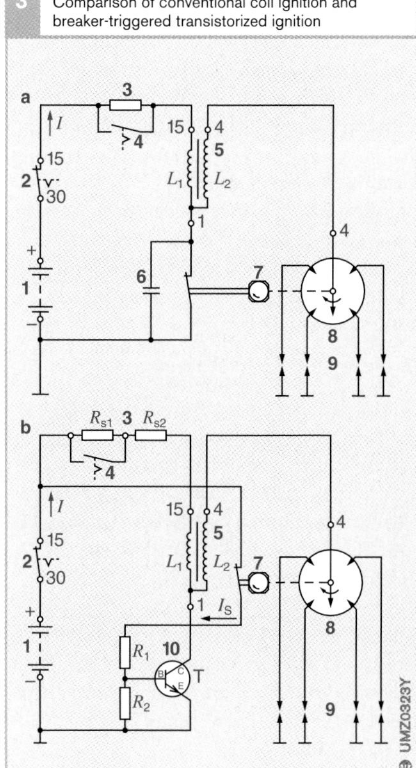

Fig. 3
a Circuit diagram for conventional coil ignition
b Circuit diagram for breaker-triggered transistorized ignition

1 Battery
2 Ignition/starter switch
3 Series resistor
4 Resistor bypass switch for starting
5 Coil with primary winding L_1 and secondary winding L_2
6 Ignition capacitor
7 Contact breakers
8 Distributor
9 Spark plugs
10 Electronic circuits with resistors for voltage distributor R_1, R_2 and transistor T
1, 4, 15, 30 Terminals

Since the number of shutters corresponds to the number of cylinders, this voltage signal thus corresponds to the signal from the contact breaker in the breaker-triggered transistorized ignition system. One system relies on the distributor shaft's cam lobe to define the dwell angle, while the other uses the pulse factor of the voltage signal produced by the shutters. Depending on the particular ignition trigger box, the width *b* of the individual shutters can determine the maximum dwell angle. This angle thus remains constant throughout the Hall sensor's entire life, at least on systems without separate dwell-angle control. Dwell adjustments of the kind required with contact-breaker points thus become redundant.

4 Hall-effect trigger in the ignition distributor

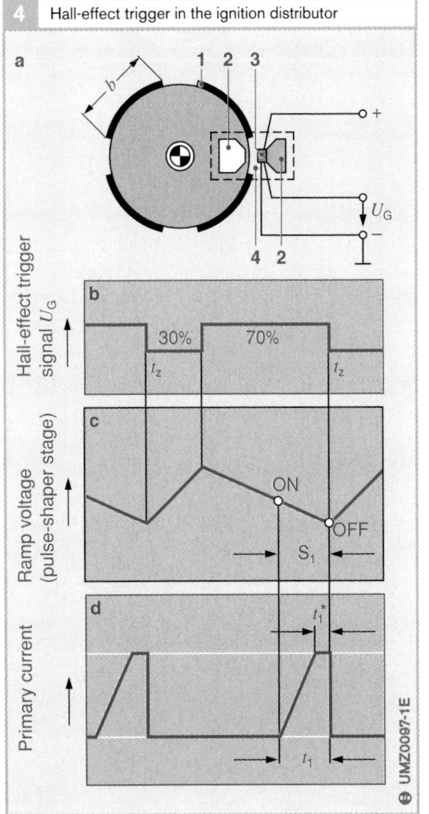

Fig. 4
a Schematic illustration of rotor design
b Hall sensor voltage output
c Ramp voltage for dwell control
d Primary current in coil

1 Shutter with width *b*
2 Soft-magnetic conductive element with permanent magnet
3 Hall IC
4 Gap

t_1 Dwell period
t_1^* Current reduction period
t_Z Firing point

Current and dwell-angle control

The application of rapid-charging, low-resistance coils made it necessary to limit primary current and power losses. The corresponding functions are integrated within the ignition system's trigger box.

Current control

The primary current is regulated to restrict flow within the coil and limit energy build-up to a defined level. Because the transistor enters its active range in its current-control phase, the voltage loss through the transistor is greater than in the switching mode. The result is high power loss in the circuit.

Dwell-angle control

An arrangement to regulate dwell to a suitable duration period is needed to minimize this power loss. Because it is possible to execute control operations by shifting the voltage threshhold using analog technology, the Hall-effect trigger's square-wave signal is converted to ramp voltage by charging and discharging a capacitor (Fig. 4c).

The firing point defined by the distributor's adjustment angle lies at the end of the shutter width, correlating with 70 %. The dwell-angle control is set to provide a current control period t_1^* that gives exactly the phase lead required for dynamic operation. The t_1 parameter is used to generate a voltage for comparison with the ramp's falling ramp. The primary current is activated to initiate the dwell period at the "ON" intersection. This voltage can be varied to shift the intersection on the ramp voltage curve to adjust the dwell period's start for any operating conditions.

Transistorized ignition with induction-type pulse generator

Only minor differences distinguish transistorized ignition with a distributor containing an inductive trigger from the system with a Hall-effect sensor (Fig. 5a). The permanent magnet (1), induction winding and core (2) on the inductive pulse generator form a fixed unit, the stator. A reluctor or "rotor" located opposite this stationary arrangement rotates to trigger the pulses. The rotor and core are manufactured in soft-magnetic material and feature spiked ends (stator and rotor spikes).

The operating concept exploits the continuous change in the gap between the rotor and stator spikes that accompanies rotation. This variation is reflected in the magnetic-flux field. The change in the flux field induces AC voltage in the induction winding (Fig. 5b). Peak voltage varies according to engine speed: approximately 0.5 V at low rpm, and roughly 100 V at high revs. The frequency f is the number of sparks per minute.

Control of current and dwell angle with inductively triggered ignition are basically the same as with Hall-effect transistorized ignition. In this case no generation of a ramp voltage is required, as the AC induction voltage can be used directly for dwell-angle control.

5 Inductive trigger in the ignition distributor

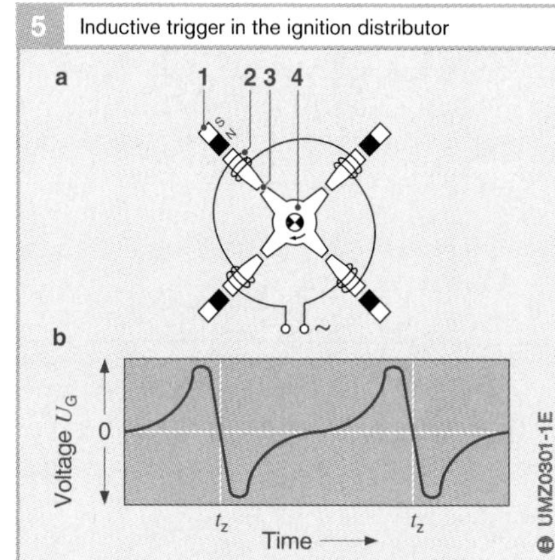

Fig. 5
a Design concept
b Inductive voltage curve

1 Permanent magnet
2 Inductive winding with core
3 Variable gap
4 Rotor

t_z Firing point

Electronic ignition

As demands for precise engine management grew, the very basic ignition timing curves offered by the centrifugal and vacuum mechanisms in conventional distributors proved unable to satisfy the requirements. In the early 1980s the introduction of automotive microelectronics opened up new options for ignition-system design.

Design and operation

Electronic ignition requires neither centrifugal nor vacuum-based timing adjustment. Instead, sensors monitor engine speed and load factor and then convert these into electrical signal data for processing in the ignition control unit. The microcontroller is essential for achieving the functionality associated with electronic ignition.

Engine speed is registered by an inductive pulse sensor than scans the teeth of a reluctor mounted on the crankshaft. An alternative is to monitor rpm using a Hall-effect sensor in the ignition distributor.

A hose connects the atmosphere within the intake manifold to a pressure sensor in the control unit. If the engine is equipped with electronic injection then the load signal employed to govern the mixture-formation process can also be tapped for ignition purposes.

The control unit uses these data to generate the control signal for the ignition's coil driver. The corresponding circuitry can be integrated within the control unit or mounted externally on the ignition coil, etc.

The most pronounced asset of electronic ignition is its ability to use a program map for ignition timing. The program map contains the ideal ignition timing for range of engine operating coordinates as defined by engine rpm and load factor; the timing is defined to provide the best compromise for each performance criterion during the engine's design process (Fig. 6a). Ignition timing for any given operating coordinates is selected based on

- Torque
- Fuel economy
- Exhaust-gas composition
- Margin to knock limit
- Driveability, etc.

Designs assign priority to specific individual parameters based on the optimization criteria. This is why 3D representations of program maps for systems with electronic control show a craggy and variegated landscape, as opposed to the smooth slopes of mechanical timing-adjustment systems (Fig. 6b).

A map based on engine speed and battery voltage is available for dwell angle. This ensures that the energy stored in the ignition coil can be regulated just as precisely as with separate dwell control.

A number of other parameters can also have an effect upon the ignition angle, and if these are to be taken into account this entails the use of additional sensors to monitor

- Engine temperature
- Intake-air temperature (optional)
- Throttle-plate aperture (at idle and at WOT)

It is also possible to monitor battery voltage – important as a correction factor for dwell angle – without a sensor. An analog-digital converter transforms the analog signals into digital information suitable for processing in the microcontroller.

Advantages

The step from mechanically-adjusted ignition timing to systems featuring electronic control brought decisive assets:

- Improved adaptation of ignition timing
- Improved starting, more stable idle and reduced fuel consumption
- Extended monitoring of operational data (such as engine temperature)
- Allows integration of knock control

Distributorless (fully-electronic) ignition

Fully-electronic ignition includes the functionality of basic electronic systems. As a major difference, the distributor used for the earlier rotating high-voltage distribution has now been deleted in favour of stationary voltage distribution governed by the control unit. The fully-electronic ignition system generates a separate, dedicated control signal for the individual cylinders, each of which must be equipped with its own ignition coil. Dual-spark ignition uses one coil for two cylinders.

Assets

The advantages of distributorless ignition are

- Substantially reduced electromagnetic interference, as there are no exposed sparks
- No rotating parts
- Less noise
- Lower number of high-tension connections and
- Design benefits for the engine manufacturer

6 Ideal electronic ignition-advance map with map for a mechanical adjustment system

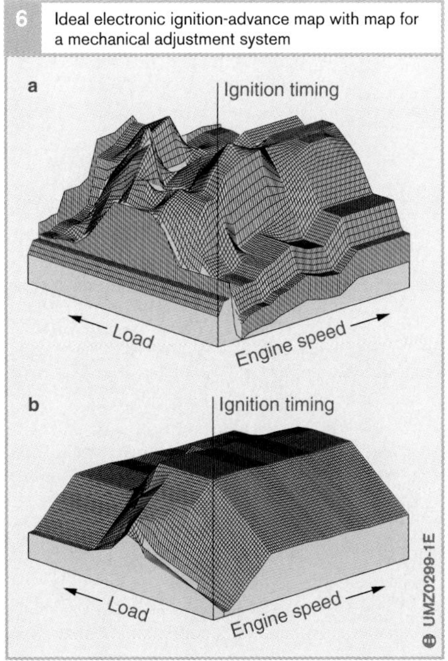

Fig. 6
a Ignition-advance map for electronic ignition
b Ignition-advance response with conventional coil ignition

Inductive ignition system

Ignition of the air-fuel (A/F) mixture in the gasoline engine is electric; it is produced by generating an arc bridging the gap between electrodes on a spark plug. The energy contained within the spark ignites the compressed mixture immediately adjacent to the spark plug, creating a flash front which then spreads to ignite the mixture in other areas of the combustion chamber. Inductive ignition systems generate the high-tension voltage required to produce flashover on each ignition stroke. The electrical energy is drawn from the battery for temporary storage in the ignition coil.

Design

Figure 1 illustrates inductive ignition's basic design concept in a distributorless system – as used in all current applications – with single-spark coils. The components in the ignition circuit are

- The ignition driver stage (5), which can be located in the Motronic control unit as well as in the ignition coil
- Compact ignition coil (3) designed to generate one (as illustrated) or two sparks, or as a rod-type ignition coil
- Spark plug (4) and
- Connection elements and interference-suppression components

Older ignition systems still feature the high-tension voltage distributor to direct the coil's voltage to the individual spark plugs.

1 Layout of an inductive ignition circuit with stationary voltage distribution and single-spark coils

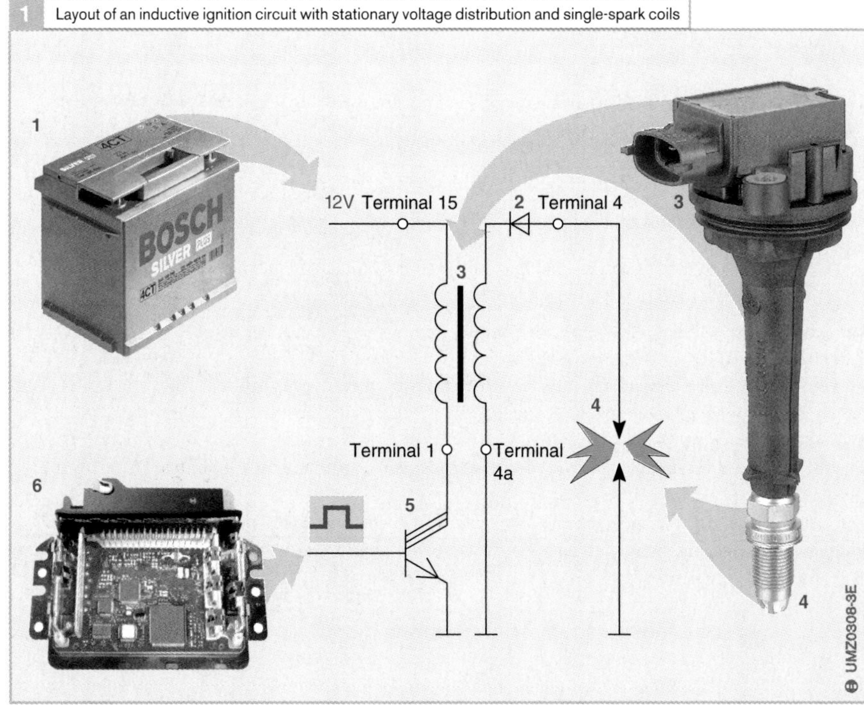

Fig. 1
1 Battery
2 EFU diode (switch-on-arc suppression), integrated in the coil
3 Ignition coil
4 Spark plug
5 Ignition driver stage (integrated in engine-management ECU or coil)
6 Engine-management ECU Motronic

Terminal 15, Terminal 1, Terminal 4, Terminal 4a Terminals
⎍ Trigger signal for ignition driver stage

Operating concept

The instant at which the spark ignites the air-fuel mixture within the combustion chamber must be selected with extreme precision. Ignition timing matched to the engine's current operating status is essential for

- High torque and
- Low emissions

The main parameters that determine ignition timing are the engine's speed and its load factor (torque). Additional parameters such as engine temperature are also used to define the optimal firing point. These parameters are registered by sensors and then relayed to the engine-management ECU (Motronic), where they are converted to digital form for further processing. The ECU uses program maps and stored response curves to calculate ignition timing and dwell angle, which are then converted to actual firing points and dwell periods. The engine-management ECU employs these data to generate the output signals for the ignition driver stage.

Since high currents flow through the coil's primary circuit, an ignition driver stage is needed. This is available in various versions for integration in either the engine-management ECU or the coil.

A magnetic field is generated as current flows through the coil's primary winding. At the firing point the current flow is interrupted and the magnetic field collapses. This rapid change in the field induces high voltage in the ignition coil's secondary circuit. When this reaches ignition voltage a spark is generated at the spark plug to ignite the compressed air-fuel mixture.

The designs and specifications of all ignition components are selected to harmonize with the demands of the overall system.

Applications

Inductive ignition systems are found in all applications where vehicles or machinery are powered by gasoline engines:

- In passenger cars
- In some specialized commercial-vehicle applications (in Europe, commercial vehicles are usually diesel powered)
- On single-track vehicles (motorcycles)
- In some ships and small craft

The most significant application for these ignition systems is in gasoline-engine passenger cars. While diesel engines are self-igniting, an externally generated ignition pulse is required to ignite the air-fuel mixtures in gasoline-burning powerplants.

Inductive ignition systems for passenger cars are available for powerplants with between 3 and 12 cylinders. The most common are 4-stroke, 4-cylinder powerplants.

In the US, large commercial vehicles are powered primarily by gasoline engines. Special inductive ignition systems are designed for this more demanding environment.

An essential requirement for inductive ignition is an onboard electrical system. Oscillating magneto ignitions are used on small engines without their own electrical system. These ignition systems are found in

- Garden machinery (lawn mowers, etc.)
- Chain saws and
- Mopeds

Ignition driver stage

Function

The ignition driver stage controls the flow of primary current in the coil.

Design and operation

Such driver stages usually consist of three-stage power transistors. The functions of "primary-current limitation" and "primary-voltage limitation" are incorporated in the driver stage in monolithic form and protect ignition components from overload.

The driver stage and the coil itself heat up in operation. Design features are required to conduct the thermal energy into the surrounding environment to prevent overheating. The primary-current limitation feature's only remaining function is its "safety net" mode, in which current is limited in response to defects (short circuit, etc.).

Driver stages can be internal or external. Internal driver stages are integrated on the engine-management ECU's circuit board. External driver stages are located in separate housings outside the ECU. Owing to cost considerations, external driver stages are not a part of new development projects.

Yet another, increasingly popular option is to integrate the driver stages in the ignition coil.

Ignition coil

Function

The ignition coil stores the requisite ignition energy and generates the high-tension voltage required to produce a spark at the firing point.

Design

State-of-the-art ignition coils consist of two magnetically coupled copper windings (primary and secondary), an iron core with individual laminations, and a plastic casing. Some designs use an enclosed core (compact coils) or a rod core (pencil coils). The arrangement and exact positions of the primary and secondary windings vary according to the specific design. A disc or chamber winding may be used on the secondary side.

The casing is filled with epoxy resin to insulate the windings from each other and from the core. Various coil designs are employed for individual applications.

2 Schematic diagrams of ignition coils

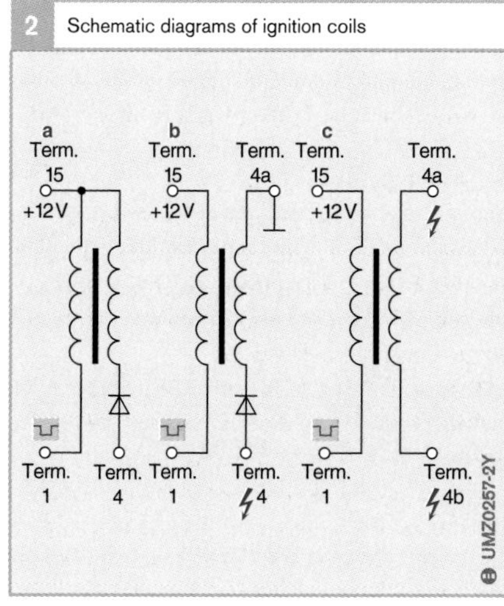

Fig. 2
a Single-spark coil in economy circuit (EFU diode not required on ignition systems with rotating high-tension voltage distribution)
b Single-spark ignition coil
c Dual-spark ignition coil

Operating concept

Operation of the ignition coil relies on induction. Magnetic induction is the means by which the energy stored in the primary winding is transferred to the coil's secondary winding. The amplification of current and voltage that occurs during the transfer from primary to secondary side is determined by the relative numbers of coil windings (turns ratio) of the two.

In the basic economy circuit (Fig. 2a) one terminal from the primary is connected to one terminal on the secondary winding, and these are both linked to Terminal 15 (ignition switch). The terminal at the other end of the primary winding is connected to the ignition driver stage (Terminal 1). The secondary winding's second connection (Terminal 4) is joined to the distributor (with rotating distribution) or directly to the spark plug (distributorless ignition). Deletion of an extra terminal makes the coil in this basic circuit less expensive. But because there is no mutual electrical insulation between the two electrical circuits, electrical interference from the coil can be propagated into the vehicle's electrical system.

The primary and secondary windings are not connected in Figures 2b and 2c. On the single-spark coil one side of the secondary winding is connected to ground (Terminal 4a), while the other side (Terminal 4) leads directly to the spark plug. Each of the connections on the dual-spark ignition coil (Terminals 4a and 4b) leads to a spark plug.

High-voltage generation

The engine-management ECU activates the ignition driver stage for the calculated dwell period. During this period the coil's primary current climbs to the specified level to generate a magnetic flux field.

The level of the primary current and the coil's primary inductance determine the amount of energy stored in the coil.

The driver stage interrupts the current flow at the firing point. The resulting shift in the magnetic flux field induces secondary voltage in the coil's secondary winding.

The factors that define the maximum available secondary voltage are the energy stored in the coil, the winding's capacity and the coil's turns ratio, the secondary draw (spark plug) and the limit on primary current imposed by the driver stage.

It is imperative that the secondary voltage exceeds the voltage level needed to create an arc between the spark plug's electrodes (ignition voltage). The energy contained in the spark must ensure that the mixture ignites, even if secondary arcing occurs. These secondary arcs occur when turbulence in the mixture deflects and interrupts the spark.

When the primary current is activated an undesired voltage of roughly 1...2 kV is induced in the secondary winding; its polarity is opposed to that of the required high-tension ignition voltage. Here it is necessary to avoid arcing at the spark plug (activation arc).

In systems with rotating high-voltage distribution the activation arc is effectively suppressed in the distributor's spark gaps. In stationary-distribution systems with single-spark coils the activation arc is suppressed by a diode (EFU diode, Figures 2a and 2b). This EFU diode can be installed on the "hot" side (toward the spark plug) or the "cold" side (away from the spark plug). In systems with dual-spark coils the activation arc is suppressed by the high flashover voltage in the series circuit feeding the two spark plugs.

Deactivating the primary current produces a self-induction voltage of several hundred volts in the primary winding. To protect the driver stage this is limited to 200 ... 400 V.

Voltage distribution

Function

The high-voltage generated in the coil must be present at the correct spark plug when the firing point arrives. This function is controlled by the high-voltage distribution system.

Rotating high-voltage distribution

In this system the high-voltage energy is generated by a single coil (Figure 3a, Pos. 2). The distributor (3) then relays the voltage to the individual spark plugs (5).

This type of voltage distribution is no longer significant for modern engine-management systems.

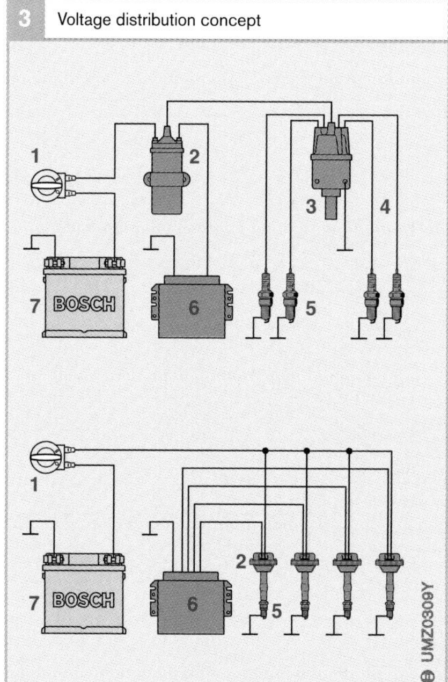

Fig. 3
- a Rotating distribution
- b Distributorless ignition with single-spark coils

1. Ignition switch
2. Coil
3. Distributor
4. Spark-plug cable
5. Spark plug
6. ECU
7. Battery

Distributorless (stationary) voltage distribution

The mechanical components have been abolished in the distributorless, or stationary electronic voltage-distribution system (Fig. 3b). The ignition coils are connected directly to the spark plugs, and voltage distribution is on their primary sides. Component wear and power losses are avoided with this distribution concept. Two versions are available.

System with single-spark coils

One driver stage and one coil are assigned to each cylinder. The engine-management ECU triggers the driver stages in the specified firing order.

As there are no distributor losses with this system, the coils can be very small. The preferred location is directly on the spark plugs.

Stationary voltage distribution with single-spark coils can be used with any number of cylinders. There are no restrictions on the dwell-angle range. However, a camshaft sensor is needed to synchronize the system with the camshaft.

System with dual-spark coils

Each combination of driver stage and coil serves two cylinders. Each end of the secondary winding is attached to a spark plug at a different cylinder. In these cylinder pairs the compression stroke in one corresponds to the exhaust stroke in the other (with even number of cylinders). A spark is produced at both spark plugs during ignition. Because it is important to avoid sparks and ignition of residual exhaust gas or fresh induction gas during the exhaust stroke, the latitude for varying ignition timing is limited with this system. However, it does not need to be synchronized with the camshaft. Owing to their intrinsic limitations, dual-spark systems cannot be recommended.

Spark plug

Function

The spark plug produces an arc between its electrodes to ignite the air-fuel mixture in the combustion chamber.

Design and operation

The spark plug (Figure 4) is a gas-tight conductor in a ceramic insulator, used to convey high voltage into the combustion chamber. It includes one center electrode (1) and one or several ground electrodes (2).

The spark type is determined by the configuration of the ground electrode(s). A ground electrode located directly across from the center electrode characterizes the air-gap plug (a, b). Laterally located ground electrodes are found on the semi-surface (c) and surface-gap plugs (d).

When the primary current is interrupted at the firing point, the energy in the coil's secondary winding quickly rises to ignition voltage (roughly 30 µs, s. Fig 5). When the ignition voltage is exceeded the spark path between the plug's center and ground electrodes becomes conductive. Having charged to ignition voltage, the capacitive elements in the secondary circuit (spark plug, high-tension cable and coil) now abruptly discharge to form a spark head. During the subsequent spark-duration period, typically 1...2 ms, the energy stored within the coil forms a glow discharge. Residual energy remaining in the coil is discharged in the postoscillation phase.

Spark-plug wear

During engine operation the spark plug's electrodes are subject to wear in the form of erosion from the spark current and corrosion in the hot gases in the combustion chamber. The electrode gap grows wider. This raises the required ignition voltage. The ignition system's secondary circuit must furnish adequate voltage reserves to ensure that the spark plug continues to fire until the end of its service life.

4 Spark plug (cutaway) and spark path

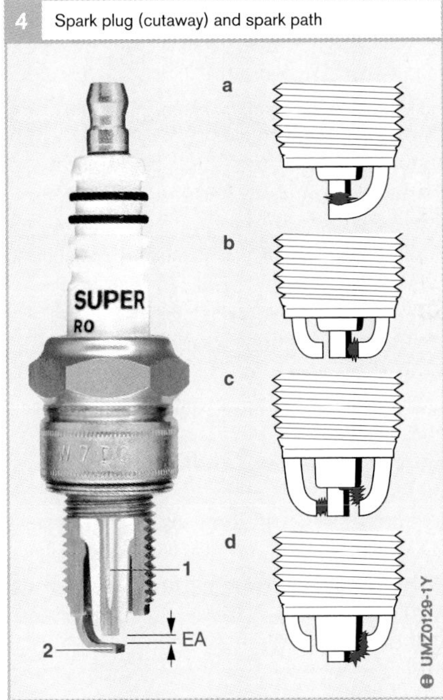

Fig. 4
1 Center electrode
2 Ground electrode

a Spark air gap with front electrode
b Spark air gap with side electrode
c Semi-surface spark gap (air gap or surface gap possible)
d Surface gap path

EA Electrode gap

5 Voltage rise and fall at the spark plug electrodes

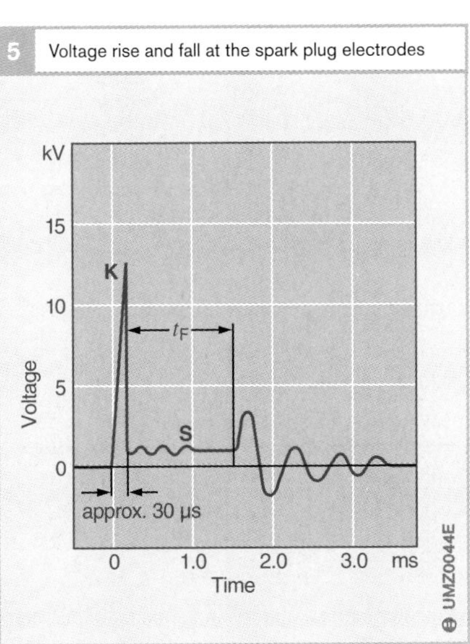

Fig. 5
K Spark head
S Spark tail

t_F Spark duration

Connecting devices and interference-suppressor equipment

Spark-plug cables

A means is required to convey the high voltage generated in the coil to the spark plugs. In systems where the coil is not mounted directly on the plug this function is discharged by plastic-insulated cables designed to withstand high voltages, with plugs on the ends for connection to the high-voltage components.

Because all high-voltage conductors tend to reduce secondary voltage by acting as capacitive resistors within the ignition system, these cables should be as short as possible.

Interference-suppression resistors, shielding

The pulsed discharge that accompanies every arcing event at the spark plug or in the distributor (with rotating distribution) is a source of radio interference (EMI/RFI). Interference-suppression resistors in the high-voltage circuit limit the maximum current during discharge. To effectively inhibit EMI these suppression resistors should be mounted as close as possible to the interference source.

Resistors for suppressing EMI are typically installed in the connector plugs and spark-plug cable sockets, as well as in the rotor with rotating spark distribution. There are also spark plugs with integral resistors. However, increasing resistance in the secondary circuit leads to energy loss within the ignition system, with lower ignition energy at the spark plug as the ultimate result.

Further reductions in EMI radiation can be achieved by partially or completely shielding the ignition system. This action includes the spark-plug cables. This effort is justified only in special cases (official government and military vehicles, special radio equipment).

Ignition voltage

The ignition voltage is the energy level at which flashover occurs between the electrodes on the spark plug. Among the factors that affect it are

- The density of the air-fuel mixture within the combustion chamber, and the ignition timing
- The composition of the air-fuel mixture (lambda excess-air factor)
- Flow velocity and turbulence
- Electrode geometry
- Electrode material and
- Electrode gap

It is vital that the ignition system ensures adequate voltage under all conditions.

Accident hazard

All ignition systems carry high electrical voltage. To avoid accidents, always disconnect the battery or switch off the ignition before working on any ignition system. These precautions apply to

- replacement of components such as ignition coils, spark plugs, ignition cables, etc.
- connecting engine testers such as timing light, dwell meter and tachometer, oscilloscope, etc.

When checking the ignition system remember that dangerously high levels of voltage are present within the system whenever the ignition is on. Always refer all tests and inspections to qualified professional personal.

Ignition energy

The deactivation current and the ignition coil's design parameters determine the amount of energy that the coil stores for application as ignition energy in the spark. The level of ignition energy is decisive for mixture ignition, and efficient ignition is vital for high performance and low emissions. The demands placed on the ignition system are considerable.

Ignition-system energy balance

The energy stored in the ignition coil is discharged when the spark is generated. This energy consists of two separate components.

Spark head

The energy E stored in the secondary circuit's capacitance C is abruptly discharged at the firing point. It rises as the square of the applied voltage U to ($E = 1/2\ CU^2$). This is reflected in the curve's quadratic progression in Figure 6.

Spark tail

The remainder of the energy stored in the ignition coil (inductive component) is then discharged. This energy represents the difference between the total energy stored in the ignition coil and the energy released during capacitive discharge.

This means: the proportion of the overall energy in the spark head increases as a function of the requirement for ignition energy.

Under high demand for ignition energy, as encountered when the spark plugs are worn, the energy in the spark tail may not be sufficient to generate a stable flame front in the mixture following initial ignition, or to produce secondary arcs capable of restarting the process should the initial spark be extinguished.

Further increases in the demand for ignition voltage lead to the miss limit. Here the available energy is no longer sufficient to generate an arc, and instead disappears in a series of low-intensity oscillations (ignition miss).

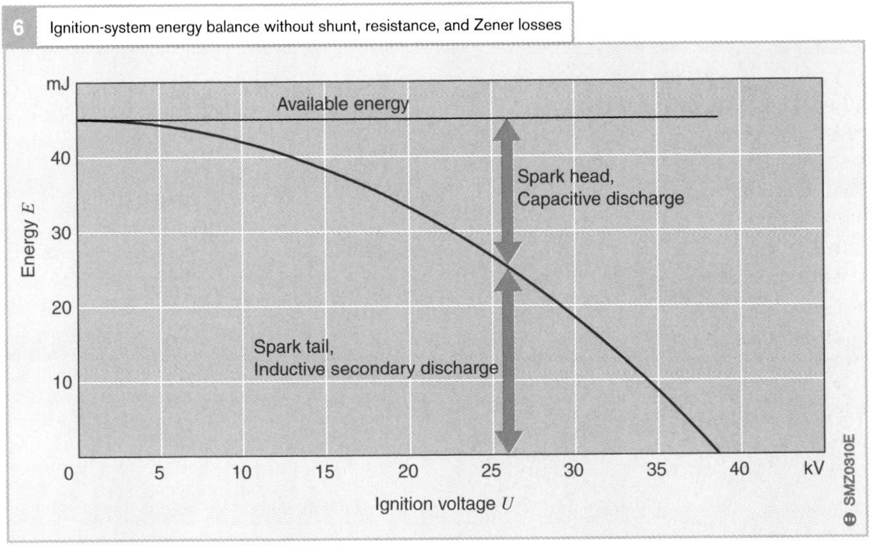

Fig. 6
The energy figures are for a sample ignition system with a coil capacitance of 35 pF, an external load of 25 pF (total capacitance C = 60 pF) and secondary inductance of 15 H.

Shunt losses

Figure 6 (previous page) is a simplified presentation of the situation. The electrical resistance in the coil and the ignition cables combines with the EMI suppression resistors to reduce the available ignition energy.

Additional losses are produced by shunt resistance. While this can result from contamination on high-tension cables, the primary source is soot and deposits on the spark plugs in the combustion chamber.

The level of shunt losses is also affected by the demand for ignition voltage. The flow of lost current through the elements acting as shunts increases as a function of the voltage at the spark plug.

7 Ignition spark on an engine with direct gasoline injection

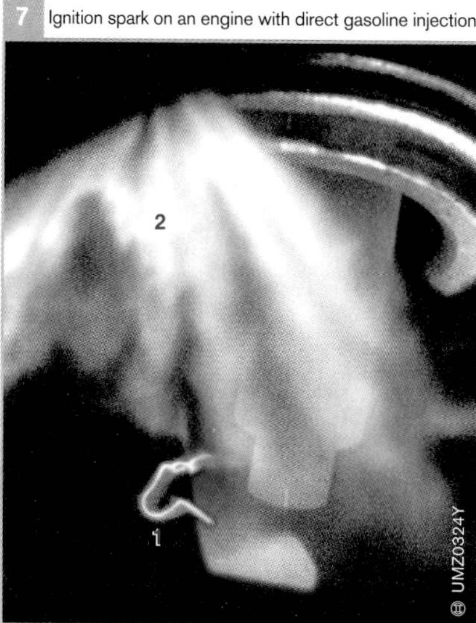

Fig. 7
Photograph of ignition spark: taken in a transparent engine using a high-speed camera

1 Ignition spark
2 Fuel spray

Mixture ignition

Under ideal conditions the energy required to ignite a stationary, homogeneous and stoichiometric air-fuel mixture with an electrical spark is roughly 0.2 mJ. Under these same conditions rich and lean mixtures require more than 3 mJ.

The energy required to ignite the mixture is only a fraction of the spark's total energy, the ignition energy. With conventional ignition systems and high breakdown voltages, energy levels in excess of 15 mJ are needed to generate a high-voltage arc at the firing point. Additional energy must be available to maintain a specific spark duration and to compensate for losses such as shunt resistance from contamination on spark plugs. These factors lead to a minimum demand for ignition energy of 30 ... 50 mJ. The corresponding energy to be stored in the coil is 60 ... 120 mJ.

Turbulence within the mixture of the kind encountered when engines with direct gasoline injection are operated in the stratified-charge mode can deflect the ignition spark to such an extent that it breaks down (Fig. 7). Secondary sparks are then needed to re-ignite the mixture; the energy for these sparks must also be furnished by the coil.

When mixtures are lean, particularly high levels of energy are needed to satisfy the increased demand for ignition voltage. At the same time, spark duration must be extended to compensate for the lower flammability of mixtures with a high proportion of air.

If only inadequate ignition energy is available the mixture will fail to ignite. No flame front is established, and combustion miss occurs.

This is why the system must furnish adequate reserves of ignition energy: to ensure reliable detonation of the air-fuel mixture, even under unfavorable conditions. It may be enough to ignite just a small portion of the mixture directly with the spark plug. The ignited mixture then initiates combustion in the remaining mixture in the combustion chamber.

Factors affecting ignition properties

Efficient formation of a mixture with unobstructed access to the spark plug fosters effective ignition, as do extended spark durations and the long arcs furnished by large electrode gaps. Turbulence in the combustion charge is also a positive factor, provided that sufficient energy is available to generate secondary sparks. Turbulence promotes more rapid expansion of the flame front for more rapid combustion of the mixture throughout the entire chamber.

Another significant factor is spark-plug contamination. Extreme contamination (deposits) causes the spark plug to act as a shunt resistor, bleeding energy from the coil while high-tension voltage is accumulating. This reduces the available high voltage while simultaneously limiting spark duration. It affects exhaust emissions, and can even lead to ignition miss under extreme conditions, as when the spark plugs are severely contaminated or wet.

Ignition miss produces combustion miss, which increases both fuel consumption and exhaust emissions, and can also lead to catalytic-converter damage.

Ignition timing

Approximately 2 milliseconds elapse between arc-over and complete combustion of the mixture. This period remains invariable as long as mixture ratio remains constant. As a result, the firing point must be shifted forward – relative to crankshaft position – by progressively larger increments as engine speed increases.

The air-fuel mixture does not ignite as well at low charge densities. This leads to extended ignition lag, making it necessary to advance ignition timing even further. Optimal torque generation is obtained by inducing peak combustion and peak pressure just after TDC while avoiding ignition knock (Fig. 8).

During operation in the stratified-charge mode (direct gasoline injection), the latitude for shifted ignition timing is limited by the end of the injection period and by the time required for mixture formation during the compression stroke.

8 Pressure curves in combustion chamber with different advance angles (ignition timing)

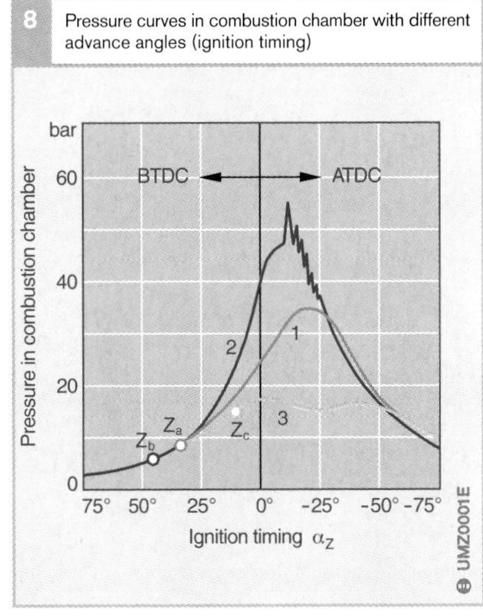

Fig. 8

1 Ignition Z_a at correct firing point

2 Ignition Z_b over-advanced (combustion knock)

3 Ignition Z_c over-retarded

Ignition coils

Within the inductive ignition system, the coil is the component responsible for converting the modest battery voltage to the high-tension energy required to generate an arc at the spark plug. It draws current from the vehicle's DC electrical system to furnish ignition pulses with the required voltage level and arc-over energy to the spark plug. In the course of the years, coil design has been continually improved to satisfy the growing demands imposed by the gasoline engine.

Function

The high-tension voltage and ignition energy required to ignite the air-fuel mixture must be generated and stored prior to spark formation. The coil acts as a dual-function device by serving as both transformer and energy accumulator. It stores the magnetic energy built up in the primary winding's magnetic flux field and then discharges this energy when the primary current is interrupted at the firing point.

The coil must be precisely matched to the other components in the ignition system (driver stage, spark plugs). Essential parameters are:

- The flashover energy W_{Fu} available to the spark plug
- The spark current I_{Fu} through the spark plug at the firing point
- The duration of the arc at the spark plug t_{Fu} and
- Ignition voltage U_Z adequate for all operating conditions

Important considerations in designing individual ignition systems include the interaction of various parameters in the driver stage, the coil and the spark plugs as well as the specific demands associated with the engine's design concept.

1 Main Bosch ignition coil types

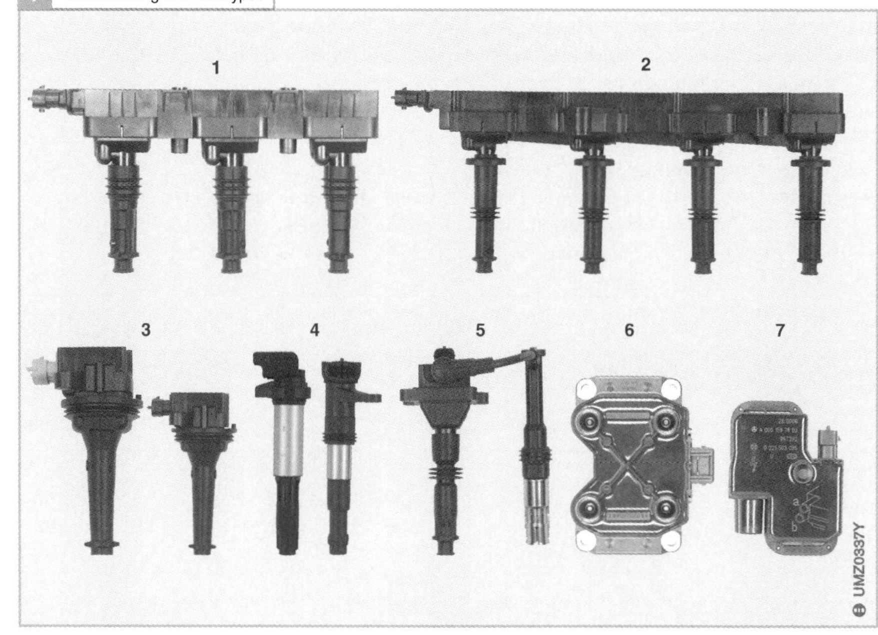

Fig. 1
1 Module with three single-spark coils
2 Module with four single-spark coils
3 Single-spark coil (compact coil)
4 Single-spark coil (pencil coil)
5 Dual-spark coil (one magnetic circuit)
6 Dual-spark coil with two magnetic circuits (four high-voltage domes)
7 Module with two single-spark coils

Examples:
- To ensure secure and reliable ignition of the mixture under all conditions, turbocharged engines need more spark energy than engines with manifold injection; powerplants with direct gasoline injection have the highest energy requirement of all.
- Spark current has a relatively limited effect on the life expectancy of modern spark plugs.
- Engines with turbochargers and superchargers need consistently higher ignition voltages than non-charged powerplants.
- The driver stage and coil must be mutually matched for correct control of primary current.
- Connections between the coil and spark plug must be designed for secure and reliable performance under all conditions (voltage, temperature, vibration, resistance to aggressive substances).

Applications
Ignition coils made their debut in Bosch ignition systems when battery ignition replaced magnetos in the 1930s. They were then subject to ongoing improvements while being adapted to new applications. Ignition coils are used in all vehicles and machines equipped with inductive ignition systems.

Requirements

Emissions regulations impose strict limits on pollution from internal-combustion engines. It is absolutely imperative to avoid the ignition miss and incomplete combustion that lead to drastic rises in HC emissions. It is thus vital to have coils that consistently provide adequate levels of ignition energy throughout their service lives. Steady increases in the levels of electrical performance provided by ignition coils have led to an ongoing series of improvements and advances.

In addition to these considerations, coils must also suit the geometry and design configuration of the powerplant. Earlier ignition systems with rotating high-tension distribution (ignition distributor, (asphalt) coil, cables) featured coils in a standardized versions for mounting on body or engine.

The ignition coil is subject to severe performance demands – electrical, chemical and mechanical – yet still expected to furnish flawless and maintenance-free operation for the life of the vehicle. For today's ignition coils, frequently installed directly in the cylinder head, this means operating under the following conditions:
- Operating temperature range of –40...+150 °C
- Secondary voltages up to 30,000 V
- Primary currents between 7...15 amperes
- Dynamic vibration up to 50 g
- Durable resistance to various substances (gasoline, oil, brake fluid, etc.)

Design concept and operation

Operating concept

Primary and secondary windings

Conceptually, the ignition coil (Fig. 1, Pos 3) is an electrical transformer. Two windings surround a shared iron core.

The primary winding consists of thick wire with a relatively low number of loops. One end of the winding is connected to the battery's positive terminal (1) via the ignition switch (Terminal 15). The other end (Terminal 1) is connected to the ignition driver stage (4) to control the flow of primary current.

Although contact-breaker points were still being used to control primary current as late as the end of the 1970s, this arrangement is now obsolete.

The secondary winding consists of thin wire looped in numerous coils. The turns ratio is usually between 1:50...1:150.

Generating the magnetic field

A magnetic-flux field is generated in the primary winding as soon as the driver stage closes the electrical circuit. Self-induction creates an inductive voltage in this winding, which opposes the cause – the current flow in the primary winding, which generates the magnetic field – as defined by Lenz. This rule explains why the rate at which the magnetic field is generated is always comparatively low (Fig. 2) depending on the sizes of the iron core and winding.

Up to a specific point the primary current will continue to rise while the circuit remains closed; beyond a certain current flow magnetic saturation occurs. The actual level is determined by the composition of the ferro-magnetic material. Inductance falls and current flow rises more sharply. Losses within the ignition coil also rise steeply. This is why operation below the magnetic saturation point is desirable. This factor is controlled by the dwell angle.

1 Concept for generating high voltage in the coil

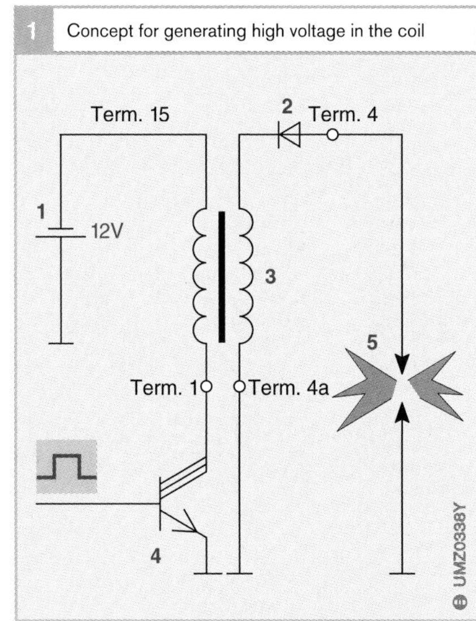

Fig. 1
1 Battery
2 EFU diode (integrated within ignition coil)
3 Coil with iron core, primary and secondary windings
4 Ignition driver stage (may be integrated in engine-management ECU or coil)
5 Spark plug

2 Primary current path in the coil

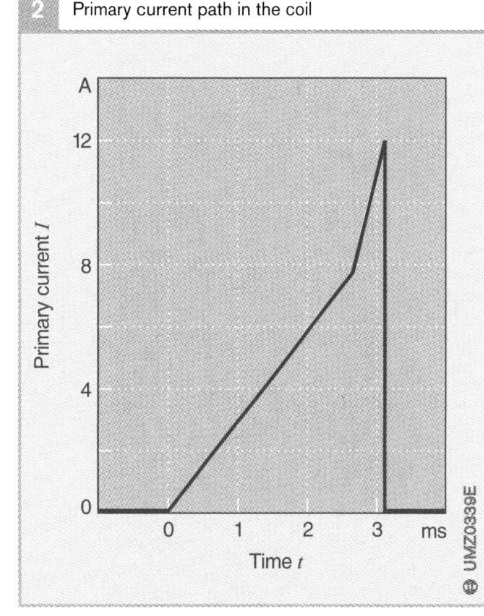

Magnetization curve and hysteresis

The ignition coil's core consists of a soft-magnetic material (permanent magnets are hard-magnetic material). This material displays a characteristic magnetization curve that defines the relationship between magnetic field strength *H* and flux density *B* within it (Fig. 3). Once maximum flux density is reached, the effect of additional increases in field strength on flux density will be minimal: saturation has occurred.

Yet another property of this material is hysteresis in the magnetization curve. While flux density remains constant, this material property makes different field strengths necessary for the magnetizing and demagnetizing processes. The intrinsic losses in the material are proportional to the level of hysteresis. Under these conditions the area encompassed by the hysteresis curve is large.

Magnetic circuit

The material most commonly used in ignition coils is electrical sheet steel, processed in various layer depths and to various specifications. Depending upon performance profile the material is either grain-oriented (high maximum flux density, expensive) or against the grain (low maximum flux density).

Sheet metal with layer depths of 0.3...0.5 mm is most commonly used. Mutually insulated plates are used to reduce eddy-current losses. The plates are stamped, combined in plate packs and joined together; this process provides the required thickness and geometrical configuration.

The best possible geometry for the magnetic circuit must be defined to obtain the desired electrical performance from any given coil geometry.

A gap is necessary to meet electrical demands and raise the magnetic energy available for storage (Fig. 4, Pos. 1). The gap promotes shear in the ferrous circuit. This substantially raises the current levels at which magnetic saturation occurs in the magnetic circuit. Without this gap, saturation would occur at low currents, and subsequent rises

3 Magnetization curve with hysteresis curve

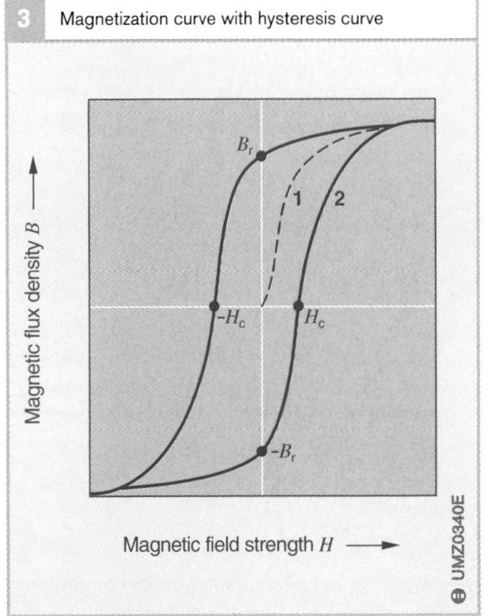

4 Iron core in compact coil with O and I core

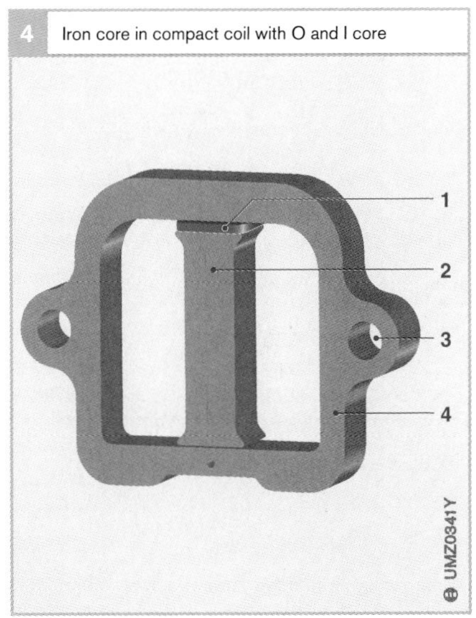

Fig. 3
1 New curve (magnetization curve of demagnetized iron core)
2 Hysteresis curve

H_c Coercive field strength
B_r Residual magnetism

Fig. 4
1 Gap or permanent magnet
2 I core
3 Attachment socket
4 O core

in current flow would produce only negligible increases in levels of stored energy (Fig. 5).

Important is that by far the largest proportion of the magnetic energy is stored in the gap.

In the coil development process, FEM simulation is employed to define the dimensions for the magnetic circuit and the gap that will provide the required electrical performance. The object is to obtain ideal geometry for maximum storable magnetic energy for a given current flow without saturating the magnetic circuit.

It is also possible to respond to the requirements associated with limited installation space, especially important with pencil coils, by installing permanent magnets (Fig. 4, Pos. 1) to increase the magnetic energy available for storage. The permanent magnets' poles are arranged so that it generates a magnetic field opposed to the field in the winding. The advantage of premagnetized materials is that they increase the magnetic circuit's energy-storage capacity.

5 Shear in magnetic circuit

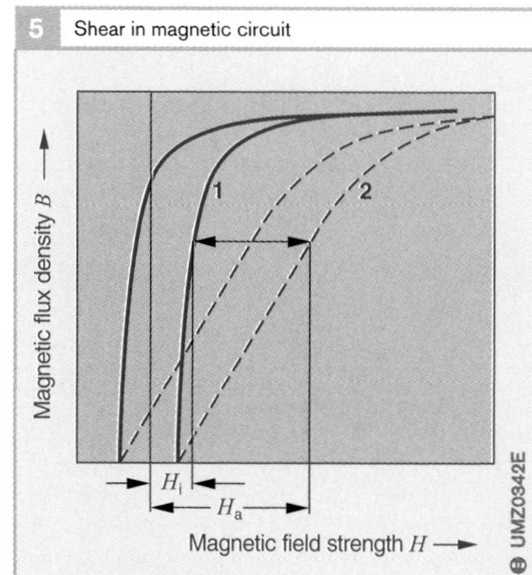

Fig. 5

1 Hysteresis with iron core and no gap

2 Hysteresis with iron core and gap

H_i Saturation control with iron core and no gap

H_a Saturation control with iron core and gap

Generating the high voltage

At the firing point the ignition driver stage interrupts the flow of current through the winding. The rapidly collapsing magnetic field induces secondary voltage in the secondary winding. In physical terms the secondary winding acts as a resonating circuit consisting of inductive, capacitive and ohmic components.

The maximum available voltage depends on

- The relative numbers of loops in the primary and secondary windings (turns ratio), together with the limit on primary voltage imposed by the coil driver
- The capacitance of the windings and the capacitance relative to adjacent components in the coil
- The stored energy
- The secondary loads (spark plug, etc.) and
- The plates in the ferromagnetic circuit

Switch-on arcs (activation arcs)

Switching on the primary current changes the current gradients to produce a sudden shift in magnetic flux in the ferrous core. This induces voltage in the secondary winding. Because the gradient for the current change is positive, the switch-on voltage's polarity is opposed to that of the induced voltage when the circuit is switched off. Because this gradient is quite small relative to the gradients that occur when the primary current is switched off, the induced voltage is also relatively low, despite the large turns ratio between the two windings. It lies within a range of 1...2 kV, and could be enough to promote spark generation and mixture ignition under some conditions. To prevent possible engine damage, preventing a switch-on arc at the spark plug is vital.

In systems with rotating high-voltage distribution the switch-on arc is effectively suppressed by the spark path within the distributor. The rotor contact is not directly across from the cap's contact when switch-on occurs.

In distributorless ignition systems with single-spark coils switch-on arcs are inhibited by the switch-on-arc suppression diode (see Fig. 1, Pos 2). With dual-spark coils the switch-on arc is suppressed by the high flashover voltage of the series circuit with its two spark plugs, and no supplementary measures are required.

Heat generation in the coil

The efficiency level, defined as the available secondary energy relative to the stored primary energy, is on the order of 50...60 %. Under certain conditions high-performance coils for special applications can achieve efficiency levels as high as 80 %.

Essentially, due to the resistance losses in the windings, and the remagnetization and eddy-current losses, the difference in energy is converted to heat.

A driver stage integrated directly in the coil can represent yet another source of thermal loss. The primary current causes a voltage drop in the semiconductor material, leading to lost efficiency. A further and thoroughly significant energy loss is attributable to the switching response when the primary current is switched-off, especially when the driver stage is “slow” in its dynamic response.

Levels of secondary voltage are usually limited by the restriction on primary voltage in the driver stage, where a portion of the energy stored in the coil is dissipated as thermal loss.

Capacitive load

Parasitic capacitance in the ignition coil cables, spark-plug well, spark plug and adjacent engine components is low in absolute terms, but remains a factor of not inconsiderable significance in view of the high voltages and voltage gradients. The increased capacitance attenuates the rise in secondary voltage. Resistive losses in the windings are higher, voltage is reduced. In the end, all of the potential secondary energy is not available to ignite the mixture.

Spark energy

The electrical energy available for the spark plug in the coil is called spark energy.
It is an essential criterion in ignition-coil design; together with the winding configuration it determines such factors as flashover current and the spark duration at the spark plug.

Spark energies of 30...50 mJ are the norm for igniting mixtures in naturally-aspirated and turbocharged powerplants. A higher energy (up to 100 mJ) is needed for reliable ignition at all operating points in gasoline engines with direct fuel injection.

Ignition-coil types

Single-spark ignition coil

Each spark plug has its own ignition coil in systems with single-spark ignition coils.

The single-spark coil generates one spark per ignition stroke. It is thus necessary to synchronize operation with a camshaft in these systems

Dual-spark ignition coil

Single-spark ignition (one spark plug per cylinder)

The dual-spark coil generates ignition voltage for two spark plugs simultaneously. The cylinders are paired

- To provide ignition for the air-fuel mixture in one cylinder at the end of the compression stroke
- To generate a spark in the other cylinder during the valve overlap phase at the end of the exhaust stroke

The dual-spark coil generates a spark for every crankshaft rotation, corresponding to twice for each ignition stroke. This means that no synchronization with the camshaft is required. However, this system is suitable for use only on powerplants with an even number of cylinders.

When the valves overlap, there is no compression pressure in the cylinder and the arcing voltage at the spark plug electrodes is thus very low. This "back-up spark" thus requires only negligible amounts of energy to bridge the gap.

Twin-spark ignition

In ignition systems with two spark plugs per cylinder, the ignition voltage from each coil is distributed to two cylinders. The resulting advantages are

- Emissions reductions
- A slight increase in power
- Two sparks at different points in the combustion chamber
- The option of using ignition offset to achieve "softer" combustion
- Good emergency-running characteristics when one ignition coil fails

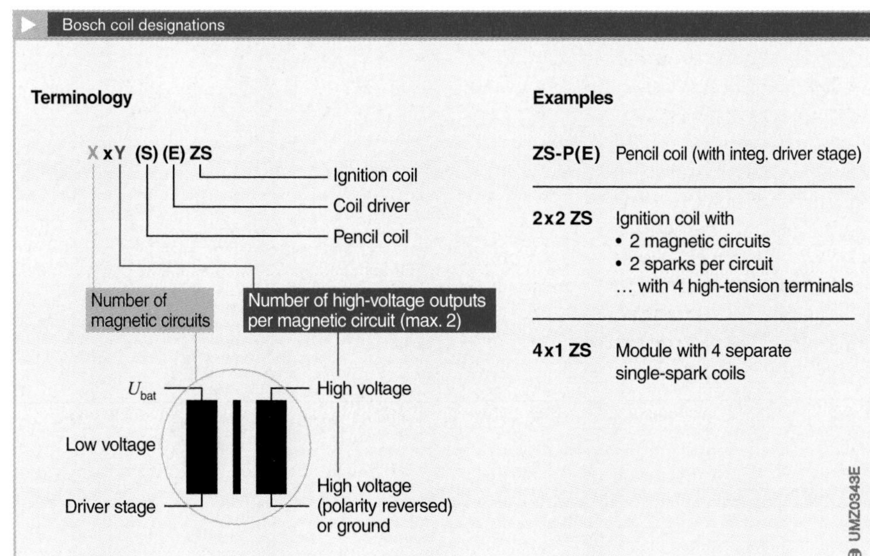

Bosch has introduced the following designations to rationalize its type definitions

Design variations

Virtually all of the coils in the ignition systems being designed today are either
- Compact coils or
- Pencil coils

It is also possible to integrate the ignition driver stage within the casing on some of the coil models described in the following section.

Compact coil

Design

The compact coil's magnetic circuit consists of the O core and the I core (Fig. 1), onto which the primary and secondary windings are plugged. This arrangement is installed in the coil casing. An electrical and mechanical link joins the primary winding (with I core carrying wire winding) to the primary socket. Also connected is the end of the secondary winding (coil body wrapped with wire). The connection between the secondary winding and the spark plug is also located in the casing, and the electrical connection is formed when the windings are manufactured.

Integrated within the casing is the high-voltage dome. This contains the contact for the spark-plug terminal and also includes a silicone insulation layer to prevent high voltage from arcing to adjacent components or the spark-plug well.

Following component assembly, resin is vacuum-injected into the inside of the casing, where it is allowed to harden. This embedding process provides
- High resistance to mechanical loads
- Effective protection against environmental factors and
- Excellent insulation against high voltage

The silicone insulation jacket is then pushed onto the high-voltage contact dome for permanent attachment.

Following 100 % testing to ensure compliance with all relevant electrical specifications the ignition coil is ready to use.

1 Compact coil design

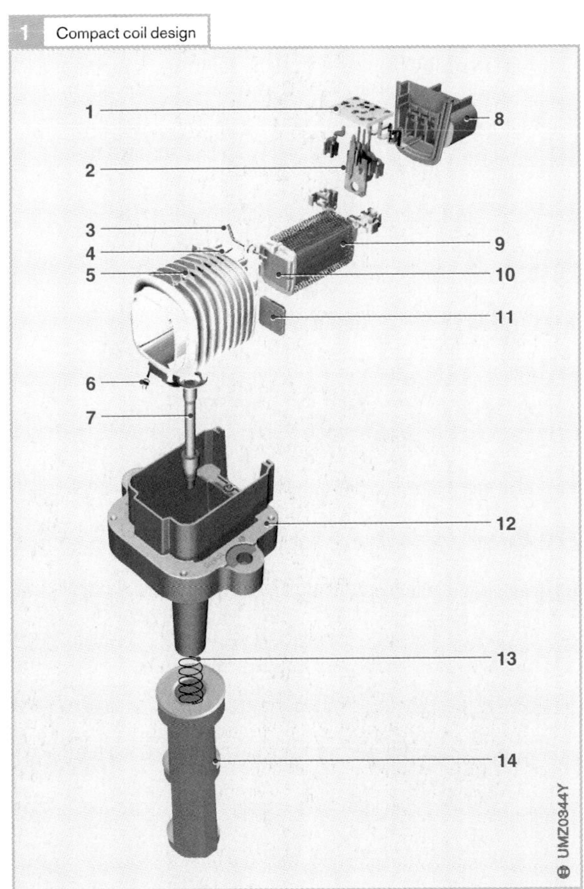

Fig. 1
1 Circuit board
2 Ignition driver stage
3 EFU diode
4 Secondary coil element
5 Secondary wire
6 Contact plate
7 High-voltage pin
8 Primary plug
9 Primary wire
10 I core
11 Permanent magnet
12 O core
13 Spring
14 Silicone jacket

Remote and COP versions

The ignition coil's compact dimensions make it possible to implement the structure shown in Fig. 1. This version is called the COP, or (Coil on Plug). The ignition coil is mounted directly on the spark plug, making extra high-voltage connecting cables unnecessary. (Fig. 2a) This reduces the capacitive load on the coil's secondary winding. The reduction in the number of components also increases reliability (no rodent bites in ignition cables, etc.).

The less-common remote version is mounted in the engine bay using bolts. Attachment lugs or a supplementary clamp can be provided. A high-voltage cable relays the high-voltage energy from the coil to the spark plug.

While the COP and remote versions share a broadly similar design concept, the remote unit (body-mounted) does not have to withstand the same thermal and vibrational stresses.

Other coil versions

ZS 2x2

Rotating high-tension distribution is being rapidly replaced by stationary, distributorless ignition. It was necessary to evolve a concept that would allow vehicle manufacturers to convert existing engine designs to the newer system.

An uncomplicated means for effecting this conversion to distributorless ignition is offered by the ZS 2x2 (Fig. 3) and the ZS 3x2. These ignition coils contain two (or three) magnetic circuits, and generate two sparks per circuit. They can thus be used to replace the distributors on four and six-cylinder engines. Because the units can be mounted almost anywhere in the engine bay, the vehicle manufacturer's adaptation effort is minimal, although the engine-management ECU has to be modified. Another factor is that high-voltage ignition cables are required for layouts with remote ignition coils.

2 Single-spark coils

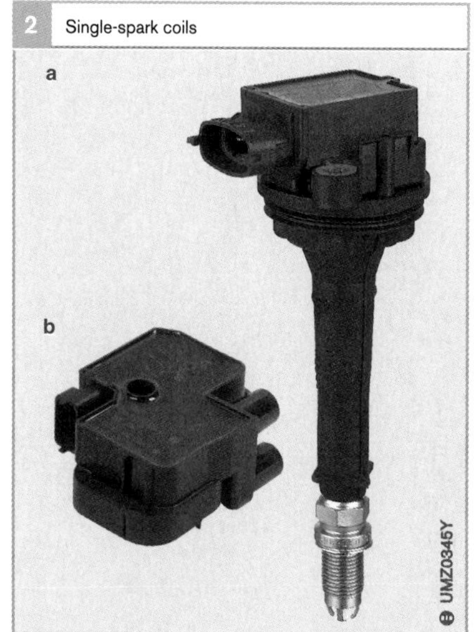

Fig. 2
- a COP version of a single-spark compact coil
- b Remote version: two single-spark coils in module, spark plugs connected with two ignition cables

3 2x2 coil for conversions to distributorless ignition

Ignition-coil modules

Coil modules combine several coils in a shared housing to form a single assembly (Fig. 4). These coils operate independently of each other.

The advantages furnished by ignition-coil modules are

- simplified installation (just a single operation for three or four coils), with
- fewer bolted/threaded connections,
- connection to the engine wiring harness with just one plug, and
- cost savings thanks to more rapid assembly and simplified wiring.

Problem issues are:

- It is essential to adapt the module's geometry to fit the engine and
- Modules must be designed to fit individual cylinder heads; no universal designs

Pencil coil

The pencil coil makes optimal use of the space available in the engine compartment. Its cylindrical form makes it possible to use the spark-plug well as a supplementary installation area, thus making for optimal use of the available space in the cylinder head (Fig. 5).

Because pencil coils are always mounted directly on the spark plug, no additional high-voltage cables are required.

Design structure and magnetic circuit

Pencil coils use the same inductive principle as compact coils. However, the spherical symmetry results in a design structure that differs considerably from that of the "compact" unit.

The most conspicuous difference is found in the magnetic circuit. Although the materials are the same, the central rod core (Fig. 6, Pos. 5) consists of laminations in

4 Ignition-coil module with compact ignition coils

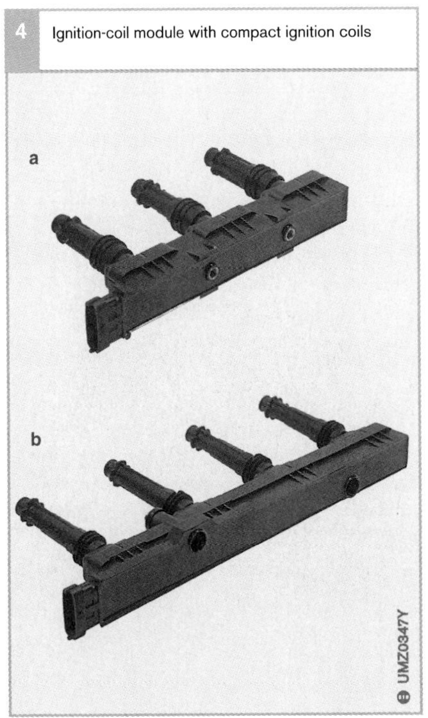

5 Installation in spark-plug well: relative dimensions of compact and pencil coils

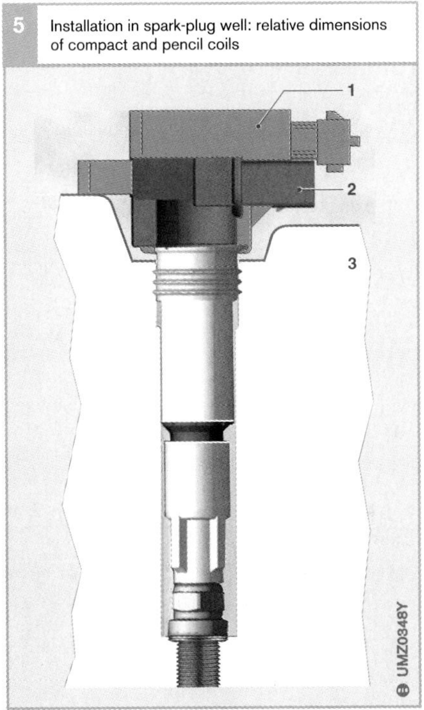

Fig. 4
a ZS 3x1M
b ZS 4x1M

Fig. 5
1 Compact coil
2 Pencil coil
3 Cylinder head

various widths stacked in packs that are virtually spherical. The yoke plate (9) that provides the magnetic circuit is a rolled and slotted shell – also in electric plate, sometimes in multiple layers.

Another difference relative to compact coils is the primary winding (7), which has a larger diameter and is located around the secondary winding (6), while the body of the winding also supports the rod core (5). This arrangement brings benefits in the areas of design and operation.

6 Design of pencil coil

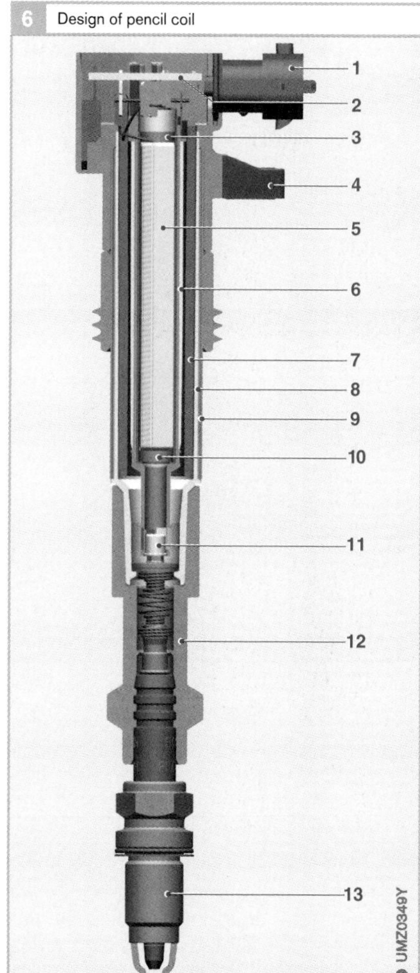

Fig. 6
1 Plug connection
2 PCB with driver stage
3 Permanent magnet
4 Attachment arm
5 Laminated electric plate core (rod core)
6 Secondary winding
7 Primary winding
8 Casing
9 Yoke plate
10 Permanent magnet
11 High-voltage dome
12 Silicone jacket
13 Attached spark plug

Cavities filled with cast resin

Owing to restrictions imposed by their geometrical configuration and compact dimensions, pencil coils allow only limited latitude for varying the magnetic circuit (core, yoke plate) and windings. Use of FEM simulation and other computer-supported design tools is absolutely essential in designing units that meet the geometrical and electrical criteria with optimal efficiency.

In the majority of pencil-coil applications, the limited installation space makes it necessary to insert permanent magnets in the coil in order to increase the spark energy (refer to Fig. 6 and the section on "operating concept/magnetic circuit").

The arrangements for electrical contact with the spark plug and the engine wiring harness are comparable to those used with compact coils.

Versions

An extended array of versions (different diameters and lengths, etc.) is available to provide pencil coils for a range of applications. The driver stage can also be integrated within the casing as an option.

A frequent diameter, as measured at the cylindrical center section (yoke plate, casing), is roughly 22 mm. This dimension is defined by the diameter of the spark-plug well within the cylinder head as used with standard spark plugs featuring a 16 mm socket fitting. The length of the pencil coil is determined by the installation space in the cylinder head and the specified or potential electrical performance specifications. Due to the accompanying increase in parasitic capacity and the reductions in magnetic-circuit efficiency, there are limits on the latitude for lengthening the active component (transformer). Optimum electrical performance becomes impossible above a certain length.

Ignition coil retrospective: the asphalt coil

A relict from an earlier era, but still on the job: the asphalt ignition coil (illustration) is still in use in a number of older vehicles.

The asphalt coil consists of a casing (5) containing magnetic yoke plates (7). The secondary winding (9) is wound directly around the laminated iron core (12), through which it is electrically connected to the centre terminal in the coil's cover (3).

Because the iron core conducts high voltage, it must be insulated by the cover as well as an additional insulation element (11) at the bottom. The primary winding (8) surrounds the secondary. Asphalt casting provides insulation for the windings while also holding them in place.

The insulated coil cover also contains the terminal for battery voltage (Terminal 15) and the wire to the distributor (Terminal 1), which are arranged symmetrically on either side of the high-voltage socket (Terminal 4).

Most of the thermal loss is generated in the primary winding, and is conducted to the casing by the yoke plates. Broad attachment clamps (6) serve to transfer maximum heat to the vehicle body.

In the 1980s the asphalt coil gradually ceased to appear in new system designs as it was supplanted by the plastic coil. However, the older type of coil is still being manufactured in Brazil to satisfy the remaining global demand for spare parts.

Structure of the asphalt coil

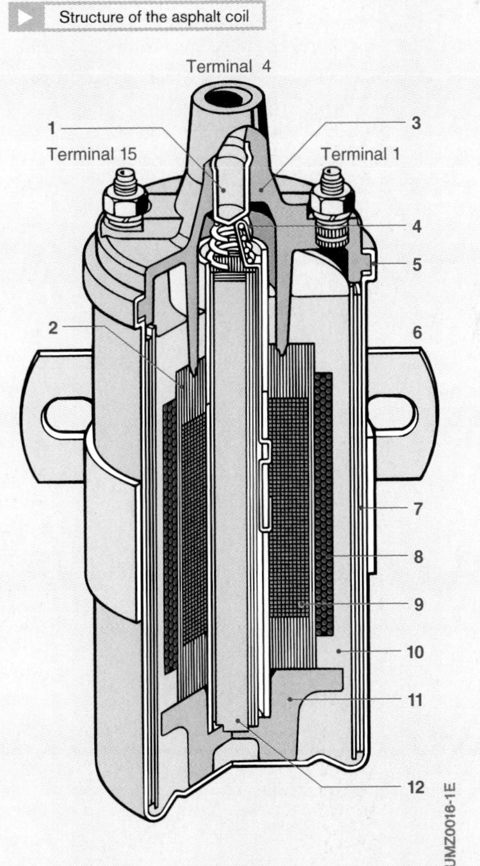

1 High-voltage connection to the outside
2 Paper insulation wrapping
3 Insulated cover (coil cover)
4 Internal high-voltage connection with spring-loaded contact
5 Casing
6 Attachment clamp
7 Magnetic metal jacket
8 Primary winding
9 Secondary winding
10 Casting (asphalt)
11 Insulator
12 Iron core

Ignition-coil electronics

In earlier designs the driver stage was usually incorporated within a separate module, and attached to the coil or the distributor in the engine bay of a vehicle with rotating voltage distribution. The conversion to distributorless ignition combined with increasing electronic-component miniaturization fostered the development of compact ignition driver stages in the form of integrated circuits (ICs) for integration in the ignition ECU or engine-management ECU.

Engine-management ECUs (Motronic) and new powerplant concepts (direct fuel injection, etc.) have increased thermal stresses (heat loss in driver stages) and reduced installation space. These factors were behind the trend toward remote driver stages located outside the engine-management control unit. One option is integration within the ignition coil, which also makes it possible to use a shorter primary wire to reduce line loss.

1 Driver-stage installation in a pencil-coil casing

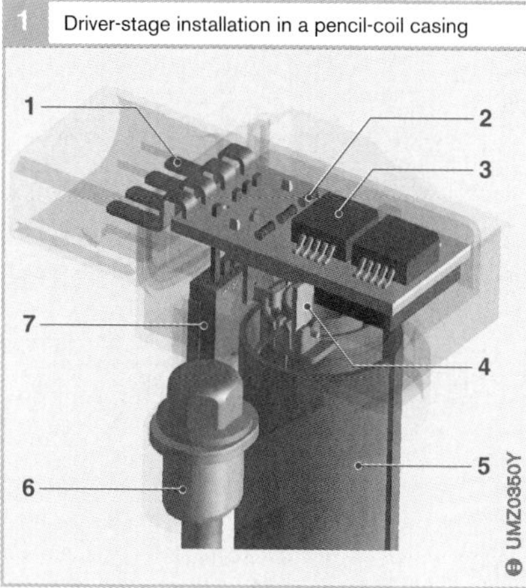

Fig. 1
1 Primary plug
2 SMD components
3 Electronics for multispark ignition and ionic current measurement
4 Primary winding contacts
5 Pencil-coil transformer
6 Attachment socket
7 Driver stage

Design

The driver stage can be integrated in both compact and pencil coils. Figure 1 shows installation in a pencil coil. The driver-stage module – which can also incorporate supplementary functions – is built on a small PCB. SMD (Surface Mounted Device) componentry is used to save space.

The driver-stage transistors (7) are integrated in standard TO housings and connected to the circuit board or conductor paths. Other electronic chips (3) can be included to provide supplementary functionality for monitoring, diagnosis, etc. (can include ionic-current signal amplification, multispark ignition, etc.). The primary terminal (1) is connected directly to the circuit board. Below the board are the contacts for the coil's primary winding (4).

Ignition driver stages

Conventional driver-stage design relies on BIP (Bosch Integrated Power) bipolar technology. The IGBT (Insulated Gate Bipolar Transistor, includes features from field-effect and bipolar transistor) opens the way to new technology offering several advantages over BIP:

- Wattless voltage control instead of current control
- Lower saturation voltage at higher currents
- Reduced switching times
- Higher active clamping voltages available
- Higher voltage-clamp energy
- No current control or limitation required
- Withstands higher temperatures in extended operation

Supplementary circuitry on the chip neutralizes the liabilities associated with the IGBT's high input impedance; this reduces sensitivity to interference.

Some driver stages transmit signals back to the control unit, allowing it to detect and assess defects (short circuit, power collapse). Some driver stages are able to avoid overheating damage by switching off in response to excessive temperatures.

Multispark ignition

Under critical conditions the energy stored in small coils will not always be enough to ensure reliable ignition of the mixture. Cost-intensive concepts (dual-energy ignition, AC ignition, etc.) are available for demand-based regulation of ignition energy. Multispark (pulsed, restrike) ignition represents a less expensive alternative.

Operating concept

At the core of the multispark concept is a fast-charging ignition coil (Fig. 2). Following initial arc-over the driver stage is repeatedly switched on and off to form a pulsed spark train consisting of numerous consecutive arcs. Recharging can start early to interrupt individual sparks. During the recharging phases no energy to ignite the mixture is transferred to the plug.

Single-spark ignition is distinguished by the following characteristics:

- Extended charge times
- Single spark
- Spark duration is defined by the amount of energy stored
- The Motronic engine-management ECU can trigger subsequent sparks only at extremely low engine speeds

Multispark ignition is characterized by

- short charging times, and
- a spark sequence of variable duration.

Beyond a certain engine speed it is not logical to apply more than a single spark; the amount of time available for a second spark is limited, and in the intervening period the crankshaft will move forward by a number of degrees.

Ionic-current measurement

When the spark collapses, the space between the electrodes is ionized. The electrical circuit – consisting of voltage source, spark plug and current monitoring device (Figure 3) – is closed. The current flow serves as an index for the quality of combustion.

Assessment of this signal entails extensive effort using considerable electronic and software resources.

The system is suitable for use in detecting combustion knock. Other, more detailed analysis options are under development.

2 Comparison of signal curves for single and dual-spark ignition

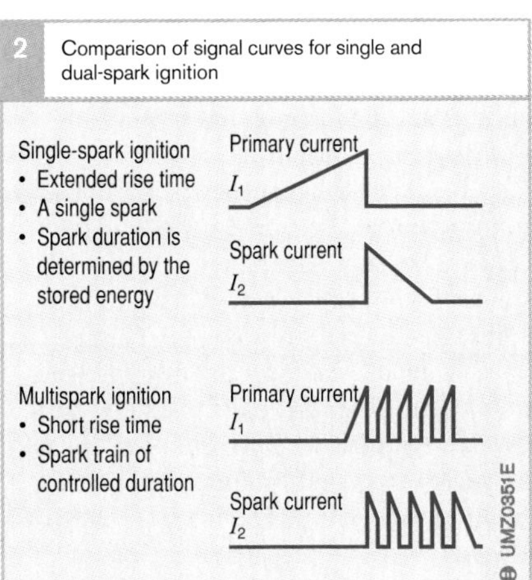

3 Measuring ionic current across spark-plug electrodes

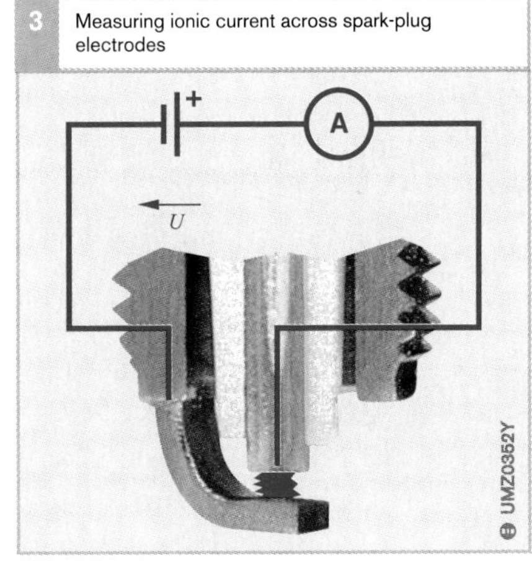

Electrical parameters

Inductance

Inductance is determined by the material and dimensions of the magnetomotive circuit, the number of windings and the geometry of the copper winding, as well as the current that flows through it.

The ignition coil includes primary and secondary inductance elements, with the secondary inductance being many times greater.

Capacitance

In ignition coils this includes self-capacitance, parasitic capacitance and load capacitance.

Self-capacitance is essentially defined by the internal capacitance in the secondary winding, as determined by the winding configuration. With pencil coils, the limited amount of insulation area between the two windings and to the core is also a significant factor in coil-on-plug units. Simulation tools are used to reduce the capacitances and endow the coil with the best possible electrostatic characteristics and improve its electrical performance.

Table 1

1	Ignition-coil specifications	
I_1	Primary current	6.5...9.0 amperes
T_1	Rise time	1.5...4.0 ms
U_2	Secondary voltage	29...35 kV
T_{Fu}	Spark duration	1.3...2.0 ms
W_{Fu}	Spark energy	30...50 mJ, up to 100 mJ for direct fuel injection
I_{Fu}	Spark current	80...115 mA
R_1	Primary winding resistance	0.3...0.6 Ω
R_2	Secondary winding resistance	5...15 kΩ
N_1	Number of coil loops in primary winding	150...200
N_2	Number of coil loops in secondary winding	8,000...22,000

External load capacitance is determined by the installation environment (metallic spark-plug well, etc.), the spark plug and any high-voltage connections that may be present. Because these factors are not usually subject to modification, appropriate coil design is used to deal with them.

Stored energy

The amount of magnetic energy that can be stored depends upon numerous factors such as the coil's design (geometry, material in the magnetic circuit, additional magnets) and the driver stage. Once a certain point is reached any additional increases in primary current will deliver only negligible rises in stored energy, while losses grow disproportionately, ultimately leading to destruction of the coil.

While continuing to satisfy numerous criteria, the ideal coil thus operates just below the magnetic circuit's magnetic saturation point.

Resistance

The resistance of the windings is determined by the temperature-sensitive specific resistance of copper.

To ensure good cold starts (with reduced battery voltage), the primary winding's resistance should be as low as possible. It is in the order of a few hundred milliohms.

The resistance in the secondary winding, at a number of kilo-ohms, has negative effects on the ignition coil's intrinsic losses.

Power loss

The coil's losses are determined by resistance in the windings, capacitive losses and remagnetization losses. Also present are losses stemming from deviations from the ideal configuration for a magnetic circuit that arise from compliance with other specifications. The efficiency level of 50...60 % translates into a relatively high net loss at elevated engine speeds. Coils specially designed for minimal losses combine with thermally resistant materials employed in suitable concepts to provide some degree of compensation.

Turns ratio
The turns ratio is the arithmetical ratio of the number of copper loops in the primary winding to the number in the secondary. On standard coils it is on the order of 1:50...1:150. Selection of suitable turns ratios can be used in combination with driver-stage specifications to affect such factors as the level of spark current and – to some degree – the maximum secondary voltage.

High-voltage and spark-generation properties
The ideal coil remains relatively impervious to load factors while producing as much high voltage as possible within an extremely brief rise period. These properties ensure flashover at the spark plug for reliable mixture ignition under all conditions encountered in operation.

At the same time, the real-world properties of the windings, the magnetic circuit and the coil driver all unite to impose limits on performance.

The polarity of the high-voltage circuit ensures that the spark plug's center electrode maintains negative potential relative to chassis ground. This negative polarity counteracts the tendency of the spark plug's electrodes to erode.

Dynamic internal resistance
Yet another important parameter is the coil's dynamic internal resistance (impedance). Because impedance combines with internal and external capacitance to help determine voltage rise times, it serves as an index of the amount of energy that can flow from the coil and through shunt resistance elements when ignition is triggerd. Low internal resistance is an asset when spark plugs are contaminated or wet. Internal resistance depends on secondary inductance.

Ignition-coil development using simulation tools

Ever-increasing technical demands mean that the efficiency limits of trial-and-error design methods are reached early. Product development processes using CAE provide a solution. CAE (Computer Aided Engineering) is a generic designation encompassing all computer-based design activities. It includes all aspects of CAD (Computer Aided Design) as well as calculation routines.

The advantages offered by CAE include:

- Informed decisions at early stages in the design-engineering process (possible without prototypes)
- Generation of specifications for prototypes that are ready to test and
- Enhanced understanding of physical interrelationships

Calculation programs are employed in various simulation sectors during ignition-coil development:

- Structural mechanics (analysis of mechanical and thermal stress factors)
- Fluid mechanics (analysis of fluidic charging processes)
- Electromagnetics (analysis of the system's electromagnetic performance)

Electromagnetic simulation tools are especially important in engineering ignition coils. They are used for geometrical and performance simulation.

The Finite Element Method (FEM) is used for geometrical analysis. In this process the coil's geometry is defined based on a CAD model. Various specifications (current density, electrical potentials, etc.) are then defined for incorporation in an FEM model. Transfer to a calculation model and derivation of specifications from the corresponding equation series follow. The result is a clear calculated solution to the problem.

This method allows simulation-based, 100 % "virtual" coil design at the computer. Depending upon the objective of the analysis exercise, the ideal coil geometry can also be defined (optimal magnetic properties in the ferromagnetic circuit, best-possible electrostatic properties in electrically conductive contours).

Following geometrical analysis, performance simulation can be employed to examine the coil's electrical characteristics within the overall system, consisting of driver stage, coil and spark plug, under conditions reflecting the actual, real-world environment. This calculation method provides the initial specification data for the coil. It also supports subsequent calculation of electrical parameters such as spark energy and spark current.

Electromagnetic simulation tools make "virtual" development of ignition coils possible. The simulation results define geometrical data and winding design to furnish the basis for prototype construction. The electrical performance of these sample coils will approximate the simulation results. This substantially reduces the number of time-consuming recursion processes that occur when coils are produced using conventional design-engineering methods.

Manufacturing ignition coils

Bosch manufactures its ignition coils using fully-automated production processes. The following section provides an overview of the sequence.

Injected thermoplastic castings
1. Primary and secondary windings
2. Coil casing
3. Primary terminal

Stamped and packed electric material
1. O core and I core (compact ignition coil)
2. Core and yoke plate (pencil coil)

Stamped connectors
1. Current rails
2. Contacts

Production steps
1. Apply coating to O core (casing) and I core (primary winding) or core rod (primary winding)
2. Winding of primary and secondary windings (Fig. 1)
3. Attach secondary wire to contact element
4. Assembly of individual components
4.1 Attach current rails and primary terminals
4.2 Connect primary terminal to primary winding
4.3 Insert pre-assembled primary winding in secondary winding
4.4 Install high-voltage socket in casing
4.5 Pencil coil: Mount yoke plate on casing
4.6 Install completed winding assembly in casing

Casting
1. Preheating
2. Vacuum casting
3. Curing

Final assembly
1. Silicone jacket and any remaining components

Final inspection
1. Check for conformity with electrical and mechanical specifications
2. Labeling

Packaging and shipping

1 Winding coil body

Fig. 1
1 Wire (gauge 0.05 mm, shown as bold line in illustration)
2 Wire guide
3 Spindle
4 Secondary winding
5 Secondary winding body
6 Contact element

2 Inserting winding body in coil casing

Fig. 2
1 Coil winding
2 Coil casing

Spark plugs

The air-fuel mixture in the spark-ignition engine is ignited electrically. Energy drawn from the battery is temporarily stored in the coil for this purpose. The high voltage generated within the coil produces a spark bridging the electrodes in the engine's combustion chamber. The energy contained in the spark then ignites the compressed air-fuel mixture.

Function

The spark plug's function is to convey energy into the gasoline engine's combustion chamber and produce a spark between its electrodes to initiate combustion in the air-fuel mixture.

Spark plugs must be designed to ensure positive insulation between spark and cylinder head while also sealing the combustion chamber.

Along with other engine components, such as the ignition and induction systems, the spark plug is decisive in defining engine operation. It must

- provide reliable cold starts,
- furnish consistent operation with no ignition miss throughout its service life,
- not overheat under extended operation at or near top speed.

To ensure this kind of performance throughout the spark plug's service life, the correct plug concept must be evolved early in the engine-design process. Research investigating the ignition process is employed to determine the spark-plug concept that will provide the best emissions and the smoothest-running engine operation.

An important spark-plug parameter is the heat range. The right heat range prevents the spark plug from overheating and inducing the thermal auto-ignition that could lead to engine damage.

1 Spark plug in gasoline engine

Application

Installation environments

The first spark plug in a passenger car came from Bosch in 1902, when it was installed in a system featuring magneto ignition. The spark plug then went on to become an unparalleled success in automotive technology.

Spark plugs are used in all vehicles and machinery powered by spark-ignition powerplants, both 2-stroke and 4-stroke. They can be found in

- Passenger cars
- Commercial vehicles
- Single-track vehicles (motorcycles, scooters, mopeds)
- Ships and small craft
- Agricultural and construction machinery
- Chainsaws
- Garden tools (lawn mowers, etc.)

Because multicylinder passenger-car powerplants require at least one plug per cylinder, it is in this sector that most spark plugs are used.

Since only small power outputs are involved, motorized machinery usually relies on a single-cylinder engine needing only a single spark plug.

Within Europe most commercial vehicles – at least in heavy-duty applications – are powered by diesel engines, which limits the demand for spark plugs in this segment. In the US, however, spark-ignition gasoline engines are also the most prevalent powerplants in heavy vehicles.

Type range

1902 engines produced only about 6 bhp for each 1,000 cm^3 of displacement. Today's comparable figure is 100 bhp, with up to 300 bhp available from competition engines. The technical resources invested in engineering and producing the spark plugs that allow this performance is enormous.

The first spark plug was expected to ignite 15...25 times per second. Today's spark plugs must ignite five times as often. Peak temperatures have risen from 600 °C to approximately 900 °C, while ignition voltage has increased from 10,000 V to as much as 30,000 V. Finally: while initially the spark-plug had to be removed every 1,000 km, today's spark plugs are expected to perform for a minimum of 15,000 km.

Although the spark plug's basic concept has changed little in the course of 100 years, in this period Bosch has designed more than 20,000 different types to meet the needs of various engine configurations.

The current spark-plug range continues to embrace a wide array of models. The spark plug is subject to immense demands in the areas of

- electrical and
- mechanical performance, as well as resistance to
- chemical and
- thermal attack.

In addition to satisfying these performance criteria the spark plug must also be matched to the geometrical conditions defined by the individual engine (plug position in the cylinder head, etc.). Combined with the extensive range of engines being manufactured, this demand catalog makes it necessary to offer a wide variety of spark plugs. Bosch currently supplies more than 1,250 different spark-plug versions, all of which must be available to service centers and commercial distributors.

Requirements

Electrical-performance requirements

During operation in electronic ignition systems, spark plugs must handle voltages as high as 30,000 V with no tracking across the insulation. Residue from the combustion process, such as soot, carbon and ash from fuel and oil additives, can be electrically conductive under some conditions. Yet under these same conditions it remains imperative that tracking through the insulation be avoided.

1 Temperature and pressure stresses on spark plugs

Two-stroke engine				
Cycle phase	Bypass flow	Com-pression	Combustion and work	Exhaust
Gas temp.	...120 °C	200... 400 °C	2,000... 2,800 °C	500... 1,200 °C
Gas pressure	1 bar	5...8 bar	15...30 bar	1...3 bar
Piston position				
Crank-shaft angle	0° BDC – 90°	90° – 180° TDC	180° TDC – 270°	270° – 360° BDC

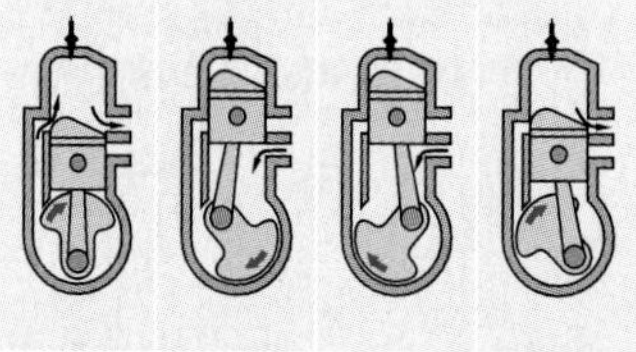

Four-stroke engine				
Cycle phase	Com-pression	Combustion and work	Exhaust	Intake
Gas temp.	300... 600 °C	2,000... 3,000 °C	1,300... 1,600 °C	...120 °C
Gas pressure	8...15 bar	30...50 bar	1...5 bar	0.9 bar
Piston position				
Crank-shaft angle	0° TDC – 180° BDC	180° BDC – 360° TDC	360° TDC – 540° BDC	540° BDC – 720° TDC

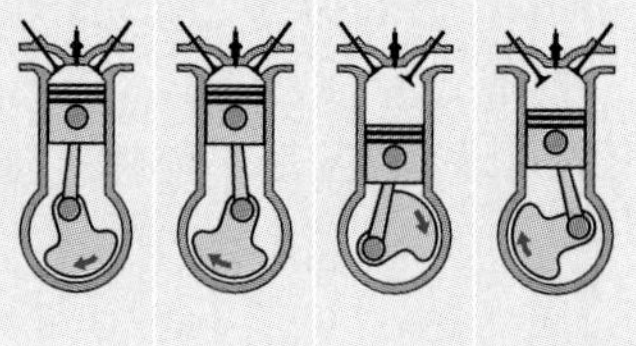

UMZ0325E

The insulator must continue to display adequate electrical resistance at up to 1,000 °C with only very minor reductions in this figure throughout its service life.

Mechanical-performance requirements

The spark plug must be capable of withstanding periodic pressure peaks (up to about 100 bar) in the combustion chamber while still providing an effective gas seal. High resistance to mechanical stresses is also required from the ceramic insulator which is highly loaded during spark-plug installation. During engine operation, the insulator is also loaded by the spark-plug connector and the ignition cable. The shell must absorb the torque applied during installation with no permanent deformation.

Chemical-performance requirements

Because the spark plug extends into a combustion chamber hot enough to make its tip glow, it is exposed to the chemical reactions that occur at extreme temperatures. Substances within the fuel can form aggressive residue deposits on the spark plug, affecting its performance characteristics.

Thermal-performance requirements

In operation the spark plug must alternately absorb heat from hot combustion gases and then withstand the cold incoming air-fuel mixtures in rapid sequence. This is why insulators must display immense resistance to thermal shock.

The spark plug must also transfer heat to the engine's cylinder head with maximum efficiency; the top of the spark plug should remain as cool as possible.

Design

The essential components of the spark plug are (Fig. 1) the

- Terminal post (1)
- Insulator (2)
- Shell (3)
- Gasket and seat (6) and
- Electrodes (8, 9)

Terminal post

The steel terminal post is mounted in the insulator with the cast-glass conductor that also provides the electrical path to the center electrode. A threaded stud on the terminal end protrudes from the insulator to serve as a connector for the spark plug cables. Some terminal posts designed to ISO/DIN standards can feature internally-threaded connection fittings (with the specified contours) installed on the stud; others are equipped with a massive ISO/DIN fitting on the stud during manufacture.

Insulator

The insulator is cast in a special ceramic material. Its function is to insulate the center electrode and terminal post from the shell. The demand for a combination of good thermal conductivity and effective electrical insulation is in stark contrast with the properties displayed by most insulating substances. Bosch employs aluminum oxide (Al_2O_3) along with minute quantities of other substances. Following firing, this special ceramic meets all requirements for mechanical and chemical strength, while its dense microstructure provides high resistance to electrical breakdown.

The outer contours of the insulator nose can also be modified to improve the plug's heating characteristic for better response during repeated cold starts.

The surface of the insulator's terminal end is coated with a lead-free glaze. The glazing helps prevent moisture and contamination from adhering to the surface and ensures good resistance to electrical tracking.

1 Spark plug design

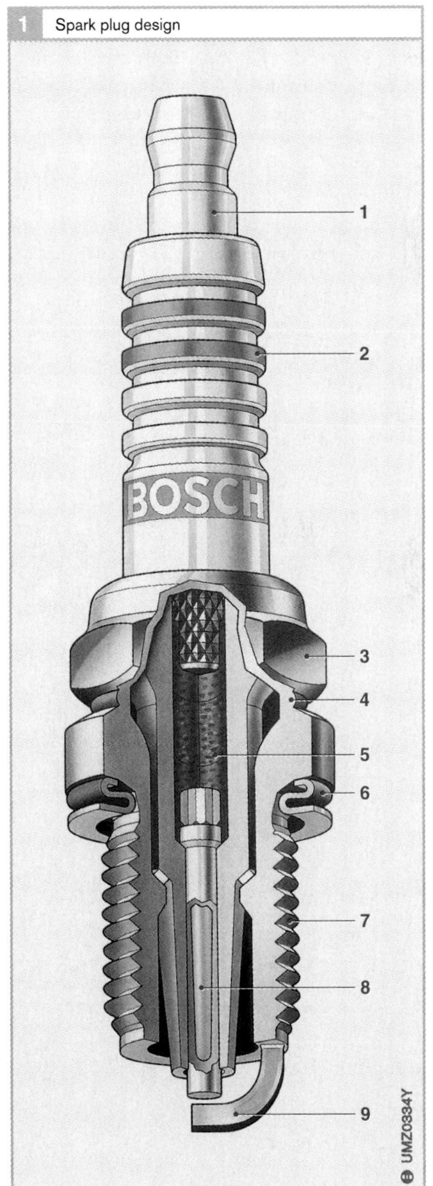

Fig. 1
1 Terminal post with nut
2 Al_2O_3 ceramic insulator
3 Shell
4 Heat-shrinkage zone
5 Conductive glass
6 Captive gasket
7 Threads
8 Composite centre electrode (Ni/Cu)
9 Ground electrode (here Ni/Cu composite)

Shell

The shell is manufactured in steel in cold-forming processes. The shell castings emerge from the pressing tool with their final contours, limiting subsequent machining operations to just a few areas. The bottom end of the shell includes threads (Fig. 1, Pos. 7) for installing the plug in the cylinder head and removing it at the end of its service life. Depending upon the specific design, as many as four ground electrodes are be welded to the shaft's end.

An electroplated nickel coating is applied to the surface to protect the shell against corrosion and prevent it from seizing aluminum cylinder heads.

To accomodate the spark-plug wrench the upper section of the conventional shell is provided with a hexagon; newer shell designs may have a double-hexagon which makes it possible to reduce the A/F size of the wrench without modifying insulator geometry. This reduces the spark plug's space demands in the cylinder head and allows the engine designer greater latitude in locating the water passages.

The plug assembly, including insulator along with the securely mounted center electrode and terminal post, is inserted in the shell. The shell is then crimped to hold this assembly in position. The subsequent shrink-fitting process – inductive heating under high pressure – produces a gas-tight connection between insulator and shell to ensure effective thermal conductivity.

2 Spark plug seal

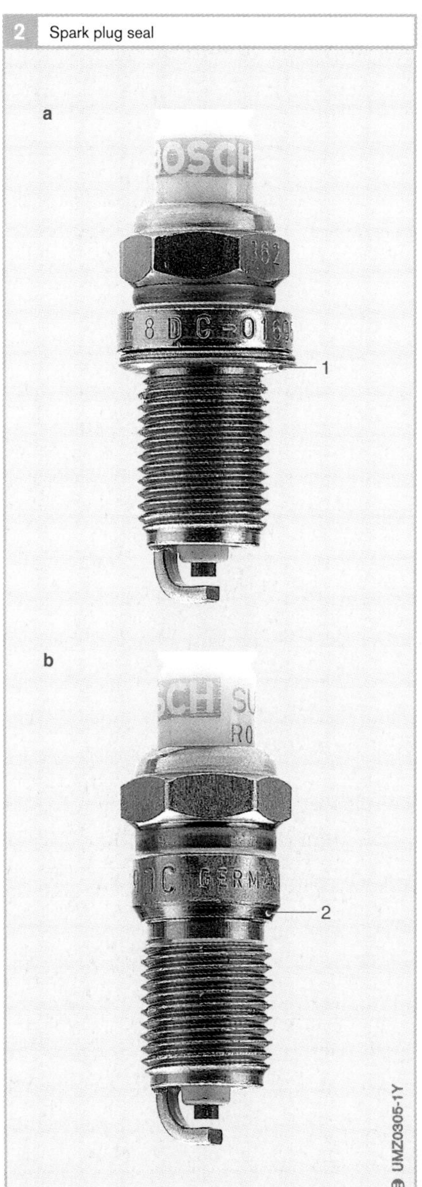

Fig. 2
a Flat seat with gasket
b Conical seat without gasket

1 Gasket
2 Conical seat

Seal seat

Depending upon engine design, either a flat or tapered seal seat (Fig. 2) provides the leak-free union between spark plug and cylinder head.

The flat seat is equipped with a captive gasket (1) which remains permanently attached to the spark plug. Its special contours adapt to form a durable yet flexible seal when the spark plug is installed.
In the case of conical-seat plugs, a rounded surface on the spark plug's shell (2) mates directly with the cylinder head to provide a seal without an additional gasket.

Electrodes

The spark's flashover combines with the stresses of high-temperature operation to promote erosion on the electrodes – the gap widens. To satisfy demands for extended replacement intervals, electrode materials must effectively resist erosion (burning from arcs) and corrosion (wear from aggressive thermochemical processes). These properties are achieved primarily through the use of temperature-resistant nickel alloys.

Center electrode

The center electrode (Fig. 1, Pos. 8), which includes a copper core for improved thermal dissipation, is anchored at one end in the glass casting.

In "extended-life" spark plugs the center electrode serves as the carrier for a noble-metal rod, permanently welded to the base electrode by laser. Other spark-plug designs still rely on electrodes formed from a single thin platinum wire which is then sintered to the ceramic base for good thermal conductivity.

Ground electrodes

The ground electrodes (9) are attached to the shell and are usually rectangular in section. Depending upon the particular configuration, one differentiates between the front electrode and the side electrode (Fig. 3). The ground electrode's long-term durability is determined by its ability to transfer heat. As with center electrodes, composite materials can be used to improve thermal transfer, but it is the length and the cross section that ultimately determine the ground electrode's temperature, and thus its resistance to wear.

Spark-plug life can be extended through the use of greater end-surface areas and multiple ground electrodes.

3 Electrode configurations

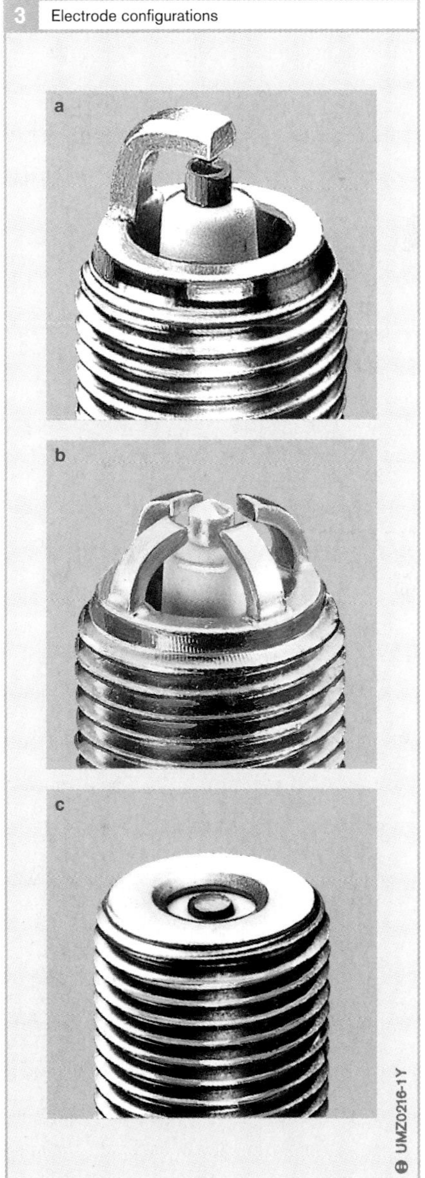

Fig. 3
a Front electrode
b Side electrodes
c Surface-gap spark plug without ground electrode (special application for racing engines)

Electrode materials

As a basic rule, pure metals display higher thermal conductivity than alloys. Yet pure metals – such as nickel – are also more sensitive than alloys to chemical attack from combustion gases, with less resistance to combustion residue. Manganese and silicon can be added to nickel to produce alloys with enhanced resistance to aggressive chemicals, especially sulfur-dioxide (sulfur is found in lubricating oils and fuels). Aluminum and yttrium additives enhance resistance to scaling and oxidation.

Compound electrodes

Corrosion-resistant nickel alloys are now the most widely used option in spark-plug manufacture. A copper core can be used for further increases in thermal conductivity, producing composite electrodes that satisfy stringent requirements for thermal conductivity and corrosion (Fig. 1).

The ground electrodes, which must be flexible enough to bend when the gap is set, may also be manufactured from a nickel-based alloy or using a composite design.

Silver center electrodes

Silver has the best electrical and thermal conductivity of any material. It also displays extreme resistance to chemical attack, provided that it is not exposed to either leaded fuels or to high temperatures in reducing atmospheres (rich air-fuel mixture).

Composite particulate materials with silver as their basic substance can substantially enhance thermostability.

Platinum electrodes

Platinum (Pt) and platinum alloys display high levels of resistance to corrosion, oxidation and thermal erosion. This is why platinum is the substance of choice for use in extended-life spark plugs.

On some spark-plug models the Pt rod is cast in the ceramic body early in the manufacturing process. In the subsequent sintering process the ceramic material shrinks onto the Pt rod to permanently fix it in position.

For other types of spark plug thin Pt rods are welded onto the center electrode (Fig. 2). Bosch relies on continuous-operation lasers to produce a durable bond.

1 Spark plugs with composite electrodes

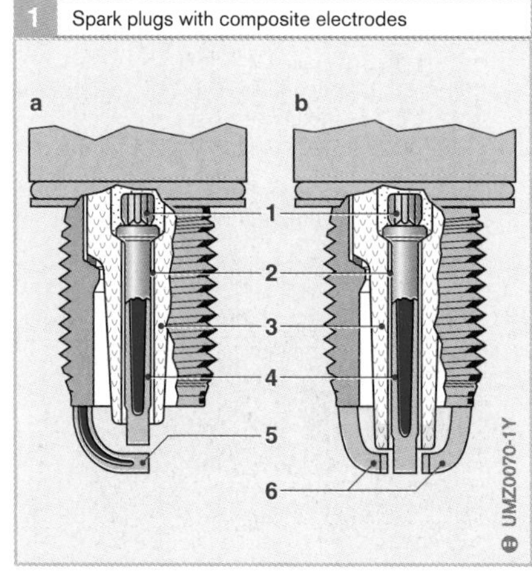

Fig. 1
a with front electrode
b with side electrode

1 Conductive glass
2 Gap
3 Insulator nose
4 Composite centre electrode
5 Composite ground electrode
6 Ground electrodes

2 Laser-welded rods

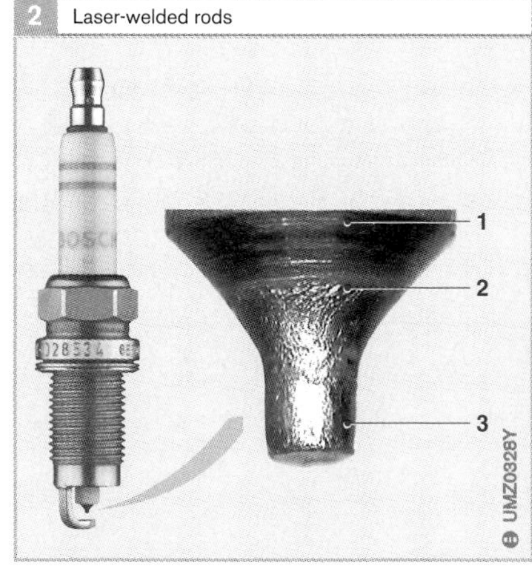

Fig. 2
1 Composite electrode (Ni/Cu)
2 Laser-welded seam
3 Platinum rod

Spark-plug concepts

The relative positions of the electrodes and the locations of the ground electrodes relative to the insulator determine the spark-plug concept (Fig. 1).

Air-gap concept

Center and ground electrodes are configured to produce a linear spark to ignite the air-fuel mixture located in the space between them.

Surface-gap concept

The design creates an arc that travels from the center electrode and across the insulator's ceramic surface before traversing the gas-filled gap to the ground electrode. Because the ignition voltage required to produce discharge across the surface is less than that needed to arc across an air gap of equal dimensions, at any given voltage surface-gap plugs can ignite over wider gaps than air-gap plugs. This produces a larger core flame for more effective creation of a stable flame front.

The surface gap also promotes self-cleaning during repeated cold starts, preventing soot deposits from forming on the insulator tip. This improves performance on engines exposed to frequent cold starts under low-temperature conditions.

Semi-surface gap concepts

On these spark plugs the ground electrodes are arranged at a specific distance from the center electrode and the end of the ceramic insulator. This produces two alternate spark paths with two different voltage requirements. Both discharge options, air and semi-surface – are available. Depending on operating conditions and spark-plug condition (wear), the arc travels along the insulator surface or directly across the gap between the electrodes.

1 Spark-plug concepts

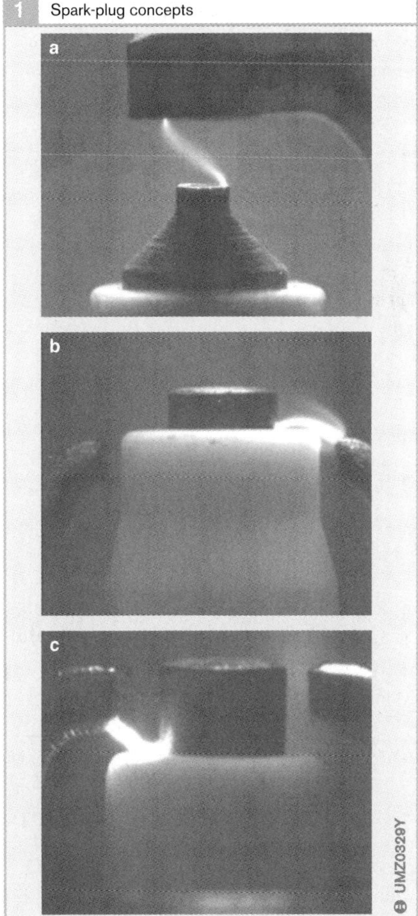

Fig. 1
a Air gap
b Surface gap
c Semi-surface gap

Electrode gap

As the shortest distance between the center and ground electrode(s), the electrode gap determines the length of the spark (Fig. 1) The voltage required to bridge this distance with an ignition spark is proportional to the gap's width, smaller gaps need less voltage.

However, narrow gaps produce only a small core flame at the electrodes. Because this flame also sacrifices energy through the electrode's contact surfaces, the core's propagation rate is limited. Under extreme conditions the energy loss from the corresponding quenching effect can be high enough to produce ignition miss.

As electrode gaps increase (from wear, etc.) lower quenching losses lead to improved conditions for ignition, but larger gaps also demand higher ignition voltages (Fig. 2). The reserves afforded by any given level of ignition voltage in the coil are reduced and the danger of ignition miss increases.

Engine manufacturers use various test procedures to determine the ideal electrode gap for each engine. The first step is to conduct ignition tests under various characteristic operating conditions to determine the minimum electrode gap. Important considerations include exhaust emissions, smooth operation and fuel economy.

The subsequent extended test runs provide insights into the spark plug's wear characteristics and their relationship to ignition-voltage requirements. The specified electrode gap is then defined at a point providing an adequate safety margin to the miss limit. Gap specifications are available in vehicle owner's manuals as well as in Bosch reference charts.

Bosch spark plugs are set to the specified gap at the factory.

1 Electrode gap (EA)

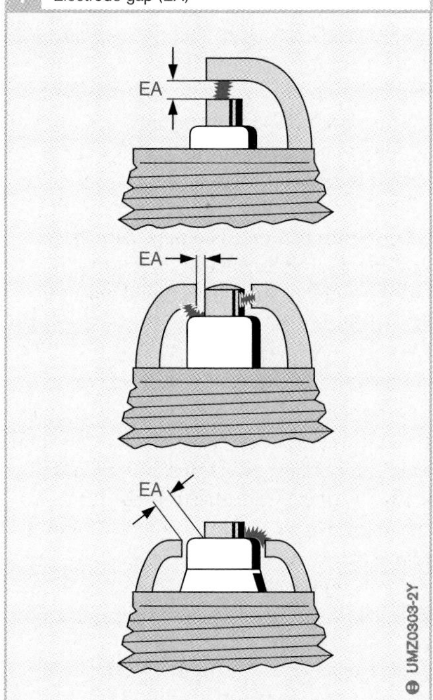

Fig. 1
- a Spark plug with front electrode (air gap)
- b Spark plug with side electrode (air gap or semi-surface gap)
- c Semi-surface gap spark plug

2 Relationship between electrode gap and ignition voltage

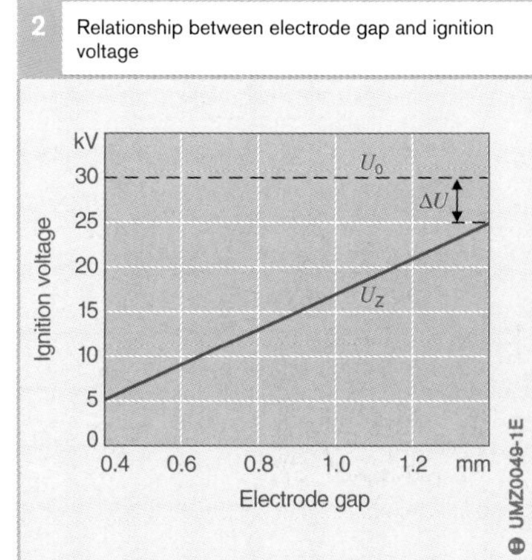

Fig. 2
- U_0 Available ignition voltage
- U_Z Ignition voltage
- ΔU Reserve ignition voltage

Spark position

The spark position (Fig. 1a) is the location of the spark path relative to the walls of the combustion chamber. Spark position has a substantial effect on combustion in modern engines (especially gasoline direct-injection powerplants). The criterion for defining the quality of the combustion process is the engine's operating consistency, or smoothness, which is in turn based on a statistical evaluation of mean effective pressures. The level of the standard deviation, described by the variation coefficient ($cov = s/p_{mi} \cdot 100\,[\%]$), provides an index for the evenness of the combustion process. The data also provide information on any major effects that delayed or missed combustion will have on engine operation. A figure of 5 % is defined for *cov* at the misfire limit.

Figure 1 illustrates the effects of leaning out the induction mixture and varying ignition timing on operational smooth running at two spark-plug positions. The lines describe constant smooth-running levels, while the 5 % boundary is in bold. Figures above this curve (<5 % range) correlate with smooth engine operation, with cyclically even combustion processes that are devoid of marked fluctuations. Results below the curve (>5 %) correspond to rough engine operation – combustion is inconsistent, with isolated miss or delayed ignition occuring under extreme conditions.

Comparison of the two diagrams for this engine indicates that shifting the spark further into the combustion chamber would substantially improve ignition, as the ignition timing sector increases above the 5 % curve and the misfire limit is pushed toward higher excess-air factors.

However, extending the length of the ground electrodes leads to higher temperatures, which produce increased electrode wear. The self-resonant frequency falls, which can lead to ruptures and fissures due to vibration. When the spark position is shifted forward, a number of other measures are needed to ensure adequate service life:

- The spark-plug shell is extended inward beyond the wall of the combustion chamber. The shoulder reduces the danger of electrode rupture.
- Copper cores are inserted in the earth electrodes. Placing copper in direct contact with the shell can reduce temperatures by roughly 70 °C.
- Electrodes are manufactured using materials designed to resist extreme temperatures.

1 Smoothness and roughness with different spark positions (f)

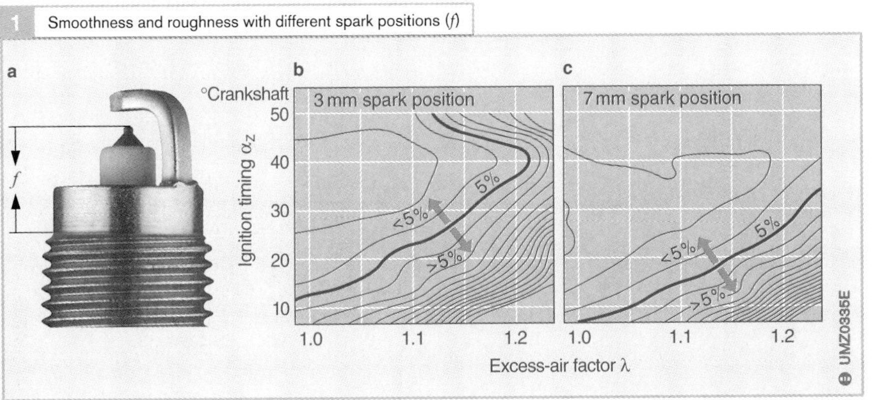

Fig. 1
a Definition of spark position f
b Diagram for $f = 3$ mm
c Diagram for $f = 7$ mm

Curves indicate operating coordinates at constant *cov* figures

$cov = s/p_{mi} \cdot 100$ [%]
s Standard deviation
p_{mi} Mean recorded pressure

5 % curve: misfire limit
<5 % range: good smooth running
>5 % range: rough running

Spark-plug heat ranges

Spark-plug operating temperatures

Range

Engines run on a rich air-fuel mixture when cold. This can lead to incomplete combustion and formation of soot deposits on spark plugs and combustion-chamber surfaces. These deposits contaminate the insulator nose to form a partially conductive link between the center electrode and the spark-plug shell (Fig. 1). This shunt effect allows a portion of the ignition energy to escape as "shunt current", reducing the overall energy available for actual ignition. As contamination increases, so does the probability that no spark will be generated.

The tendency for combustion residue to form deposits on the insulator nose is largely a function of temperature. This phenomenon is encountered primarily at temperatures below approximately 500 °C. At higher temperatures the carbon-based residue is burned from the insulator – the spark plug cleans itself.

The objective is thus to heat the insulator nose to a temperature of roughly 500 °C (with unleaded fuel) as soon as possible after starting and then to maintain it at this self-cleaning level.

It is also possible to observe the upper temperature limit, which lies at approximately 900 °C. Above this limit, hot combustion gases start to promote major corrosion and oxidation on the electrodes.

If temperatures rise even further, auto-ignition will become a potential problem (Fig. 2). In this process, the air-fuel mixture ignites on superheated spark-plug components to produce uncontrolled ignition which can damage or even destroy the engine.

Resistance to extreme temperatures

When the engine runs, the spark plug is warmed by the heat generated in the combustion process. A portion of the heat absorbed by the spark plug is transferred to the fresh induction gas. Most is conducted through the center electrode and the insulator to the spark-plug shell, from which it is transferred to the cylinder head (Fig. 3). The

Fig. 1
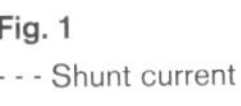
--- Shunt current

Fig. 2
1 Spark plug with correct heat range
2 Spark plug with heat range too low (cold plug)
3 Spark plug with heat range too high (hot plug)

The temperature in the operating range should be 500...900 °C at the insulator nose, varying according to engine output

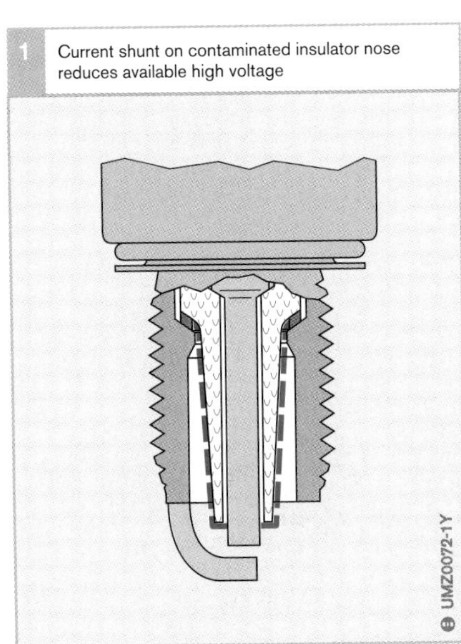

1 Current shunt on contaminated insulator nose reduces available high voltage

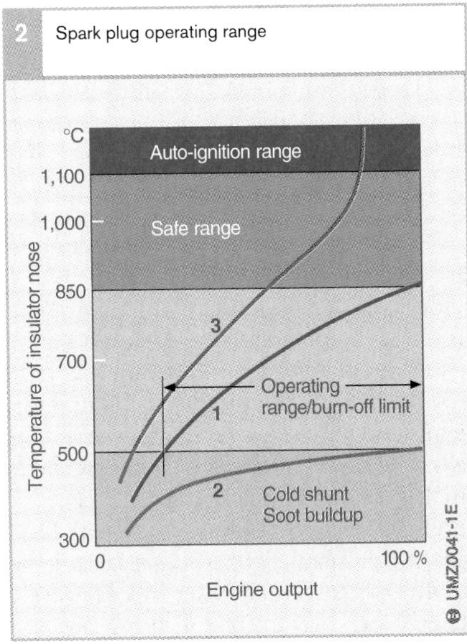

2 Spark plug operating range

ultimate operating temperature is the point at which absorption of heat from the engine and its dissipation into the cylinder head reach a state of equilibrium.

Heat transfer from the engine depends upon powerplant configuration. Engines with high specific outputs generally operate with higher combustion chamber temperatures than those with a low specific output.

The design of the insulator nose is the primary determinant of heat dissipation. Insulator-surface size determines thermal absorption, while cross-sectional area and center electrode affect thermal transfer.

The spark plug's thermal performance must thus be matched to the individual engine type. The index indicating a spark plug's resistance to thermal extremes is its heat range.

Heat ranges and code numbers

Each spark plug's heat range is determined using calibration plugs prior to assigning it a heat-range code number. A low heat range (such as 2...5) indicates a "cold" spark plug with a short insulator nose affording only limited thermal absorption. Higher heat ranges (codes numbers of 7...10 and so on) identify "hot" spark plugs with extended insulator noses to absorb more heat. Recognition of each spark plug's heat range and the engine applications for which it is intended is facilitated by the heat-range code number forming an integral part of the spark plug's designation.

Because WOT operation subjects the spark plug to maximum heat, it is under just these conditions that the correct heat range is determined. During operation, the spark plugs should never become so hot as to represent a source of thermally-induced auto-ignition. The heat-range recommendation is always defined with a safety margin relative to the auto-ignition limit to accomodate production variations in both plugs and engines. This margin is also important in view of the fact that an engine's thermal properties can vary over the course of time. One example is the potential increase in compression ratio caused by ash deposits within the combustion chamber, which in turn results in higher temperature loading for the spark plug. If the selected spark plug absolves the subsequent cold-starting tests without failure due to soot contamination, it is then defined as the correct plug for the specific application.

Because individual powerplants display a wide array of properties in the areas of operating loads, operating concept, compression, rpm, cooling and fuels, it is impossible to use just one spark plug for all engines. A plug that overheats in one engine would run at relatively cold temperatures in another.

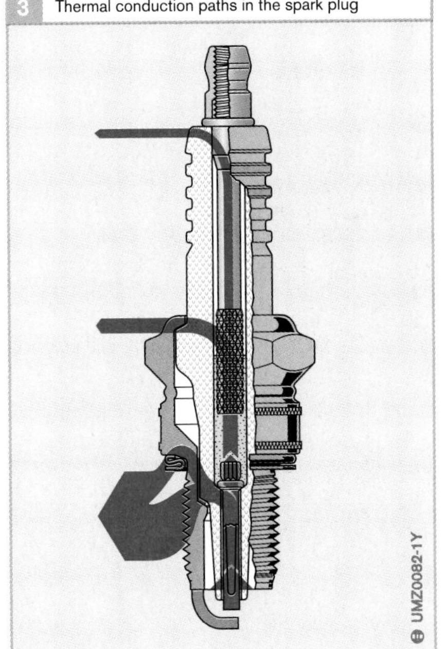

Fig. 3
A large proportion of the heat absorbed from the combustion chamber is dissipated via thermal conductivity (small contribution to cooling of approximately 20 % from flow of fresh induction mixture is not included)

Spark-plug selection

Bosch works together with engine manufacturers in jointly defining the ideal spark plugs for each powerplant.

Temperature measurements

A special spark plug designed and produced for thermal monitoring (Fig. 1) provides initial information on the right choice of plug. A thermocouple (2) embedded in the test plug's center electrode (3) makes it possible to monitor temperatures in relation to the engine's speed and load factor. This process represents a simple means of identifying the hottest cylinder and operating conditions for subsequent measurements as well as assisting in reliable designation of the correct plug for any specific application.

Ionic-current measurement

The Bosch ionic-current measurement employs the combustion characteristics as a factor for determining the correct heat range. The progress of combustion over time as reflected in the ionizing effect of flames can be assessed based on conductivity measurement in the spark path (Fig. 2). Because the electrical ignition spark produces a large number of charge carriers in the spark path, ionic current rises sharply at the firing point. Although the current flow falls once the ignition coil is discharged, the number of charge carriers maintained by the combustion process is large enough to allow continued monitoring. If the combustion-chamber pressure is monitored at the same time, normal combustion can be seen with an even pressure increase, with the pressure peaking after ignition TDC. If the spark plug's heat range is also varied during these trials, the resulting combustion characteristic will display typical shifts owing to the change in the plug's sensitivity to heat when the range is changed (Fig. 4).

1 Thermocouple spark plug

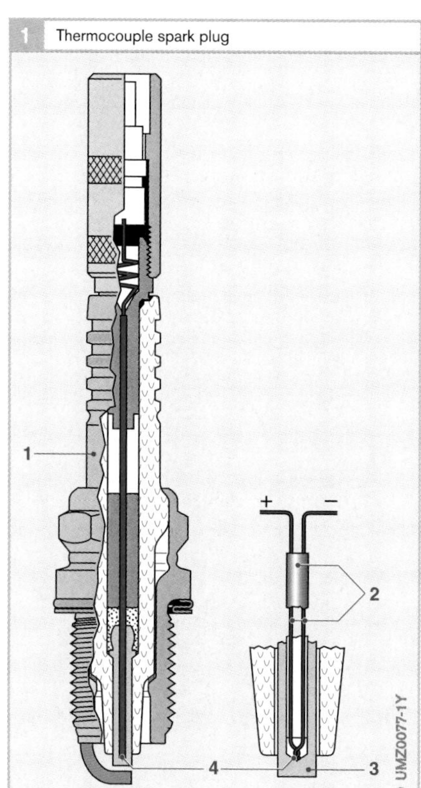
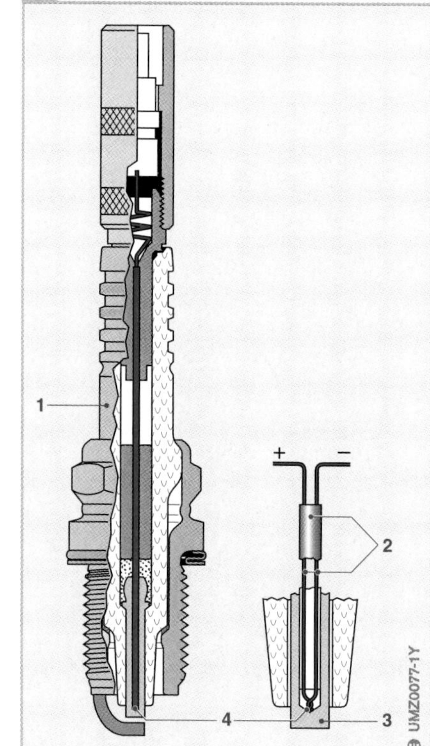

Fig. 1
1 Insulator
2 Thermocouple sleeve
3 Center electrode
4 Measurement point

2 Ionic-current test circuit diagram

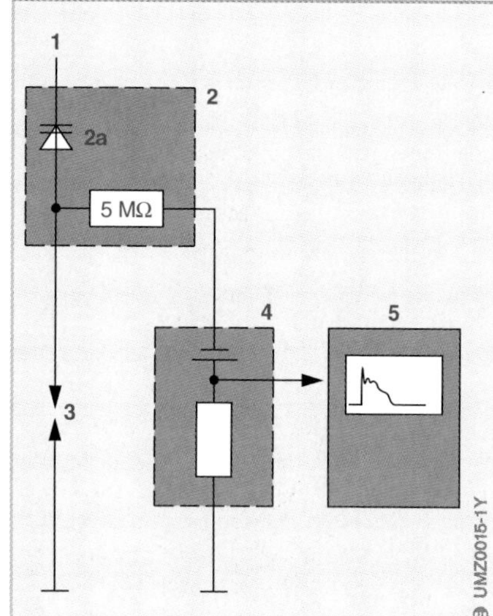

Fig. 2
1 From ignition distributor
2 Ionic-current adapter
2a Break-over diode (BOD)
3 Spark plug
4 Ionic-current device
5 Oscilloscope

The advantage of this method over internal measurements focusing exclusively on temperatures in the combustion chamber is that it indicates ignition probability, which is not solely dependent on temperature, but is also affected by the engine's design parameters and the spark plug itself.

Definition of terminology

Terminology and definitions for uncontrolled ignition in air-fuel mixtures for specifying spark-plug heat ranges have been defined in an internationally valid document (ISO 2542 – 1972, Fig. 3).

Thermal auto-ignition

Auto-ignition is defined as a process that results in ignition of the air-fuel mixture without an ignition spark, usually starting on a superheated surface (for instance, on the insulator nose of a spark plug with too high a heat range). These events can be classified in one of two categories, according to the point at which they occur relative to the firing point.

3 Terms in heat-range definition

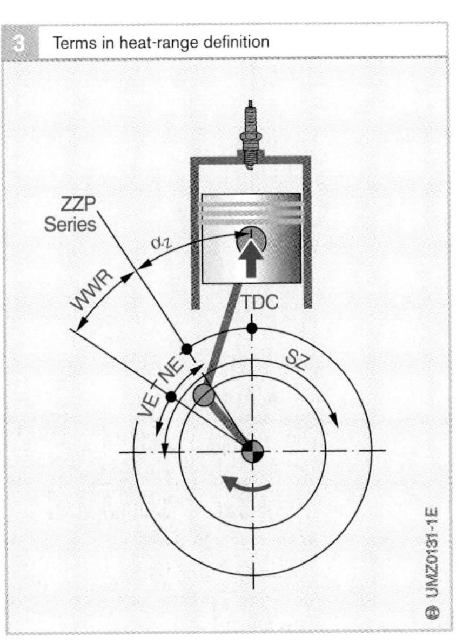

4 Characteristic ionic-current oscillogram

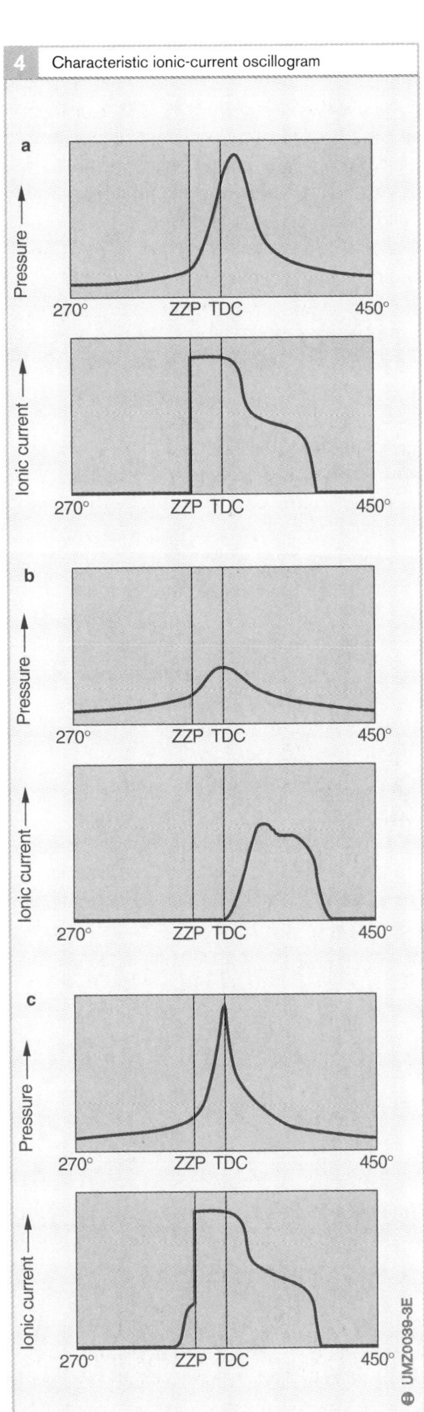

Fig. 3

SZ	Auto-ignition
TDC	Top Dead Center
VE	Pre-ignition
NE	Post-ignition
WWR	Heat-range reserve in °Crankshaft
ZZP	Ignition point in °Crankshaft BTDC
α_Z	Ignition angle

Fig. 4

a	Normal combustion
b	Suppressed ignition with post-ignition
c	Pre-ignition

Post-ignition

Because post-ignition takes place after electrically-induced ignition, it is not a critical factor in actual engine operation. Conducting measurements to determine whether the spark plug is producing thermal ignition of this kind entails suppressing the electrical spark. When post-ignition occurs the major rise in ionic current does not occur until after the firing point. Because it initiates a combustion process, a pressure rise and the corresponding torque generation are registered (Fig. 4b).

Pre-ignition

Pre-ignition occurs before the electrical firing point (Fig. 4c). The resulting uncontrolled combustion can lead to serious engine damage. Premature initiation of the combustion process shifts the pressure peak to TDC while also increasing maximum pressure levels in the combustion chamber and causing additional thermal stress on components inside it. In selecting spark plugs it is thus vital to ensure that no pre-ignition can take place.

Assessment of measurement results

The Bosch ionic-current measurement procedure can be used to detect both types of thermal ignition, whereby the ignition spark must be suppressed at specific intervals for detection of post-ignition. The point at which post-ignition occurs relative to the electrical firing point combines with the percentage of post-ignition events relative to the suppression rate to furnish insights regarding the stresses to which the spark plug is being subjected within the engine. Because spark plugs with extended insulator noses (hot spark plugs) absorb more heat from the combustion chamber and dissipate that heat less effectively, they are more likely to induce pre-ignition than spark plugs with shorter insulator noses. The application trials employed to select the correct heat range for a given engine environment thus rely on mutual comparisons of spark plugs with various heat ranges and analysis of their tendency to produce pre-ignition or post-ignition.

The preferred environments for conducting spark-plug selection testing are thus the engine test bed and the chassis dynamometer. Safety considerations prevent using extended WOT operation on public highways to determine the hottest operating parameters under full load.

Spark-plug selection

The object is to select a spark plug which will not produce pre-ignition and which possesses adequate heat-range reserves. This means that pre-ignition should not occur with spark plugs that are not hotter by at least two heat range numbers.

As this section has indicated, selection and application of spark plugs is a finely-tuned process. The procedure for choosing the ideal spark plug generally includes close cooperation between the spark-plug supplier and the engine manufacturer.

5 Bosch Spark Plug Catalog (German edition)

Fig. 5
Spark-plug selection: The engine manufacturer's specifications and the recommendations contained in the Bosch reference material are mandatory.
Bosch supplies the ideal spark plug for every engine – You can find the right plug in this catalog.

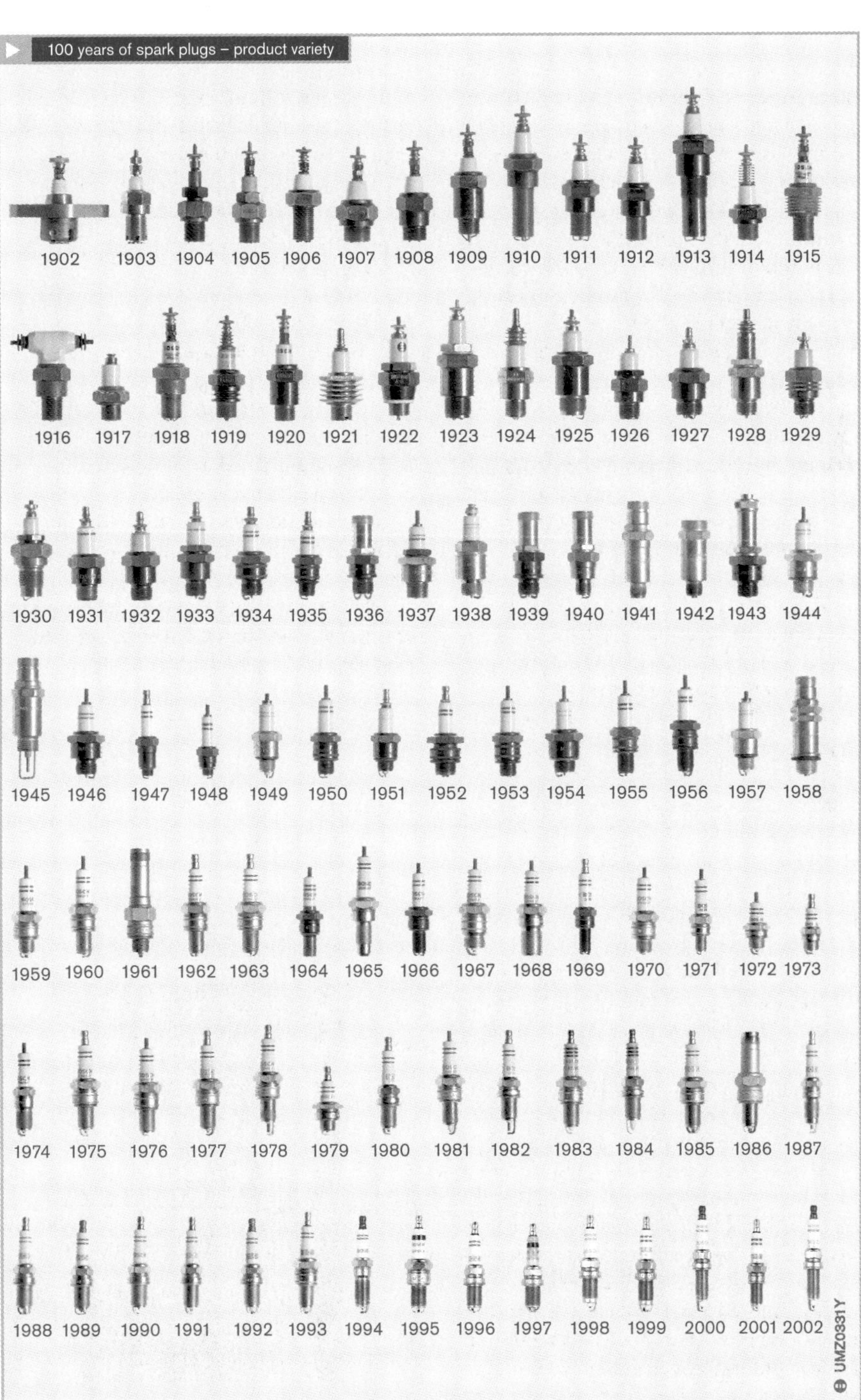
100 years of spark plugs – product variety
1902 1903 1904 1905 1906 1907 1908 1909 1910 1911 1912 1913 1914 1915
1916 1917 1918 1919 1920 1921 1922 1923 1924 1925 1926 1927 1928 1929
1930 1931 1932 1933 1934 1935 1936 1937 1938 1939 1940 1941 1942 1943 1944
1945 1946 1947 1948 1949 1950 1951 1952 1953 1954 1955 1956 1957 1958
1959 1960 1961 1962 1963 1964 1965 1966 1967 1968 1969 1970 1971 1972 1973
1974 1975 1976 1977 1978 1979 1980 1981 1982 1983 1984 1985 1986 1987
1988 1989 1990 1991 1992 1993 1994 1995 1996 1997 1998 1999 2000 2001 2002
UMZ0331Y

Spark-plug performance

Changes in the course of service life

Spark plugs operate in an aggressive atmosphere, sometimes at extremely high temperatures. This leads to electrode erosion, which raises the demand for ignition energy. When the situation finally reaches the point at which the demand for ignition voltage exceeds the coil's supply capacity, ignition miss occurs.

Spark-plug operation can also be negatively affected by changes in an aging engine and by contamination. As engines age, blowby and leakage increase, raising the amount of oil in the combustion chamber. This, in turn, leads to more deposits of soot, ash and carbon on the spark plug, with the danger of electrical tracking and ignition miss. Yet another factor is the use of antiknock additives in fuels, which form deposits that can become conductive at high temperatures and produce hot shunts. These cause ignition miss, characterized by a substantial increase in harmful emissions along with potential catalytic-converter damage. This is why spark plugs should be replaced at regular intervals.

1 Wear on center and ground electrodes

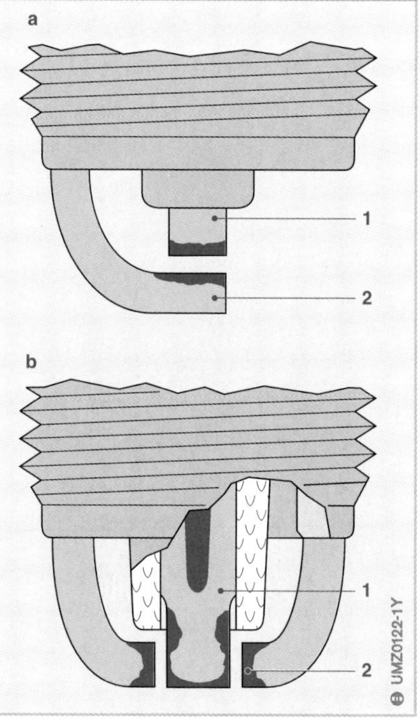

Fig. 1
a Spark plug with front electrode
b Spark plug with side electrodes

1 Center electrode
2 Ground electrode

Electrode wear

Electrode wear is synonymous with electrode erosion, a material loss which causes the gap to grow substantially over the course of time. This phenomenon essentially arises from two sources:

- Spark erosion and
- Corrosion in the combustion chamber

Spark erosion and corrosion

Arcing electrical sparks heat the electrodes to their melting point. The microscopically small melted surface areas react with oxygen or other elements in the combustion gases. This produces material erosion, widening the electrode gap and raising the demand for voltage from the coil (Fig. 1).

Electrode wear is minimized by using materials with high resistance to thermal extremes (such as platinum and platinum alloys). It is also possible to reduce erosion without limiting service life using suitable electrode geometry (smaller diameter, thin cores) and alternate spark-plug designs (surface-gap plugs).

The electrical resistance in the glass casting also reduces erosion and wear.

Abnormal operating conditions

Unusual operating conditions can destroy both the spark plugs and the engine.
Such conditions include:

- Auto-ignition
- Knock and
- High oil consumption (ash and carbon deposits)

Engine and spark plugs can also be damaged by incorrect ignition-system settings, spark plugs with the wrong heat range and unsuitable fuels.

Auto-ignition

Full-throttle operation can produce localized hot spots and induce auto-ignition in the following areas:

- At the spark-plug insulator nose
- On exhaust valves
- On protruding sections of the head gasket and
- On flaking deposits

Auto-ignition is an uncontrolled ignition process accompanied by increases in combustion-chamber temperatures severe enough to cause serious damage to both spark plugs and engine.

Combustion knock

Knocking is defined as uncontrolled combustion featuring extremely steep pressure spikes. Knock is caused by spontaneous ignition of the mixture in areas which the advancing flame front, initiated by the electrical spark, has not yet reached. Combustion proceeds at a considerably faster rate than normal. High-frequency pressure oscillations with extreme pressure peaks are then superimposed on the normal pressurization characteristic (Fig. 3). The severe pressure gradients expose components (cylinder head, valves, pistons and spark plugs) to extreme thermal loads capable of damaging one or numerous components.

The damage is similar to that associated with cavitation damage from ultrasonic flow currents. On the spark plug, pitting on the ground electrode's surface is the first sign of combustion knock.

2 Ground electrode damaged by excessive knock

3 Cylinder pressure characteristic

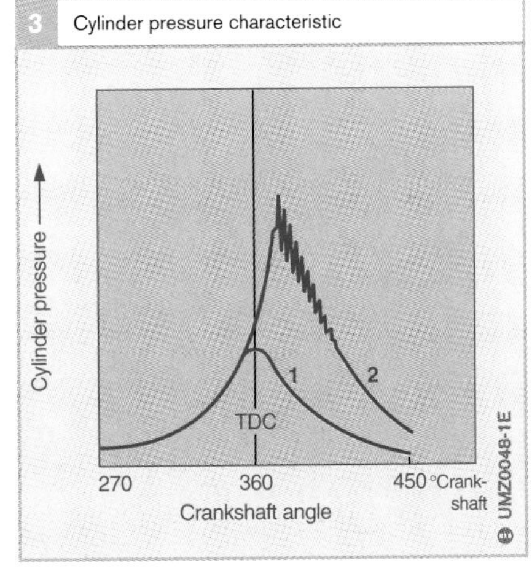

Fig. 3
1 with normal combustion
2 with combustion knock

Versions

To accomodate the wide array of potential applications, more than 1,400 different spark-plug designs are available for:

- Passenger cars
- Commercial vehicles
- Single-track vehicles
- Ships and small craft
- Agricultural and construction machinery
- Chainsaws
- Gardening machinery, etc.

SUPER spark plug

SUPER spark plugs (Fig. 1) make up most of the Bosch spark-plug range, and serve as the basis for various derivative spark-plug types and concepts. A suitable version with precisely the right heat range is available for virtually every engine and application.

A cutaway view of the SUPER spark plug is provided in the "design" section (Fig. 1). The most significant characteristics of the SUPER spark plug are:

- A composite center electrode consisting of nickel-chrome alloy and featuring a copper core
- As an option, a composite ground electrode designed to operate at lower temperatures for reduced erosion and
- Electrode gaps that are preset for the individual application at the factory

Various spark-plug profiles are employed to satisfy specific demands. The spark plug illustrated in Figure 2b varies in a number of points from the classic SUPER (Fig. 2a). The spark position projects further into the combustion chamber, while a thinner center electrode and optimized isolator-nose geometry provide for improved performance in repeated cold starts.

The version in Figure 2c features a noble-metal center pin welded into place by laser. This not only extends service life, but also improves ignition and flame propagation thanks to its small diameter.

1 The SUPER spark plug from Bosch

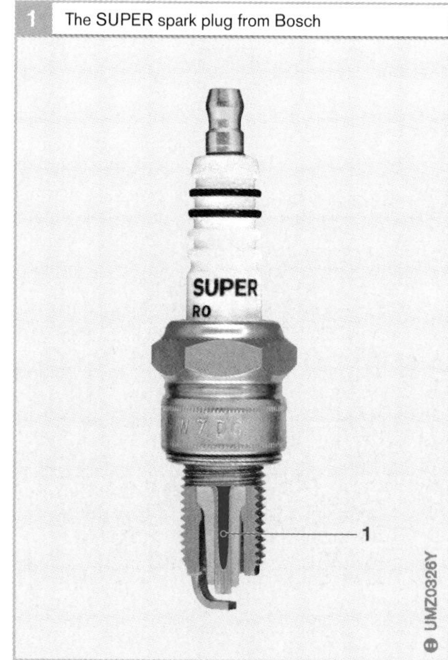

Fig. 1
1 Composite center electrode with copper core

2 SUPER spark-plug electrode shapes

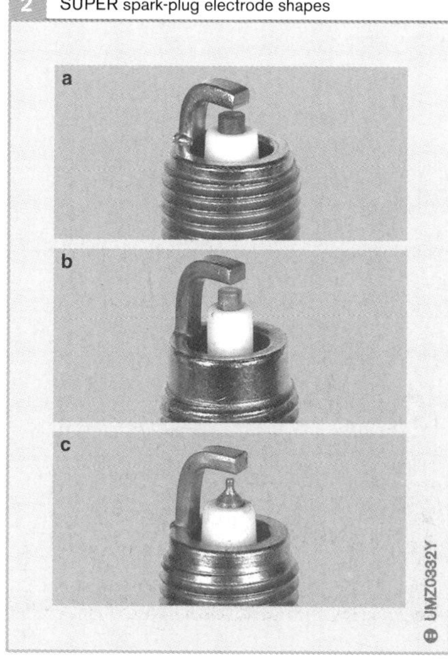

Fig. 2
a Front electrode
b Front electrode and forward spark
c Front electrode and platinum center electrode

SUPER 4 spark plug

Design

The special features that distinguish the Bosch SUPER 4 spark plug from conventional SUPER plugs include

- Four symmetrically arranged ground electrodes (Fig. 3)
- A silver-coated center electrode and
- A preset gap requiring no adjustment during the plug's service life

Operating concept

The four ground electrodes are manufactured in a thin profile to ensure good ignition and flash-front propagation. The defined gap separating them from the center electrode and the insulator nose allows the spark to arc either directly between electrodes (air gap) or along a semi-surface path. The result is a total of eight potential spark paths. Each spark's actual trajectory is usually determined by random factors.

Uniform electrode wear

Because the probability of spark propagation is the same for all electrodes, the arcs are evenly distributed across the insulator nose. Thus wear is balanced among the four electrodes.

Heat range

The silver-coated center electrode provides effective heat dissipation. This reduces susceptibility to auto-ignition, extending the secure operating range. These assets mean that each SUPER 4 has a heat range corresponding to at least two ranges in a conventional spark plug. A relatively limited number of spark plugs thus furnishes ideal replacements for an extended range of vehicles.

Spark-plug efficiency

The SUPER 4's thin ground electrodes absorb less energy from the ignition spark than the electrodes on conventional spark plugs. The SUPER 4 thus offers higher operating efficiency by providing up to 40 % more energy to ignite the air-fuel mixture (Fig. 4).

Ignition probability

Higher excess air (lean mixture, $\lambda > 1$) reduces the probability that the energy transfered to the gas will be sufficient for reliable ignition. In laboratory testing, the SUPER 4 has demonstrated the ability to provide reliable ignition of mixtures as lean as $\lambda = 1.55$, even though half of all ignition attempts will fail under these conditions when standard spark plugs are installed (Fig. 5).

3 Electrodes on SUPER 4 spark plug from Bosch

4 Spark-plug efficiency

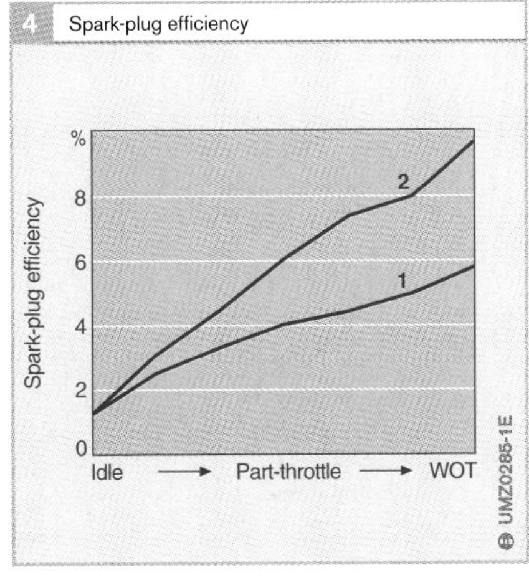

Fig. 4
1 Conventional spark plug
2 SUPER 4 spark plug from Bosch

Cold-start performance

Arcs with a surface-gap spark track ensure effective self-cleaning, even at low temperatures. This means that up to three times as many cold starts (without allowing the engine to warm completely in the intervals) are available as with conventional plugs.

Protection of environment and catalytic converter

Improved cold-start performance and more reliable ignition under all conditions including the warm-up phase, reduce uncombusted fuel for lower levels of HC emissions.

Advantages

The special features that set the SUPER 4 apart from conventional spark plugs include:

- Eight potential spark paths for reliable ignition
- Self-cleaning from surface-gap technology and
- Extended heat range

Fig. 5
1 Conventional spark plug
2 SUPER 4 spark plug from Bosch

5 Effect of mixture composition on ignition probability

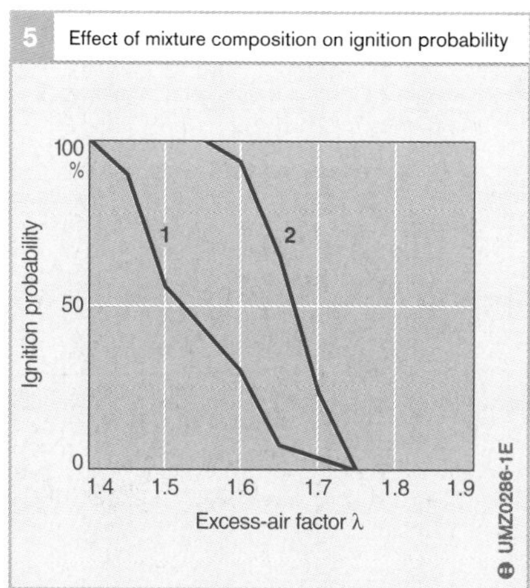

Platinum +4 spark plug

Design

The Platinum +4 spark plug (Fig. 6) is a surface-gap spark plug designed for extended replacement intervals. It is distinguished from conventional spark plugs by

- Four symmetrically arranged, double-bent ground electrodes (9)
- A thin sintered center electrode in platinum (8)
- An internal conductor pin (7) featuring improved geometry and a special alloy
- A ceramic insulator (2) with high breakdown resistance and
- An insulator nose redesigned for improved performance

6 Design of Platinum+4 spark plug

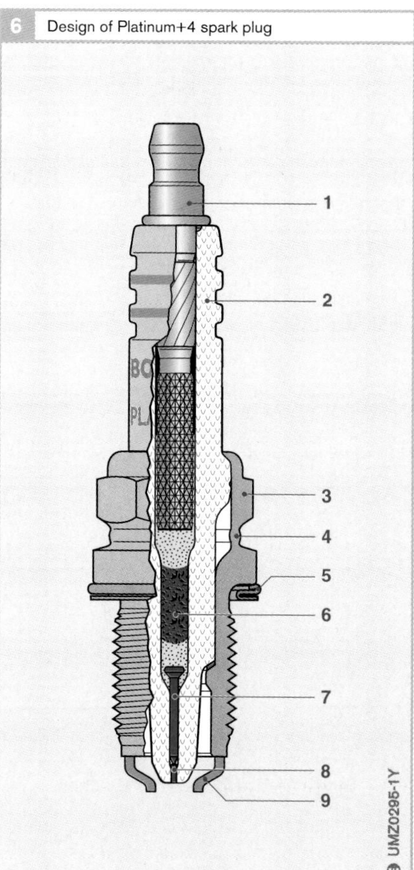

Fig. 6
1 Terminal stud
2 Insulator
3 Shell
4 Heat-shrinkage zone
5 Captive gasket
6 Conductive glass seal
7 Conductor pin
8 Platinum pin (center electrode)
9 Ground electrodes (only two of four electrodes shown)

Operating concept

Reliable ignition

The extended gap of 1.6 mm endows the Platinum +4 with the ability to supply superbly reliable ignition, while the four ground electrodes are ideally positioned within the combustion chamber to ensure that the ignition spark has unobstructed access to the mixture. This allows the core flame to spread into the combustion chamber with virtually no interference, ensuring complete ignition of the entire air-fuel mixture.

Response to repeated cold starts

The surface-gap concept provides substantial improvements over air-gap plugs in repeated cold starting.

Electrode wear

There are also advantages regarding electrode wear, thanks to the erosion-resistant platinum rod in the center electrode and improved materials in the four ground electrodes. The resistance in the conductive glass seal reduces capacitive discharge, making a further contribution to reduced arcing erosion.

The comparison in Figure 7 shows the rise in demand for ignition energy during 800 h on an engine test stand (corresponding to 100,000 km of highway use). The Platinum +4 spark plug's lower electrode wear leads to substantial reductions in the rate at which voltage demand increases relative to conventional spark plugs. Figures 8 and 9 illustrate the condition of a Platinum +4 spark plug when new and after 800 h of in-engine operation; the minimal electrode wear at the end of the endurance test is obvious.

8 Appearance of new Platinum+4 spark plug

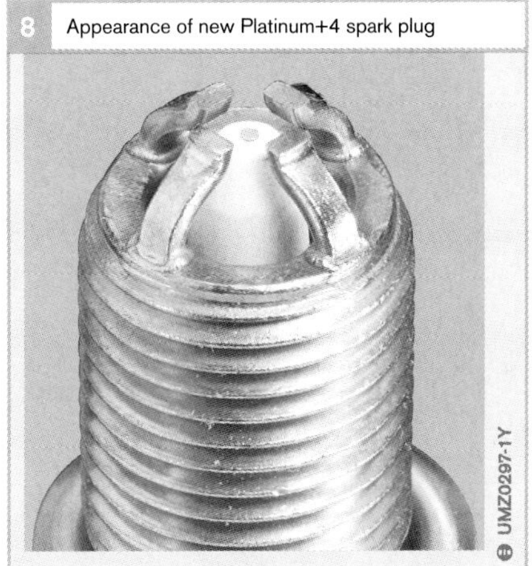

7 Increase in ignition-voltage demand over time

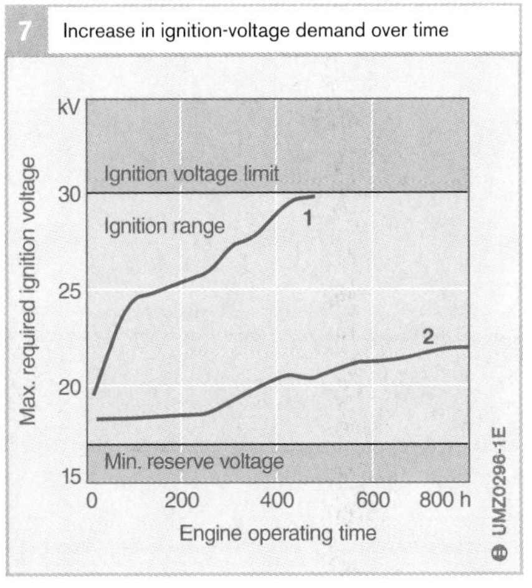

9 A Platinum+4 spark plug after 800 hours of operation

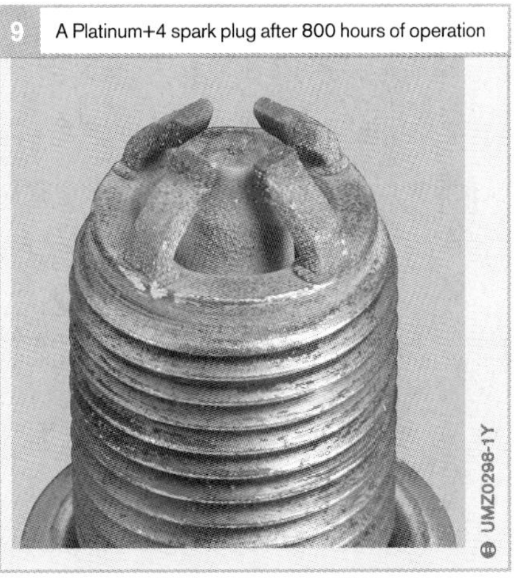

Fig. 7

1 Spark plug with air gap (gap = 0.7 mm)

2 Platinum+4 spark plug with surface gap (gap = 1.6 mm)

Advantages of the Platinum+4 spark plug

The Platinum+4 spark plug incorporates an array of assets making it ideal for extended-duty applications:

- Durable electrodes and ceramic components extend plug replacement intervals to as much as 100,000 km
- Higher numbers of repeated cold starts possible
- Extremely good ignition and flash-front propagation for major improvements in running smoothness

Spark plugs for direct-injection engines

These powerplants contrast with manifold-injection units by relying upon a pronounced swirl effect to transport the air-fuel mixture to the spark plug. Because both the mass and direction of the flow vary under different engine operating conditions, a plug that projects a spark well into the combustion chamber is of distinct advantage. This forward-spark concept has negative consequences on the temperature of the ground electrode though, meaning that countermeasures are required. Efficient concepts rely on extended shells to reduce the length of the ground electrodes and thereby lower their temperatures. Both air-gap and surface-gap spark plugs may be employed to meet the needs of specific combustion concepts (Fig. 10).

10 Spark plugs for engines with direct gasoline injection

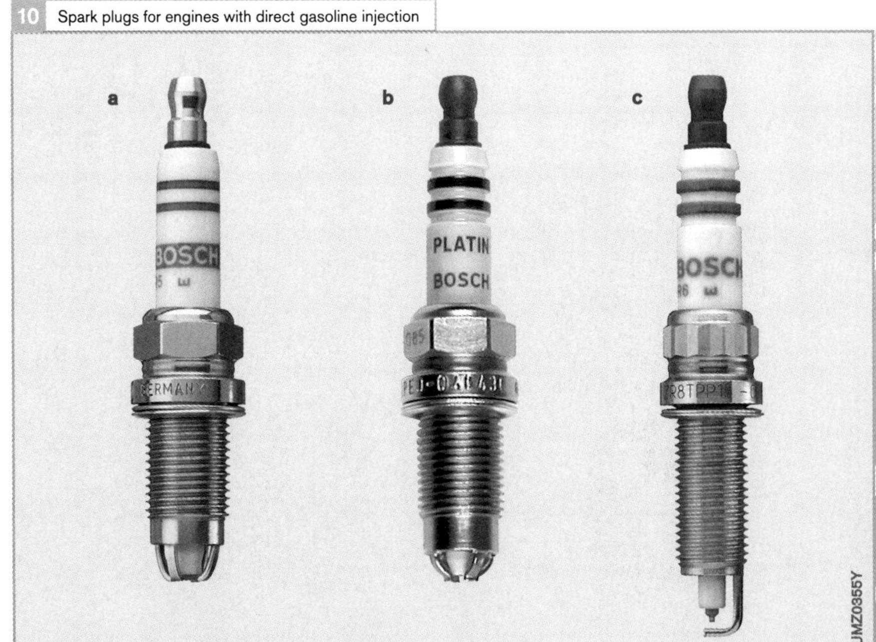

Fig. 10
- a Surface-gap spark plug without noble metal
- b Surface-gap spark plug with platinum center electrode
- c Air-gap spark plug with platinum on center electrode

Special-purpose spark plugs

Applications

Special-purpose spark plugs are available for use in certain applications. These plugs feature unique designs dictated by the operating conditions and installation environments in individual engines.

Competition spark plugs

Extended full-throttle operation subjects the spark plugs in competition vehicles to extreme thermal loads. The spark plugs produced for this operating environment usually have noble-metal electrodes (silver, platinum) and a short insulator nose. This type of insulator nose keeps thermal absorption extremely low, while the heat dissipation through the center electrode is high (Fig. 11).

Spark plugs with resistors

A resistor can be installed in the spark plug's internal conductor path to suppress transmission of interference pulses to the ignition cable and reduce RFI. The reduced current in the ignition spark's arcing phase also leads to lower electrode erosion. The resistor consists of a special conductive glass seal between the center electrode and the terminal stud. Appropriate additives endow the seal glass with the desired level of resistance.

Fully-shielded spark plug

Shielded spark plugs may be required in applications marked by extreme demands in the area of interference suppression (cell phones, communications equipment).

In fully-shielded spark plugs the insulator is surrounded by a metal sleeve. The connection is inside the insulator. A union nut attaches the shielded ignition cable to the sleeve. Fully-shielded spark plugs are also watertight (Fig. 12).

11 Competition spark plug

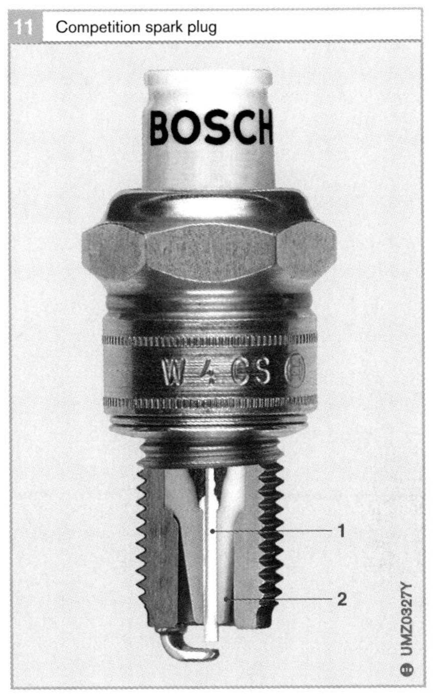

12 Fully-shielded spark plug

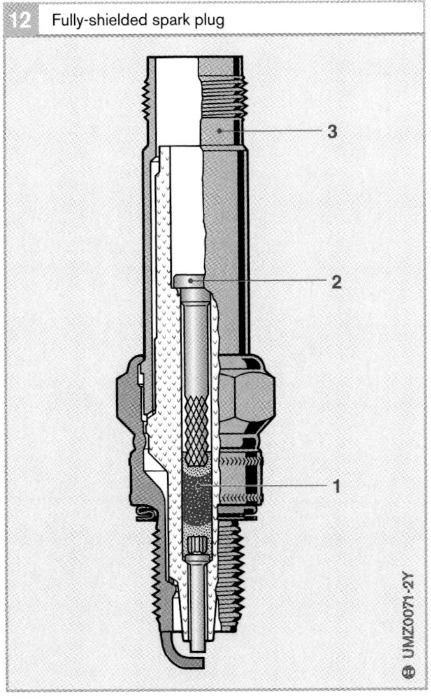

Fig. 11
1 Silver center electrode
2 Short insulator

Fig. 12
1 Special glass seal (RFI-suppression resistor)
2 Ignition-cable terminal
3 Shield

Spark-plug type designations

Spark plugs are identified by a designation code (Fig. 1). This designation code includes all of the spark plug's vital data with the exception of the gap. This is provided on the package. Spark-plug specifications for individual engine applications are defined by Bosch and the engine manufacturers.

1 Type code key for Bosch spark plugs

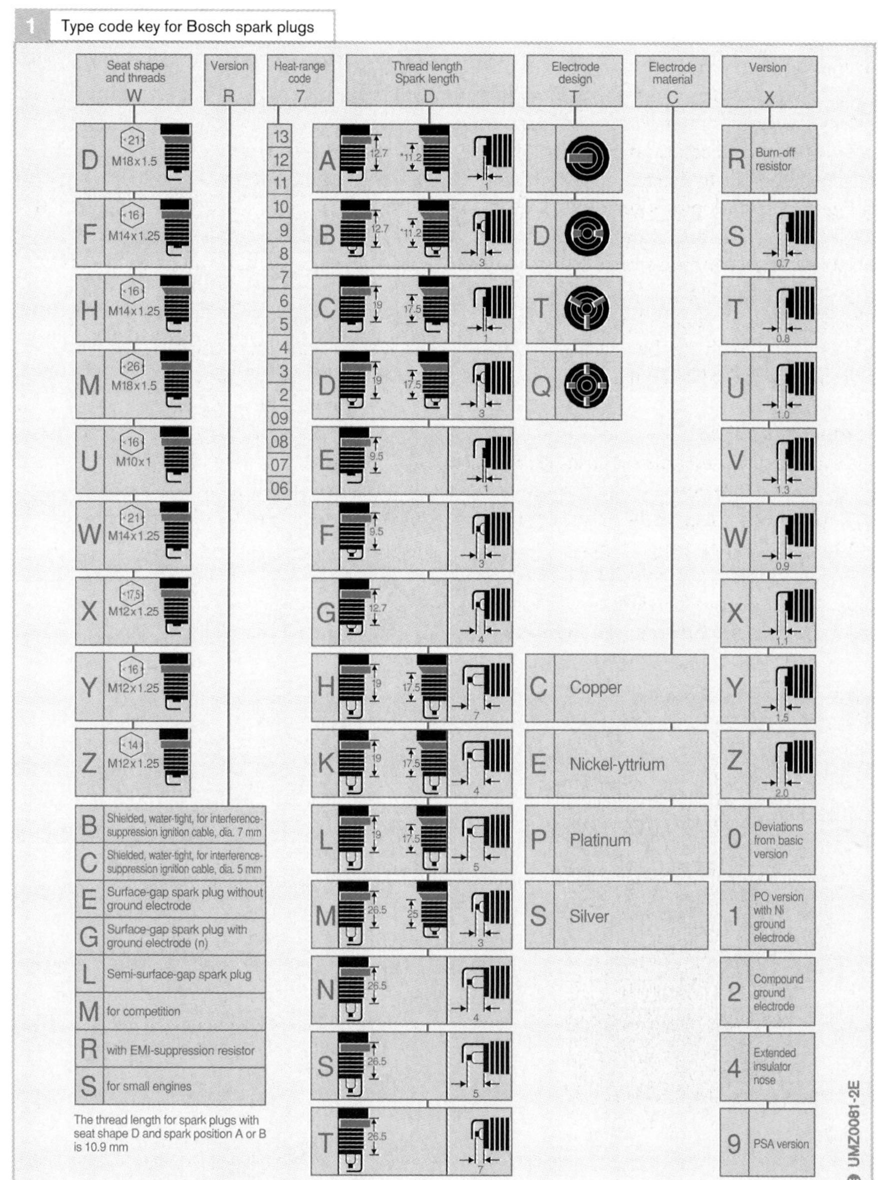

Simulation-based spark-plug development

The Finite Element Method (FEM) is a mathematical approximation procedure for solving differential equations and describing the properties of physical systems. The process entails dividing structures into individual sectors, or finite elements.

In spark-plug design, FEM is employed to calculate ranges for temperature, electrical properties and structural mechanics. It makes it possible to determine the effects of changes to a spark plug's geometry and constituent materials, and variations in general environmental conditions, in advance, without extensive testing. The results furnish the basis for precisely focused production of the test samples which are then used for verification of the calculated data.

Temperature field

The maximum temperatures of the ceramic insulator and the center electrode within the combustion chamber are decisive factors for the spark plug's heat range. Figure 1a is an axisymmetrical representation of a spark plug along with a section of the cylinder head. The temperature ranges as indicated in the colored sections show that the highest temperatures occur at the nose of the ceramic insulator.

Electric field

The high voltage applied at the firing point is intended to create an arc at the electrodes. Breakdown in the ceramic material or current tracking between the ceramic insulator and the spark plug's shell can lead to delayed combustion and ignition miss. Figure 1b shows an axisymmetrical model with the corresponding field strength vectors between center electrode and shell. The electrical field permeates the nonconductive ceramic material and the intermediate gas.

Structural mechanics

High pressures within the combustion chamber make a gas-tight union between the spark plug's shell and the insulator essential. Figure 1c shows a symmetrical, axially split model of a spark plug after the shell has been crimped and heat-shrunk. The retention force and the mechanical stress in the spark-plug shell are measured.

1 FEM application on spark plug

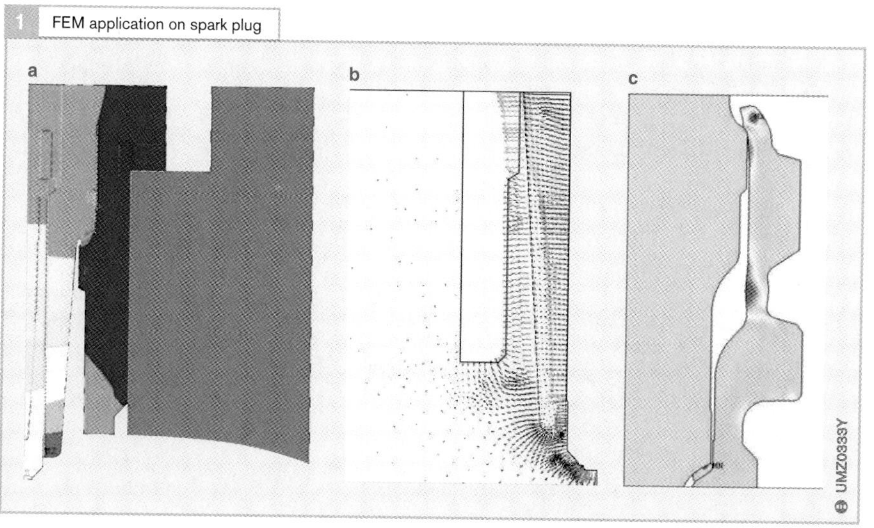

Fig. 1
Axially split model of a spark plug
- a Temperature distribution in ceramic insulator and center electrode
- b Electric field strength adjacent to center electrode and shell
- c Retaining force and mechanical stress in spark-plug shell

Spark-plug manufacture

Each day roughly one million spark plugs emerge from our Bamberg plant, the only Bosch facility manufacturing these products in Europe. Spark plugs conforming to the universal Bosch quality standards are also produced for local OEM and replacement markets in India, Brazil, China and Russia. Bosch has now produced a total of far more than seven billion spark plugs.

The individual components joined to form the finished spark plug in final assembly are produced in three parallel manufacturing processes.

Insulator

The basic substance in the high-quality ceramic insulator is aluminum oxide. Aggregate materials and binders are added to this aluminum oxide, which is then ground to a fine consistency. The granulate is poured into moulds and processed at high pressure. This gives the raw castings their internal shape. The outer contours are ground to produce the soft core, which already displays a strong similarity to the later plug core. The ceramic castings are run through a sintering oven where they receive their final form at roughly 1,600 °C. The soft core must be manufactured to compensate for the contraction that occurs in the sintering process, which is approximately 20 %.

Once the insulators have been fired the labeling is applied to the insulator body, which is then coated with a lead-free glaze.

Plug core

Center electrodes are manufactured using blanks containing a mixture of copper and nickel alloy. These are joined and then shaped to form a center electrode in an extrusion process. A lug at the rear of the electrode is staked in place to ensure that the unit will remain firmly attached to the plug core later on. Once the center electrode is inserted in the insulator, paste is filled into the hole. The *paste* consists of glass particles to which conductive particles are added to produce an electrical connection to the terminal stud after casting. The individual components can also be varied to manipulate the paste's resistance. Resistances of up to 10 kΩ are possible.

The *terminal stud* is manufactured from wire and formed by flattening and edge knurling. It receives a protective nickel surface and is inserted in the core. The core then travels through an oven where it is heated to over 850 °C. At these temperatures the paste assumes a molten texture. It flows around the center electrode, and the terminal post can be pressed into the material. The core cools to form a gas-tight and electrically conductive connection between the center electrode and the terminal stud.

Shell

The shell is manufactured in steel using extrusion processing. A section several centimeters in length is cut from the coil and then cold-formed in several pressing operations until the spark-plug shell assumes its final contours. Only a limited number of machining operations (to produce shrinkage and threaded sections) are then required. After the ground electrodes (up to four depending upon spark-plug type) are welded to the shell, the threads are rolled and the entire shell is nickel-coated for protection against corrosion.

Spark-plug assembly

During spark-plug assembly a seal ring and the plug core are installed in the shell. The upper shell is crimped and beaded to position the plug core. A subsequent shrinking process (induction is used to heat parts of the spark-plug shell to over 900 °C) provides a gas-tight union between spark-plug shell and core. Then an outer seal is mounted on flat-seat spark plugs in an operation that reshapes the material to form a captive gasket. This ensures that the combustion chamber will be effectively sealed when the spark plug is subsequently installed in a cylinder head.

On some spark-plug versions an SAE nut must then be installed on the terminal stud's M4 threads and staked several times to form a firm attachment.

The electrode gap is then adjusted to the engine manufacturer's specifications, and the spark plugs are inserted in the packages specified for the intended customer and commercial application.

1 Spark-plug production sequence

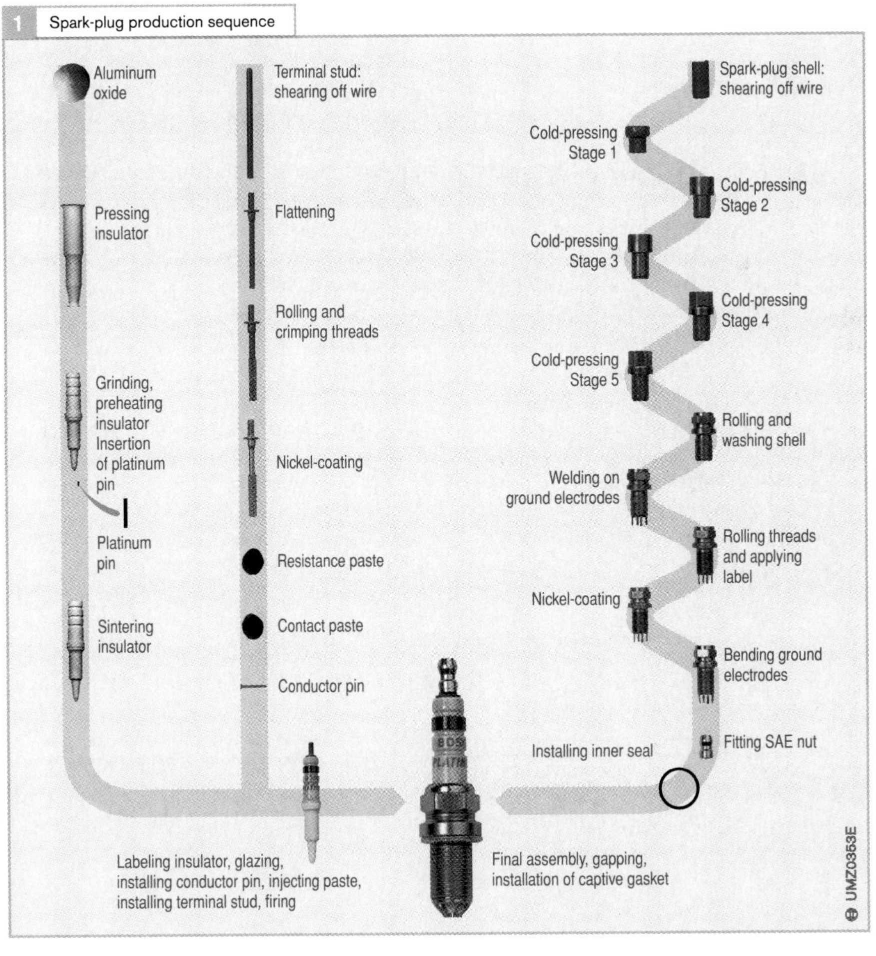

Motronic engine management

"Motronic" is synonymous with an engine-management concept that unites an extensive array of open and closed-loop control functions for gasoline power plants within a single electronic control unit. Bosch started series production of the first Motronic system in 1979. The system's focus was on controlling the electronic ignition and fuel injection. Progress in microelectronics technology has spurred an unbroken string of advances in performance levels. Increasingly complex systems incorporating progressively more extensive levels of functionality have allowed Motronic to keep pace with present-day performance demands.

System overview

The first Motronic system combined L-Jetronic's electronic intermittent-injection system with map-controlled electronic ignition timing. Ignition still relied on rotating high-voltage distribution (ROV). The mechanical distributor required for ROV was subsequently rendered redundant by stationary voltage distribution (RUV).

Although cost considerations limited application of early Motronic versions to the automotive upper class, progressively more stringent demands for clean emissions gradually led to widespread use of this system. Currently, Bosch bases all engine-management development projects on the Motronic system.

Monitoring operational data

Sensors and setpoint generators

Motronic employs sensors and setpoint generators to gather the operational data required for open and closed-loop control of the engine (Figure 1).

The setpoint generators (switches, etc.) register the selections made by the vehicle operator. Monitored parameters include

- The position of the key in the ignition switch (Terminal 15)
- The positions of the air-conditioning switches and
- Settings for the cruise-control system

By monitoring physical and chemical parameters the sensors are able to furnish information on the engine's current operating status. Examples of these monitored parameters include:

- Engine temperature
- Induction air mass
- Intake-manifold pressure
- Throttle-plate angle
- Excess-air factor λ
- Crankshaft rotation rate
- Camshaft position
- Vehicle speed

Various sensors relay data in digital, pulse or analog form.

Signal processing in the ECU

Input circuits within the electronic control unit – or, increasingly, local circuitry within the sensors – prepare raw signal data for processing. These circuits transform voltages to the levels required for subsequent processing in the control unit's microprocessor.

The microcontroller records digital input signals directly for storage in digital form. Analog signals must be transformed into digital data in the analog-digital converter, or ADC.

1 Components used for electronic control in Motronic systems

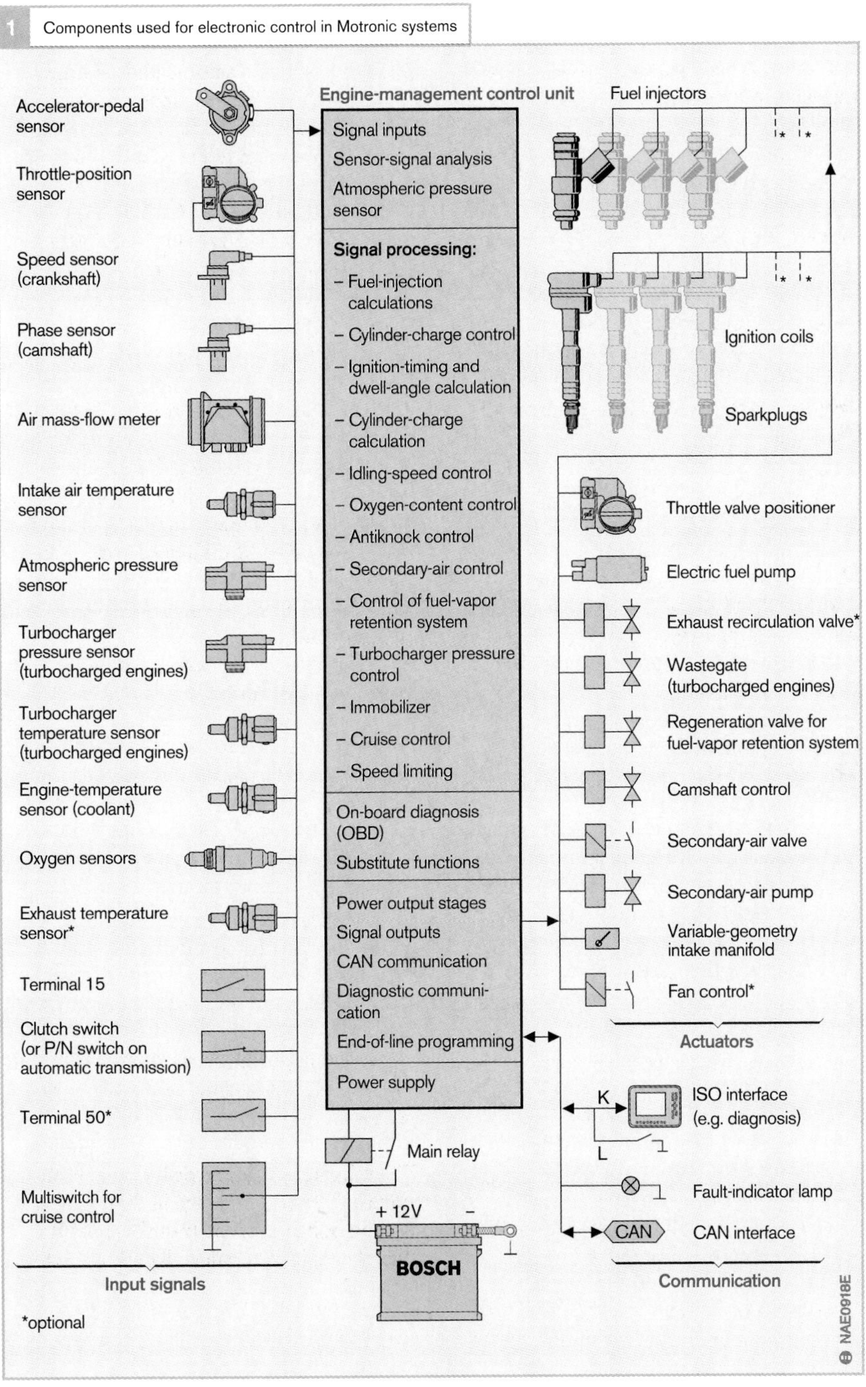

Processing operational data

By allowing the engine-management ECU to monitor the engine's current status, these input signals combine with the direct-demand monitors (registering demand from driver and ancillary equipment) to provide the basis for processing operations in which the ECU generates control signals for the actuators.

The engine-management ECU's activities are divided into specific functions. The corresponding algorithms are programmed into software in the ECU's program memory.

Basic functions

Motronic has two basic functions. Firstly, metering the correct mass of fuel in accordance with the air mass drawn into the engine, and secondly, the triggering of the ignition point at the most appropriate moment in time. Integrating these functions within a single system makes it possible to coordinate the injection and ignition functions for optimal performance.

Supplementary functions

The ongoing expansion in the computational capacity of available microcontrollers is powering a drive to integrate ever-increasing functionality within Motronic's open and closed-loop control processes. Progressive tightening of emissions limits simultaneously spurs the demand for functions capable of improving the composition of the engine's exhaust gases. The functions with the potential to support this objective include:

- Idle-speed control
- Closed-loop lambda control
- Regulation of the evaporative-emissions control system (tank purge system)
- Knock control
- Exhaust-gas recirculation for reduced NO_X emissions
- Control of the secondary-air injection system to ensure rapid catalyst response

The following functions can also be added to meet special drivetrain demands:

- Closed-loop control of turbocharger operation
- Control of variable-geometry intake manifold for enhanced torque and power
- Camshaft control allowing variable valve-timing systems to reduce exhaust emissions while increasing both power and fuel economy
- Torque and speed limiting functions to protect engine and vehicle

Ever-increasing priority is being assigned to the driver's comfort and convenience. This development has pronounced implications for the engine-management sector. Examples of typical comfort and convenience functions include:

- Cruise control
- ACC (adaptive cruise control)
- Torque adaptation during upshifts with automatic transmissions
- Load transition control (for smoother response to driver demand)

Actuators

The componentry in the output driver circuits furnishes the current that triggers the actuators (ignition coils, injectors, EGR valve, etc.). These components, in turn, are controlled by activation signals processed within the ECU's microcontroller.

Electronic diagnosis

The electronic control unit's integrated diagnostic functions monitor the Motronic system (sensors and actuators as well as ECU) for signs of defects and malfunctions. The system responds to detected problems by storing error codes in the malfunction log, and can also initiate activation of default control strategies as required. A diagnosis lamp or a display within the instrument cluster alerts the vehicle's operator to the problem.

The diagnostic interface provides access to any error codes and complementary status data stored within the ECU.

The diagnostic function was originally intended to assist technicians in providing service in the field. Then, with the promulgation of laws requiring OBD (**On**-**B**oard **D**iagnosis) in America, it evolved into a utility for detecting and warning of specific emissions-relevant problems. The mandatory European EOBD is an adapted version of the American OBD.

Vehicle management

Bus systems such as the CAN (**C**ontroller **A**rea **N**etwork) support data communications between Motronic ECU and other electronic systems within the vehicle. Fig. 2 illustrates several examples. The control units can integrate data from other systems as supplementary input signals for use in generating their own open and closed-loop control algorithms. For example, in order to achieve smoother shifts, Motronic reduces engine torque in response to data indicating that the transmission is changing gear.

Motronic versions

Motronic systems have undergone a process of continuing evolution reflecting the increasingly demanding performance requirements of vehicle systems. The following Motronic systems are currently available:

- M-Motronic, with the basic and supplementary functions described above.
- ME-Motronic, which is based on M-Motronic, expands on the base system by providing integral EGAS (electronic throttle control, EDC).
- MED-Motronic further expands on ME-Motronic by furnishing closed-loop control of direct gasoline injection.

Motronic systems (such as MEG-Motronic) are also available with integrated transmission control. However, these are not in extensive use, as the demands on hardware are considerable.

2 Components for data communications with Motronic (examples)

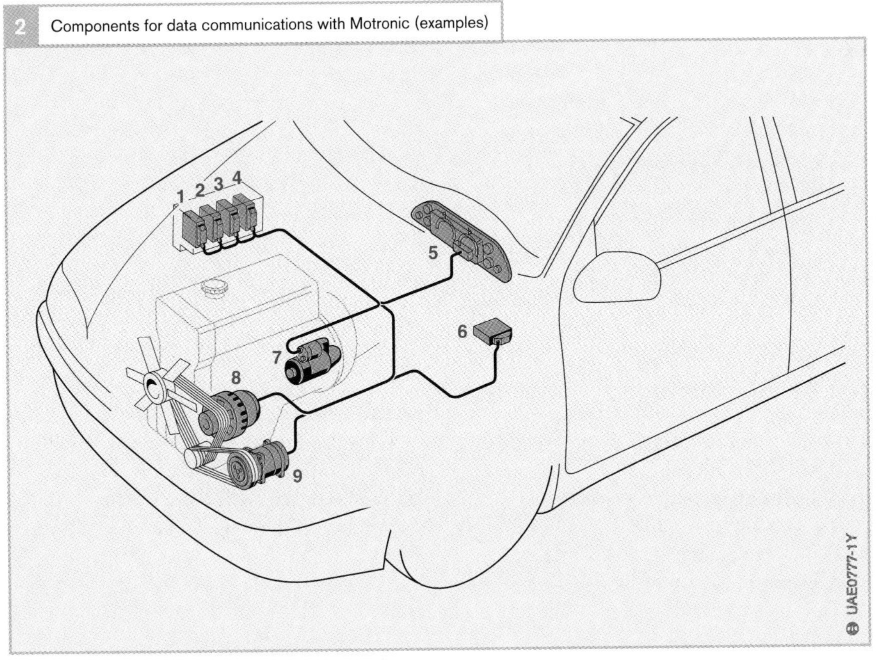

Fig. 2
1 Engine-management ECU (Motronic)
2 ESP control unit (with ABS and ASR)
3 Transmission control unit
4 Climate control ECU
5 Instrument cluster module with on-board computer
6 Immobilizer ECU
7 Starter
8 Alternator
9 A/C compressor

System structure

A few years ago it was still possible to represent the functionality of Motronic systems by "simple" system and function descriptions. Now the control and regulation operations for gasoline engines have become so complex that a structured system description is necessary.

The introduction of a torque structure can be seen as a milestone in the development of Motronic systems. All torque requirements placed on the engine are handled by the Motronic system as torque data and centrally coordinated. This structure was first introduced on the ME-Motronic. This system calculates the required torque and delivers it by making adjustments to:

- The electrically controlled throttle valve (air system)
- The ignition angle (ignition system)
- The fuel quantity in the case of direct-injection engines (fuel system)
- The use of injection suppression and
- Controlling the wastegate on exhaust-gas turbocharged engines

Figure 3 shows the system structure used for new Motronic systems and their various subsystems. Some subsystems are purely software constructs in the ECU (e.g. torque structure), while others incorporate hardware components (e.g. fuel system and fuel injectors). The various subsystems are interconnected by defined interfaces.

This system structure is also used for new M-Motronic systems that incorporate a torque structure.

3 Motronic system structure

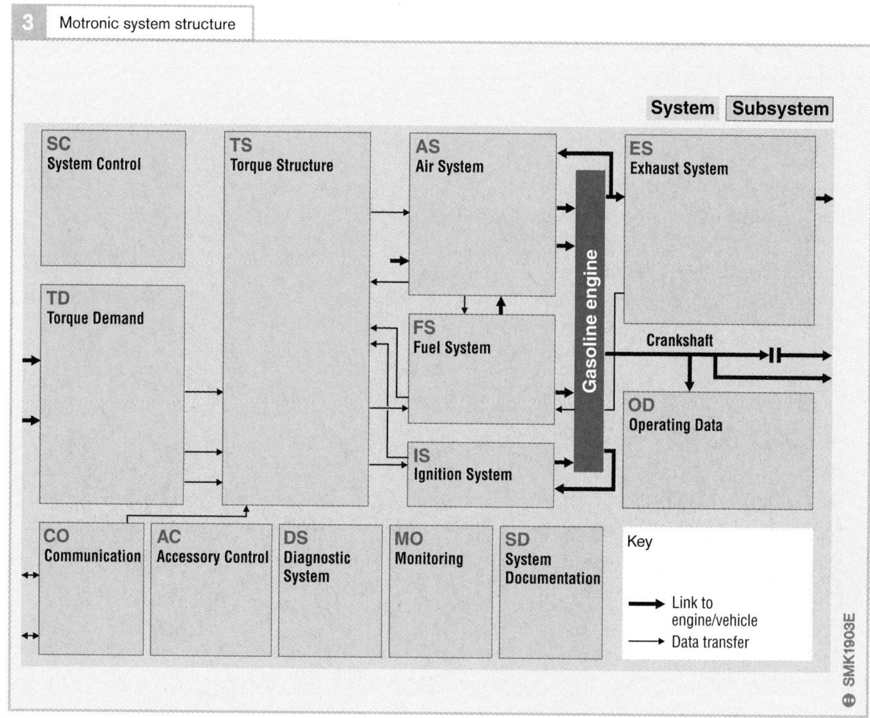

Motronic systems in motorsport

Simultaneously with the introduction of the Motronic systems on production vehicles, modified versions were also used on racing engines. Whereas the development objectives for production versions are aspects such as convenience, safety, reliability, emission limits and fuel consumption, the main focus in motor-racing applications is on maximum performance over a short period. The production costs with regard to choice of materials and dimensioning of components are a secondary consideration.

But the production and racing versions of the Motronic system are still based on identical principles because in both cases similar functions achieve contrasting aims. The excess-air factor and knock control systems are examples of this.

Environmental protection regulations are increasingly a consideration even in motorsport. The cars in the German Touring Car Championship, for example, are now fitted with catalytic converters. Noise and fuel-consumption levels have to be limited in more and more classes of racing. Consumption-reducing developments used on production cars quickly transfer to motor racing where shorter or less frequent refueling stops can make the difference between winning and losing. The 2001 Le Mans 24-hour race, for example, was won for the first time by a car with a Bosch gasoline direct-injection system.

The high revving speed of racing engines minimizes the time available during each operating cycle. The vast amount of process data requires high processor clock frequencies and the use of multiprocessor systems to a greater extent.

Not only the ECU but also the ignition and fuel-injection components have to operate at extremely high speeds. This requires ignition coils with fast charging times and fuel-system components that are capable of quicker throughput and higher pressures. Spark plugs with smaller thread diameters made of materials adapted to the operating temperatures encountered allow higher compression ratios.

During the race, data can be transmitted by radio from the car to the pits. Known as telemetry, this technology allows constant monitoring of operating parameters such as pressures and temperatures.

UAV0059Y

M-Motronic

The M-Motronic includes all components that are necessary to manage an engine with intake-manifold fuel injection and conventional throttle valve. The system diagram (Fig. 1) shows an example of an M-Motronic system. The system scope is determined by the specifications for engine performance and the restrictions imposed by the prevailing emission-control legislation. The control center of the M-Motronic system is the engine ECU (21) which detects all input signals and generates control signals for the actuators.

In the course of the development history of Motronic systems, successive Motronic generations (e.g. M1, M3, M7) have differed mainly in the design of the hardware. The basic distinguishing features are the microcontroller family, the peripheral modules and the output-stage modules (chipset). The hardware variations arising from the requirements of different vehicle manufacturers are distinguished by manufacturer-specific identification numbers (e.g. M4.3).

Air-system components

Throttle valve

The accelerator pedal is connected to the throttle valve (16) by a linkage or a Bowden cable. The position of the accelerator pedal determines how far the throttle valve opens. This controls the air mass flowing through the intake manifold to the cylinders.

Idle actuator

The idle actuator (15) allows air to bypass the throttle valve at a defined mass air flow. This provides a means of holding the engine speed at a constant level when idling, for example (idle-speed control). To do so, the engine ECU controls the aperture of the bypass channel.

Sensors for detecting engine load

The engine load is an important operating parameter for electronic engine control. It is a measure of the air charge enclosed in the cylinder at the point of combustion and therefore of the mass air flow of the intake air as well. It is the decisive parameter for calculating injection time.

Fig. 1

1 Activated charcoal canister
2 Tank-leakage diagnosis module [1])
3 Regeneration valve
4 Secondary-air pump
5 Secondary-air valve
6 Air mass sensor with integrated temperature sensor
7 Intake-manifold pressure sensor [1])
8 Variable-geometry intake manifold with controllable flaps
9 Fuel rail
10 Fuel injector
11 Actuators and sensors for variable valve timing
12 Ignition coil with spark plug mounted
13 Camshaft phase sensor
14 Throttle-valve angle sensor
15 Idle actuator
16 Throttle valve
17 Exhaust-gas recirculation valve
18 Knock sensor
19 Engine-temperature sensor
20 Lambda sensor upstream of catalytic converter
21 Engine ECU
22 Speed sensor
23 Three-way catalytic converter (in some cases separate primary and secondary catalytic converters)
24 Diagnostics interface
25 Fault indicator lamp [1])
26 Interface with immobilizer ECU
27 Interface with transmission ECU [2])
28 CAN interface
29 Fuel tank
30 Tank pressure sensor [1])
31 Fuel line
32 In-tank unit comprising electric fuel pump, fuel filter and fuel-pressure regulator
33 Lambda sensor downstream of catalytic converter [1])

[1]) Components used specifically for on-board diagnostics (system configuration illustrated for CARB-OBD)
[2]) Communication also possible via CAN

1 Components used for electronic control in an M-Motronic system (example)

There are various methods of determining engine load. The M-Motronic system uses the following sensors (Fig. 2):

- Air-flow sensor
- Hot-wire air-mass meter
- Hot-film air-mass sensor
- Intake-manifold pressure sensor and
- Throttle-valve sensor

Depending on the configuration of the Motronic system, certain sensors or combinations of sensors may or may not be used.

Air-flow sensor

The air-flow sensor is located between the air filter and the throttle valve. It detects the volumetric flow rate [m³/h] of the intake air. The air flow acts on the baffle plate in the air-flow sensor, forcing it to deflect against the return force of the spring. The angle of deflection of the baffle plate is a measure of the volumetric air flow. It is converted to electrical voltage by a potentiometer.

In order to be able to take account of variations in air density due to changes in intake-air temperature, the air-flow sensor has an integrated temperature sensor. Its signal is used by the control unit to calculate a correction factor.

The air-flow sensor was previously a standard component of Motronic systems but has gradually been superseded by air-mass meters.

Air-mass meter

The air-mass meter (Fig. 1, Pos. 6) is fitted between the air filter and the throttle valve and detects the mass flow rate [kg/h] of the intake air.

The hot-wire and hot-film air-mass meters are thermal load sensors. They both operate according to the same principle – the air flow passing over a heated sensor element removes heat from it. The heating current dependent on the mass air flow keeps the sensor element at a constant temperature. An analyzer circuit processes the heating current and generates an analyzable voltage signal.

In the *hot-wire air-mass meter*, the heated sensor element is a thin platinum filament – the "hot wire". In the *hot-film air-mass meter*, the heated sensor element is a platinum-film resistor – the "hot film". It is

2 Examples of load sensors used in the course of M-Motronic system history

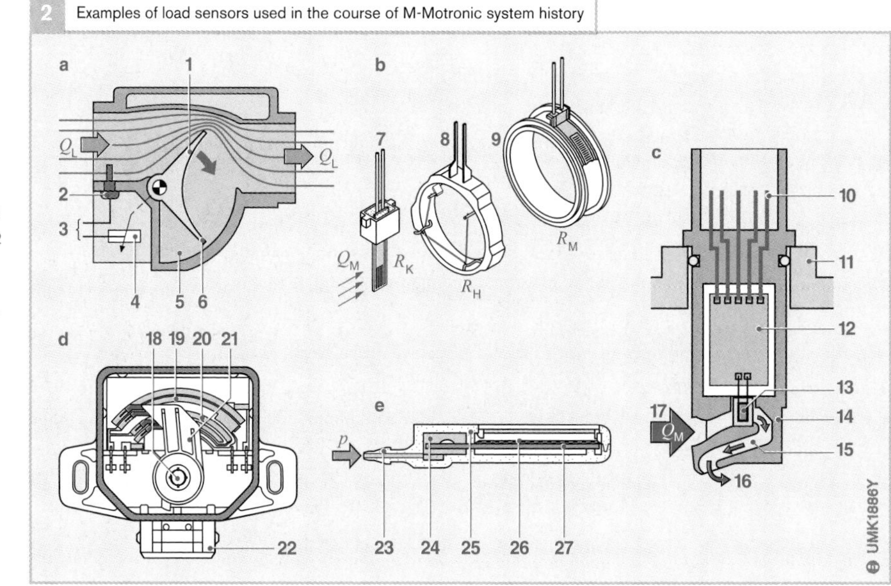

Fig. 2
- a Air-flow sensor
- b Hot-wire air-mass meter
- c Hot-film air-mass sensor (type HFM5)
- d Throttle-valve sensor
- e Intake-manifold pressure sensor (thick-film pressure sensor)

1. Sensor plate
2. Air-temperature sensor
3. Connection to ECU
4. Potentiometer
5. Compression volume
6. Compensation flap
7. Temperature-compensation resistor R_K
8. Sensor ring with hot-wire element (resistance R_H)
9. Measuring resistor R_M
10. Electrical connections
11. Sensor-tube housing
12. Electronic analyzer (hybrid circuit)
13. Sensor measurement cell
14. Sensor housing
15. Flow-component metering channel
16. Outlet for flow component Q_M
17. Inlet for flow component Q_M
18. Throttle-valve shaft
19. Potentiometer track 1
20. Potentiometer track 2
21. Rotating arm with sliding contacts
22. Electrical connection
23. Pressure connection
24. Pressure-measuring cell
25. Sealing partition
26. Analyzer circuit
27. Thick-film hybrid circuit

- Q_L Intake-air flow
- Q_M Mass-air flow
- p Pressure

located together with other sensor components on a ceramic plate.

The air density is taken into account from the outset with this type of sensor because it is a determining factor in the amount of heat removed from the sensor element by the air flow. The hot-film air-mass meter also detects the rate of air backflow and takes it into account.

Current M-Motronic systems only use the hot-film air-mass meter.

Intake-manifold pressure sensor

The intake-manifold pressure sensor (Fig. 1, Pos. 7) is connected by an air line to the intake manifold and detects the absolute pressure in the intake manifold. From the intake-manifold pressure, the intake-air temperature and the engine speed, the mass of the air drawn into the cylinders is calculated.

Throttle-valve sensor

The throttle-valve angle sensor (14) uses a potentiometer to detect the position of the throttle valve and produces an analog signal. This signal is generally used only as a secondary load signal. It provides supplementary information for dynamic functions, for engine operating-range detection (idle, medium throttle, wide-open throttle). It serves as a backup signal if the primary load sensor fails.

If the throttle-valve sensor is used as the primary load sensor, the level of measurement accuracy required is greater. This greater degree of accuracy is achieved by the use of two potentiometers and higher-quality valve bearings.

The air mass of the intake air is calculated from the throttle valve angle and the engine rotational speed. Temperature-related changes in air density are taken into account by measuring the intake-air temperature.

Exhaust-gas recirculation

Exhaust-gas recirculation allows an increase in the inert-gas content of the cylinder charge. The exhaust-gas recirculation system forms a connection between the exhaust-gas system and the intake manifold. The variable aperture of the exhaust-gas recirculation valve (Fig. 1, Pos. 17) controls the volume of exhaust that is drawn back into the cylinders (exhaust-gas recirculation rate). The recirculated exhaust gas has an effect on combustion. It reduces the peak temperature and lowers the NO_X emission levels. By reducing the degree to which the intake-air flow is restricted, it also lowers fuel consumption.

The aperture of the exhaust-gas recirculation valve is indirectly controlled by a PWM signal from the engine ECU via an electropneumatic exhaust-gas recirculation valve actuator. More recent exhaust-gas recirculation valves can be directly controlled by electrical means and have a position feedback facility provided by a potentiometer. Position feedback allows extremely precise setting of the valve aperture.

The permissible exhaust-gas recirculation rate depends essentially on the engine operating status and the operating conditions (e.g. engine temperature). If the exhaust-gas recirculation rate is too high, delayed combustion and even misfiring can occur and this causes uneven engine running. Such considerations have to be taken account when calibrating the system for a particular engine.

Variable-geometry intake manifold

Variable-geometry intake manifolds (Fig. 1, Pos. 8) can be used to enhance the utilization of speed-dependent dynamic turbocharging effects. Systems with variable-geometry intake manifolds achieve better cylinder charging and therefore improved torque curves.

The intake-manifold geometry is varied by the engine ECU by means of electrically or electropneumatically operated flaps, depending on engine operating status (in particular engine speed and load).

Exhaust-gas turbocharging

Exhaust-gas turbocharging is another means of increasing cylinder charge and therefore engine torque (Fig. 3). The exhaust-gas turbocharger is located in the exhaust-gas system so that the flow of gas can drive the turbine (14). In the compressor (12), the impeller, which is mounted on a common shaft with the turbine, compresses the intake air, thereby increasing the cylinder charge. In the process of being compressed, the intake air is heated up. As a result, the intake air passes through an intercooler (5) to reduce its temperature again.

In order to prevent the compressor from pumping when the throttle valve (2) is closed and generating unwanted additional noise or even suffering damage, the divert-air valve (9) in the compressor bypass passage is opened.

The charge-air pressure has to be adjusted to the engine operating status. At high engine speeds and high levels of torque demand, the exhaust-gas mass flow is so high that the engine would overload if no measures were taken. Therefore, a proportion of the exhaust-gas mass flow is diverted past the exhaust-gas turbine by a bypass valve called a wastegate (11). This limits the turbine speed and therefore the turbocharger pressure generated in the intake manifold. The aperture of the bypass passage is controlled by the pneumatically operated boost-pressure control valve (10). This valve is controlled by the pulse valve (13), which in turn is controlled by a PWM (pulse-width modulation) signal from the engine ECU.

Variable valve timing

Variable valve timing that changes the engine operating status provides a means of increasing torque output and reducing exhaust-gas emissions.

Systems with variable-valve timing can change valve timing. By varying the phase of the inlet-valve camshaft (Fig. 1, Pos. 11) relative to the crankshaft, the valve overlap can be adjusted and therefore the inert-gas rate.

3 Exhaust-gas turbocharging system

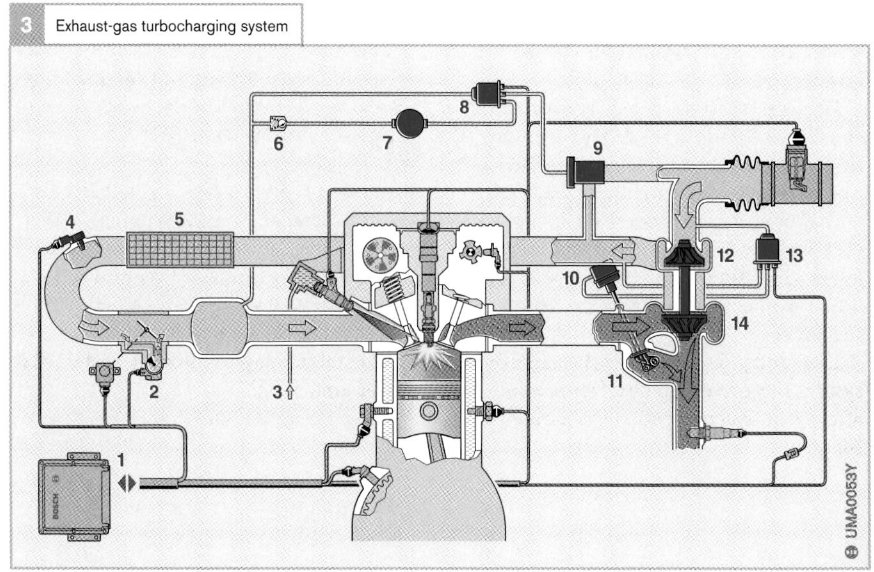

Fig. 3
1 Engine ECU
2 Throttle valve with idle actuator and throttle-valve angle sensor
3 Fuel supply
4 Charge-air pressure and charge-air temperature sensor
5 Charge-air cooler
6 Non-return valve
7 Vacuum accumulator
8 Solenoid valve (pulse valve)
9 Divert-air valve (dump valve)
10 Boost-pressure control valve
11 Wastegate (bypass valve)
12 Exhaust-gas turbocharger compressor
13 Solenoid valve (pulse valve)
14 Exhaust-gas turbine

Fuel-system components

Electric fuel pump

The electric fuel pump (Fig. 1, Item 32) pumps fuel from the fuel tank (29) through the fuel pipe (31) at the pressure determined by the pressure regulator. On older systems, the electric fuel pump is fitted outside the fuel tank in the fuel pipe ("inline"). On more recent systems, the electric fuel pump is integrated in the in-tank unit inside of the fuel tank.

Fuel filter

The fuel filter removes contamination from the fuel. In so doing, it protects the components of the fuel-injection system from blockage and damage. The fuel filter is usually fitted in the fuel pipe outside of the fuel tank, though it may also be integrated in the in-tank unit.

Fuel-pressure regulator

The fuel-pressure regulator keeps the pressure in the fuel supply system at a predetermined level by diverting excess fuel to the fuel-return pipe and back to the fuel tank.

In fuel systems with a fuel-return pipe, the fuel-pressure regulator is usually mounted on the fuel rail. The air pipe that leads to the intake manifold ensures that there is a constant pressure differential between the fuel injectors and the intake manifold.

In fuel systems without a return-fuel pipe (returnless fuel systems, RLFS), the pressure in the fuel system is kept constant relative to atmospheric pressure. The variable difference between fuel pressure and intake-manifold pressure requires adjustment to calculate the injection time. In such systems, the fuel-pressure regulator is integrated in the fuel tank.

Depending on the type of fuel-injection system, there may also be a fuel-pressure attenuator to compensate for pressure fluctuations.

Evaporative-emissions control system

The evaporative-emissions control system ensures that fuel vapor does not escape to the atmosphere. Fuel vapor evaporating from the fuel tank is absorbed by the activated charcoal canister (1) in the evaporative-emissions control system. The stored fuel vapor is entrained by the air flow drawn in through the shutoff valve (2) and directed to the intake manifold so that it can flow into the cylinders. This process regenerates the activated charcoal canister. The engine ECU regulates the regenerating flow of air by controlling the regeneration valve (3) depending on the engine operating status.

Fuel rail

The fuel delivered by the electric fuel pump passes into the fuel rail (9) which is connected to the fuel injectors.

Electromagnetic fuel injectors

The electromagnetic fuel injectors (10) protrude into the intake manifold in such a way that the injection jet is directed at the area just in front of it or on the inlet-valve plate (or valves, if there are more than one). The fuel mixes with the intake air in the intake manifold at that point to form the air-fuel mixture. There is a fuel injector for each cylinder.

The fuel injectors are clocked by the engine ECU. The length of the injection signal determines the injected-fuel quantity. Depending on the type of fuel injection, the fuel injectors are operated simultaneously or individually. New systems use only sequential or cylinder-specific fuel injection. The fuel is injected into each cylinder individually in precisely the right quantity at precisely the right moment.

The output-stage modules are integrated in the control unit.

Ignition-system components

Ignition coil

The ignition coil stores the necessary electrical energy to ignite the air-fuel mixture and generates the high voltage required (ignition voltage requirement) to produce the spark across the spark-plug electrodes.

In the ignition system illustrated in the system diagram (Fig. 1), the ignition coil (12) is attached directly to the spark plug. In other words, this is an ignition system with stationary voltage distribution and single-spark ignition coils. It has a separate ignition coil to generate the high ignition voltage for each cylinder. Systems with dual-spark ignition coils are also in widespread use. They have pairs of spark plugs served by a single coil in each case. Systems with only one ignition coil also require a mechanical rotary high-voltage distributor. Such configurations are no longer used in more recent ignition systems.

Spark plugs

The spark plug ignites the air-fuel mixture inside of the cylinder by means of an electric spark that arcs across its electrodes.

4 Ignition coils for Motronic systems with stationary voltage distribution (examples)

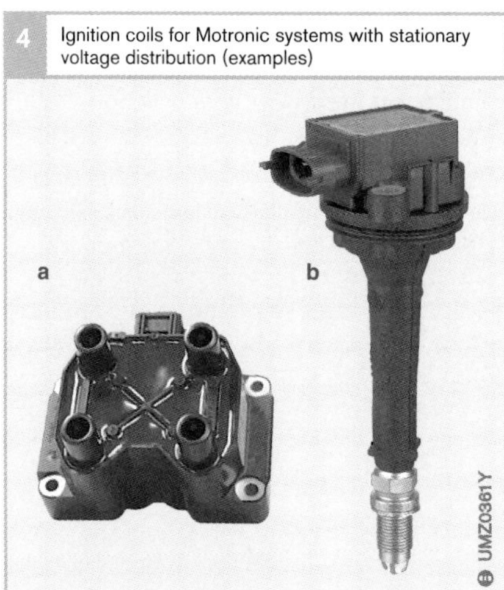

Fig. 4
a ZS 2x2: Module with two dual-spark ignition coils, i.e. with four HT outlets for four spark plugs (spark plugs connected by HT leads)
b ZS 1x1: Single-spark ignition coil (compact-coil version) with plug-on spark plug

Emission-control system components

Three-way catalytic converter

The three-way catalytic converter (Fig. 1, Item 23) converts the harmful emissions produced as a byproduct of combusting the air-fuel mixture, i.e. carbon monoxide (CO), hydrocarbons (HCs) and nitrogen oxides (NO_X). It produces water vapor (H_2O), nitrogen (N_2) and carbon dioxide (CO_2).

Lambda sensors

The lambda sensor (20) measures the oxygen content in the exhaust gas. It provides information about the composition of the air-fuel mixture entering the cylinders. The engine ECU uses the information from the lambda sensor to maintain the stoichiometric ratio ($\lambda = 1$) of the air-fuel mixture. This mixture composition demonstrates the best possible exhaust-gas treatment for the three-way catalytic converter.

In M-Motronic systems, only two-point lambda sensors are used. Depending on the system, an additional sensor (33) may be fitted downstream of the catalytic converter to allow twin-sensor control and monitoring of catalytic-converter aging.

Secondary-air system

Injecting air in the exhaust pipe for a short period after the engine is started helps in afterburning unburned hydrocarbons (HCs) in the exhaust gas. This not only reduces HC emissions, it also reduces the warm-up time of the catalytic converter so that it reaches its operating temperature more quickly.

The secondary-air pump (4) injects secondary air and the air pipe is closed off by the secondary-air valve (5) when the pump is inactive. The two components are controlled by the engine ECU.

On-board diagnosis components
The components indicated by [1]) in Figure 1 are used for on-board diagnosis. Californian emission-control legislation (Californian Air Resources Board (CARB)) places particularly stringent requirements on the diagnostic capabilities of the Motronic system. Some emissions-related systems can only be diagnostically assessed with the aid of supplementary components.

Operating data
In addition to the sensors already mentioned, there is a whole series of additional sensors and desired-value generators for engine operating data. Examples include:

- The engine-speed sensor (Fig. 1, Item 22) for detecting the crankshaft position and calculating engine speed
- The phase sensor (13) for detecting the camshaft position/phase (i.e. the phase of the engine operating cycle)
- The engine-temperature sensor (19) and intake-air temperature sensor for calculating temperature-dependent adjustments
- The knock sensor (18) for detecting engine knock

Auxiliary systems
Radiator fan
The ECU switches on the radiator fan depending on the engine temperature in order to assist engine cooling and reduce engine temperature. Depending on the system, the radiator fan may also have several speeds.

In conventional control systems, the radiator fan is controlled by a thermostat.

Air-conditioner compressor
A large amount of power is required to drive the air-conditioner compressor and this has to be supplied by the engine. This proportion of engine performance is not available to drive the vehicle. In situations where full engine power is required (e.g. when overtaking), the engine management system can temporarily switch off the compressor.

Communication
CAN interface
Depending on the vehicle's equipment level, the M-Motronic may be fitted with a CAN bus system. The CAN interface (Fig. 1, Item 28) provides the facility for data exchange with other electronic systems (e.g. the transmission ECU, ABS, etc.).

Malfunction indicator lamp (MIL)
The malfunction indicator lamp (25) is integrated in the instrument panel or instrument cluster. It indicates to the driver if there is a malfunction in the Motronic system.

Fuel-consumption signal
Motronic calculates fuel consumption from the injection time and passes the information on to the on-board computer in the form of a digital signal. Fuel-consumption data can also be sent via the CAN interface.

Diagnostic interface
The diagnostic interface (24) is used to connect system testing equipment (e.g. KTS500) when the vehicle is being serviced or repaired at the Customer Service Workshop. This equipment can read all faults recorded by the diagnostics system while the engine is running.

In the past, fault data could only be read by blink codes detected by a diagnosis lamp. Now the information can be viewed on an engine tester capable of displaying text. On vehicles fitted with an OBD II on-board diagnosis system, a “scan tool” (tester) can read off emission-related fault data including the “ambient conditions” that were present at the time the fault occurred.

ME-Motronic

An electronic engine-management system with electronic throttle control (EGAS) is characterized by the fact that there is no longer a mechanical link (linkage or Bowden cable) between the accelerator pedal and the throttle valve. The position of the accelerator pedal, i.e. the driver command, is detected by a potentiometer attached to the accelerator pedal (pedal-travel sensor in the accelerator-pedal module, Fig. 1 Pos. 23) and sent to the engine ECU (12) in the form of an analog voltage signal. In response, the ECU generates output signals that set the aperture of the electrically controlled throttle valve (3) so that the engine produces the desired torque.

A system that regulates engine performance in this way was first introduced by Bosch in 1986. In addition to the engine ECU, the original system also had a separate control unit for engine performance.

The increasingly higher packaging density of electronic circuits allowed the combination of Motronic functions and engine performance control in a single control unit (1994). Nevertheless, functions remained divided between two microcontrollers.

The next step was taken in 1998 with the launch of the new Motronic generation, the ME7, which incorporates all engine-management functions in a single microcontroller. This advance was made possible by the ever-increasing processing capacity of microchips. The chip used in the ME7 is a 16-bit processor.

The system diagram (Fig. 1) shows an example of an ME-Motronic system. The system scope is defined by the requirements for engine performance and the restrictions imposed by the prevailing exhaust-gas and diagnostic-system legislation.

The essential features that distinguish the ME-Motronic from the M-Motronic are the electronically controlled throttle valve and the torque-based function structure of the application software.

Fig. 1
1 Activated charcoal canister
2 Hot-film air-mass sensor with integrated temperature sensor
3 Throttle device (electronic throttle control)
4 Regeneration valve
5 Intake-manifold pressure sensor
6 Fuel rail
7 Fuel injector
8 Actuators and sensors for variable valve timing
9 Ignition coil and spark plug
10 Camshaft phase sensor
11 Lambda sensor upstream of primary catalytic converter
12 Engine ECU
13 Exhaust-gas recirculation valve
14 Speed sensor
15 Knock sensor
16 Engine-temperature sensor
17 Primary catalytic converter (three-way catalytic converter)
18 Lambda sensor downstream of primary catalytic converter
19 CAN interface
20 Fault indicator lamp
21 Diagnostics interface
22 Interface with immobilizer ECU
23 Accelerator-pedal module with pedal-travel sensor
24 Fuel tank
25 In-tank unit comprising electric fuel pump, fuel filter and fuel-pressure regulator
26 Main catalytic converter (three-way)

The on-board diagnostics system configuration illustrated by the diagram reflects the requirements of EOBD

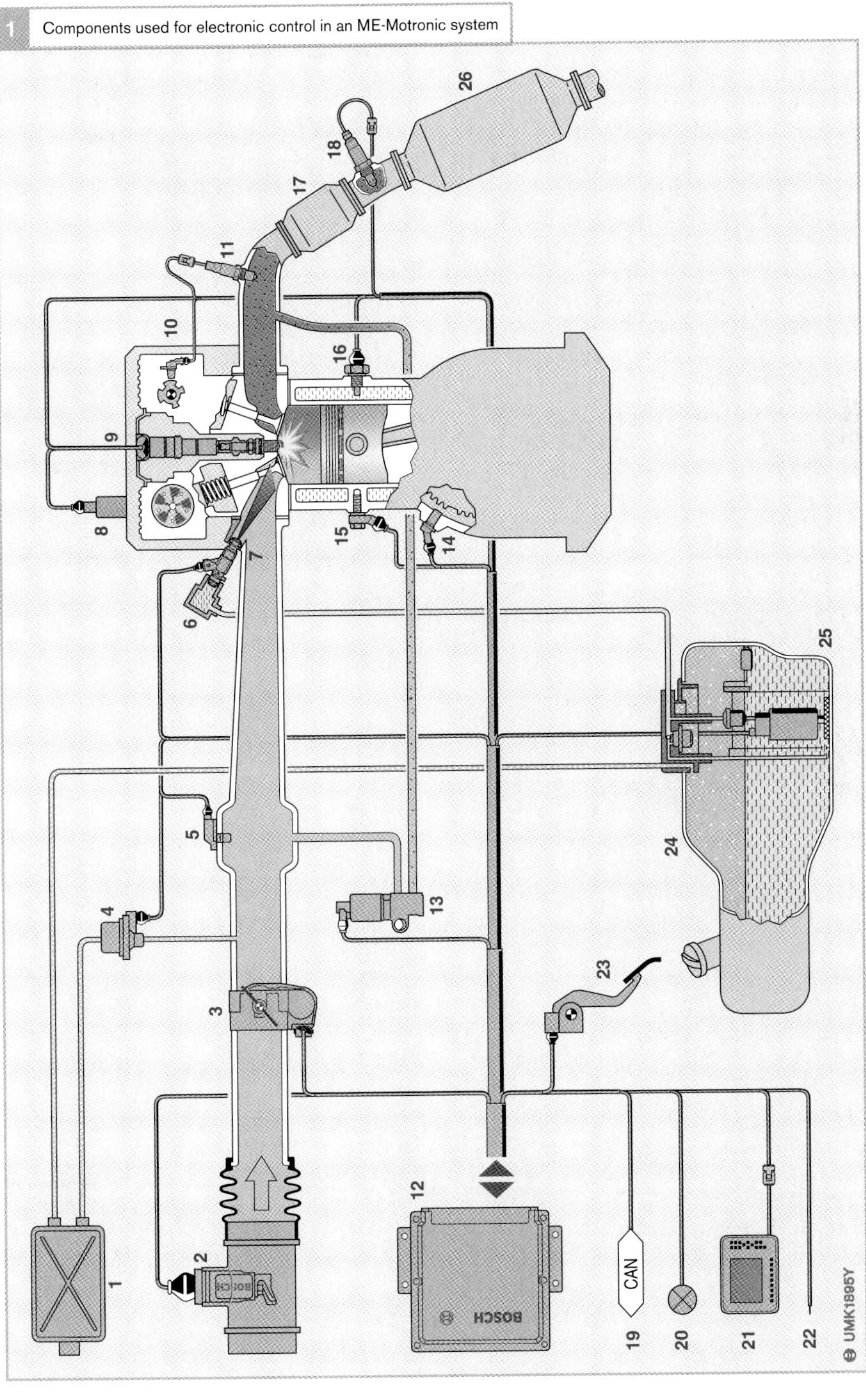

1 Components used for electronic control in an ME-Motronic system

Air-system components

Throttle valve

On the ME-Motronic, the throttle-valve position is electrically adjusted by the servomotor integrated in the throttle device (Fig. 1 Pos. 3), i.e. engine performance is controlled electronically. Consequently, the mass-air flow through the intake manifold can be adjusted independently of the accelerator-pedal position. The accelerator-pedal position detected by the pedal-travel sensor (23) serves as the input variable for the engine ECU; it is the indicator of the driver's throttle position.

When the engine is idling, the throttle valve opens to a position at which the engine is running at the required idle speed. The supplementary air bypass provided by the idle actuator on the M-Motronic system is no longer required.

Sensors for detecting engine load

The ME-Motronic system uses the following load sensors:
- The hot-film air-mass meter (2) and/or
- The intake-manifold pressure sensor (5)

The primary load sensor is the air-mass meter which senses the mass air flow drawn in by the engine. From this information, the Motronic calculates the mass of air enclosed inside of the cylinder. In systems with an air mass-flow sensor, the intake-manifold pressure sensor is used only as a secondary sensor for analyzing the exhaust-gas recirculation system.

As an alternative to the above arrangement, there are also *p* systems (*p* stands for pressure) which determine engine load by means of the intake-manifold pressure sensor. From the vacuum prevailing in the intake manifold, the temperature of the intake air and the engine speed, it is possible to calculate the mass of air available inside the cylinder for combustion.

On turbocharged engines, an additional sensor which detects charge-air pressure is also required.

Fuel-system components

The fuel system for the ME-Motronic does not differ from the arrangement used for the M-Motronic. The timing of the fuel injectors (7) is controlled sequentially, i.e. individually for each cylinder, except under certain operating conditions.

Ignition-system components

The ME-Motronic is exclusively a stationary-voltage distribution system which uses single-spark (9) or dual-spark ignition coils.

Exhaust-gas treatment system components

As with the M-Motronic, the exhaust-gas treatment system components used by the ME-Motronic are the following:
- One or two three-way catalytic converters (primary catalytic converter and main catalytic converter, 17 and 26 respectively)
- The lambda sensors upstream and downstream of the primary and secondary catalytic converters (11 and 13 respectively)

Other components that may also be used are:
- The secondary-air system comprising secondary-air pump and secondary-air valve for rapid heating of the catalytic converter and
- An exhaust-gas temperature sensor for monitoring the exhaust-gas temperature (for component protection purposes) on vehicles with exhaust-gas turbocharged engines

Figure 1 shows a system with a primary catalytic converter fitted close to the engine and a main converter further downstream. As well as the two-point lambda sensors used to measure the oxygen concentration in the exhaust gas upstream of the catalytic converter, broadband lambda sensors are also used (for lambda control). This provides better dynamic response for the lambda closed-loop control. The lambda sensors used for dual-sensor control and those fitted downstream of the catalytic converter for diagnostic purposes are always two-point sensors.

Monitoring concept

It is imperative that, when the vehicle is in motion, it is never able to accelerate when the driver does not want it to. Consequently, the safety concept for the electronic engine-performance control system must meet exacting requirements. To this end, the ECU includes a monitoring processor in addition to the main processor, and the two processors monitor one another.

Torque structure

The torque-based system structure was first introduced on the ME-Motronic. All performance demands (Fig. 2) placed on the engine are always converted into a torque requirement. The torque coordinator prioritizes the torque demands from internal and external power consumers and other requirements relating to engine efficiency. The resulting required torque is allocated proportionately to the air, fuel and ignition systems.

The air-system proportion is implemented by varying the throttle-valve aperture and – in the case of turbocharged engines – the wastegate valve control. The fuel-system proportion is essentially determined by means of the injected fuel, taking account of fuel-tank ventilation (evaporative-emissions control system).

The torque is adjusted via two channels. The air-system channel (main channel) involves calculating the required cylinder charge on the basis of the required torque. From the required cylinder charge, the required throttle-valve aperture is calculated. The required injected-fuel mass is directly related to the cylinder charge due to the fixed oxygen level specified. The air-system channel only permits gradual changes in torque output (e.g. idle-speed control integral component).

The crankshaft-synchronized channel uses the cylinder charge currently available to calculate the maximum possible torque output for the current operating status. If the desired torque is less than the maximum possible torque, a rapid reduction in torque (e.g. idle-speed control differential component, torque reduction for gear shifting, surge damping) can be achieved by retarding the ignition or shutting down one or more cylinders altogether (injection interludes, e.g. intervention by traction control or when the engine is overrunning).

On earlier M-Motronic systems, a reduction in torque (e.g. at the request of the automatic transmission when changing gear) is performed directly by the function concerned, for example by retarding the ignition angle. There is no coordination between individual requests or of command implementation.

2 Torque-based system structure of ME-Motronic

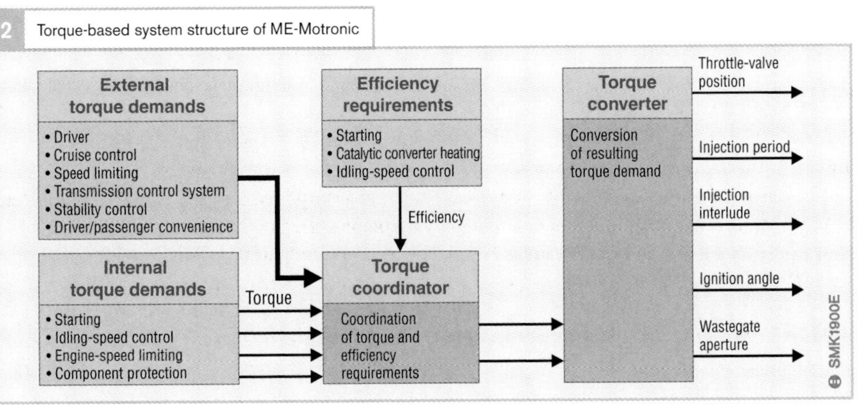

MED-Motronic

With the introduction of direct injection for gasoline engines, a new control concept was required. The fuel injector must be capable of producing both a homogeneous mixture distribution (homogeneous-charge mode) – as implemented on the M-Motronic and ME-Motronic systems for intake-manifold fuel injection – and a locally limited stratified charge (stratified-charge mode) in the combustion chamber. Homogeneous mixture distribution is achieved by injecting the fuel during the induction stroke, while the stratified mixture is produced by injecting the fuel shortly before the end of the compression stroke shortly before ignition. It is only by the use of the stratified-charge mode, which is used in the lower to medium engine-speed and torque ranges, that the fuel-saving benefits of direct injection can be fully exploited. However, there are also direct-injection systems which use a homogeneous, stoichiometric composition mixture ($\lambda = 1$) across the entire engine operating range.

The engine-management system that meets these requirements is called the MED-Motronic. In comparison with the ME-Motronic, it requires a substantially larger processing capacity.

The system diagram (Fig. 1) shows an example of an MED-Motronic system. The first MED-Motronic was launched on the Volkswagen Lupo in 2000.

The essential features that distinguish the MED-Motronic from the ME-Motronic are in the fuel system and in exhaust-gas system, which incorporates a NO_X-accumulator catalytic converter. The descriptions that follow essentially concentrate on the differences from the ME-Motronic.

Fig. 1
1 Activated charcoal canister
2 Regeneration valve
3 Type HDP2 high-pressure pump with integrated fuel-quantity control valve
4 Actuators and sensors for variable valve timing
5 Ignition coil and spark plug
6 Hot-film air-mass meter with integrated temperature sensor
7 Throttle device (electronic throttle control EGAS with position sensor)
8 Intake-manifold pressure sensor
9 Fuel-pressure sensor
10 High-pressure fuel rail
11 Camshaft phase sensor
12 Lambda sensor upstream of primary catalytic converter
13 Exhaust-gas recirculation valve
14 High-pressure fuel injector
15 Knock sensor
16 Engine-temperature sensor
17 Primary catalytic converter (three-way catalytic converter)
18 Lambda sensor downstream of primary catalytic converter (optional)
19 Speed sensor
20 Engine ECU
21 CAN interface
22 Fault indicator lamp
23 Diagnostics interface
24 Interface with immobilizer ECU
25 Accelerator-pedal module with pedal-travel sensor
26 Fuel tank
27 In-tank unit comprising electric fuel pump, fuel filter and fuel-pressure regulator
28 Exhaust-gas temperature sensor
29 Main catalytic converter (NO_X accumulator plus three-way catalytic converter)
30 Lambda sensor downstream of main catalytic converter

1 Components used for electronic control in an MED-Motronic system

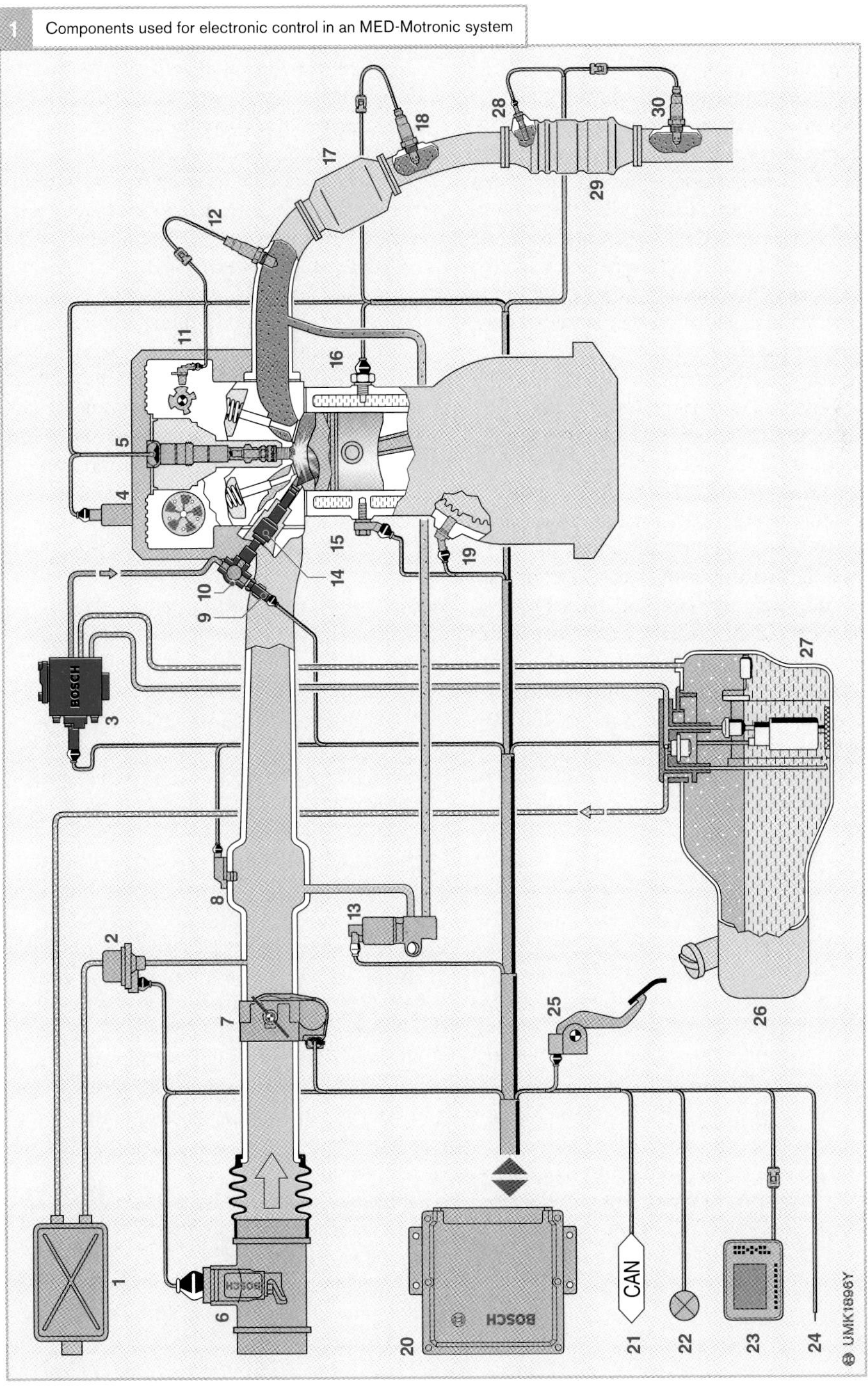

Air-system components

Throttle valve

The throttle device (7) is identical in design to the device used on the ME-Motronic. However, the engine is operated in stratified-charge mode at low speeds (< 3,000 rpm) and when the torque demand is low so that the throttle valve remains wide open. In this operating mode, the torque is not controlled by means of the intake-air mass but by the quantity of fuel injected. Consequently, the cylinder charge is not proportional to the torque output.

Sensors for detecting cylinder charge

The cylinder-charge sensing equipment of the gasoline direct-injection system is more complex than that of the intake-manifold injection system since exhaust-gas recirculation is used to reduce NO_X emissions in stratified-charge mode. In order to be able to exploit the fuel-economy benefits and keep emission levels low, precise control and measurement of the engine's mass flows (air and recirculated exhaust gas) are necessary. For this reason, gasoline direct-injection systems use two cylinder-charge sensors to determine the mass flows. There are two ways in which this can be done:

Hot-film air-mass sensor and intake-manifold pressure sensor

The hot-film air-mass sensor (6) measures the mass-air flow entering the intake manifold. This information can be used to calculate the partial pressure of the air in the intake manifold. The difference between the intake-manifold pressure (measured by the intake-manifold pressure sensor, 8) and the partial air pressure indicates the mass flow of the recirculated exhaust gas.

When the engine is running on a lean mixture, the recirculated exhaust gas contains a proportion of unused oxygen that can be determined from the oxygen level measured by the lambda sensor in the exhaust gas.

Intake-manifold pressure sensor and ambient-pressure sensor

Together with the pressure above the throttle valve and the temperature of the intake air, the measured throttle-valve aperture can be used to calculate the intake-air mass across the throttle valve. Using the same algorithm, the mass flow of the recirculated exhaust gas via the exhaust-gas recirculation valve is calculated using the imputed exhaust-gas back pressure. This demands accurate data concerning the position of the exhaust-gas recirculation valve.

Both systems employ temperature sensors to detect the temperature of the intake air. If a hot-film air-mass sensor is fitted, the temperature sensor is integrated in it; in a pressure system, a separate temperature sensor has to be fitted upstream of the throttle valve.

Fuel-system components

In the MED-Motronic system, the fuel system consists of a low-pressure and a high-pressure system. The MED-Motronic is therefore significantly different from the ME-Motronic, in which the fuel injectors are connected to the low-pressure fuel system. Figure 2 shows the standard gasoline direct-injection system with the type HDP2 demand-controlled high-pressure injection pump.

The low-pressure system may be configured in different ways depending on the requirements of the vehicle manufacturer. In this example, it consists of:

- The electric fuel pump with integrated pressure-limiting valve (Fig. 2, Pos. 2)
- The low-pressure regulator (4) and
- The shutoff valve (5)

The high-pressure system consists of:

- The high-pressure pump (6) which generates injection pressures of up to 12 MPa
- The fuel-quantity control valve (7) which regulates the amount of fuel delivered by the pump

- The fuel rail (9) which serves as a pressure accumulator for the injected fuel
- The fuel-rail pressure sensor (11) which provides feedback on the actual pressure in the fuel rail
- And the pressure-limiting valve which limits the pressure in the fuel rail to the permissible maximum

Connected to the fuel rail are the high-pressure fuel injectors (12). In comparison with the fuel injectors used on intake-manifold injection systems, these injectors have to meet much more demanding standards. An essential consideration is that the fuel has to be injected within a much shorter period of time. This requires a complex control signal which is generated by the output stage integrated in the ECU (Fig. 1, Pos. 20).

Ignition-system components

In the MED system, the high voltage required for the ignition spark is generated by single-spark ignition coils that are directly connected to the spark plugs (Fig. 1, Pos. 5). In order to be able to ignite the air-fuel mixture reliably, the ignition coil has to deliver a greater amount of energy than with intake-manifold injection systems. For this reason, special ignition coils are required.

Emission-control system components

The emission-control components in the MED-Motronic system comprise:

- A primary catalytic converter fitted close to the engine (three-way catalytic converter, Fig. 1, Item 17)
- The main catalytic converter fitted further downstream (NO_X-accumulator and three-way catalytic converter, 29) and
- The lambda sensors upstream of the primary catalytic converter (12) and downstream of the main catalytic converter (30) and the exhaust-temperature sensor (28)

The MED-Motronic is able to dispense with the secondary-air system because the exhaust-gas section can be rapidly brought up to operating temperature by means of a special engine operating mode.

In contrast with intake-manifold injection systems, a gasoline direct-injection system requires a NO_X-accumulator catalytic converter. This stores the nitrogen oxides that cannot be converted by the three-way catalytic converter when the engine is operating with excess air. As the accumulator catalytic converter gets closer to its maximum capacity, its ability to bind nitrogen oxides diminishes. It then has to be regenerated. This is

2 Schematic diagram of fuel system of a gasoline direct-injection system with type HDP2 high-pressure pump

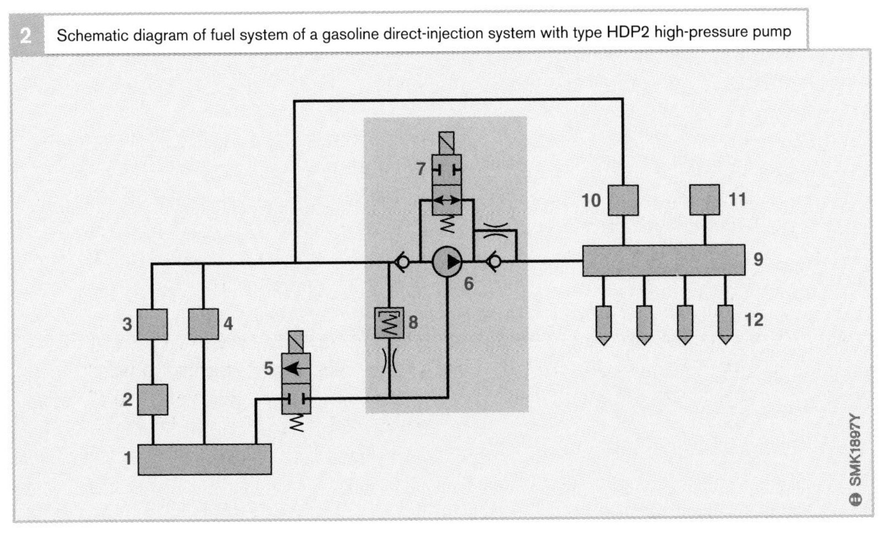

Fig. 2
1 Fuel tank
2 Electric fuel pump
3 Fuel filter
4 Low-pressure regulator
5 Shutoff valve
6 Type HDP2 high-pressure pump with integrated
7 Fuel-quantity control valve and
8 Fuel-pressure attenuator
9 Fuel rail
10 Pressure-limiting valve
11 Fuel-rail pressure sensor
12 High-pressure fuel injectors

achieved by the MED-Motronic switching to homogeneous-charge mode with excess fuel. The stored nitrogen oxides are released from the accumulator converter and converted by the three-way converter stage.

Since the direct-injection engine runs on a lean mixture in stratified-charge mode, the use of a broadband lambda sensor to measure the exhaust composition upstream of the catalytic converter is absolutely essential.

3 Changeover from stratified-charge to homogeneous-charge mode

Operating data

In addition to the operating data known to the M-Motronic and ME-Motronic systems, the MED-Motronic also detects and analyzes fuel-rail pressure by means of the fuel-rail pressure sensor. In this way, it is able to regulate the fuel pressure present at the high-pressure fuel injectors depending on the engine operating status.

Operating-mode coordination and selection

Whereas systems with intake-manifold fuel injection can only operate in homogeneous-charge mode, gasoline direct-injection systems have a number of other possible operating modes in addition to homogeneous mode. The MED-Motronic can inject fuel in such a way that, at the moment of ignition, a mixture layer forms in the vicinity of the spark plug. The remainder of the combustion chamber contains only air and inert gases (stratified-charge mode). Another possibility is to charge the engine with a lean mixture (homogeneous-layer mode). In addition, there is a homogeneous knock-

4 Operating mode selection

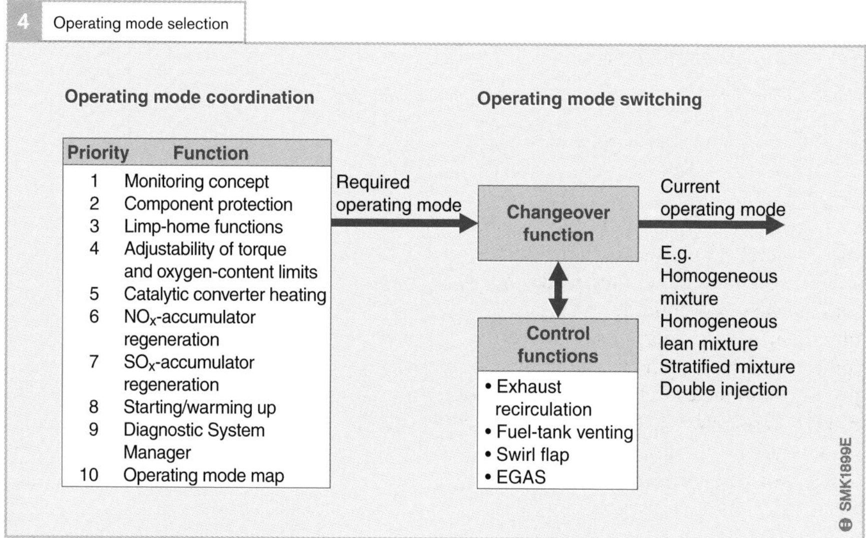

control mode and a stratified catalytic-converter-heating mode for specific operating conditions.

In homogeneous-charge mode, torque control is performed using a specified cylinder charge and air-fuel mixing ratio (λ). The injected-fuel mass is regulated to suit the available intake air. In stratified-charge mode and in homogeneous-layer mode, torque is controlled by means of the injected-fuel mass, i.e. by the opening period of the fuel injector. The injected-fuel mass is specified and the engine operates with as little restriction of intake air-flow as possible.

The MED-Motronic incorporates an operating-mode coordinator which enables changeover to a different operating mode when engine requirements demand. The basis for selecting an operating mode is the operating-mode map, which plots operating mode against engine speed and torque. The priority list (Fig. 4) lists operating-mode requirements in the order of their priority. This produces the required operating mode. But before ignition and fuel injection can be changed over to the new operating mode, control functions for exhaust-gas recirculation, tank ventilation, charge-movement flap and throttle-valve setting are initiated if required. The system then waits for acknowledgement.

In stratified-charge mode with $\lambda > 1$, the throttle valve is fully open and the intake air can enter the intake manifold without restriction. The torque output is proportional to the injected-fuel mass. In order to change over to homogeneous-charge mode, the mass air flow, which is now the means of varying torque, must be reduced very quickly and the oxygen concentration adjusted to a suitable level for a stoichiometric mixture ratio of $\lambda = 1$. The torque produced by the engine now varies in relation to the accelerator-pedal position (Fig. 3), though the driver is unaware of any change.

Brake-booster vacuum control

When the engine is operating with an unrestricted intake air-flow, there is insufficient vacuum in the intake manifold to provide the vacuum required by the brake booster. A vacuum switch or pressure sensor is used to detect whether there is sufficient vacuum in the brake booster. If necessary, the engine has to be switched to a different operating mode in order to provide the required vacuum for the brake booster.

On-board diagnosis

The essential difference from the ME-Motronic is in the additional diagnostic functions required for the high-pressure fuel system. All components in the high-pressure system have to be monitored for plausible response. In the event of a malfunction, a safe emergency mode must be ensured and an entry recorded in the fault memory.

Sensors

Sensors register operating states (e.g. engine speed) and setpoint/desired values (e.g. accelerator-pedal position). They convert physical quantities (e.g. pressure) or chemical quantities (e.g. exhaust-gas concentration) into electric signals.

Automotive applications

Sensors and actuators represent the interfaces between the ECUs, as the processing units, and the vehicle with its complex drive, braking, chassis, and bodywork functions (for instance, the Engine Management, the Electronic Stability Program ESP, and the air conditioner). As a rule, a matching circuit in the sensor converts the signals so that they can be processed by the ECU.

The field of mechatronics, in which mechanical, electronic, and data-processing components are interlinked and cooperate closely with each other, is rapidly gaining in importance in the field of sensor engineering. These components are integrated in modules (e.g. in the crankshaft CSWS (Composite Seal with Sensor) module complete with rpm sensor).

Since their output signals directly affect not only the engine's power output, torque, and emissions, but also vehicle handling and safety, sensors, although they are becoming smaller and smaller, must also fulfill demands that they be faster and more precise. These stipulations can be complied with thanks to mechatronics.

Depending upon the level of integration, signal conditioning, analog/digital conversion, and self-calibration functions can all be integrated in the sensor (Fig. 1), and in future a small microcomputer for further signal processing will be added. The advantages are as follows:

- Lower levels of computing power are needed in the ECU
- A uniform, flexible, and bus-compatible interface becomes possible for all sensors
- Direct multiple use of a given sensor through the data bus
- Registration of even smaller measured quantities
- Simple sensor calibration

1 Sensor integration levels

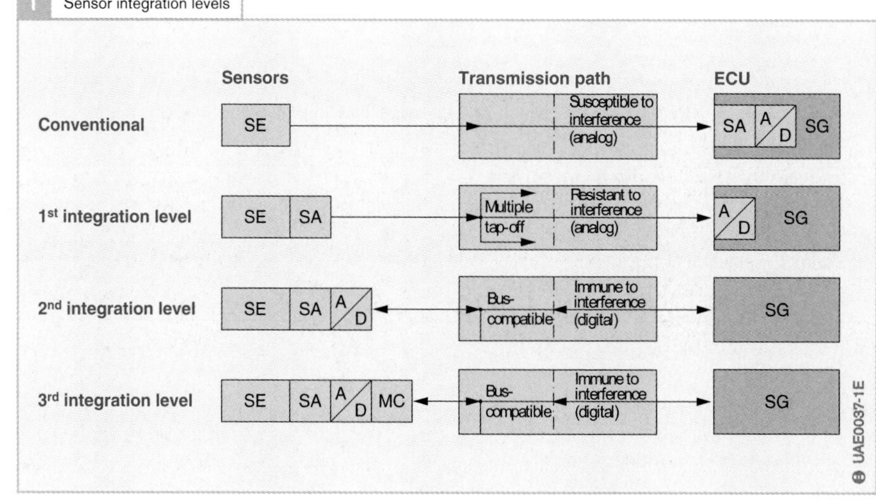

Fig. 1
SE Sensor(s)
SA Analog signal conditioning
A/D Analog-digital converter
SG Digital ECU
MC Microcomputer (evaluation electronics)

Miniaturization

Thanks to micromechanics it has become possible to locate sensor functions in the smallest possible space. Typically, the mechanical dimensions are in the micrometer range. Silicon, with its special characteristics, has proved to be a highly suitable material for the production of the very small, and often very intricate mechanical structures. With its elasticity and electrical properties, silicon is practically ideal for the production of sensors. Using processes derived from the field of semiconductor engineering, mechanical and electronic functions can be integrated with each other on a single chip or using other methods.

Bosch was the first to introduce a product with a micromechanical measuring element for automotive applications. This was an intake-pressure sensor for measuring load, and went into series production in 1994. Micromechanical acceleration and yaw-rate sensors are more recent developments in the field of miniaturisation, and are used in driving-safety systems for occupant protection and vehicle dynamics control (**E**lectronic **S**tability **P**rogram ESP). The illustrations below show quite clearly just how small such components really are.

Micromechanical acceleration sensor

Electric circuit

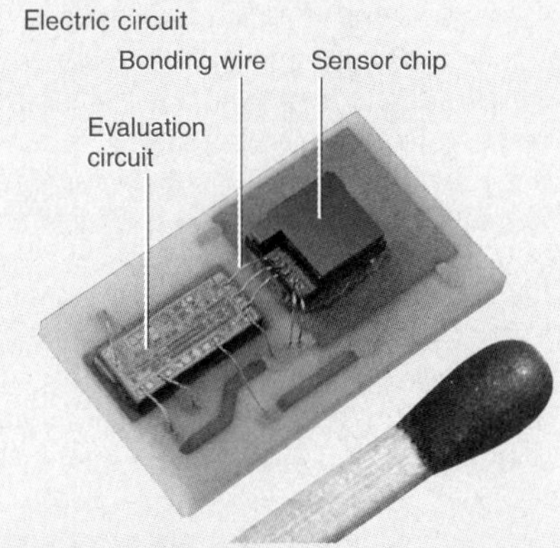

Comb-like structure compared to an insect's head

UAE0787E

Micromechanical yaw-rate sensor

DRS-MM1 vehicle-dynamics control (ESP)

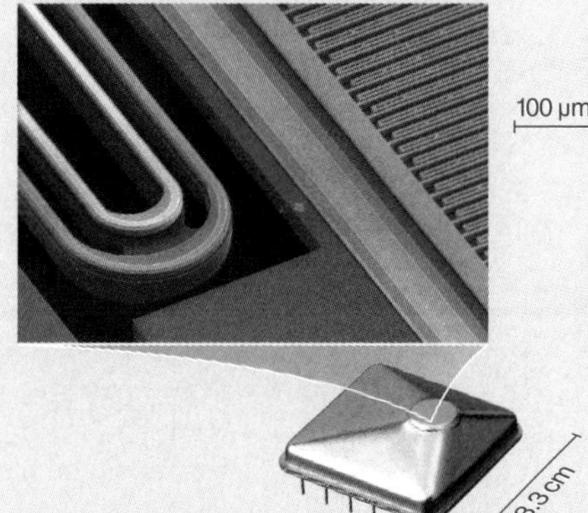

DRS-MM2 roll-over sensing, navigation

UAE0788Y

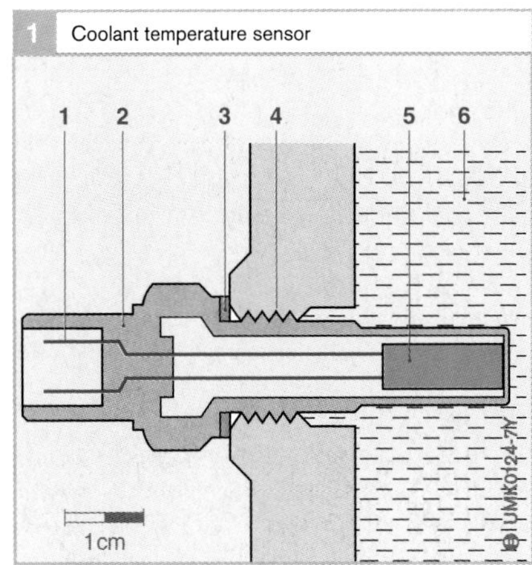

1 Coolant temperature sensor

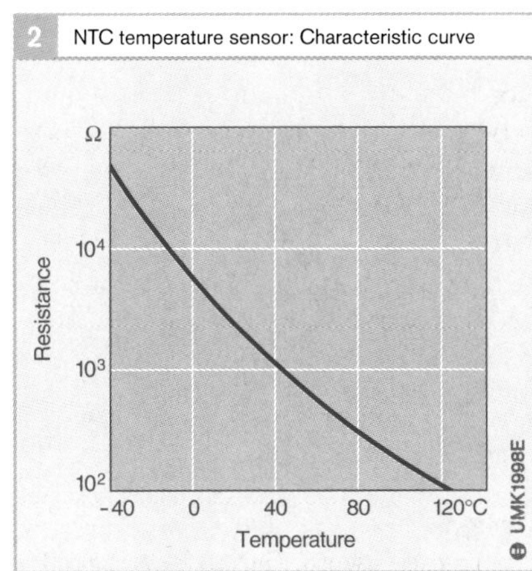

2 NTC temperature sensor: Characteristic curve

Fig. 1
1 Electrical connections
2 Housing
3 Gasket
4 Thread
5 Measuring resistor
6 Coolant

Temperature sensors

Applications

Engine-temperature sensor

This is installed in the coolant circuit (Fig. 1). The engine management uses its signal when calculating the engine temperature (measuring range – 40...+130 °C).

Air-temperature sensor

This sensor is installed in the air-intake tract. Together with the signal from the boost-pressure sensor, its signal is applied in calculating the intake-air mass. Apart from this, desired values for the various control loops (e.g. EGR, boost-pressure control) can be adapted to the air temperature (measuring range –40...+120 °C).

Engine-oil temperature sensor

The signal from this sensor is used in calculating the service interval (measuring range –40...+170 °C).

Fuel-temperature sensor

Is incorporated in the low-pressure stage of the diesel fuel circuit. The fuel temperature is used in calculating the precise injected fuel quantity (measuring range –40...+120 °C).

Exhaust-gas temperature sensor

This sensor is mounted on the exhaust system at points which are particularly critical regarding temperature. It is applied in the closed-loop control of the systems used for exhaust-gas treatment. A platinum measuring resistor is usually used (measuring range –40...+1,000 °C).

Design and operating concept

Depending upon the particular application, a wide variety of temperature sensor designs are available. A temperature-dependent semiconductor measuring resistor is fitted inside a housing. This resistor is usually of the NTC (Negative Temperature Coefficient, Fig. 2) type. Less often a PTC (Positive Temperature Coefficient) type is used. With NTC, there is a sharp drop in resistance when the temperature rises, and with PTC there is a sharp increase.

The measuring resistor is part of a voltage-divider circuit to which 5 V is applied. The voltage measured across the measuring resistor is therefore temperature-dependent. It is inputted through an analog to digital (A/D) converter and is a measure of the temperature at the sensor. A characteristic curve is stored in the engine-management ECU which allocates a specific temperature to every resistance or output-voltage.

Fuel-level sensor

Application

It is the job of the fuel-level sensor to register the level of the fuel in the tank and send the appropriate signal to the ECU or to the display device in the vehicle's instrument panel. Together with the electric fuel pump and the fuel filter, it is part of the in-tank unit. These are installed in the fuel tank (gasoline or diesel fuel) and provide for an efficient supply of clean fuel to the engine (Fig. 1).

Design

The fuel-level sensor (Fig. 2) is comprised of a potentiometer with wiper arm (wiper spring), printed conductors (twin-contact), resistor board (pcb), and electrical connections. The complete sensor unit is encapsulated and sealed against fuel. The float (fuel-resistant Nitrophyl) is attached to one end of the wiper lever, the other end of which is fixed to the rotatable potentiometer shaft (and therefore also to the wiper spring). Depending upon the particular version, the float can be either fixed in position on the lever, or it can be free to rotate). The layout of the resistor board (pcb) and the shape of the float lever and float are matched to the particular fuel-tank design.

Operating concept

The potentiometer's wiper spring is fixed to the float lever by a pin. Special wipers (contact rivets) provide the contact between the wiper spring and the potentiometer resistance tracks, and when the fuel level changes the wipers move along these tracks and generate a voltage ratio which is proportional to the float's angle of rotation. End stops limit the rotation range of 100° for maximum and minimum levels as well as preventing noise. Operating voltage is 5...13 V.

1 Fuel-level sensor installed in a fuel tank

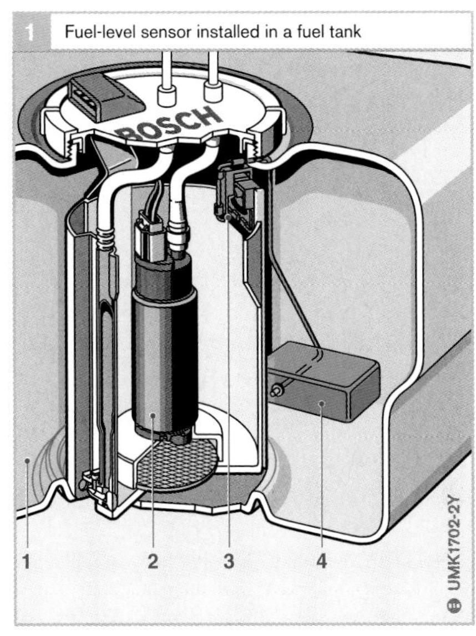

Fig. 1
1 Fuel tank
2 Electric fuel pump
3 Fuel-level sensor
4 Float

2 Fuel-level sensor

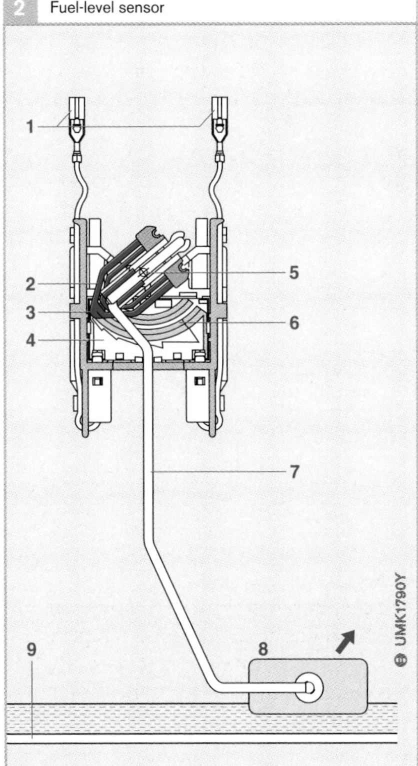

Fig. 2
1 Electrical connections
2 Wiper spring
3 Contact rivet
4 Resistor board
5 Bearing pin
6 Twin contact
7 Float lever
8 Float
9 Fuel-tank floor

Sensor-plate potentiometer

Application

The sensor-plate potentiometer is used in the air-flow sensor of the KE-Jetronic fuel-injection system to register the position (angle of rotation) of the sensor flap. The rate at which the driver presses the accelerator pedal is derived from the sensor plate's movement, which is only slightly delayed with respect to the throttle-valve movement. This signal corresponds to the change in intake air quantity as a function of time, in other words approximately engine power. The potentiometer inputs it to the ECU which applies it when triggering the electro-hydraulic pressure actuator (Fig. 2).

1 Sensor-plate potentiometer (highly simplified)

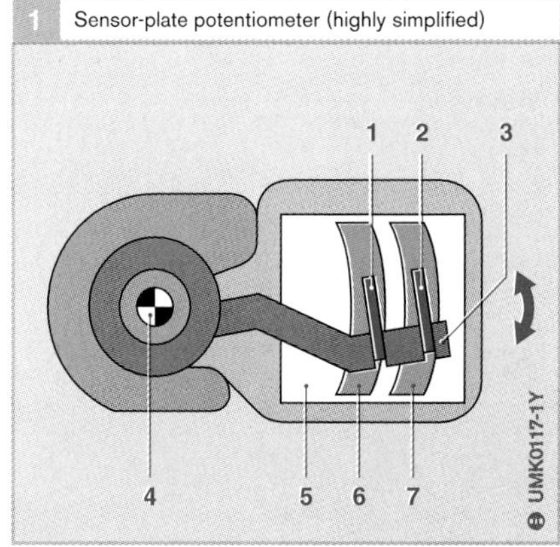

Fig. 1
1 Pick-off wiper brush
2 Main wiper brush
3 Wiper lever
4 Air-flow sensor shaft
5 Potentiometer board
6 Pick-off track
7 Measurement track

2 Sensor-plate potentiometer in the KE-Jetronic air-flow sensor (schematic)

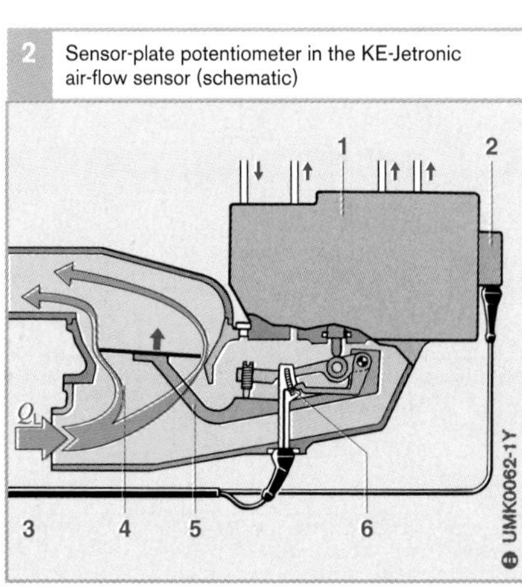

Fig. 2
1 Fuel distributor
2 Electrohydraulic pressure actuator
3 To the ECU
4 Air-flow sensor
5 Sensor plate
6 Potentiometer

Q_L Air quantity

Depending upon the engine's operating state and the corresponding current signal generated by the ECU, the pressure actuator changes the pressure in the vacuum chambers of the differential-pressure valves in the fuel distributor, and with it the amount of fuel metered to the injectors.

Design and operating concept

The potentiometer in the air-flow sensor is produced on a ceramic substrate using film techniques. It is a potentiometer-type angle-of-rotation sensor, which for measurement purposes applies the relationship which exists between the length of a film resistor (printed conductor) and its resistance. The printed-conductor width is varied in order to make the potentiometer characteristic non-linear so that the highest acceleration signal is generated when sensor-plate movement originates from the idle setting. The signal decreases along with increasing engine power output.

The brush wiper slides over the potentiometer tracks (pick-off track and wiper track) and is comprised of a number of very fine wires welded to a lever which is mechanically connected to the sensor-plate shaft (from which it is electrically insulated). The individual wires only apply very light pressure to the potentiometer tracks so that wear remains at a very low level. The large number of wires leads to good electrical contact in case the track surface is very rough and also when the brush is moved very quickly over the track. The wiper voltage is picked-off by a second brush wiper which is connected electrically to the main wiper (Fig. 1).

Damage due to air blowback in the intake manifold is ruled out since the wiper is free to travel far enough beyond the measurement range at both ends of the track. Protection against electrical short circuit is provided by a fixed film resistor connected in series with the wiper.

Throttle-valve sensor

Application

The throttle-valve sensor registers the angle of rotation of the gasoline-engine throttle valve. On M-Motronic engines, this is used to generate a secondary-load signal which, amongst other things, is used as auxiliary information for dynamic functions, as well as for recognition of operating range (idle, part load, WOL), and as a limp-home or emergency signal in case of failure of the primary-load sensor (air-mass meter). If the throttle-valve sensor is used as the primary-load sensor, the required accuracy is achieved by applying two potentiometers for two angular ranges.

The ME-Motronic adjusts the required engine torque via the throttle valve. In order to check that the throttle valve moves to the required position, the throttle-valve sensor is used to evaluate the valve's position (closed-loop position control). As a safety measure, this sensor is provided with two parallel-operation (redundant) potentiometers with separate reference voltages.

Design and operating concept

The throttle-valve sensor is a potentiometer-type angle-of-rotation sensor with one (or two) linear characteristic curve(s).

The wiper arm is connected mechanically with the throttle-valve shaft, and with its brushes slides across the respective potentiometer tracks. In the process, it converts the rotation of the throttle valve shaft into a voltage ratio U_A/U_V which is proportional to the valve's angle of rotation (Fig. 2). The operating voltage is 5 V. The electrical wiper connection is usually through a second potentiometer track. This has the same surface, but the track itself is formed of a low-resistance printed-conductor material (Figs. 1 and 3).

As a protection against overload, the voltage is applied to the measurement (potentiometer) tracks through small series resistors (also used for zero-point and slope calibration). The shape of the characteristic curve can be adapted by varying the width of the potentiometer track (variation can also apply to sections of the track).

1 Throttle-valve sensor

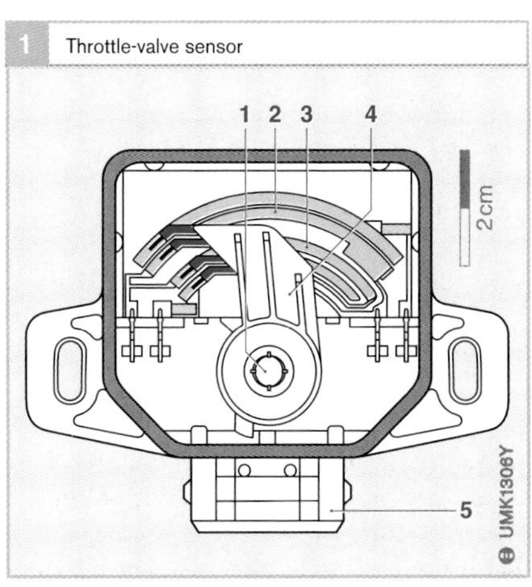

Fig. 1
1 Throttle-valve shaft
2 Resistance track 1
3 Resistance track 2
4 Wiper arm with wipers
5 Electric connection (4-pole)

2 Throttle-valve sensor with two curves

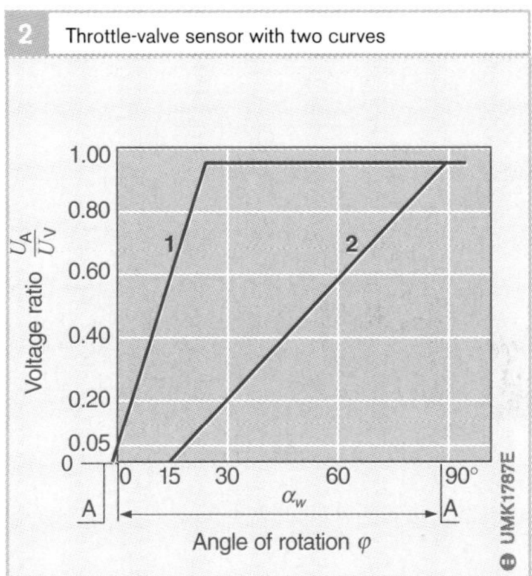

Fig. 2
A Internal stop

1 Curve for high resolution in angular range 0°...23°
2 Curve for angular range 15°...88°

U_0 Supply voltage
U_A Measurement voltage
U_V Operating voltage
α_W Effective measured angle

3 Throttle-valve sensor (circuit)

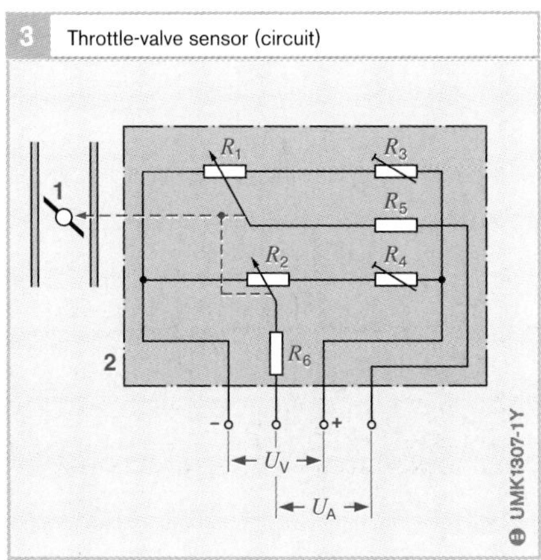

Fig. 3
1 Throttle valve
2 Throttle-valve sensor

U_A Measurement voltage
U_V Operating voltage
R_1, R_2 Resistance tracks 1 and 2
R_3, R_4 Calibration resistors
R_5, R_6 Protective resistors

Accelerator-pedal sensors

Application

In conventional engine-management systems, the driver transmits his/her wishes for acceleration, constant speed, or lower speed, to the engine by using the accelerator pedal to intervene mechanically at the throttle plate (gasoline engine) or at the injection pump (diesel engine). Intervention is transmitted from the accelerator pedal to the throttle plate or injection pump by means of a Bowden cable or linkage.

On today's electronic engine-management systems, the Bowden cable and/or linkage have been superseded, and the driver's accelerator-pedal inputs are transmitted to the ECU by an accelerator-pedal sensor which registers the accelerator-pedal travel, or the pedal's angular setting, and sends this to the engine ECU in the form of an electric signal. This system is also known as "drive-by-wire". The accelerator-pedal module (Figs. 2b, 2c) is available as an alternative to the individual accelerator-pedal sensor (Fig. 2a). These modules are ready-to-install units comprising accelerator pedal and sensor, and make adjustments on the vehicle a thing of the past.

Design and operating concept

Potentiometer-type accelerator-pedal sensor

The heart of this sensor is the potentiometer across which a voltage is developed which is a function of the accelerator-pedal setting. In the ECU, a programmed characteristic curve is applied in order to calculate the accelerator-pedal travel, or its angular setting, from this voltage.

A second (redundant) sensor is incorporated for diagnosis purposes and for use in case of malfunctions. It is a component part of the monitoring system. One version of the accelerator-pedal sensor operates with a second potentiometer. The voltage across this potentiometer is always half of that across the first potentiometer. This provides two independent signals which are used for trouble-shooting (Fig. 1). Instead of the second potentiometer, another version uses a low-idle switch which provides a signal for the ECU when the accelerator pedal is in the

1 Characteristic curve of an accelerator-pedal sensor with redundant potentiometer

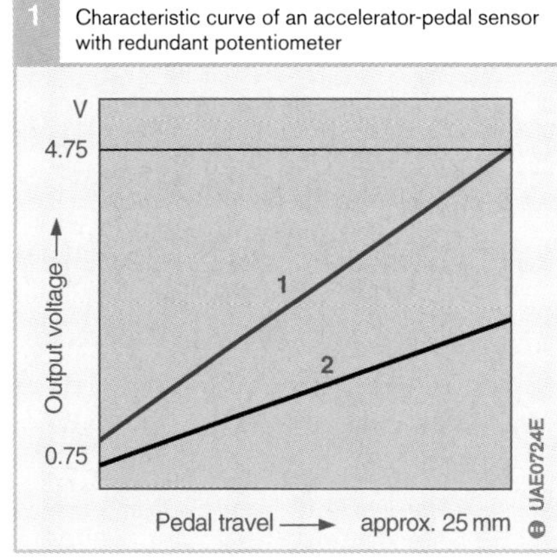

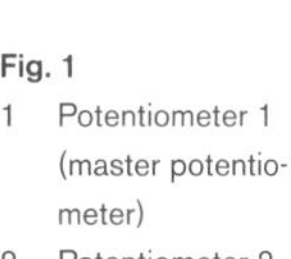
Fig. 1
1 Potentiometer 1 (master potentiometer)
2 Potentiometer 2 (50 % of voltage)

2 Accelerator-pedal-sensor versions

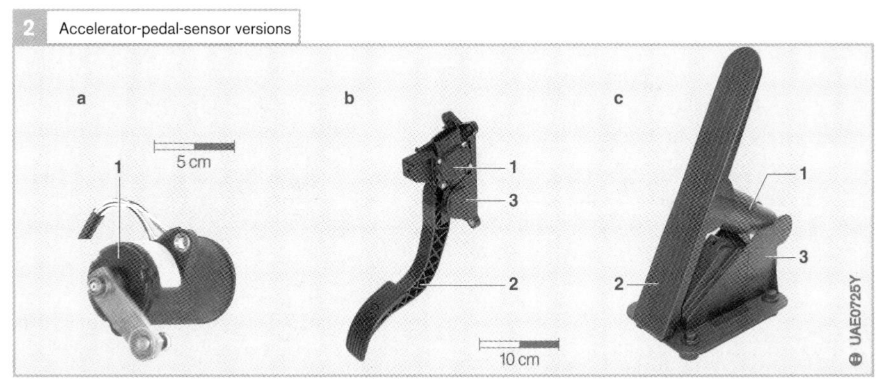

Fig. 2
a Individual accelerator-pedal sensor
b Top-mounted accelerator-pedal module
c Bottom-mounted accelerator-pedal module FMP1

1 Sensor
2 Vehicle-specific pedal
3 Pedal bracket

idle position. For automatic transmission vehicles, a further switch can be incorporated for a kick-down signal.

Hall-effect angle-of-rotation sensors

The ARS1 (**A**ngle of **R**otation **S**ensor) is based on the movable-magnet principle. It has a measuring range of approx. 90° (Figs. 3 and 4).

A semicircular permanent-magnet disc rotor (Fig. 4, Pos. 1) generates a magnetic flux which is returned back to the rotor via a pole shoe (2), magnetically soft conductive elements (3) and shaft (6). In the process, the amount of flux which is returned through the conductive elements is a function of the rotor's angle of rotation φ. There is a Hall-effect sensor (5) located in the magnetic path of each conductive element, so that it is possible to generate a practically linear characteristic curve throughout the measuring range.

The ARS2 is a simpler design without magnetically soft conductive elements. Here, a magnet rotates around the Hall-effect sensor. The path it takes describes a circular arc. Since only a small section of the resulting sinusoidal characteristic curve features good linearity, the Hall-effect sensor is located slightly outside the center of the arc. This causes the curve to deviate from its sinusoidal form so that the curve's linear section is increased to more than 180°.

Mechanically, this sensor is highly suitable for installation in an accelerator-pedal module (Fig. 5).

3 Hall-effect angle-of-rotation sensor ARS1

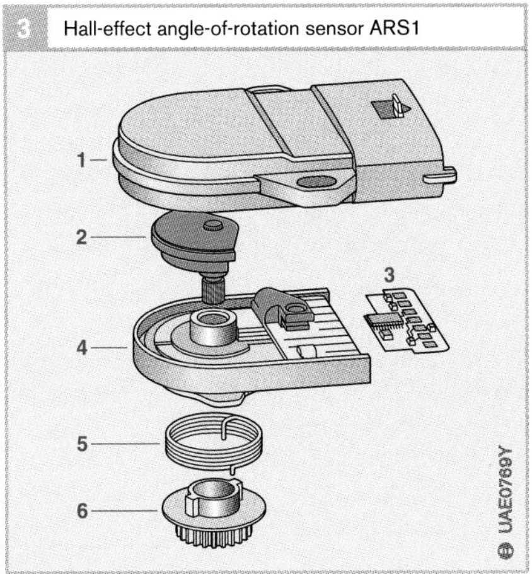

Fig. 3
1 Housing cover
2 Rotor (permanent magnet)
3 Evaluation electronics with Hall-effect sensor
4 Housing base
5 Return spring
6 Coupling element (e.g. gear)

4 Hall-effect angle-of-rotation sensor ARS1 (shown with angular settings a...d)

Fig. 4
1 Rotor (permanent magnet)
2 Pole shoe
3 Conductive element
4 Air gap
5 Hall-effect sensor
6 Shaft (magnetically soft)

φ Angle of rotation

5 Hall-effect angle-of-rotation sensor ARS 2

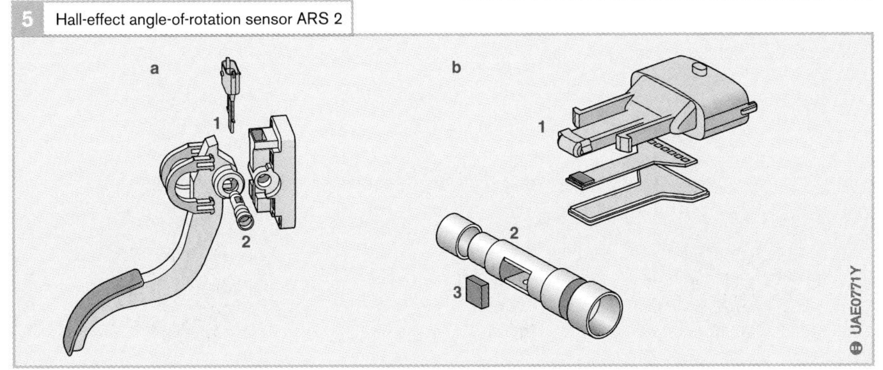

Fig. 5
a Installation in the accelerator-pedal module
b Components

1 Hall-effect sensor
2 Pedal shaft
3 Magnet

Hall-effect phase sensors

Application

The engine's camshaft rotates at half the crankshaft speed. Taking a given piston on its way to TDC, the camshaft's rotational position is an indication as to whether the piston is in the compression or exhaust stroke. The phase sensor on the camshaft provides the ECU with this information.

Design and operating concept

Hall-effect rod sensors

As the name implies, such sensors (Fig. 2a) make use of the Hall effect. A ferromagnetic trigger wheel (with teeth, segments, or perforated rotor, Pos. 7) rotates with the camshaft. The Hall-effect IC is located between the trigger wheel and a permanent magnet (Pos. 5) which generates a magnetic field strength perpendicular to the Hall element.

If one of the trigger-wheel teeth (Z) now passes the current-carrying rod-sensor element (semiconductor wafer), it changes the magnetic field strength perpendicular to the Hall element. This causes the electrons, which are driven by a longitudinal voltage across the element to be deflected perpendicularly to the direction of current (Fig. 1, angle α).

This results in a voltage signal (Hall voltage) which is in the millivolt range, and which is independent of the relative speed between sensor and trigger wheel. The evaluation electronics integrated in the sensor's Hall IC conditions the signal and outputs it in the form of a rectangular-pulse signal (Fig. 2b "High"/"Low").

Differential Hall-effect rod sensors

Rod sensors operating as per the differential principle are provided with two Hall elements. These elements are offset from each other either radially or axially (Fig. 3, S1 and S2), and generate an output signal which is proportional to the difference in magnetic flux at the element measuring points. A two-track perforated plate (Fig. 3a) or a two-track trigger wheel (Fig. 3b) are needed in order to generate the opposing signals in the Hall elements (Fig. 4) as needed for this measurement.

Such sensors are used when particularly severe demands are made on accuracy. Further advantages are their relatively wide air-gap range and good temperature-compensation characteristics.

1 Hall element (Hall-effect vane switch)

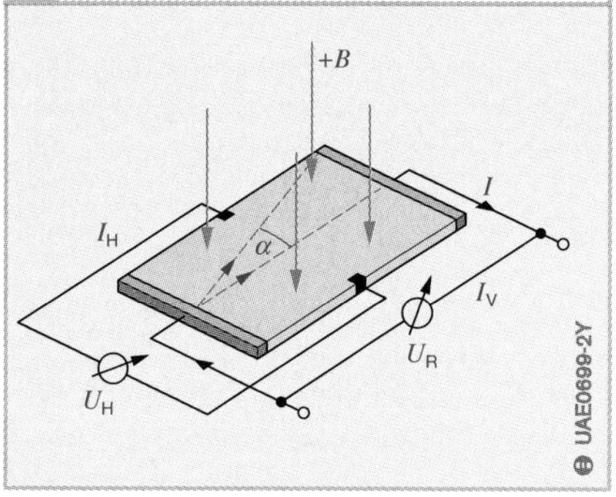

Fig. 1
- I Wafer current
- I_H Hall current
- I_V Supply current
- U_H Hall voltage
- U_R Longitudinal voltage
- B Magnetic induction
- α Deflection of the electrons by the magnetic field

2 Hall-effect rod sensor

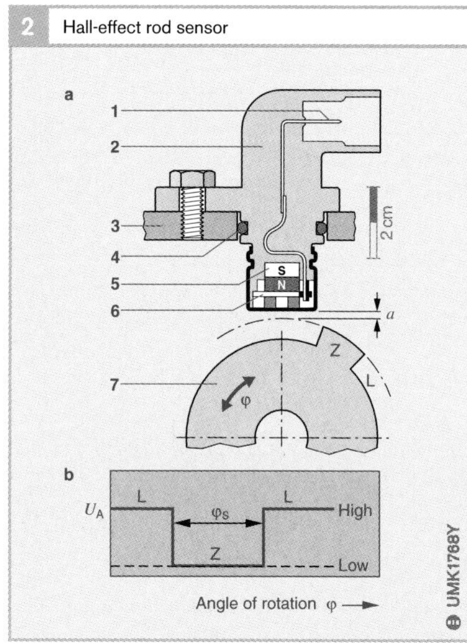

Fig. 2
- a Positioning of sensor and single-track trigger wheel
- b Output signal characteristic U_A

1. Electrical connection (plug)
2. Sensor housing
3. Engine block
4. Seal ring
5. Permanent magnet
6. Hall-IC
7. Trigger wheel with tooth/segment (Z) and gap (L)

- a Air gap
- φ Angle of rotation

3 Differential Hall-effect rod sensors

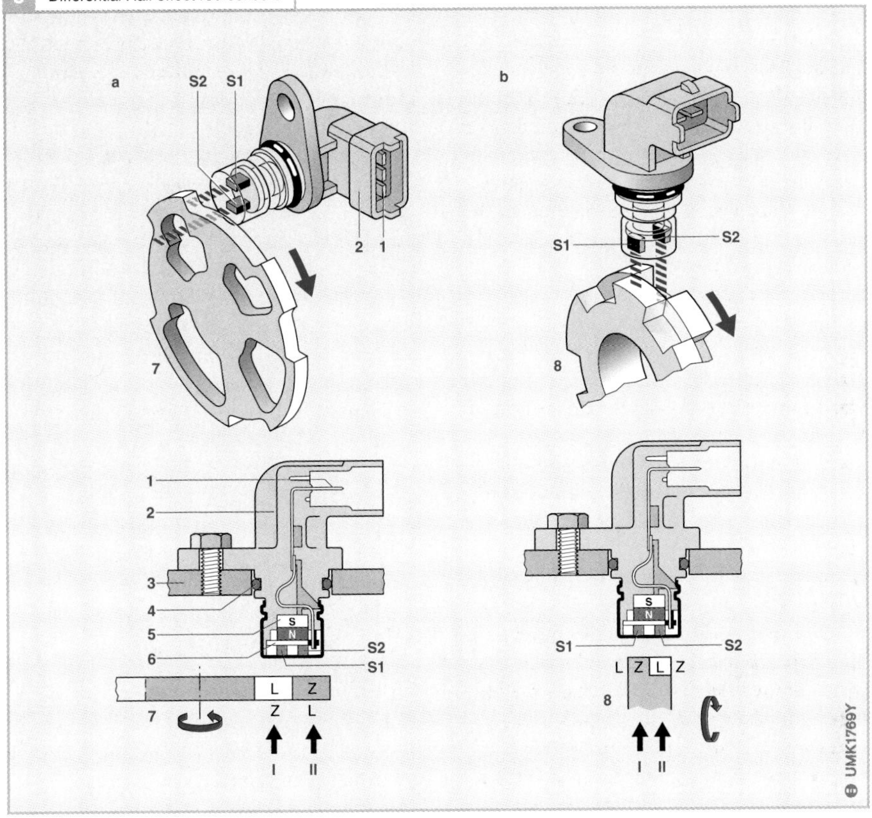

Fig. 3
a Axial tap-off (perforated plate)
b Radial tap-off (two-track trigger wheel)

1 Electrical connection (plug)
2 Sensor housing
3 Engine block
4 Seal ring
5 Permanent magnet
6 Differential Hall-IC with Hall elements S1 and S2
7 Perforated plate
8 Two-track trigger wheel

I Track 1
II Track 2

4 Characteristic curve of the output signal U_A from a differential Hall-effect rod sensor

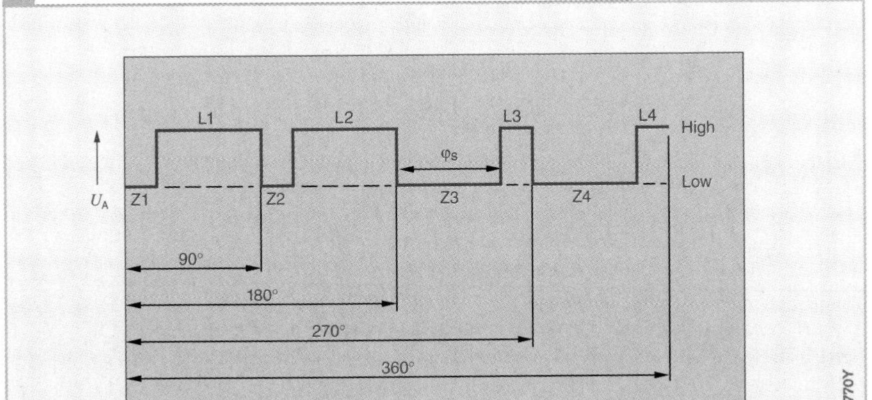

Fig. 4
Output signal “Low”: Material (Z) in front of S1, gap (L) in front of S2

Output signal “High”: Gap (L) in front of S1, material (Z) in front of S2

φ_S signal width

Induction-type sensors for transistorized ignition

Applications

For ignition-triggering purposes, the TC-I transistorized ignition uses an induction-type sensor which serves as an AC generator. The switch-on point for the dwell angle is defined by comparing its AC signal with that of a voltage signal which corresponds to the current-control time.

Design and construction

The induction-type sensor is incorporated in the ignition-distributor housing in place of the former contact-breaker points (Fig. 1).

The soft-magnetic core of the induction winding is disc-shaped, and together with the permanent magnet and the induction winding, forms a fixed, enclosed subassembly, the stator.

The rotor (trigger wheel) on the distributor shaft rotates past the ends of the stators. Similar to the distributor cam for the former contact breaker assembly, it is firmly attached to the hollow shaft surrounding the distributor shaft.

Core and rotor are produced from soft-magnetic material and have toothed extensions (stator teeth and rotor teeth). The stator teeth are at the ends of the stator "limbs" and bent upwards at right angles. The rotor has similar teeth, but these are bent downwards at right angles.

As a rule, the number of teeth on rotor and stator correspond to the number of cylinders in the engine. The fixed and rotating teeth are separated by a mere 0.5 mm when directly opposite to each other.

Operating concept

The principle of functioning depends upon the air gap between the rotor teeth and the stator teeth, and thus the magnetic flux, changing periodically along with rotation of the rotor. This change in magnetic flux induces an AC voltage in the induction winding whose peak voltage $\pm \hat{U}_S$ is proportional to the rotor's speed of rotation. At low speeds it is approx. 0.5 V and at high speeds approx. 100 V. The frequency f of this AC voltage (Fig. 2) corresponds to the number of ignition sparks per minute (sparking rate). The following applies

$$f = z \cdot n/2$$

where

f Frequency or sparking rate (rpm)
z Number of engine cylinders
n Engine speed (rpm)

1 Induction-type sensor in the ignition distributor (principle)

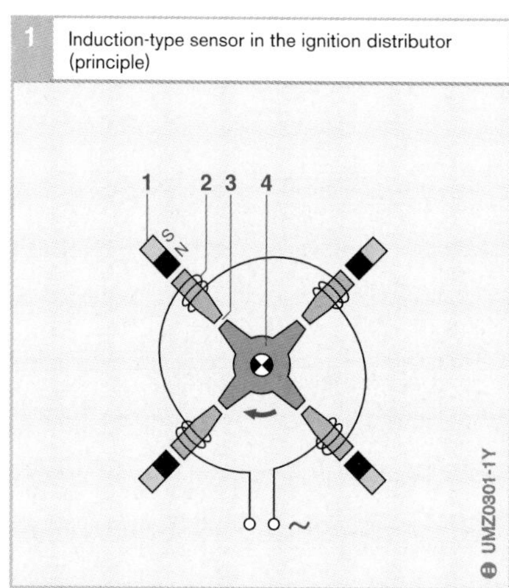

Fig. 1
1 Permanent magnet
2 Induction winding with core
3 Variable air gap
4 Rotor

2 Induction-type sensor in the ignition distributor (characteristic)

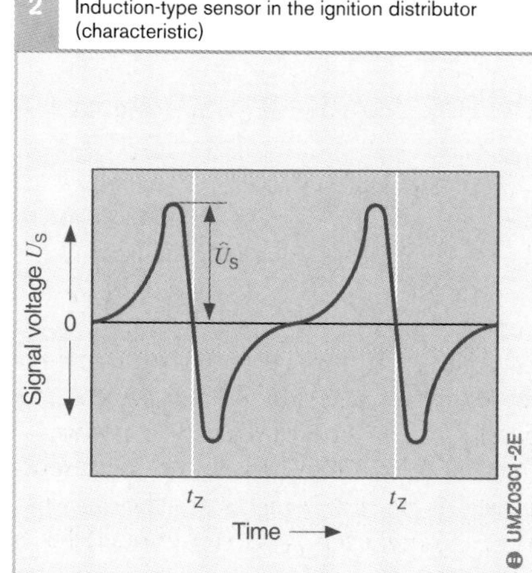

Fig. 2
U_S Signal voltage
$\hat{U}_S$ Peak voltage
t_Z Ignition point

Hall-effect sensors for transistorized ignition

Application

The Hall-effect sensor is also used as the ignition-triggering sensor for the TI-H transistorized ignition system. The information contained in the signal from the Hall generator located in the ignition distributor corresponds to that in the signal generated by the breaker points in a conventional breaker-triggered coil-ignition system. Whereas with the conventional ignition system the distributor cam defines the dwell angle via the contact-breaker points, on the transistorized system the Hall-effect sensor in the ignition distributor defines the on/off ratio by means of the rotor (trigger-wheel) vane.

Design and construction

The Hall-effect sensor (Fig. 1) is installed in the ignition distributor, and its vane switch is attached to the movable mounting plate. The Hall IC is mounted on a ceramic substrate and in order to protect it against moisture, dirt, and mechanical damage is encapsulated in plastic at one of the conductive elements. The conductive elements and the rotor are made of a soft-magnetic material. The number of vanes on the rotor corresponds to the number of cylinders in the engine. Depending on the type of ignition trigger box, the width b of the rotor's individual conductive elements can define the ignition system's maximum dwell angle. The dwell angle therefore remains practically constant throughout the Hall sensor's service life and dwell-angle adjustment is unnecessary.

Operating concept

When the ignition-distributor shaft rotates, the rotor vane's pass through the Hall IC air gap without making contact. If the air gap is not occupied by a vane, the magnetic field is free to permeate the Hall IC and the Hall-effect sensor element (Fig. 1). The magnetic flux density is high, the Hall voltage is at its maximum, and the Hall-IC is switched on. As soon as a rotor vane enters the air gap, the majority of the magnetic flux is diverted through the vane and is isolated from the Hall-IC. The magnetic flux density at the Hall sensor element reduces to a negligible level which results from the leakage field, and the Hall voltage drops to a minimum. The dwell angle is defined by the rotor vane's shape as follows: A ramp voltage is generated from the signal voltage U_S (converted Hall voltage, Fig. 2). The switch-on point for the dwell angle is shifted as required along this ramp. The Hall-effect sensor's priniple of operation and its construction permit the ignition to be adjusted with the engine at standstill provided no provision is made for peak-coil-current cut-off.

1 Hall-effect sensor in the ignition distributor (principle of operation)

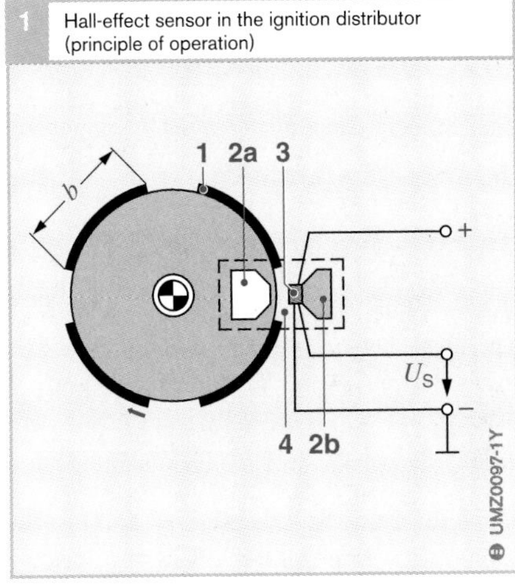

Fig. 1
1 Vane with width b
2a Permanent magnet
2b Soft-magnetic conductive element
3 Hall-IC
4 Air gap

U_S Signal voltage (converted Hall voltage)

2 Hall-effect sensor in the ignition distributor (characteristic curve)

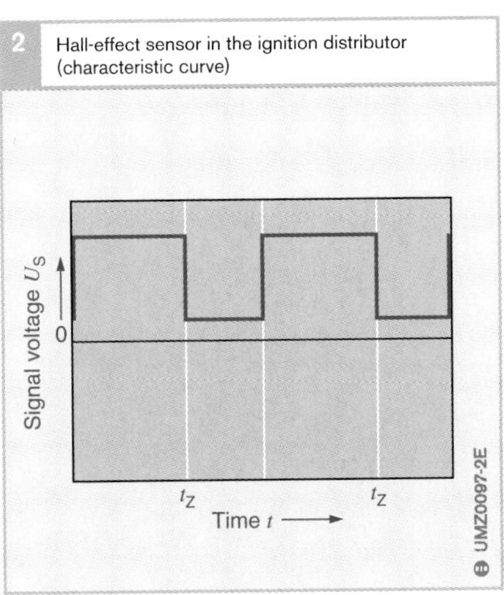

Fig. 2
U_S Signal voltage (converted Hall voltage)
t_z Ignition point

Inductive engine-speed sensors

Applications

Such engine-speed sensors are used for measuring:

- Engine rpm
- Crankshaft position (for information on the position of the engine pistons)

The rotational speed is calculated from the sensor's signal frequency. The output signal from the rotational-speed sensor is one of the most important quantities in electronic engine management.

1 Inductive rpm sensor

Fig. 1
1 Permanent magnet
2 Sensor housing
3 Engine block
4 Pole pin
5 Solenoid winding
6 Air gap
7 Trigger wheel with reference-mark gap

Design and operating concept

The sensor is mounted directly opposite a ferromagnetic trigger wheel (Fig. 1, Pos. 7) from which it is separated by a narrow air gap. It has a soft-iron core (pole pin) (4), which is enclosed by the solenoid winding (5). The pole pin is also connected to a permanent magnet (1), and a magnetic field extends through the pole pin and into the trigger wheel. The level of the magnetic flux through the winding depends upon whether the sensor is opposite a trigger-wheel tooth or gap. Whereas the magnet's stray flux is concentrated by a tooth and leads to an increase in the working flux through the winding, it is weakened by a gap. When the trigger wheel rotates therefore, this causes a fluctuation of the flux which in turn generates a sinusoidal voltage in the solenoid winding which is proportional to the rate of change of the flux (Fig. 2). The amplitude of the AC voltage increases strongly along with increasing trigger-wheel speed (several mV...>100 V). At least about 30 rpm are needed to generate an adequate signal level.

The number of teeth on the trigger wheel depends upon the particular application. On solenoid-valve-controlled engine-management systems for instance, a 60-pitch trigger wheel is normally used, although 2 teeth are omitted (7) so that the trigger wheel has 60 – 2 = 58 teeth. The very large tooth gap is allocated to a defined crankshaft position and serves as a reference mark for synchronizing the ECU.

There is another version of the trigger wheel which has one tooth per engine cylinder. In the case of a 4-cylinder engine, therefore, the trigger wheel has 4 teeth, and 4 pulses are generated per revolution.

The geometries of the trigger-wheel teeth and the pole pin must be matched to each other. The evaluation-electronics circuitry in the ECU converts the sinusoidal voltage, which is characterized by strongly varying amplitudes, into a constant-amplitude square-wave voltage for evaluation in the ECU microcontroller.

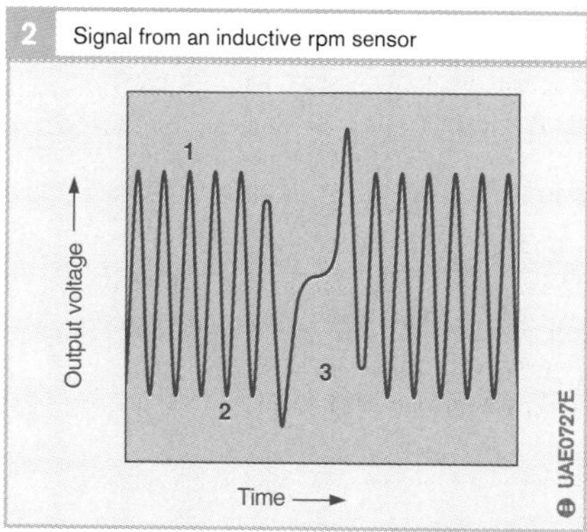

2 Signal from an inductive rpm sensor

Fig. 2
1 Tooth
2 Tooth gap
3 Reference mark

Piezoelectric knock sensors

Application

Regarding their principle of functioning, knock sensors are basically vibration sensors and are suitable for detecting structure-borne acoustic oscillations. These occur as "knock" for instance in a vehicle engine when uncontrolled ignition takes place, and are converted into electrical signals by the sensor and inputted to the ECU. As a rule, 4-cylinder in-line engines are equipped with *one* knock sensor; 5 and 6-cylinder engines, with *two;* and 8 and 12-cylinder engines have *two or more.* They are switched in accordance with the ignition sequence.

Design and operating concept

Due to its inertia, a mass excited by a given oscillation or vibration exerts a compressive force on a toroidal piezoceramic element at the same frequency as the excitation oscillation. Inside the ceramic element, these compressive forces cause a charge transfer so that a voltage appears across the ceramic element's two outer faces which is picked-off by contact discs and inputted to the ECU for processing. Sensitivity is defined as the output voltage per unit of acceleration [mV/*g*].

1 Knock-sensor signal

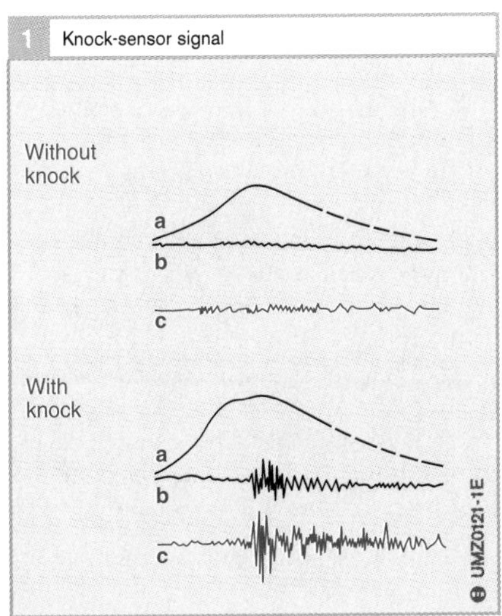

Fig. 1
a Cylinder-pressure curve
b Filtered pressure signal
c Knock-sensor signal

2 Knock sensor (design and mounting)

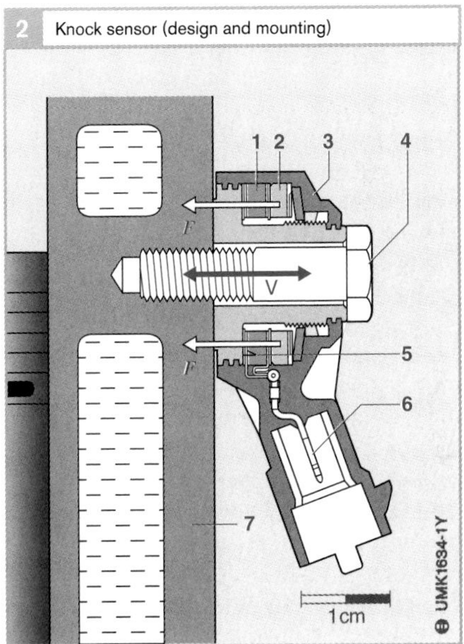

Fig. 2
1 Piezoceramic element
2 Seismic mass with compressive forces F
3 Housing
4 Fastening screw
5 Contact surface
6 Electrical connection
7 Cylinder block

V Vibration

The sensor's voltage output is evaluated by a high-resistance AC amplifier in the ECU of the ignition or Motronic engine-management system (Figs. 1 and 2).

Mounting

Depending on the particular engine, the knock-sensor installation point is selected so that knock can be reliably detected from each cylinder. The sensor is usually screwed to the side of the engine-cylinder block. In order that the resulting signals (structure-borne oscillations) can be transferred from the measuring point on the engine block and into the sensor without resonant-frequency effects and in agreement with the stipulated characteristic curve, the following points must be observed:

- The fastening bolt must have been tightened with a defined torque
- The sensor's contact surface and bore in the engine block must comply with certain quality requirements and
- No washers of any type may be used

Micromechanical pressure sensors

Application

Manifold-pressure or boost-pressure sensor

This sensor measures the absolute pressure in the intake manifold between the supercharger and the engine (typically 250 kPa or 2.5 bar) and compares it with a reference vacuum, not with the ambient pressure. This enables the air mass to be precisely defined, and the boost pressure exactly controlled in accordance with engine requirements.

Atmospheric-pressure sensor

This sensor is also known as an ambient-pressure sensor and is incorporated in the ECU or fitted in the engine compartment. Its signal is used for the altitude-dependent correction of the setpoint values for the control loops. For instance, for the exhaust-gas recirculation (EGR) and for the boost-pressure control. This enables the differing densities of the surrounding air to be taken into account. The atmospheric-pressure sensor measures absolute pressure (60...115 kPa or 0.6...1.15 bar).

Oil and fuel-pressure sensor

Oil-pressure sensors are installed in the oil filter and measure the oil's absolute pressure. This information is needed so that engine loading can be determined as needed for the Service Display. The pressure range here is 50...1,000 kPa or 0.5...10.0 bar. Due to its high resistance to media, the measuring element can also be used for pressure measurement in the fuel supply's low-pressure stage. It is installed on or in the fuel filter. Its signal serves for the monitoring of the fuel-filter contamination (measuring range: 20... 400 kPa or 0.2...4 bar).

Version with the reference vacuum on the component side

Design and construction

The measuring element is at the heart of the micromechanical pressure sensor. It is com-

1 Pressure-sensor measuring element with reference vacuum on the components side

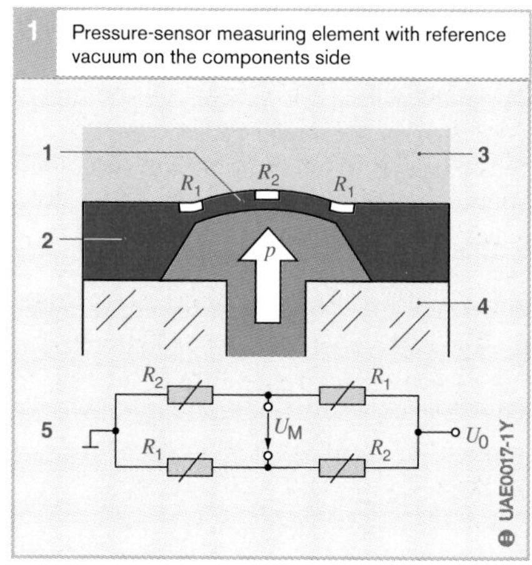

Fig. 1
1 Diaphragm
2 Silicon chip
3 Reference vacuum
4 Glass (Pyrex)
5 Bridge circuit

p Measured pressure
U_0 Supply voltage
U_M Measured voltage
R_1 Deformation resistor (compressed)
R_2 Deformation resistor (extended)

2 Pressure-sensor measuring element with cap and reference vacuum on the components side

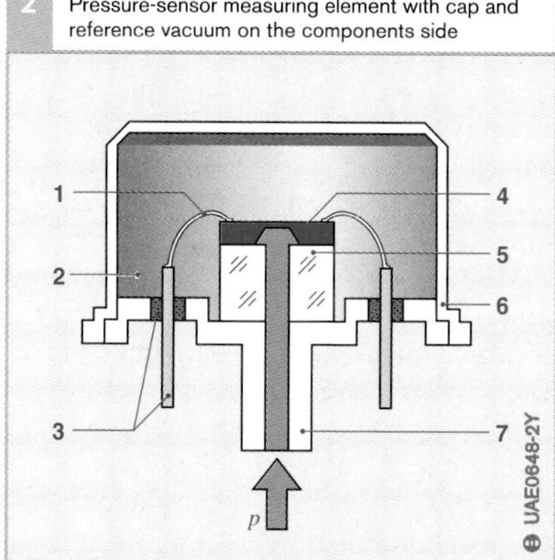

Fig. 2
1, 3 Electrical connections with glass-enclosed lead-in
2 Reference vacuum
4 Measuring element (chip) with evaluation electronics
5 Glass base
6 Cap
7 Input for measured pressure p

3 Pressure-sensor measuring element with cap and reference vacuum on the components side

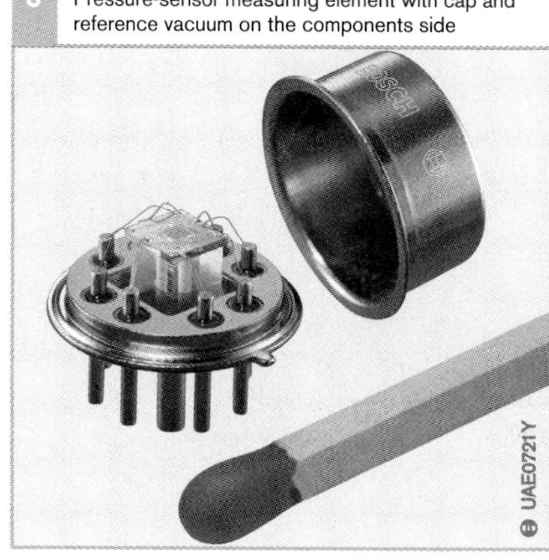

prised of a silicon chip (Fig. 1, Pos. 2) in which a thin diaphragm has been etched micromechanically (1). Four deformation resistors (R_1, R_2) are diffused on the diaphram. Their electrical resistance changes when mechanical force is applied. The measuring element is surrounded on the component side by a cap which at the same time encloses the reference vacuum (Figs. 2 and 3). The pressure-sensor case can also incorporate an integral *temperature sensor* (Fig. 4, Pos. 1) whose signals can be evaluated independently. This means that at any point a single sensor case suffices to measure temperature and pressure.

Method of operation

The sensor's diaphragm deforms more or less (10 ... 1,000 μm) according to the pressure being measured. The four deformation resistors on the diaphragm change their electrical resistances as a function of the mechanical stress resulting from the applied pressure (piezoresistive effect).

The four measuring resistors are arranged on the silicon chip so that when diaphragm deformation takes place, the resistance of two of them increases and that of the other two decreases. These deformation resistors form a Wheatstone bridge (Fig. 1, Pos. 5), and a change in their resistances leads to a change in the ratio of the voltages across them. This leads to a change in the measurement voltage U_M. This unamplified voltage is therefore a measure of the pressure applied to the diaphragm.

The measurement voltage is higher with a bridge circuit than would be the case when using an individual resistor. The Wheatstone bridge circuit thus permits a higher sensor sensitivity.

The component side of the sensor to which pressure is not supplied is subjected to a reference vacuum (Fig. 2, Pos. 2) so that it measures the absolute pressure.

4 Micromechanical pressure sensor with reference vacuum on the components side

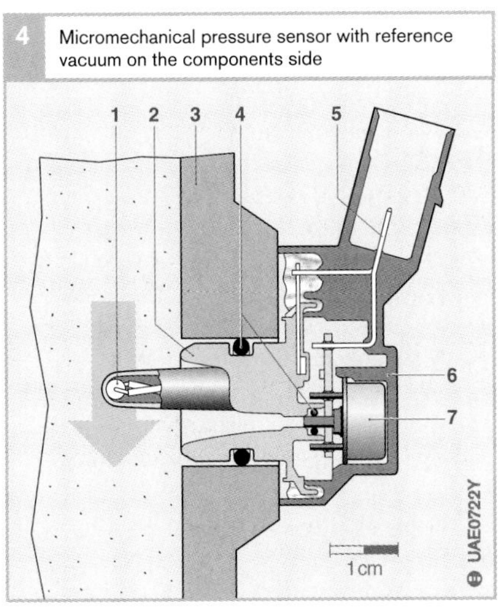

Fig. 4
1 Temperature sensor (NTC)
2 Lower section of case
3 Manifold wall
4 Seal rings
5 Electrical terminal (plug)
6 Case cover
7 Measuring element

5 Micromechanical boost-pressure sensor (example of curve)

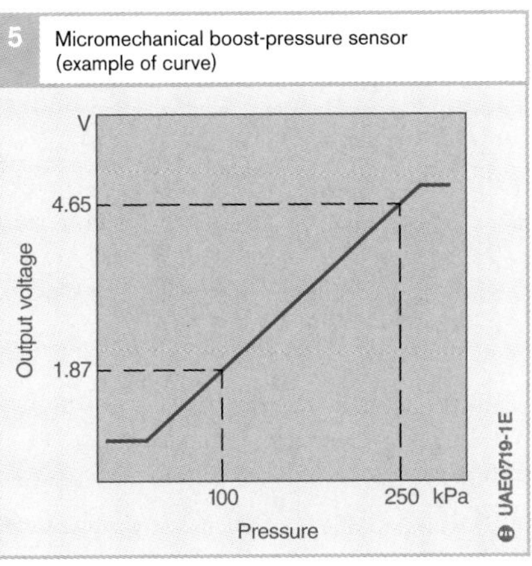

The signal-conditioning electronics circuitry is integrated on the chip. Its assignment is to amplify the bridge voltage, compensate for temperature influences, and linearise the pressure curve. The output voltage is between 0...5 V and is connected through electrical terminals (Fig. 4, Pos. 5) to the engine-management ECU which uses this output voltage in calculating the pressure (Fig. 5).

Version with reference vacuum in special chamber

Design and construction

The *manifold or boost-pressure sensor* version with the reference vacuum in a special chamber (Figs. 6 and 7) is easier to install than the version with the reference vacuum on the components side of the sensor element. Similar to the pressure sensor with cap and reference vacuum on the components side of the sensor element, the sensor element here is formed from a silicon chip with four etched deformation resistors in a bridge circuit. It is attached to a glass base. In contrast to the sensor with the reference vacuum on the components side, there is no passage in the glass base through which the measured pressure can be applied to the sensor element. Instead, pressure is applied to the silicon chip from the side on which the evaluation electronics is situated. This means that a special gel must be used at this side of the sensor to protect it against environmental influences (Fig. 8, Pos. 1). The reference vacuum is enclosed in the chamber between the silicon chip (6) and the glass base (3). The complete measuring element is mounted on a ceramic hybrid (4) which incorporates the soldering surfaces for electrical contacting inside the sensor.

A *temperature sensor* can also be incorporated in the pressure-sensor case. It protrudes into the air flow, and can therefore respond to temperature changes with a minimum of delay (Fig. 6, Pos. 4).

Operating concept

The operating concept, and with it the signal conditioning and signal amplification together with the characteristic curve, corresponds to that used in the pressure sensor with cap and reference vacuum on the sensor's structure side.The only difference is that the measuring element's diaphragm is deformed in the opposite direction and therefore the deformation resistors are "bent" in the other direction.

6 Micromechanical pressure sensor with reference vacuum in a chamber

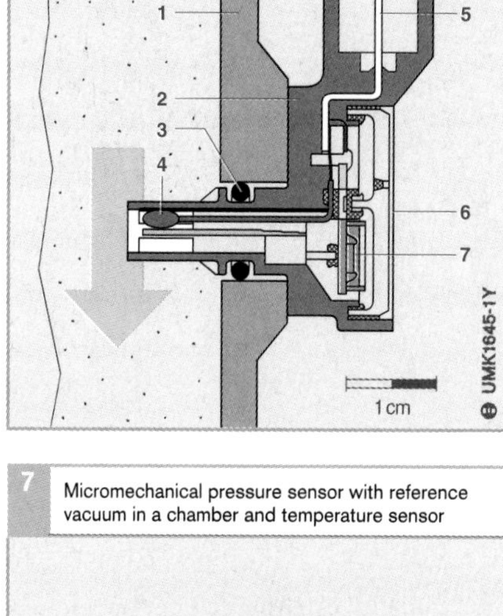

Fig. 6
1 Manifold wall
2 Case
3 Seal ring
4 Temperature sensor (NTC)
5 Electrical connection (socket)
6 Case cover
7 Measuring element

7 Micromechanical pressure sensor with reference vacuum in a chamber and temperature sensor

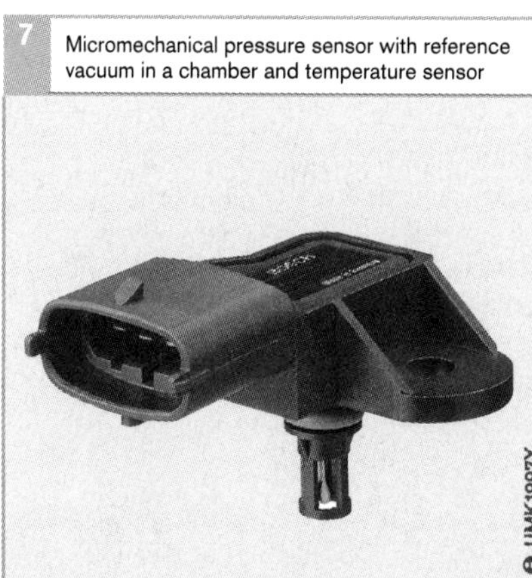
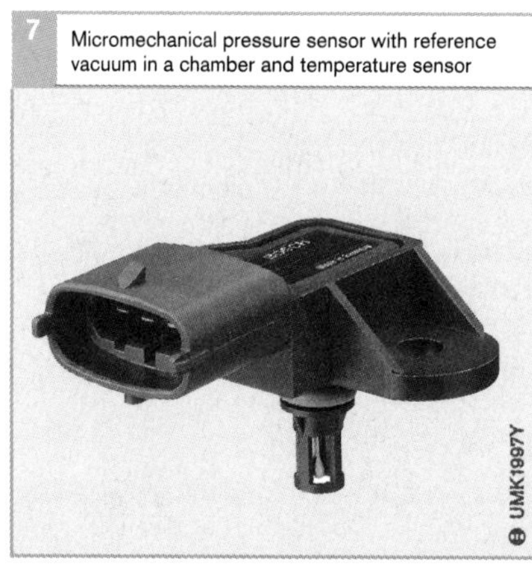

8 Measuring element of pressure sensor with reference vacuum in a chamber

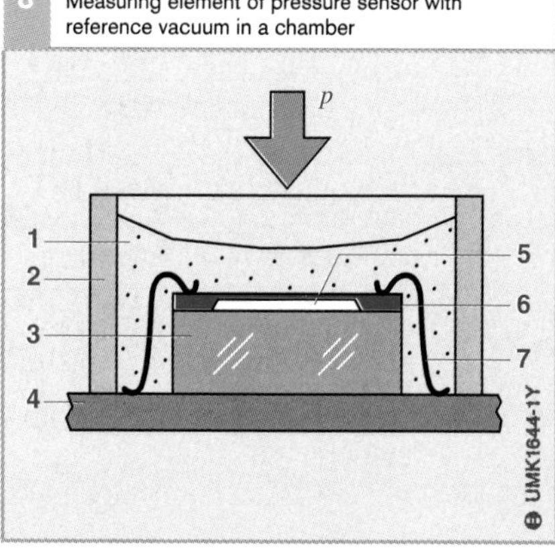

Fig. 8
1 Protective gel
2 Gel frame
3 Glass base
4 Ceramic hybrid
5 Chamber with reference volume
6 Measuring element (chip) with evaluation electronics
7 Bonded connection

p Measured pressure

Thick-film pressure sensors

Application

As an alternative to micromechanical pressure sensors, thick-film pressure sensors can sometimes be used (for instance in engine-management systems, M and ME Motronic). These are in the form of a module for installation in the ECU or a stand-alone component. They are used as:

- Manifold-pressure or boost-pressure sensor (pressure range 20...400 kPa or 0.2...4.0 bar) and
- Atmospheric-pressure sensor (pressure range 60...115 kPa or 0.6...1.15 bar)

Design and operating concept

The sensor is subdivided into a pressure-measuring cell and a chamber for the evaluation circuit. Both are arranged on a common ceramic substrate (Fig. 1).

The pressure-measuring cell (Fig. 2) comprises a "bubble-shaped" thick-film diaphragm which encloses a reference pressure of 0.1 bar. The diaphragm deforms as a function of the pressure being measured. There are four deformation resistors on the diaphragm which are connected to form a bridge circuit. Two of these active deformation resistors are located in the center of the diaphragm and change their conductivity when mechanical stress is applied (measured pressure). Two passive deformation resistors are situated on the diaphragm's periphery and function primarily as bridge resistors for temperature compensation. They have little effect upon the output signal.

When pressure is applied, the diaphragm deforms and changes the bridge-circuit balance. The bridge's measurement voltage U_M is therefore a measure of the measured pressure p (Fig. 3). The evaluation circuit amplifies the bridge voltage, compensates for the influence of temperature, and linearises the pressure curve. The evaluation circuit's output voltage U_A is inputted to the ECU.

1 Thick-film pressure sensor (for ECU installation)

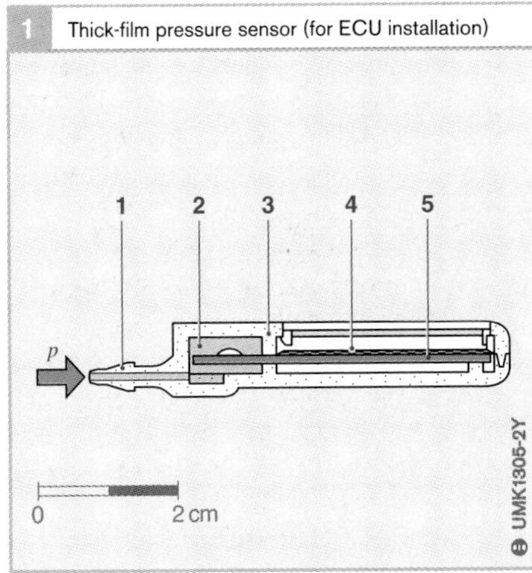

Fig. 1
Measuring range:
1 Pressure connection for the measured pressure p
2 Pressure-measuring cell
3 Sealing web

Signal conditioning:
4 Evaluation circuit
5 Thick-film hybrid on ceramic substrate

2 Thick-film pressure sensor (pressure-measuring cell)

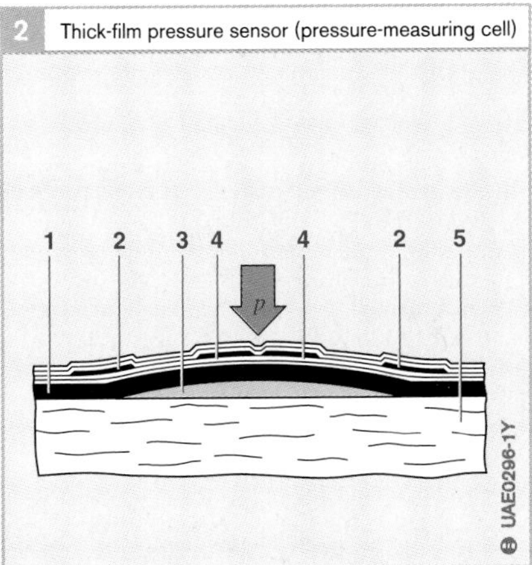

Fig. 2
1 Thick-film diaphragm
2 Passive reference deformation resistor
3 Reference-pressure chamber ("bubble")
4 Active deformation resistor
5 Ceramic substrate

p Measured pressure.

3 Thick-film pressure sensor (circuit)

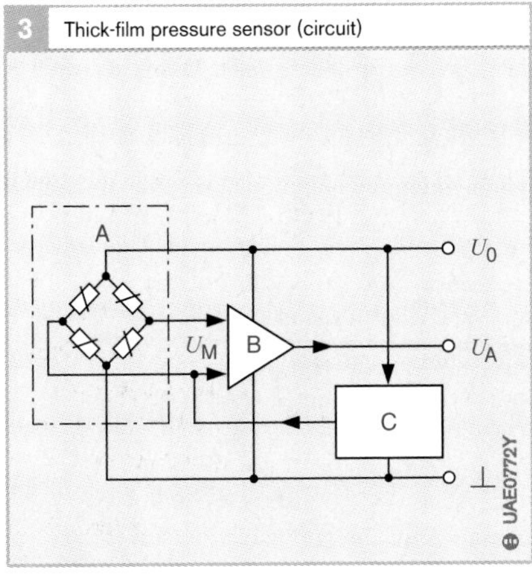

Fig. 3
A DMS pressure-measuring cell
B Amplifier
C Temperature-compensation circuit

U_0 Supply voltage
U_M Measured voltage
U_A Output voltage

Micromechanics

Micromechanics is defined as the application of semiconductor techniques in the production of mechanical components from semiconductor materials (usually silicon). Not only silicon's semiconductor properties are used but also its mechanical characteristics. This enables sensor functions to be implemented in the smallest-possible space. The following techniques are used:

Bulk micromechanics

The silicon wafer material is processed at the required depth using anisotropic (alkaline) etching and, where needed, an electrochemical etching stop. From the rear, the material is removed from inside the silicon layer (Fig. 1, Pos. 2) at those points underneath an opening in the mask. Using this method, very small diaphragms can be produced (with typical thicknesses of between 5 and 50 µm, as well as openings (b), beams and webs (c) as are needed for instance for acceleration sensors.

Surface micromechanics

The substrate material here is a silicon wafer on whose surface very small mechanical structures are formed (Fig. 2). First of all, a "sacrificial layer" is applied and structured using semiconductor processes such as etching (a). An approx. 10 µm polysilicon layer is then deposited on top of this and structured vertically using a mask and etching. In the final processing step, the "sacrificial" oxide layer underneath the polysilicon layer is removed by means of gaseous hydrogen fluoride. In this manner, the movable electrodes for acceleration sensors (Fig. 3) are exposed.

Wafer bonding

Anodic bonding and sealglass bonding are used to permanently join together (bonding) two wafers by the application of tension and heat or pressure and heat. This is needed for the hermetic sealing of reference vacuums for instance, and when protective caps must be applied to safeguard sensitive structures.

1 Structures produced by bulk micromechanics

2 Surface micromechanics (processing steps)

3 Surface micromechanics (structure details)

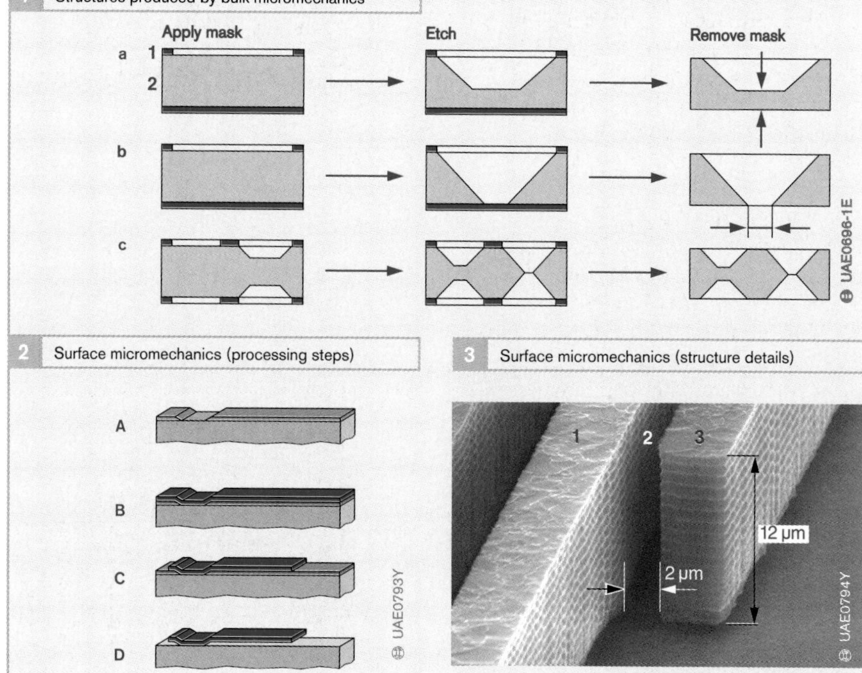

Fig. 1
a Diaphragms
b Openings
c Beams and webs

1 Etching mask
2 Silicon

Fig. 2
A Cutting and structuring the sacrificial layer
B Cutting the polysilicon
C Structuring the polysilicon
D Removing the sacrificial layer

Fig. 3
1 Fixed electrode
2 Gap
3 Spring electrodes

High-pressure sensors

Application

In automotive applications, high-pressure sensors are used for measuring the pressures of fuels and brake fluids.

Diesel rail-pressure sensor

In the diesel engine, the rail-pressure sensor measures the pressure in the fuel rail of the Common Rail accumulator-type injection system. Maximum operating (nominal) pressure p_{max} is 160 MPa (1,600 bar). Fuel pressure is controlled by a closed control loop, and remains practically constant independent of load and engine speed. Any deviations from the setpont pressure are compensated for by a pressure control valve.

Gasoline rail-pressure sensor

As its name implies, this sensor measures the pressure in the fuel rail of the DI Motronic with gasoline direct injection. Pressure is a function of load and engine speed and is 5...12 MPa (50...120 bar), and is used as an actual (measured) value in the closed-loop rail-pressure control. The rpm and load-dependent setpoint value is stored in a map and is adjusted at the rail by a pressure control valve.

Brake-fluid pressure sensor

Installed in the hydraulic modulator of such driving-safety systems as ESP, this high-pressure sensor is used to measure the brake-fluid pressure which is usually 25 MPa (250 bar). Maximum pressure p_{max} can climb to as much as 35 MPa (350 bar). Pressure measurement and monitoring is triggered by the ECU which also evaluates the return signals.

Design and operating concept

The heart of the sensor is a steel diaphragm onto which deformation resistors have been vapor-deposited in the form of a bridge circuit (Fig. 1, Pos. 3). The sensor's pressure-measuring range depends upon the diaphragm's thickness (thicker diaphragms for higher pressures and thinner ones for lower pressures). When the pressure is applied via the pressure connection (4) to one of the diaphragm faces, the resistances of the bridge resistors change due to diaphragm deformation (approx. 20 µm at 1,500 bar).

The 0...80 mV output voltage generated by the bridge is conducted to an evaluation circuit which amplifies it to 0...5 V. This is used as the input to the ECU which refers to a stored characteristic curve in calculating the pressure (Fig. 2).

1 High-pressure sensor

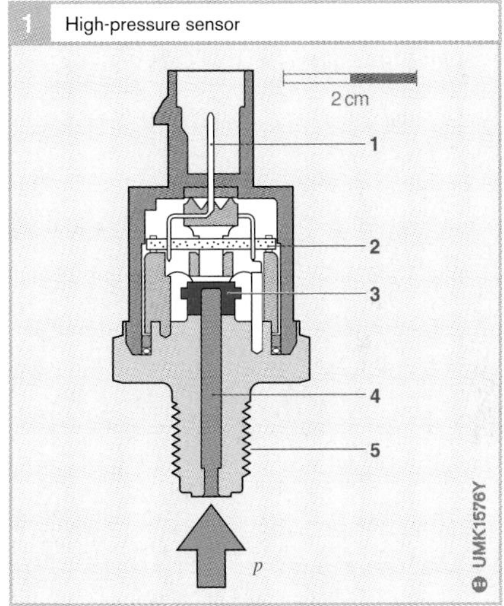

Fig. 1
1 Electrical connection (socket)
2 Evaluation circuit
3 Steel diaphragm with deformation resistors
4 Pressure connection
5 Mounting thread

2 High-pressure sensor (curve, example)

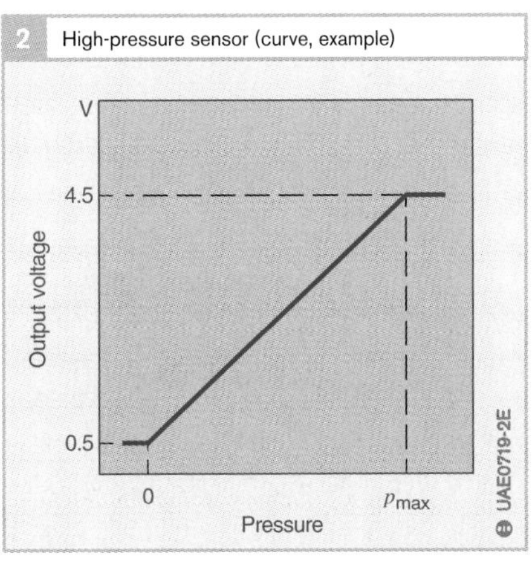

Sensor-flap (impact-pressure) air-flow sensor LMM

Application

The sensor-flap air-flow sensor is still in operation in a number of engines equipped with certain versions of the L-Jetronic or M-Motronic. It is installed between the air filter and the throttle valve and applies the sensor-flap principle in registering the air flow Q_L drawn in by the engine (Figs. 1 through 3).

Design and operating concept

The air-flow sensor's pivoting sensor flap (Fig. 1, Pos. 1) forms a variable orifice plate. The incoming air Q_L deflects the flap against the constant return force of a spring, whereby the free cross-section area increases along with increasing air flow the more the plate is deflected.

The change of the free air-flow-sensor cross section as a function of the sensor-flap setting has been selected so that there is a logarithmic relationship between the sensor-flap angle and the air quantity drawn in by the engine. This leads to high air-flow sensor sensitivity, a valuable asset in the case of small air quantities which necessitate high measuring accuracy. The stipulated measuring accuracy is 1...3 % of the measured value throughout a range defined by $Q_{max} : Q_{min} = 100:1$.

The sensor-plate angle is picked-off by a potentiometer (4) which converts it into an output voltage U_A (Fig. 4) which is used as an input to the ECU. In order to eliminate the effects of potentiometer aging and temperature coefficient on accuracy, the ECU only evaluates resistance ratios.

A further phenomenon which must be taken into account are the intake or induction strokes from the individual cylinders. These generate oscillations in the intake manifold, which the air-flow sensor can only follow up to about 10 Hz. To keep these effects down to a minimum, the measuring flap has a compensation flap attached to it which, in combination with a damping chamber (5), serves to damp the oscillations of the pulsating intake air.

Instead of the desired mass flow which is proportional to the product from $\rho \cdot v$, measurement according to the impact-pressure principle only measures a flow which is proportional to the product $\sqrt{\rho} \cdot v$. This means that density compensation (air temperature, air pressure) is required in order to achieve precise fuel metering.

The intake air's density changes along with its temperature. This fact is taken into account by the ECU calculating a correcting quantity from the temperature-dependent resistance of a temperature sensor integrated in the air-flow sensor (2). M-Motronic versions always feature barometric-pressure compensation. Here, a manifold-pressure sensor is connected pneumatically to the intake manifold so that it can pick-off the absolute manifold pressure. It is either integrated directly in the ECU (connected by hose to the intake manifold), or located in the vicinity of the intake manifold, or attached directly to it.

1 Impact-pressure airflow sensor

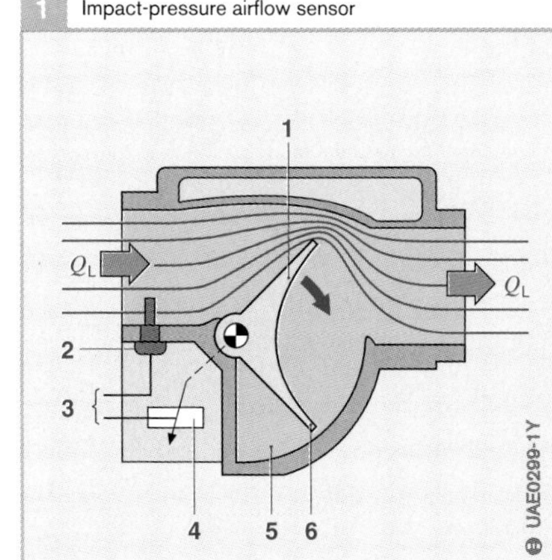

Fig. 1
1 Sensor plate
2 Air-temperature sensor
3 To ECU
4 Potentiometer
5 Damping chamber
6 Compensation flap

Q_L Intake-air flow

2 Impact-pressure air-flow sensor (air-chamber side)

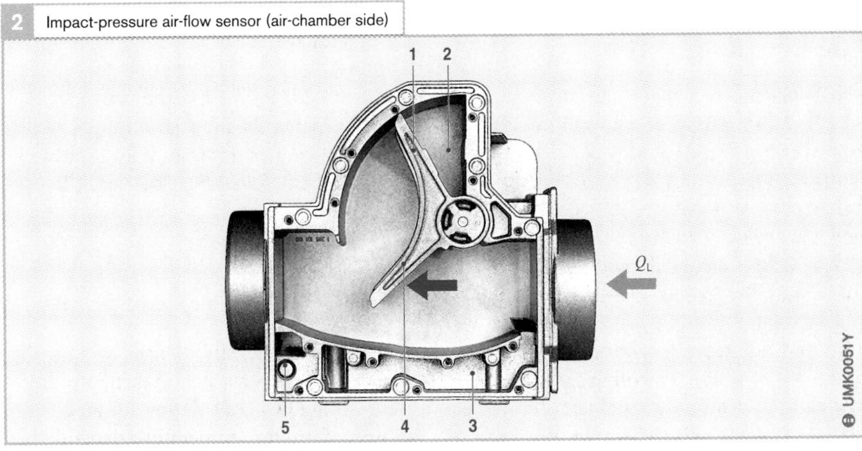

Fig. 2
1 Compensation flap
2 Damping chamber
3 Bypass
4 Sensor plate
5 Idle-mixture adjusting screw

Q_L Intake-air flow

3 Impact-pressure air-flow sensor (connection and components side)

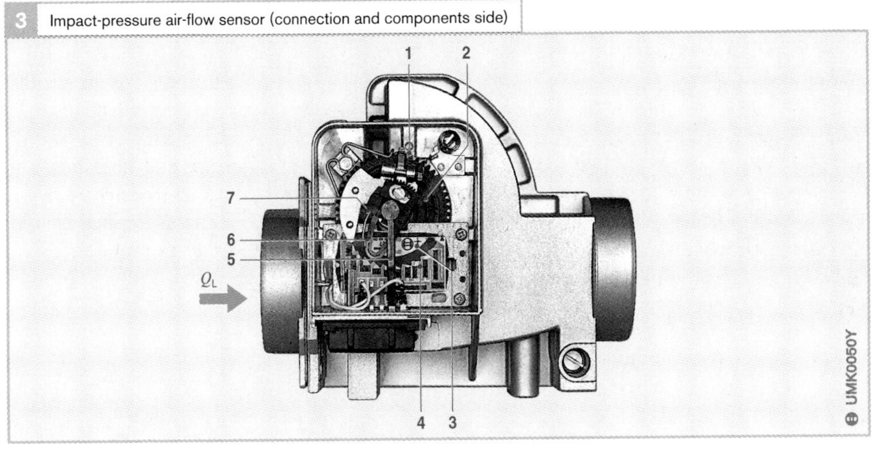

Fig. 3
1 Ring gear for spring preload
2 Return spring
3 Wiper track
4 Ceramic plate with resistors and printed conductors
5 Wiper pick-off
6 Wiper
7 Pump contact

Q_L Intake-air flow

4 Impact-pressure air-flow sensor (potentiometer circuitry and voltage curve)

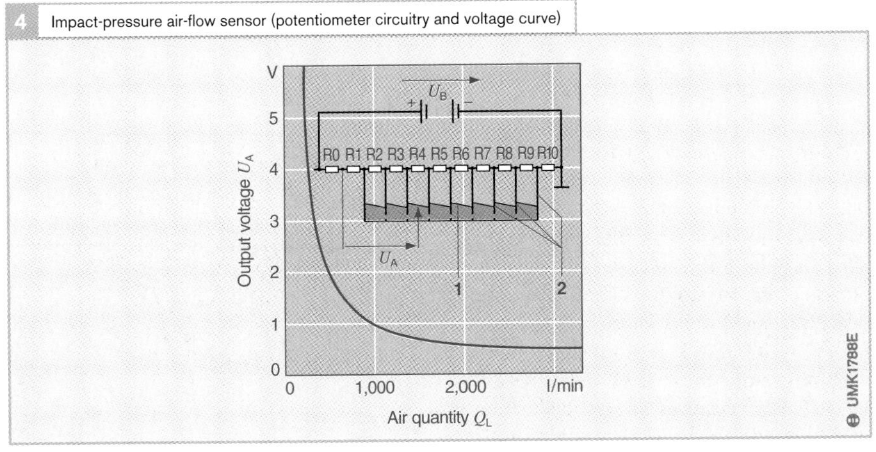

Fig. 4
1 Wiper track
2 Conductor segments (data points)

$U_A \sim 1/Q_L$ applies for the curve

Hot-wire air-mass meter HLM

Application

The HLM hot-wire air-mass meter is installed as a "thermal" load sensor between air filter and throttle plate in a number of LH-Jetronic or M-Motronic gasoline engines. It registers the air-mass flow Q_M drawn in by the engine, and applies this to determine the engine load. Being as it is able to follow average fluctuations of up to 1 Hz, the HLM is the fastest of the air flowmeters at present in use.

Design and construction

The intake air drawn in by the engines flows through the tubular HLM housing which is protected at each end by a wire mesh. A heated, 70 μm thin platinum wire element is suspended across the HLM measuring tube. It is suspended trapezoidally so that in good approximation it is able to cover the whole of the flow cross-section. A temperature-compensation (thin film) resistor projects into the air flow just upstream of the hot wire. Both of these components (hot wire and resistor) are integral parts of a closed-loop control circuit, where they function as temperature-dependent resistors. The control circuit is basically a bridge circuit and an amplifier (Figs. 1 and 2).

Operating concept

Before the air flowing through the HLM cools down the hot wire, its temperature is measured by means of the temperature-compensation resistor. A closed-loop control circuit regulates the heater current so that the hot wire assumes a temperature which is held at a constant level above that of the intake air. Since the air density influences the amount of heat dissipated to the air by the hot wire, this measuring concept must adequately take it into account. The heating current I_H is therefore a measure of the air-mass flow, and across a precision measuring resistor (R_M) it generates a voltage signal U_M, for input to the ECU, which is proportional to air-mass flow. The HLM, on the other hand, cannot detect the direction of air flow.

In order to prevent measurement-result drift due to deposits on the platinum wire, this is heated for about 1 second to a burn-off temperature of approx. 1,000 °C every time the engine is switched off. Here, the dirt deposits evaporate or flake off and leave the hot wire in a clean state.

1 Hot-wire air-mass meter (circuit)

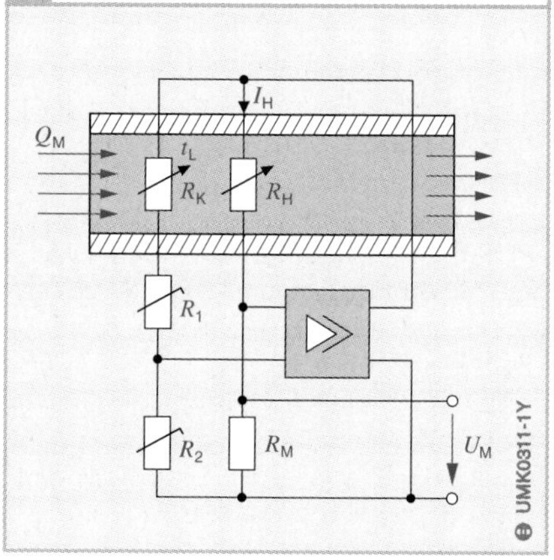

Fig. 1
R_K Temperature-compensation resistor
R_H Hot-wire heater resistor
R_M Measuring resistor
$R_{1,2}$ Bridge balance resistors
U_M Measurement voltage
I_H Heating current
t_L Air temperature
Q_M Air-mass flow

2 Hot-wire air-mass meter

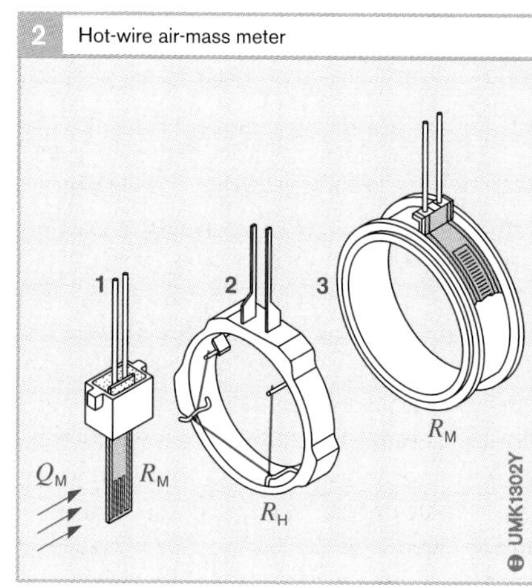

Fig. 2
1 Temperature-compensation resistor R_K
2 Sensor ring with hot wire R_H
3 Precision measuring resistor (R_M)

Q_M Air-mass flow

Hot-film air-mass meter HFM2

Application

The HFM2 **h**ot-**f**ilm air-mass **m**eter is a thick-film sensor which is installed as a "thermal" load sensor between air filter and throttle plate in a number of LH-Jetronic or M-Motronic gasoline engines. It very accurately registers the air-mass flow Q_M drawn in by the engine, and applies this to determine the engine load.

Design and construction

Together with bridge resistors, the electrically heated HFM2 platinum heater resistor R_H is located on a ceramic chip (substrate, Fig. 1).

The bridge also incorporates a temperature-dependent resistor R_S (flow sensor) which registers the heater temperature. Separation of the heater and the flow sensor is advantageous for the (closed-loop) control circuit. The heater element and the air-temperature compensation sensor (resistor R_K) are decoupled thermally by two saw cuts (Fig. 2).

Since the dirt is deposited mainly on the front edge of the sensor element, the components which are decisive for the heat transition are situated downstream on the ceramic substrate. Furthermore, the sensor is so constructed that air flow around the sensor remains unaffected by dirt deposits.

Operating concept

The electrically heated platinum heater resistor projects into the intake-air flow which cools it down. A closed-loop control circuit regulates the heater current so that the hot wire assumes a temperature which is held at a constant level above that of the intake air. Since the air density, just as much as the flow rate, is decisive regarding the amount of heat dissipated to the air by the hot wire, this measuring concept takes it into account to the appropriate degree. The heating current I_H, and the voltage at the heater, is thus a non-linear measure for the air-mass flow Q_M.

The HFM2 electronic circuitry converts this voltage into the voltage UM which it adapts to make it suitable for input into the ECU. The computer than uses this to calculate the air mass drawn in by the engine for every working cycle. The HFM2 cannot determine the direction of air flow.

The long-term measuring accuracy of ±4 % referred to the measured value applies even without the burn-off of dirt deposits.

1 Hot-film air-mass meter (circuit)

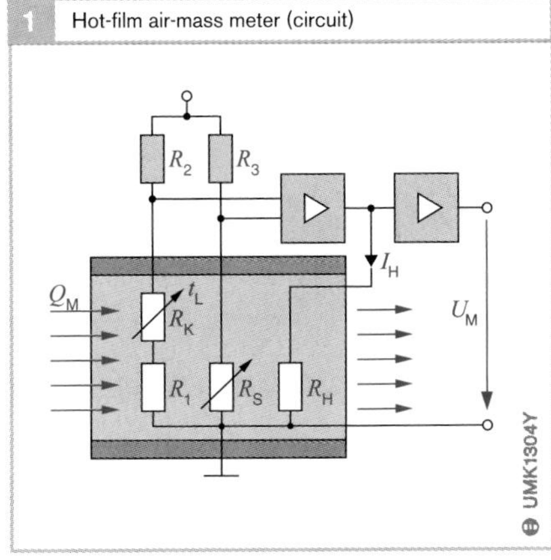

Fig. 1
R_K Temperature-compensation sensor (resistor)
R_H Heater resistor
R_S Sensor resistor
R_1, R_2, R_3 Bridge resistors
U_M Measurement voltage
I_H Heating current
t_L Air temperature
Q_M Air-mass flow

2 Hot-film air-mass meter (substrate)

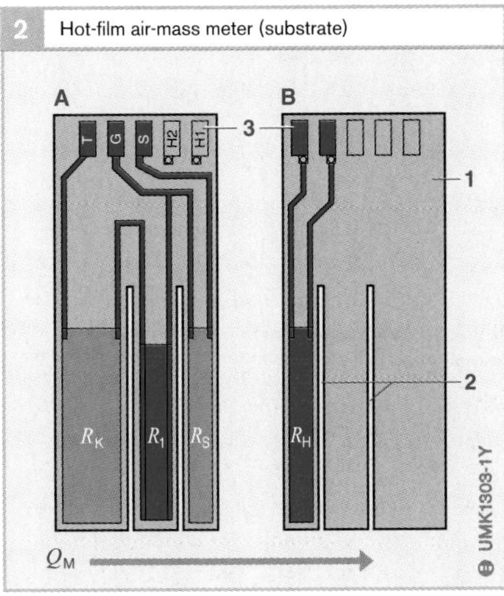

Fig. 2
A Front side
B Rear side

1 Ceramic substrate
2 Two saw cuts
3 Contacts

R_K Temperature-compensation sensor (resistor)
R_H Heater resistor
R_S Sensor resistor
R_1 Bridge resistor

Hot-film air-mass meter HFM5

Application

For optimal combustion as needed to comply with the emission regulations imposed by legislation, it is imperative that precisely the necessary air mass is inducted, irrespective of the engine's operating state.

To this end, part of the total air flow which is actually inducted through the air filter or the measuring tube is measured by a hot-film air-mass meter. Measurement is very precise and takes into account the pulsations and reverse flows caused by the opening and closing of the engine's intake and exhaust valves. Intake-air temperature changes have no effect upon measuring accuracy.

Design and construction

The housing of the HFM5 **h**ot-**f**ilm air-mass **m**eter (Fig. 1, Pos. 5) projects into a measuring tube (2) which, depending upon the engine's air-mass requirements, can have a variety of diameters (for 370...970 kg/h). This tube is installed in the intake tract downstream from the air filter. Plug-in versions are also available which are installed inside the air filter.

The most important components in the sensor are the sensor element (4), in the air intake (8), and the integrated evaluation electronics (3). The partial air flow as required for measurement flows across this sensor element.

Vapor-deposition is used to apply the sensor-element components to a semiconductor substrate, and the evaluation-electronics (hybrid circuit) components to a ceramic substrate. This principle permits very compact design. The evaluation electronics are connected to the ECU through the plug-in connection (1). The partial-flow measuring tube (6) is shaped so that the air flows past the sensor element smoothly (without whirl effects) and back into the measuring tube via the air outlet (7). This method ensures efficient sensor operation even in case of extreme pulsation, and in addition to forward flow, reverse flows are also detected (Fig. 2).

1 Hot-film air-mass meter HFM5 (circuit)

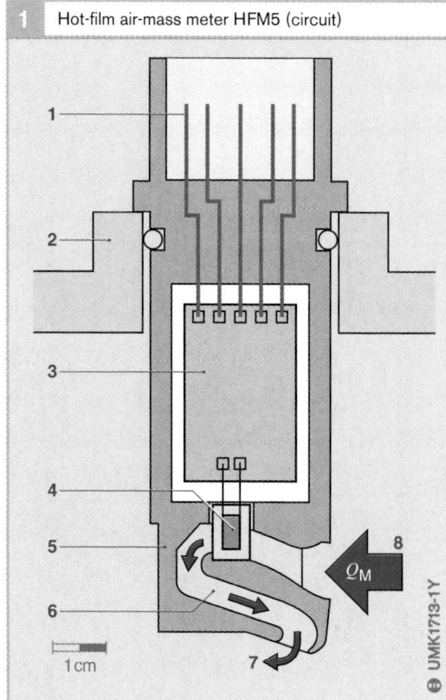

Fig. 1
1 Electrical plug-in connection
2 Measuring tube or air-filter housing wall
3 Evaluation electronics (hybrid circuit)
4 Sensor element
5 Sensor housing
6 Partial-flow measuring tube
7 Air outlet for the partial air flow Q_M
8 Intake for partial air flow Q_M

Operating concept

The hot-film air-mass meter is a "thermal sensor" and operates according to the following principle:

A micromechanical sensor diaphragm (Fig. 3, Pos. 5) on the sensor element (3) is heated by a centrally mounted heater resistor and held at a constant temperature. The temperature drops sharply on each side of this controlled heating zone (4).

The temperature distribution on the diaphragm is registered by two temperature-dependent resistors which are mounted upstream and downstream of the heater resistor so as to be symmetrical to it (measuring points M_1, M_2). Without the flow of incoming air, the temperature characteristic (1) is the same on each side of the heating zone ($T_1 = T_2$).

As soon as air flows over the sensor element, the uniform temperature distribution at the diaphragm changes (2). On the intake side, the temperature characteristic is steeper since the incoming air flowing past this area cools it off. Initially, on the opposite side (the side nearest to the engine), the sensor element cools off. The air heated by the heater element then heats up the sensor element. The change in temperature distribution leads to a temperature differential (ΔT) between the measuring points M_1 und M_2.

The heat dissipated to the air, and therefore the temperature characteristic at the sensor element is a function of the air mass flow. Independent of the absolute temperature of the air flowing past, the temperature differential is a measure of the air mass flow. Apart from this, the temperature differential is directional, which means that the air-mass meter not only registers the mass of the incoming air but also its direction.

Due to its very thin micromechanical diaphragm, the sensor has a highly dynamic response (<15 ms), a point which is of particular importance when the incoming air is pulsating heavily.

The evaluation electronics (hybrid circuit) integrated in the sensor convert the resistance differential at the measuring points M_1 and M_2 into an analog signal of 0...5 V which is suitable for processing by the ECU. Using the sensor characteristic (Fig. 2) programmed into the ECU, the measured voltage is converted into a value representing the air mass flow [kg/h].

The shape of the characteristic curve is such that the diagnosis facility incorporated in the ECU can detect such malfunctions as an open-circuit line. A temperature sensor for auxiliary functions can also be integrated in the HFM5. It is located on the sensor element upstream of the heated zone.

It is not required for measuring the air mass. For applications on specific vehicles, supplementary functions such as improved separation of water and contamination are provided for (inner measuring tube and protective grid).

2 Hot-film air-mass meter (output voltage as a function of the partial air mass flowing past it)

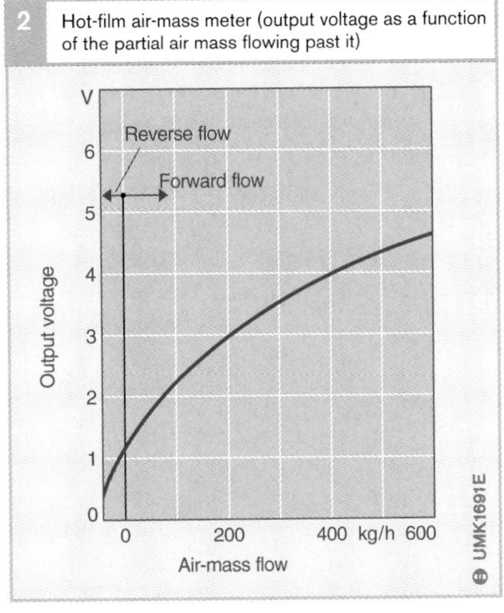

3 Hot-film air-mass meter: Measuring principle

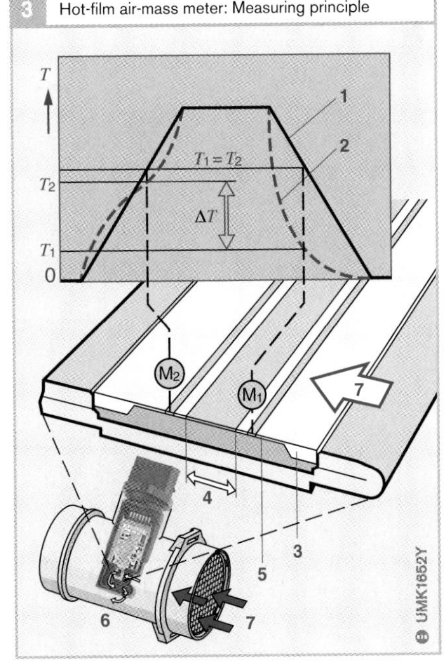

Fig. 3

1 Temperature profile without air flow across sensor element
2 Temperature profile with air flow across sensor element
3 Sensor element
4 Heated zone
5 Sensor diaphragm
6 Measuring tube with air-mass meter
7 Intake-air flow

M_1, M_2 Measuring points

T_1, T_2 Temperature values at the measuring points M_1 and M_2

ΔT Temperature differential

Two-step Lambda oxygen sensors

Application

These sensors are used in gasoline engines equipped with two-step Lambda control. They extend into the exhaust pipe between the engine's exhaust manifold and the catalytic converter, and register the exhaust-gas flow leaving each cylinder. Thanks to the Lambda sensor being heated, it can be installed further away from the engine so that long periods of full-load (WOT) engine operation are unproblematical. The LSF4 sensor is also suitable for operation with exhaust-gas control systems featuring a number of sensors (for instance with OBDII).

"Two-step sensors" compare the residual-oxygen content in the exhaust gas with that of the reference atmosphere inside the sensor. They then indicate whether the A/F mixture in the exhaust gas is "rich" ($\lambda < 1$) or "lean" ($\lambda > 1$). The sudden jump in the characteristic curve of these sensors permits A/F control to $\lambda = 1$ (Fig. 1).

Design and construction

LSH25 tube-type (finger) sensor

Sensor ceramic with protective tube

The solid-state electrolyte is a ceramic element and is impermeable to gas. It is a mixed oxide comprising the elements zirconium and yttrium in the form of a tube closed at one end (finger). The inside and outside surfaces have each been provided with a porous platinum coating which serves as an electrode.

The platinum electrode on the outside surface of the ceramic body protrudes into the exhaust pipe, and acts as a catalytic converter in miniature. Exhaust gas which reaches this electrode is processed catalytically and brought to a stoichiometric balance ($\lambda = 1$). In addition, the outside of the sensor which is in contact with the exhaust gas is provided with a porous multiple ceramic (Spinel) layer to protect it against contamination. The ceramic body is also protected against mechanical impact and thermal shocks by a slotted metal tube. A number of slots in the protective tube are specially shaped so that on the one hand they are particularly effective against extensive thermal and chemical stresses, while on

1 Two-step Lambda oxygen sensor (voltage curve for 600 °C working temperature)

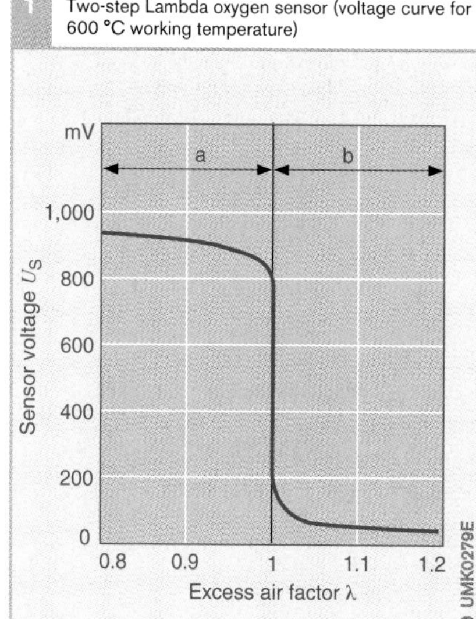

Fig. 1
- a Rich A/F mixture (air deficiency)
- b Lean A/F mixture (excess air)

2 Configuration of a tube-type (finger) Lambda oxygen sensor in the exhaust pipe

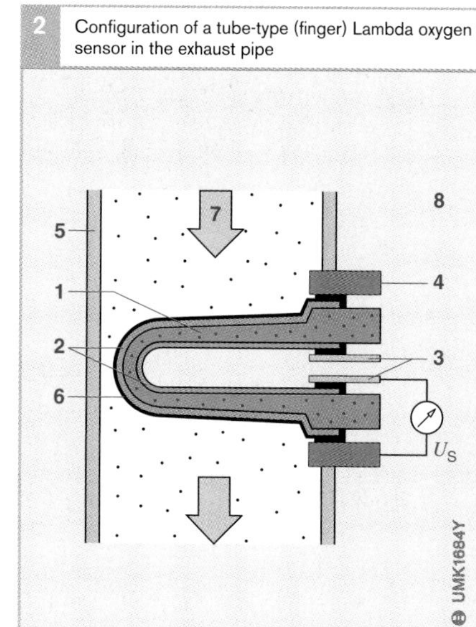

Fig. 2
1. Sensor ceramic element
2. Electrodes
3. Contacts
4. Housing contact
5. Exhaust pipe
6. Ceramic protective coating (porous)
7. Exhaust gas
8. Outside air

U_S Sensor voltage

the other hand preventing a sharp drop in sensor-ceramic temperature when the exhaust is "cool".

The sensor's "open" inner chamber is isolated from the exhaust gas and connected to the surrounding air which acts as a reference gas (Fig. 2).

Sensor body with heater element and electric connection

A ceramic support tube and a disc spring hold the active finger-shaped sensor ceramic in the sensor housing and seal it off. A contact element between the support tube and the active sensor ceramic provides the contact between the inner electrode and the connection cable.

A metal seal ring connects the outer electrode with the sensor housing. The sensor's complete internal structure is located and held in place by a protective metal sleeve which at the same time acts as the support for the disc spring. The protective sleeve also guards against contamination of the sensor's interior. The connection cable is crimped to the contact element which protrudes from the sensor, and is protected against damp and mechanical loading by means of a special temperature-resistant cap.

The tube-type (finger) Lambda sensor (Fig. 3) is also equipped with a electrical heater element. On this sensor, at low engine loads (e.g. low exhaust-gas temperatures) the ceramic element's temperature is defined by the electrical heater, and at high loads by the exhaust-gas temperature. Depending on the heater element's power rating, exhaust-gas temperatures of a low as 150...200 °C suffice to bring the sensor ceramic up to operating temperature. Thanks to its external electrical heating, the sensor heats up so quickly that it has already reached operating temperature 20...30 s after the engine has started so that the Lambda closed-loop control can come into operation. In fact, the sensor heating ensures that above the func-

3 LSH25 heated tube-type (finger) Lambda oxygen sensor (view and section)

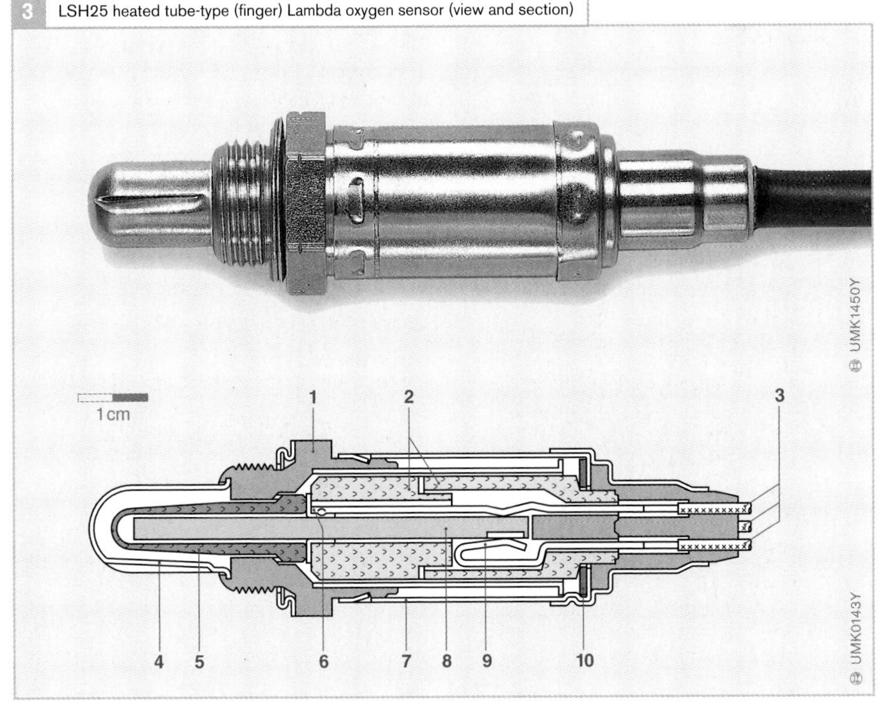

Fig. 3
1 Sensor housing
2 Ceramic support tube
3 Connection cable
4 Protective tube with slots
5 Active sensor ceramic
6 Contact element
7 Protective sleeve
8 Heater element
9 Clamp-type heater-element connections
10 Disc spring

tion limit (350 °C) the sensor is always at optimum operating temperature, and it also contributes to low and stable exhaust-gas emission figures.

Initially the LS21 Lambda sensor was unheated. This version is only rarely found in older vehicles as a replacement part.

LSF4 planar Lambda oxygen sensor
Regarding their function, planar Lambda sensors correspond to the heated finger sensors with their voltage-jump curve at $\lambda = 1$.

4 Planar Lambda oxygen sensor (functional layers)

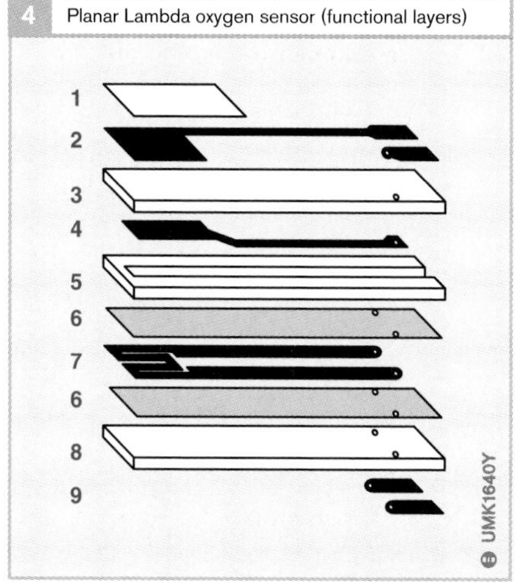

Fig. 4
1 Porous protective layer
2 Outer electrode
3 Sensor foil
4 Inner electrode
5 Reference-air-passage foil
6 Insulation layer
7 Heater
8 Heater foil
9 Connection contacts

5 LSF4 planar Lambda oxygen sensor (schematic)

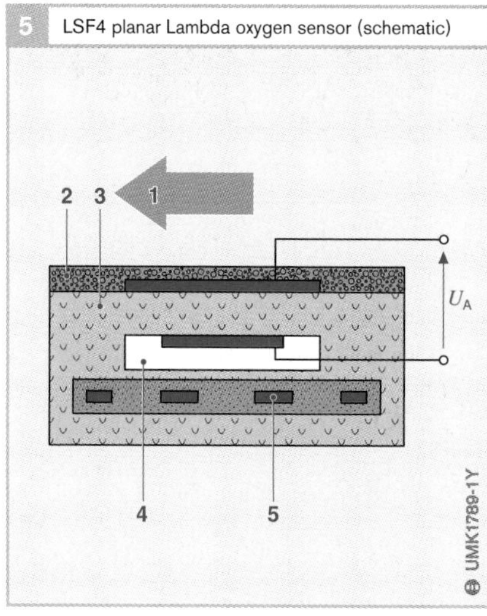

Fig. 5
1 Exhaust gas
2 Porous protective ceramic layer
3 Measuring element with microporous noble-metal coating
4 Reference-air passage
5 Heater

U_A Output voltage

On the planar sensor though, the solid-state electrolyte is comprised of a number of individual laminated foils stacked one on top of the other (Fig. 4). The sensor is protected against thermal and mechanical influences by a double-walled protective tube.

The sensor's planar ceramic element (measuring element and heater are integrated) is shaped like a long stretched-out wafer with rectangular cross-section.

The measuring element's surfaces have a microporous noble-metal coating. On the exhaust-gas side this has an exrtra microporous ceramic coating to protect it against the erosive effects of the exhaust-gas components. The heater is a wave-shaped element containing noble metal. It is integrated in the ceramic wafer and insulated from it, and even though it has a low power input ensures that the sensor heats up quickly.

The LSF4 sensor (Figs. 5 and 6) operates as a reference-gas sensor. The reference-air passage inside this sensor has a direct connection to the surrounding air. This enables it to compare the residual oxygen content of the exhaust gas with the oxygen content of the reference gas. In other words, with the air inside the sensor. This means that in the case of the planar sensor, the sensor voltage also features a characteristic jump (Fig. 1) in the area of stoichiometric A/F mixture ($\lambda = 1$).

Operating concept

The two-step sensors operate in accordance with the principle of the galvanic oxygen-concentration cell with solid-state electrolyte (Nernst principle). As from about 350 °C, the sensor ceramic becomes conductive for oxygen ions (the most efficient and reliable functioning is at temperatures >> 350 °C). Since there is a sharp change in the oxygen content on the exhaust-gas side in the area of $\lambda = 1$ (e.g. $9 \cdot 10^{-15}$ Vol% for $\lambda = 0.99$, and 0.2 Vol% for $\lambda = 1.01$), this leads to the generation of a voltage between the sensor's boundary layers due to the different oxygen concentrations inside and outside the sensor. This means that the exhaust-gas oxygen content can be applied as a measure for the A/F ratio. The integral heater ensures that the sensor is operational at exhaust-gas temperatures as low as 150 °C.

The sensor's output voltage U_S is a function of the oxygen content in the exhaust gas. In the case of a rich A/F mixture ($\lambda < 1$) it reaches 800...1,000 mV. For a lean mixture ($\lambda > 1$), only approx. 100 mV are generated. The transition from the rich to the lean area at U_{reg} is at about 450...500 mV.

The ceramic structure's temperature influences its ability to conduct the oxygen ions, and therefore also the shape of the output-voltage curve as a function of the excess-air factor λ (the values in Fig. 1 apply for about 600 °C). Apart from this, the response time for a voltage change when the A/F mixture changes is also highly dependent upon temperature.

Whereas response times at ceramic temperatures below 350 °C are in the seconds range, at optimum temperatures of around 600 °C the sensor responds in less than 50 ms. When the engine is started therefore, the Lambda closed-loop control is switched off until the minimum operating temperature of about 350 °C is reached. During this period, the engine is open-loop controlled.

Excessive temperatures reduce the sensor's useful life. This means that the Lambda sensor must be installed so that the operating temperatures 850 °C (LSH25), and 930 °C (LSF4), are not exceeded.

6 LSF4 planar Lambda oxygen sensor (view and section)

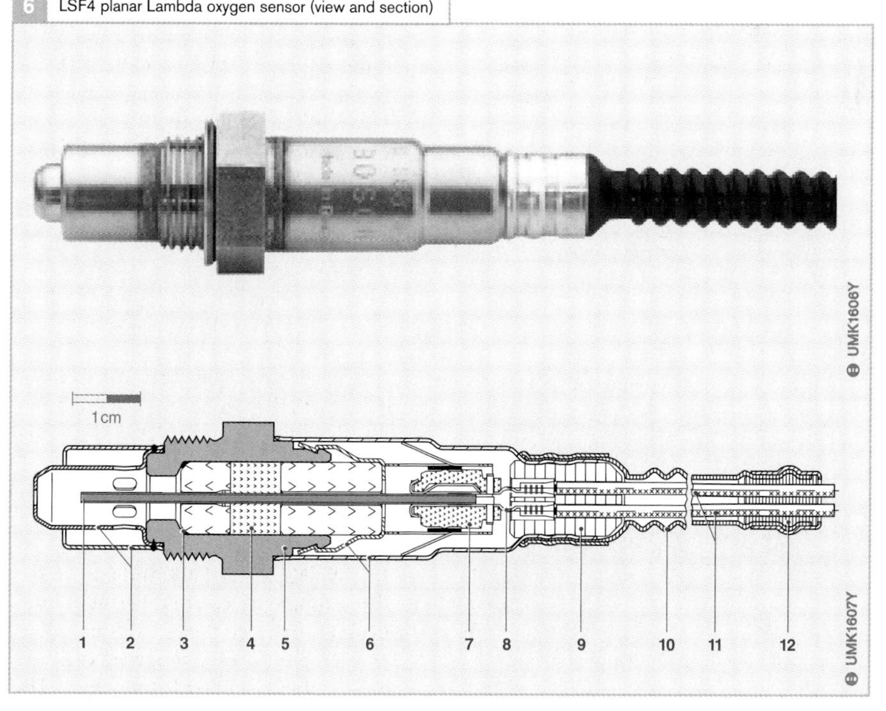

Fig. 6
1 Planar measuring element
2 Double protective tube
3 Seal ring
4 Seal packing
5 Sensor housing
6 Protective sleeve
7 Contact holder
8 Contact clip
9 PTFE sleeve
10 Shaped PTFE sleeve
11 Five connection cables
12 Seal

LSU4 planar broad-band Lambda oxygen sensor

Application

As its name implies, the broad-band Lambda oxygen sensor is used across a very extensive range to determine the oxygen concentration in the exhaust gas. The figures provided by the sensor are an indication of the air-fuel (A/F) ratio in the engine's combustion chamber. The excess-air factor λ is used when defining the A/F ratio.

The sensor protrudes into the exhaust pipe and registers the exhaust-gas mass flow from all cylinders. Broad-band Lambda sensors make precise measurements not only at the stoichiometic point $\lambda = 1$, but also in the lean range ($\lambda > 1$) and in the rich range ($\lambda < 1$). In combination with electronic closed-loop control circuitry, these sensors generate an unmistakable, continuous electrical signal (Fig. 3) in the range from $0.7 < \lambda < \infty$ (= air with 21% O_2). This means that the broad-band Lambda sensor can be used not only in engine-management systems with two-step control ($\lambda = 1$), but also in control concepts with rich and lean air-fuel (A/F) mixtures. This type of Lambda sensor is therefore also suitable for the Lambda closed-loop control used with lean-burn concepts on gasoline engines, as well as for diesel engines, gaseous-fuel engines, and gas-powered central heaters and water heaters (this wide range of applications led to the designation LSU: Lambda Sensor Universal (taken from the German), in other words Universal Lambda Sensor).

In a number of systems, several Lambda sensors are installed for even greater accuracy. Here, for instance, they are fitted upstream and downstream of the catalytic converter as well as in the individual exhaust tracts (cylinder banks).

Design and construction

The LSU4 broad-band Lambda sensor (Fig. 2) is a planar dual-cell limit-current sensor. It features a zirconium-dioxide/ceramic (ZrO_2) measuring cell which is the combination of a Nernst concentration cell (sensor cell which functions the same as a two-step Lambda sensor) and an oxygen-pump cell for transporting the oxygen ions. The oxygen-pump cell (Fig. 1, Pos. 8) is so arranged with respect to the Nernst concentration cell (7) that there is a 10...50 µm diffusion gap (6) between them in which there are two porous platinum electrodes: A pump electrode and a Nernst measuring electrode. The gap is connected to the exhaust gas through a gas-access passage (10). The porous diffusion barrier (11) serves to limit the inflow of oxygen molecules from the exhaust gas.

Fig. 1
1 Exhaust gas
2 Exhaust pipe
3 Heater
4 Control electronics
5 Reference cell with reference-air passage
6 Diffusion gap
7 Nernst concentration cell with Nernst measuring electrode (on the diffusion-gap side), and reference electrode (on the reference-cell side)
8 Oxygen-pump cell with pump electrode
9 Porous protective layer
10 Gas-access passage
11 Porous diffusion barrier

I_P Pump current
U_P Pump voltage
U_H Heater voltage
U_{Ref} Reference voltage (450 mV, corresponds to $\lambda = 1$)
U_S Sensor voltage

1 Planar broad-band Lambda oxygen sensor (installation in the exhaust pipe and schematic diagram of the measuring cell)

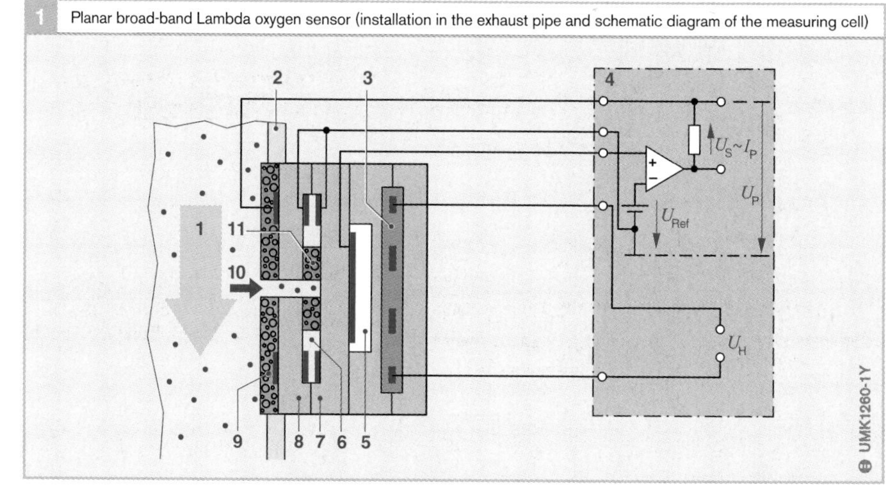

On the one side, the Nernst concentration cell is connected to the atmosphere by a reference-air passage (5), on the other it is connected to the exhaust gas in the diffusion gap.

An electronic closed-loop control circuit is needed in order to generate the sensor signal and for the sensor temperature control.

An integral heater (3) heats up the sensor quickly so that it soon reaches its operating temperature of 650...900 °C needed for generating a usable signal. This functions decisively reduces the effects that the exhaust-gas temperature has on the sensor signal.

Operating concept

The exhaust gas enters the actual measuring chamber (diffusion gap) of the Nernst concentration cell through the pump cell's gas-access passage. In order that the excess-air factor λ can be adjusted in the diffusion gap, the Nernst concentration cell compares the gas in the diffusion gap with that in the reference-air passage.

The complete process proceeds as follows: By applying the pump voltage U_P across the pump cell's platinum electrodes, oxygen from the exhaust gas is pumped through the diffusion barrier and into or out of the diffusion gap. With the help of the Nernst concentration cell, an electronic circuit in the ECU controls the voltage (U_P) across the pump cell in order that the composition of the gas in the diffusion gap remains constant at $\lambda = 1$. If the exhaust gas is lean, the pump cell pumps the oxygen to the outside (positive pump current). On the other hand, if it is rich, due to the decomposition of CO_2 and H_2O at the exhaust-gas electrode the oxygen is pumped from the surrounding exhaust gas and into the diffusion gap (negative pump current). Oxygen transport is unnecessary at $\lambda = 1$ and pump current is zero. The pump current is proportional to the exhaust-gas oxygen concentration and is this a non-linear measure for the excess-air factor λ (Fig. 3).

3 Pump current I_P of a broad-band Lambda sensor as a function of the exhaust-gas excess-air factor (λ)

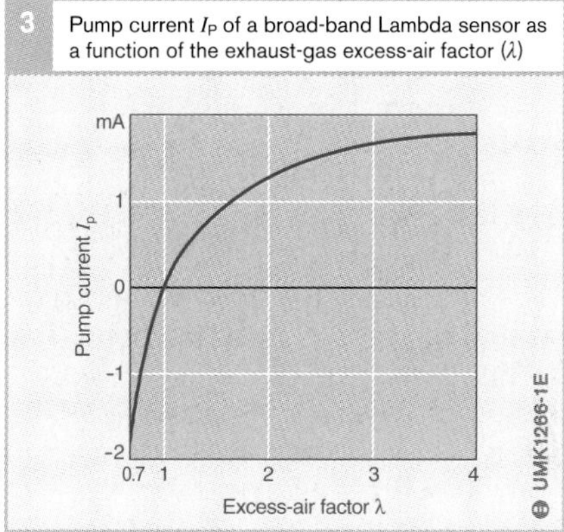

2 LSU4 planar broad-band Lambda oxygen sensor (view and section)

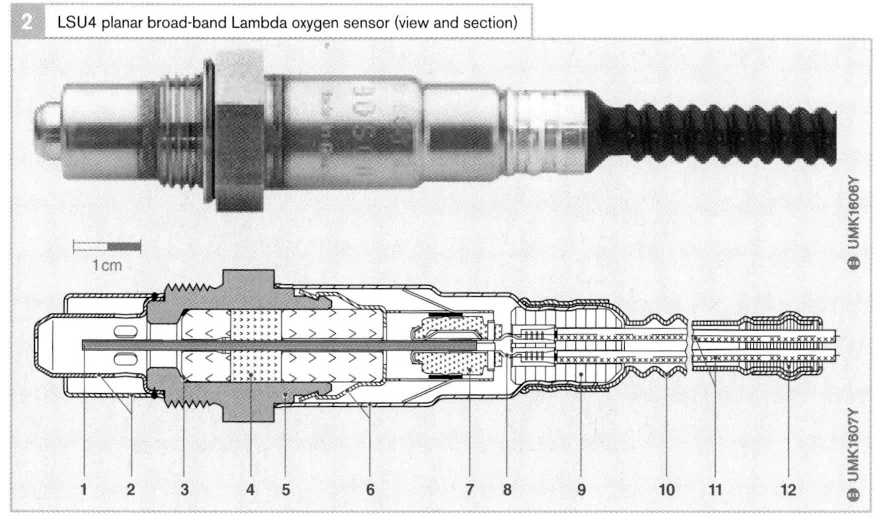

Fig. 3

1 Measuring cell (combination of Nernst concentration cell and oxygen-pump cell)
2 Double protective tube
3 Seal ring
4 Seal packing
5 Sensor housing
6 Protective sleeve
7 Contact holder
8 Contact clip
9 PTFE sleeve (Teflon)
10 PTFE shaped sleeve
11 Five connecting leads
12 Seal ring

Electronic control unit (ECU)

Digital technology furnishes an extensive array of options for open and closed-loop control of automotive electronic systems. A large number of parameters can be included in the process to support optimal operation of various systems. After receiving the electric signals transmitted by the sensors, the ECU processes these data in order to generate control signals for the actuators. The software program for closed-loop control is stored in the ECU's memory. The program is executed by a microcontroller. The ECU and its components are referred to as hardware. The Motronic ECU contains all of the algorithms for open and closed-loop control needed to govern the engine-management processes (ignition, induction and mixture formation, etc.).

Operating conditions

The ECU operates in an extremely harsh and demanding environment. It is exposed to

- Extreme temperatures (ranging from –40 to +60...+125 °C) under normal operating conditions
- Abrupt temperature variations
- Exposure to fluids (oil, fuel, etc.)
- The effects of moisture and
- Mechanical stresses such as engine vibration

The engine-management ECU must continue to perform flawlessly in the face of fluctuations in electrical supply, during starts with a weak battery (cold starts, etc.) as well as at high voltages (surges in onboard electrical system).

Other requirements arise from the need for EMC (electro-magnetic compatibility). The requirements for resistance to electromagnetic interference and for suppressing EMI emissions from the system itself are both very high.

Design

The printed-circuit board with the electrical components (Fig. 1) is installed in a housing of plastic or metal. A multipin plug (1) connects the ECU to the sensors, actuators and electrical power supply. The high-power driver circuits (3) that provide direct control of the actuators are specially integrated within the housing to ensure effective heat transfer to the housing and the surrounding air.

Most of the electronic components are SMDs (surface-mounted devices). This concept provides extremely efficient use of space in low-weight packages. Only the power elements and the plugs are mounted using conventional insertion technology.

Hybrid versions combining compact dimensions with extreme resistance to thermal attack are available for mounting directly on the engine.

Data processing

Input signals

The sensors join the actuators as the peripheral components linking the vehicle and the central processing device, the engine-management ECU. The electrical signals from the sensors travel through the wiring harness and the plug (1) to reach the control unit. These signals can be in various forms:

Analog input signals

Analog input signals can have any voltage level within a specific range. Samples of physical parameters monitored as analog data include induction air mass, battery voltage and intake-manifold pressure (including boost pressure) as well as the temperatures of the coolant and induction air. An analog/digital (A/D) converter within the control unit's microcontroller transforms the signal data into the digital form required by the microcontroller's central processing unit. The maximum resolution of these analog signals is 5 mV. This translates into roughly 1,000 incremental graduations based on an overall monitoring range of 0...5 V.

Digital input signals

Digital input signals have only two conditions: high (logical 1) and low (logical 0). Samples of digital input signals are switch control signals (on/off) and digital sensor signals such as the rotational-speed pulses from Hall-effect and magnetoresistive sensors. The microcontroller can process these signals without prior conversion.

Pulse-shaped input signals

The pulse-shaped input signals with information on rpm and reference marks transmitted by inductive sensors are conditioned in special circuitry within the ECU. In this process interference pulses are suppressed while the actual signal pulses are converted to digital square-wave signals.

Signal conditioning

Protective circuits limit the voltages of incoming signals to levels suitable for conditioning. Most of the superimposed interference signals are removed from the useful signal by filters. When necessary, the useful signals are then amplified to the input voltage required by the microcontroller (0...5 V).

Some or all of this initial conditioning can be carried out in the sensor itself, depending upon its level of integration.

Signal processing

The ECU is the switching center governing all of the functions and sequences regulated by the engine-management system. The control algorithms are executed by the microcontroller. The input signals from sensors and interfaces linking other systems (from CAN bus, etc.) serve as the input parameters. The processor runs backup plausibility checks on these data. The ECU program supports calculation of the output signals used to control the actuators.

1 ECU structure, using ME Motronic as an example

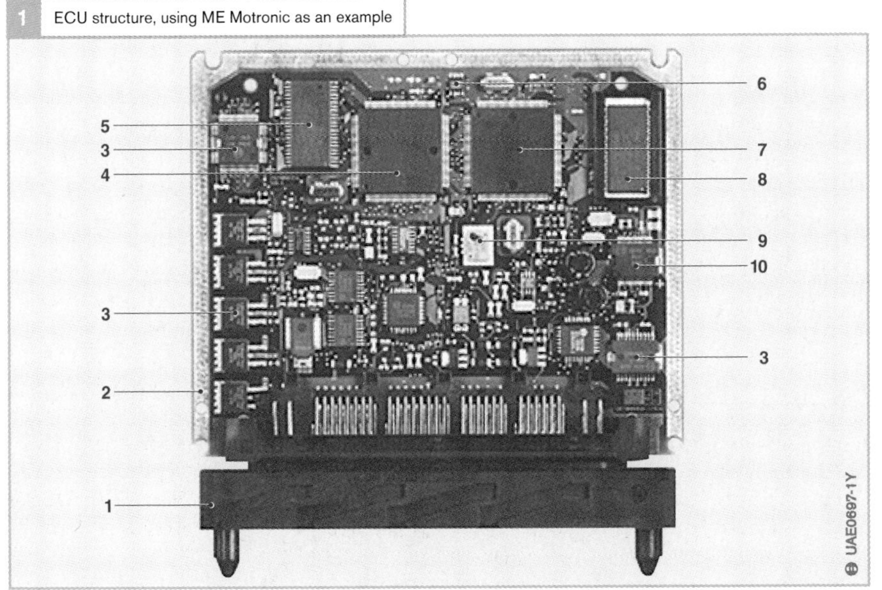

Fig. 1
1 Multipin plug
2 Printed-circuit board
3 Driver stages
4 Microcontroller with ROM (operations processor)
5 Flash EPROM (supplementary program memory for vehicle-specific program)
6 EEPROM
7 Microcontroller with ROM (extension processor, ASIC)
8 Flash EPROM (program memory for extension processor)
9 Barometric-pressure sensor
10 Peripheral module (integrated 5V voltage supply and processing circuitry for inductive sensors)

RAM is on the bottom of the PCB and thus not visible

Microcontroller

The microcontroller is the central component of a control unit and controls its operative sequence (Fig. 2). Apart from the CPU (**C**entral **P**rocessing **U**nit), the microcontroller contains not only the input and output channels, but also timer units, RAMs, ROMs, serial interfaces, and further peripheral assemblies, all of which are integrated on a single micro-chip. Quartz-controlled timing is used for the microcontroller.

Program and data memory

In order to carry out the computations, the microcontroller needs a program – the "software". This is in the form of binary numerical values arranged in data records and stored in a program memory.

These binary values are accessed by the CPU which interprets them as commands which it implements one after the other.

This program is stored in a Read-Only Memory (ROM, EPROM, or Flash-EPROM) which also contains variant-specific data (individual data, characteristic curves, and maps). This is non-variable data which cannot be changed during vehicle operation. It is used to regulate the program's open and closed-loop control processes.

The program memory can be integrated in the microcontroller and, depending upon the particular application, extended in a separate component (e.g. an external EPROM or a flash EPROM).

ROM

Program memories can be in the form of a ROM (**R**ead **O**nly **M**emory). This is a memory whose contents have been defined permanently during manufacture and thereafter remain unalterable. The ROM installed in the microcontroller only has a restricted memory capacity, which means that an additional ROM is required in case of complicated applications.

2 Signal processing in the ECU

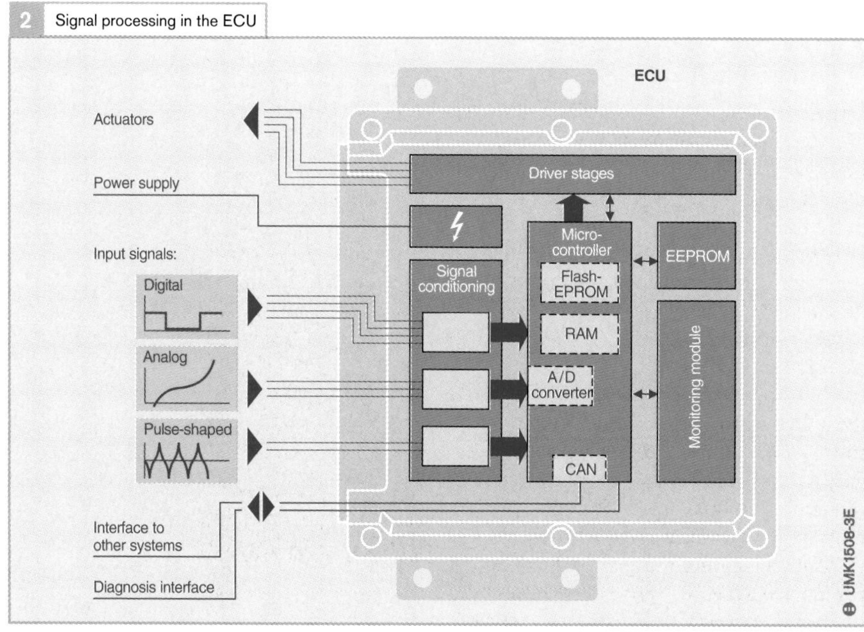

EPROM
The data on an EPROM (**E**rasable **P**rogrammable **ROM**) can be erased by subjecting the device to UV light. Fresh data can then be entered using a programming unit. The EPROM is usually in the form of a separate component, and is accessed by the CPU through the Address/Data-Bus.

Flash-EPROM (FEPROM)
The contents of the flash EPROM can be electrically erased. In the process, the ECU is connected to the reprogramming unit through a serial interface.

If the microcontroller is also equipped with a ROM, this contains the programming routines for the Flash programming. Flash-EPROMs are available which, together with the microcontroller, are integrated on a single microchip (as from EDC16).

Its decisive advantages have helped the Flash-EPROM to largely supersede the conventional EPROM.

Variable-data or main memory
Such a read/write memory is needed in order to store such variable data (variables) as the computational and signal values.

RAM
Instantaneous values are stored in the RAM (**R**andom **A**ccess **M**emory) read/write memory. If complex applications are involved, the memory capacity of the RAM incorporated in the microcontroller is insufficient so that an additional RAM module becomes necessary. It is connected to the ECU through the Address/Data-Bus.

When the control unit is disconnected from the power supply, all data stored in the RAM is lost (volatile memory). However, the next time the engine is started the control unit has to have access to adaptation data (learned data relating to engine condition and operating status). That information must not be lost when the ignition is switched off. In order to prevent that happening, the RAM is permanently connected to the power supply (continuous power supply). If the battery is disconnected, however, the information will nevertheless be lost.

EEPROM (also known as the E^{2}PROM)
Data that must not be lost even if the battery is disconnected (e.g. important adaptation data, codes for the immobilizer) must be permanently stored in a nonvolatile, nonerasable memory. The EEPROM is an electrically erasable EPROM in which (in contrast to the flash EPROM) every single memory location can be erased individually. Therefore, the EEPROM can be used as a nonvolatile random-access memory.

Some control-unit variants also use separately erasable areas of the flash EPROM as nonvolatile memories.

ASIC
The ever-increasing complexity of ECU functions means that the computing powers of the standard microcontrollers available on the market no longer suffice. The solution here is to use so-called ASIC modules (**A**pplication **S**pecific **I**ntegrated **C**ircuit). These IC's are designed and produced in accordance with data from the ECU development departments and, as well as being equipped with an extra RAM for instance, and inputs and outputs, they can also generate and transmit pwm signals (see "PWM signals" below).

Monitoring module
The ECU is provided with a monitoring module. Using a "Question and Answer" cycle, the microcontroller and the monitoring module supervise each other, and as soon as a fault is detected one of them triggers appropriate back-up functions independent of the other.

Output signals

With its output signals, the microcontroller triggers driver stages which are usually powerful enough to operate the actuators directly. For particularly high power consumers (e.g. radiator fan) some driver stages can also operate relays.

The driver stages are proof against shorts to ground or battery voltage, as well as against destruction due to electrical or thermal overload. Such malfunctions, together with open-circuit lines or sensor faults are identified by the driver-stage IC as an error and reported to the microcontroller.

Switching signals

These are used to switch the actuators on and off (for instance, for the engine fan).

PWM signals

Digital output signals can be in the form of pwm (pulse-width modulated) signals. Such "pulse-width modulated" signals are square-wave signals with a constant frequency and variable signal duration (Fig. 3). These signals can be used to move a variety of actuators (e.g. exhaust recirculation valve, turbocharger actuator) to any desired working position.

3 PWM signals

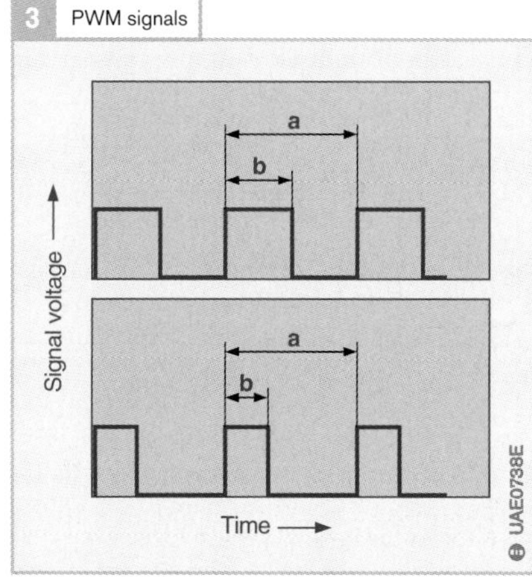

Fig. 3
a Fixed frequency
b Variable on-time

Communication within the ECU

In order to be able to support the microcontroller in its work, the peripheral components must communicate with it. This takes place using an address/data bus which, for instance, the microcomputer uses to issue the RAM address whose contents are to be accessed. The data bus is then used to transmit the relevant data. For former automotive applications, an 8-bit structure sufficed whereby the data bus comprised 8 lines which together can transmit 256 values simultaneously. The 16-bit address bus commonly used with such systems can access 65,536 addresses. Presently, more complex systems demand 16 bits, or even 32 bits, for the data bus. In order to save on pins at the components, the data and address buses can be combined in a multiplex system. That is, data and addresses are dispatched through the same lines but offset from each other with respect to time.

Serial interfaces with only a single data line are used for data which need not be transmitted so quickly (e.g. data from the fault storage).

EoL programming

The extensive variety of vehicle variants with differing control programs and data records, makes it imperative to have a system which reduces the number of ECU types needed by a given manufacturer. To that end, the flash EPROM's entire memory area can be programmed at the end of the production line with the program and the variant-specific data record (this is referred to as end-of-line, or EoL, programming).

A further means of reducing variant diversity is to have a number of data variants (e.g. gearbox variants) stored in the memory, which can then be selected by encoding at the end of the production line. This coding is stored in an EEPROM.

Performance of electronic control units

The performance of electronic control units (ECUs) goes hand-in-hand with advances achieved in the field of microelectronics. The first gasoline fuel-injection systems used analog technology and as such, they were not so versatile when it came to implementing control functions. These functions were constrained by the hardware.

Progress advanced in quantum leaps with the arrival of digital technology and the microcontroller. The entire engine management system was taken over by the universally applicable semiconductor microchip. In microcontroller-based systems, the actual control logic is accommodated on a programmable semiconductor memory chip.

From systems that initially simply controlled fuel injection, complex engine-management systems were then developed. They controlled not only fuel injection but also the ignition system including knock control, exhaust-gas recirculation and a whole variety of other systems. This continuous process of development is bound to continue in a similar vein over the next decade as well. The integration of functions and, above all, their complexity are constantly increasing. This pattern of development is only possible because the microcontrollers used are also undergoing a similar process of improvement.

For a long time microcontrollers of the Intel 8051 family were used until they were superseded at the end of the 1980s by the 80515 series which had extra input/output capabilities for timed signals and an integrated analog-digital converter. It was then possible to create relatively powerful systems. Figure 1 shows a comparison between the performance of a fuel-injection system (LH3.2) and an ignition system (EZ129K) – equipped with 80C515 controllers – and that of the succeeding Motronic systems. With a clock speed of 40 MHz, the ME7 has almost 40 times the processing power of the LH/EZ combination. With the benefit of a new generation of microcontrollers and a further increase in clock frequency on the ME9, this figure will increase to a factor of well over 50.

In the foreseeable future microcontrollers will process more than just digital control sequences. They will have integrated signal processors that will be able to process signals directly, such as signals from the engine-knock sensors, for example.

Advances in the development of semiconductor memory chips are also worthy of note. Complex control programs require an enormous amount of memory space. The capacity of memory chips at the start of the 1980s was still only 8 kilobytes. The ME7 now uses 1-megabyte chips and soon memory capacities of 2 megabytes will be required. Figure 1 shows this pattern of development and likely future trends.

1 Development of electronic control units

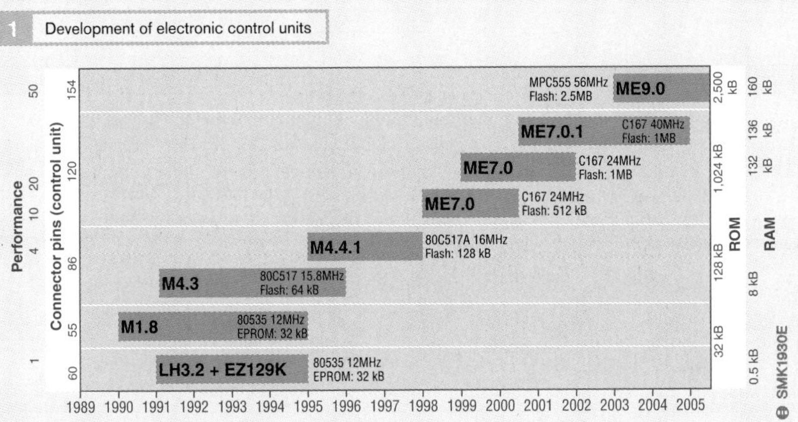

Fig. 1
Chart illustrating
- The performance of engine-management systems
- Number of connector pins on the electronic control units
- Capacity of the program memory
- Capacity of the data memories (RAM)

By way of comparison, the performance of an engine-management system with the very latest technology far exceeds the capabilities of Apollo 13.

Electronic control systems

The function of the engine electronic control unit (ECU) is to control all the actuators in the engine-management system (Motronic) to obtain optimum engine operation in terms of fuel consumption, performance, exhaust-gas emissions and driving smoothness. In order to do so, a large number of operating parameters have to be detected, e.g. with the aid of sensors, and processed using algorithms – i.e. a set of defined mathematical rules. The results obtained take the form of sequences of signals which are used to control the actuators.

Overview

The "command center" of the engine ECU is a small microcomputer (function processor) with a program memory (EPROM) which stores all the algorithms for the control processes (Fig. 1). The input variables that are derived from the data received from desired-value generators and sensors influence the algorithm calculations and, consequently, the control signals for the actuators. The actuators convert the electrical signals into physical quantities (e.g. the aperture of a valve).

The engine ECU can also exchange data via the CAN (Controller Area Network) with other electronic systems such as the ESP (Electronic Stability Program). In this way, the engine-management system can be integrated in the overall vehicle-management system.

Systems with an electronic throttle control (EGAS) have to satisfy extremely demanding requirements regarding functional safety and reliability because there is no physical link with the actuator that controls engine torque (throttle valve). A monitoring module oversees the function processor and initiates alternative measures in the event of a malfunction.

1 Components used for electronic control in an ME-Motronic system

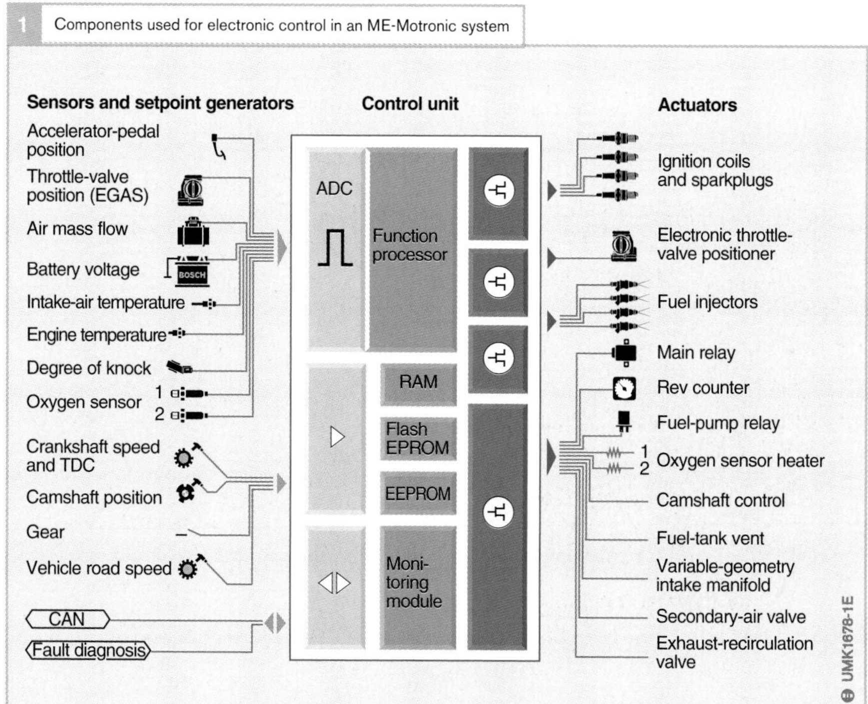

System structure

The system structure sets out the functional and static aspects of the Motronic software architecture. The Motronic software is divided into 13 subsystems (e.g. air system, fuel system). They in turn are subdivided into a total of 50 main functions (e.g. boost-pressure control, lambda closed-loop control, Fig. 2).

The functional core of the Motronic software is the torque-based structure (the subsystems "Torque Demand" and "Torque Structure"). This was first introduced with the advent of the electronic throttle control (EGAS) function on the ME7. The control of cylinder charge by the electrically controlled throttle valve allows regulation of engine torque output depending on the driver's torque command as indicated by the accelerator-pedal position. At the same time all other torque demands that arise from vehicle operation (e.g. when the air-conditioner compressor is switched on) can be coordinated within the torque structure.

On earlier M-Motronic systems, all torque demands were implemented individually through separate functions by altering (advancing or retarding) ignition timing, adjusting the idle actuator (throttle bypass) or varying mixing formation (by changing the injection time). The new generations of the M-Motronic use the torque-based structure in order to coordinate the engine torque setting.

2 Motronic system structure

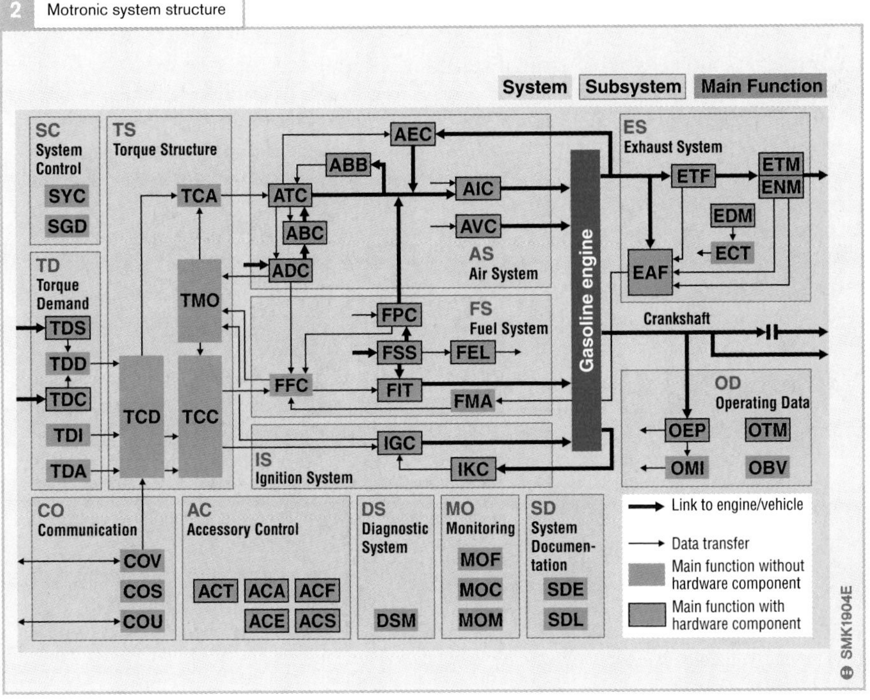

Subsystems and main functions

The description which follows provides a very general summary of the essential features of the main functions implemented on a Motronic system. A more detailed presentation is not possible within the scope of this publication.

System Documentation (SD)

System Documentation consists of technical documents which describe the customer project (e.g. description of electronic control unit, engine and vehicle data, and configuration descriptions).

System Control (SC)

The *System Control* (SC) subsystem encompasses the system control functions that affect the overall Motronic system.

The main function *System Control* (SYC) defines the microcontroller statuses, i.e.

- Initialization (system run-up)
- Running State (normal operation) – this is the status in which the main functions are executed
- Control unit run-on (e.g. for radiator-fan run-on, hardware test)

The main *System Control Gasoline Direct Injection Mode* (SGD) function coordinates and switches between the operating modes for gasoline direct injection (MED-Motronic). In order to determine the required operating mode, the requirements for various functionalities are coordinated on the basis of defined priorities by the operating-mode coordinator.

Torque Demand (TD)

Within the system structure of the ME-Motronic and MED-Motronic systems and increasingly in M-Motronic systems, all torque demands placed on the engine are systematically coordinated at torque level. The *Torque Demand (TD)* subsystem detects all torque demands and passes them to the *Torque Structure (TS)* subsystem as input variables.

The main function *Torque Demand Signal Conditioning* (TDS) function essentially consists of detecting the accelerator-pedal position. The pedal position is detected by two independent angle-position sensors and converted to a standardized accelerator-pedal angle. A number of plausibility checks are carried out to ensure that, in the event of a simple fault, the standardized accelerator-pedal angle cannot adopt a greater value than the actual accelerator-pedal position.

The main *Torque Demand Driver* (TDD) function calculates a setpoint value for the engine torque output from the accelerator-pedal position. In addition, it defines the accelerator-pedal characteristic.

The main *Torque Demand Auxiliary Functions* (TDA) function generates internal torque limitations and demands (e.g. for engine-speed limitation or engine-bucking oscillations).

Torque Demand Idle Speed Control (TDI) regulates the speed of the engine at idle when the accelerator pedal is not depressed. The setpoint value for idle speed is defined to obtain even and smooth engine running at all times. Accordingly, the setpoint idle speed is set higher than the nominal idle speed under certain operating conditions (e.g. when the engine is cold). A higher idle speed may also be used to assist catalytic-converter heating, for example to increase the output of the air-conditioner compressor, or if the battery charge level is low.

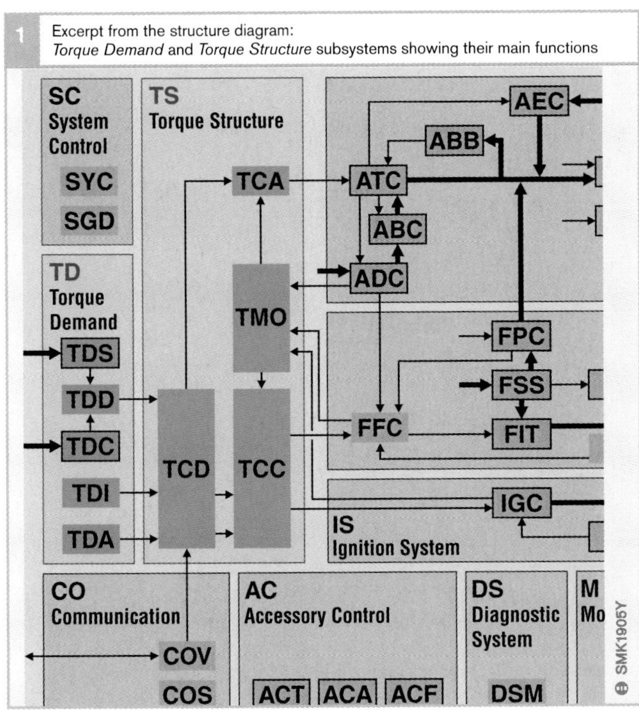

1 Excerpt from the structure diagram:
Torque Demand and *Torque Structure* subsystems showing their main functions

The *Torque Demand Cruise Control, TDC* (vehicle-speed controller) function holds the vehicle at a constant speed as long as the accelerator pedal is not depressed, assuming this is possible with the available engine torque. The most important shut-down conditions for this function include operating the "Off button" on the driver's control lever, operating the brakes or disengaging the clutch, and failure to reach the required minimum road speed.

Torque Structure (TS)

The *Torque Structure* (TS) subsystem is where all torque demands are coordinated. The required torque is then set by the air, fuel and ignition systems.

The main *Torque Coordination* (TCD) function coordinates all torque demands. The various demands (e.g. from the driver, the engine-speed limitation function, etc.) are prioritized and converted into torque setpoints for the various control channels, depending on the active operating mode.

The *Torque Conversion Air* (TCA) function calculates the setpoint value for relative air mass from the torque input variables. That setpoint value for cylinder charge is calculated so that the setpoint for the air mass/torque is obtained at precisely the moment when the specified oxygen content and the specified ignition timing are applied.

The *Torque Conversion Combustion* (TCC) function calculates the setpoint values for oxygen content, ignition timing and injection-suppression stage from the input variables for torque setpoints.

The main *Torque* (TMO) function calculates a theoretically optimum indicated engine torque from the current values for cylinder charge, oxygen content, ignition timing, reduction stage and engine speed. An indicated actual torque is determined with the aid of an efficiency chain. The efficiency chain consists of three different efficiency levels – the injection-suppression efficiency (proportional to the number of firing cylinders), the ignition-timing efficiency (resulting from the shift in the actual ignition point relative to the optimum ignition point), and the oxygen-content efficiency (obtained from plotting the efficiency characteristic against the air-fuel ratio).

2 Excerpt from the structure diagram: *Air System* and *Fuel System* subsystems showing their main functions

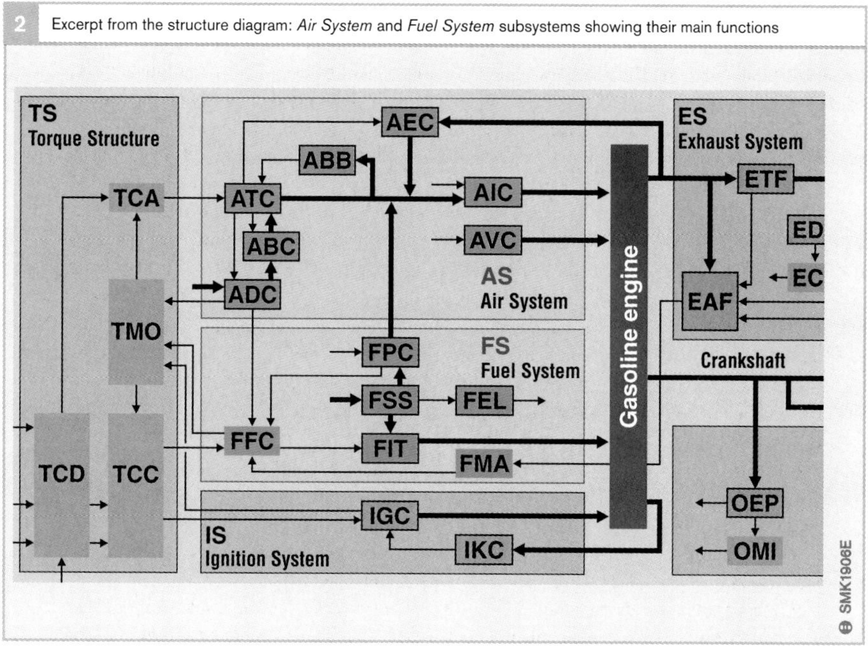

Air System (AS)

The *Air System* (AS) subsystem is where the required cylinder charge for the torque output required is set. In addition, the exhaust-gas recirculation, boost-pressure control, variable-tract intake-manifold geometry, swirl control and valve-timing functions are also part of the air system.

The main *Air System Throttle Control* (ATC) function calculates the required position for the throttle valve from the setpoint mass-air flow.

The *Air System Determination of Charge* (ADC) function determines the cylinder charge composed of intake air and inert gas using data from the available load sensors. The mass-air flows are used to model the pressure conditions in the intake manifold (intake-manifold pressure model).

The *Air System Intake Manifold Control* (AIC) function calculates the setpoint positions for the intake-manifold and swirl flaps. The vacuum in the intake manifold allows exhaust-gas recirculation which is calculated and controlled by the *Air System Exhaust-Gas Recirculation* (AEC) function.

The *Air System Valve Control* (AVC) function calculates the setpoint values for inlet and exhaust-valve timing and controls these settings. This influences the quantity of residual exhaust gas that is recirculated internally.

The *Air System Boost Control* (ABC) function is responsible for calculating the charge-air pressure on exhaust-gas turbocharged engines and controls the actuators for this system.

Gasoline engines with direct fuel injection are run in stratified-charge mode with the throttle fully open at low loads. Consequently, the pressure in the intake manifold under such conditions is virtually atmospheric pressure. The *Air System Brake Booster* (ABB) function ensures that there is sufficient vacuum in the brake booster by requesting a required amount of flow restriction.

Fuel System (FS)

The Fuel System (FS) subsystem calculates the output variables for the fuel-injection system relative to crankshaft position, i.e. the point(s) at which fuel is injected and the quantity of fuel injected.

The *Fuel System Feed Forward Control* (FFC) function calculates the setpoint fuel mass from the setpoint cylinder charge, the setpoint oxygen content and additional adjustments (e.g. transition compensation) or multiplier adjustments (e.g. for engine start, warm-up and restart). Other adjustments arise from the lambda closed-loop (oxygen-content) control, tank-ventilation and mixture-adaptation functions. On MED systems, specific data is calculated for the various operating modes (e.g. fuel injection during the induction stroke or during the compression stroke, multiple injection).

The *Fuel System Injection Timing* (FIT) function calculates the injection time and the fuel-injection position. It ensures that the fuel injectors are open at the correct time relative to crankshaft rotation. The injection time is calculated on the basis of previously calculated fuel mass and status variables (e.g. intake-manifold pressure, battery voltage, fuel-rail pressure, combustion-chamber pressure).

Fuel System Mixture Adaptation (FMA) improves the control accuracy of the oxygen content by adjusting the longer-term lambda control unit error relative to the neutral value. For smaller cylinder charges, the lambda control unit error is used to calculate an additive correction value. On systems with a hot-film air mass sensor, this normally reflects small amounts of intake-manifold leakage. On systems with an intake-manifold pressure sensor, the lambda control unit corrects the pressure-sensor residual exhaust gas or offset error. For larger cylinder charges, a multiplier correction factor is calculated. This essentially represents the hot-film air-mass sensor gain error, fuel-rail pressure regulator inaccuracies (on MED systems) and fuel-injector characteristic-gradient errors.

The *Fuel Supply System* (FSS) has the function of delivering fuel from the fuel tank to the fuel rail at the required pressure and in the required quantity. On demand-controlled systems, the pressure can be regulated between 200 and 600 kPa. A pressure sensor provides feedback of the actual level.

In a gasoline direct-injection system, the fuel supply system also includes a high-pressure section consisting of type HDP1 high-pressure pump and pressure-control valve, or type HDP2 demand-controlled high-pressure pump with fuel-quantity control valve. This allows pressure in the high-pressure system to be varied between 3 and 11 MPa depending on engine operating status. The setpoint value is calculated depending on engine operating status and the actual pressure is detected by a high-pressure sensor.

The *Fuel System Purge Control* (FPC) function controls regeneration of the fuel that evaporates from the fuel tank and that is collected in the activated charcoal canister of the evaporative-emissions control system. On the basis of the specified on/off ratio for operating the tank-ventilation valve and the pressure conditions, an actual value for the total mass-flow rate through the valve is calculated. This is taken into account by the Air System Throttle Control (ATC) function. An actual fuel-content value is also calculated and is subtracted from the required fuel mass.

The *Fuel System Evaporation Leakage Detection* (FEL) function checks the gas-tightness of the fuel tank in accordance with the requirements of the Californian OBD II regulations. The design and method of operation of the on-board diagnosis system are described in the chapter "Electronic fault diagnosis/OBD – Individual diagnoses".

Ignition System (IS)

The *Ignition System* (IS) subsystem calculates the output variables for the ignition and controls the ignition coils.

The *Ignition Control* (IGC) function calculates the current setpoint ignition angle from the engine operating conditions, taking account of intervention by the torque structure. It then generates a spark across the spark plug electrodes at the required time. The resulting ignition angle is calculated from the basic ignition angle and the ignition-timing adjustments and requirements that are determined by the operating status. When determining the engine-speed and load-dependent basic ignition angle, the consequences of variable valve timing, swirl flap setting, cylinder grouping and special gasoline direct-injection operating modes are taken into account where applicable. In order to calculate the most advanced possible ignition angle, the basic ignition angle is corrected by the advance angles for engine warm-up, knock control and – where applicable – exhaust-gas recirculation. The point at which the ignition driver stage needs to be triggered is calculated from the current ignition angle and the required charge time for the ignition coil. The driver stage is activated accordingly.

3 Excerpt from the structure diagram: *Ignition System* and *Exhaust System* subsystems showing their main functions

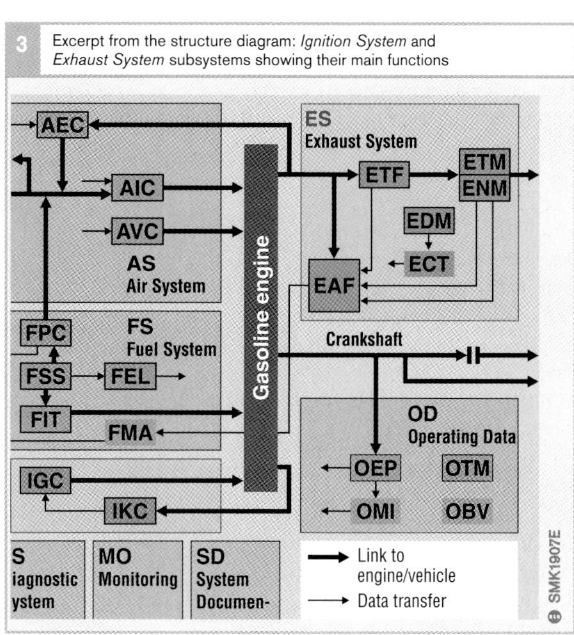

The *Ignition System Knock Control* (IKC) function runs the engine at the knock limit for optimum efficiency, but prevents potentially damaging engine knock (pinking). The combustion process in all cylinders is monitored by means of knock sensors. The structure-borne noise signal detected by the sensors is compared with a reference level that is obtained for individual cylinders via a low-pass filter from previous combustion strokes. The reference level therefore represents the background engine noise when the engine is running free of engine knock. The comparison analyzes how much louder current combustion is than the background level. Above a certain threshold, engine knock is assumed to occur. Both calculation of the reference level and detection of engine knock can take account of changes in operating conditions (engine speed, engine-speed dynamics, engine-load dynamics).

The IKC function generates an ignition timing adjustment for each individual cylinder. This is taken into account when calculating the current ignition angle (ignition retard). When engine knock is detected, that ignition timing adjustment is increased by a specified amount. The ignition timing retard is then reduced in small increments if, over a specified time period, engine knock does not occur.

If a hardware fault is detected, a safety function is activated (safety ignition-timing retard).

Exhaust System (ES)

The Exhaust System (ES) subsystem intervenes in the mixture-formation system, adjusts the excess-air factor, and controls the capacity utilization of the catalytic converters.

The prime functions of *Exhaust System Description and Modelling* (EDM) are to model physical variables in the exhaust-gas system, analyze the signals from and detect faults in the exhaust-gas temperature sensors (where present), and supply key exhaust-gas system data to the tester output. The physical variables that are modeled are temperature (e.g. for component protection purposes), pressure (primarily for residual exhaust-gas detection) and mass flow (for lambda closed-loop control and catalytic-converter fault diagnosis). In addition, the exhaust-gas excess-air factor is calculated (for NO_X accumulator catalytic converter control and fault diagnosis).

The purpose of *Exhaust System Air Fuel Control* (EAF) using the lambda sensor upstream of the front catalytic converter is to regulate the excess-air factor to a specified level. This minimizes harmful emissions, prevents engine-torque fluctuations and keeps the exhaust-gas composition on the right side of the lean-mixture limit. The input signals from the lambda closed-loop control system downstream of the main catalytic converter allows further minimization of emissions.

The main *Exhaust System Three-Way Front Catalyst* (ETF) function uses the lambda sensor downstream of the front catalytic converter (if fitted). Its signal is a measure of the oxygen content in the exhaust gas and serves as the basis for reference-value regulation and catalytic-converter fault diagnosis. Reference-value regulation can substantially improve mixture control and permit optimum exhaust-gas treatment response by the catalytic converter.

Reference-value regulation may take different forms depending on the system. A NO_X accumulator catalytic converter operated at $\lambda = 1$ displays optimum conversion response at a specific oxygen storage-capacity usage level. The reference-value regulation function sets the capacity usage to this level. Divergences are corrected by compensation components.

The main *Exhaust System Three-Way Main Catalyst* (ETM) function basically operates in the same way as the ETF function described above. In systems with a NO_X accumulator catalytic converter, changeover to a special mode allows desulfurization of the catalytic converter.

The task of the *Exhaust System NO_X Main Catalyst (ENM)* function is to ensure that NO_X emission requirements in particular are complied with when the engine is running on a lean mixture by controlling the NO_X accumulator catalytic converter.

Depending on the condition of the catalytic converter, the NO_X storage phase is terminated and the engine switched over to an operating mode ($\lambda < 1$) in which the NO_X accumulator is emptied and the stored NO_X emissions converted to N_2. Regeneration of the NO_X accumulator catalytic converter is terminated in response to the change of signal from the sensor downstream of the NO_X accumulator catalytic converter.

The *Exhaust System Control of Temperature* (ECT) function controls the temperature of the exhaust-gas system. Its aim is to speed up the time it takes the catalytic converters to reach operating temperature after the engine is started (catalytic-converter heating), prevent the catalytic converters from cooling down during operation (catalytic-converter temperature retention), heat up the NO_X accumulator catalytic converter for desulfuration, and prevent thermal damage to exhaust-gas system components (component protection). A torque reserve for the TS *(Torque Structure)* subsystem is determined from the heat flow required for a temperature increase. The temperature increase is then achieved by retarding the ignition, for example. When the engine is idling, the heat flow can also be increased by raising the idle speed.

Operating Data (OD)

The *Operating Data* (OD) subsystem records all important engine operating parameters, checks their plausibility and provides substitute data where required.

The *Operating Data Engine Position Management* (OEP) function calculates the position of the crankshaft and the camshaft using the processed input signals from the crankshaft and camshaft sensors. It calculates the engine speed from this data. The crankshaft timing wheel (two missing teeth) and the characteristics of the camshaft signal are used to synchronize the engine and the ECU and to monitor synchonization while the engine is running.

The camshaft signal pattern and the engine shutoff position are analyzed in order to optimize the start time. This allows rapid synchronization.

Operating Data Temperature Measurement (OTM) processes the temperature readings provided by the temperature sensors, performs plausibility checks on them, and provides substitute data in the event of faults. The ambient temperature and the engine-oil temperature may also be detected in addition to the temperature of the engine and the intake air. The input voltage signals are assigned to a temperature reading. This is followed by calculation of a characteristic curve.

The *Operating Data Battery Voltage* (OBV) function is responsible for providing the supply-voltage signals and performing diagnostic operations on them. The raw signal is detected at terminal 15 and, if necessary, the main relay.

The *Misfire Detection Irregular Running* (OMI) function monitors the engine for misfiring and combustion misses (see the chapter on "Electronic fault diagnosis/OBD – Individual diagnoses").

4 Excerpt from the structure diagram: *Operating Data* subsystem

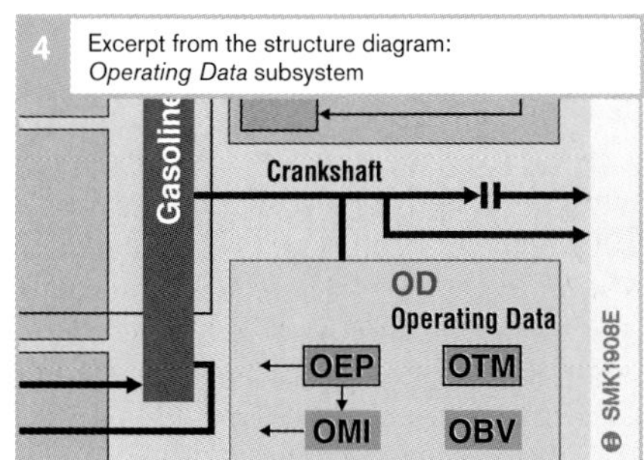

5 Excerpt from the structure diagram: *Communication* and *Accessory Control*subsystems

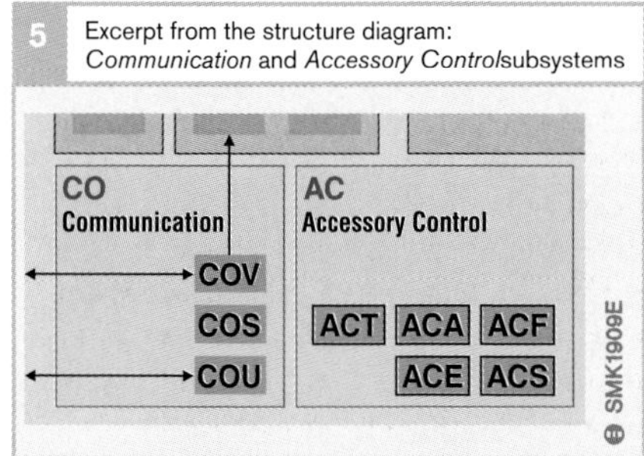

Communication (CO)

The *Communication* subsystem encompasses all Motronic main functions that communicate with other systems.

The *Communication User Interface* (COU) function provides the connection to diagnostic (e.g. engine analyzer) and calibration equipment. Communication takes place via the "K" wire, though the CAN interface can also be used for this purpose. Different communication protocols are available for various applications (e.g. KWP2000, McMess).

The *Communication Vehicle Interface* (COV) function looks after communication with other control units, sensors and actuators.

Communication Security Access (COS) provides for communication with the immobilizer and – as an option – enables access control for reprogramming the flash EPROM.

Accessory Control (AC)

The *Accessory Control* subsystem controls the auxiliary systems.

The *Accessory Control Air Condition* (ACA) function controls operation of the air-conditioner compressor and analyzes the signals from the pressure sensor in the air conditioner. The air-conditioner compressor is switched on when, for example, a request is received from the driver or the air-conditioner ECU via a switch. The air-conditioner ECU signals to the Motronic that the air-conditioner compressor needs to be switched on. It is switched on shortly afterwards. When the engine is idling, the engine management system has sufficient time to develop the required torque reserves.

Various conditions can result in the air conditioner being switched off (e.g. critical pressure in the air conditioner, fault in the pressure sensor, low ambient temperature).

The *Accessory Control Fan Control* (ACF) function controls the radiator fan in response to demand and detects faults in the fan and the control system. Under certain circumstances, the fan may be required to run on when the engine is not running.

The *Accessory Control Thermal Management* (ACT) function regulates the engine temperature according to operating conditions. The required engine temperature is determined depending on engine performance, driving speed, engine operating status, and ambient temperature. This helps the engine to reach its operating temperature more quickly and is then adequately cooled. The coolant volumetric flow through the radiator is calculated and the map-controlled thermostat is operated accordingly based on the temperature setpoint.

The *Accessory Control Electrical Machines* (ACE) function is responsible for controlling the "electrical machines", i.e. the starter motor and alternator.

The task of the *Accessory Control Steering* (ACS) function is to control the power-steering pump.

Monitoring (MO)

The *Function Monitoring* (MOF) function monitors all Motronic elements that affect engine torque and speed. The core function is torque comparison. This compares the permissible torque calculated on the basis of driver request with the actual torque calculated from the engine data. If the actual torque is too large, suitable measures are initiated to ensure that a controllable status is re-established.

The *Monitoring Module* (MOM) combines all monitoring functions that contribute to or perform reciprocal monitoring between the function processor and the monitoring module. The function processor and the monitoring module are components of the ECU. Reciprocal monitoring between them takes place by means of continuous query-and-response communication.

The *Microcontroller Monitoring* (MOC) function combines all monitoring functions that can detect a fault or malfunction in the processor and its peripherals. Examples include:

- The analog-digital converter test
- Memory tests for RAM and ROM
- Monitoring program run
- Command test

Diagnostic System (DS)

Component and system diagnosis are performed by the main functions of the subsystems. The *Diagnostic System* (DS) is responsible for coordinating the various diagnosis results.

The task of the *Diagnostic System Manager* (DSM) is to:

- Store details of faults and associated ambient conditions
- Switch on the malfunction indicator lamp
- Establish communication with the diagnostic tester
- Coordinate execution of the various diagnostic functions (taking account of priorities and bars) and verify faults

Electronic diagnosis

Integral diagnosis functions included within the electronic control unit are a standard component in electronic engine-management systems. Algorithms are used to keep track of input and output signals during normal vehicle operation. The system simultaneously monitors the overall system for signs of problems and malfunctions. It records detected malfunctions as error codes in the malfunction log. A serial interface allows technicians to access these stored error codes during the course of normal vehicle service, making it easier to localize and repair problems quickly.

Self-diagnosis

The original concept called for specific self-diagnosis utilities for each vehicle manufacturer. This feature was intended to support quick and convenient diagnosis of problems related to the engine-management system when the vehicle was serviced. Legal requirements then combined with increasingly extensive electronics to promote the adoption of engine-management systems incorporating expanded diagnostic capabilities.

Monitoring input signals

The system monitors the status of the sensors and the wiring to the control unit (Table 1) by processing input signals. In addition to registering sensor malfunctions, these tests also detect short circuits to battery voltage U_{Batt} and to ground, as well as open wiring. This system provides this functionality by:

- Monitoring the voltage supply to the sensors.
- Analyzing registered data for conformity with the specified operating ranges (such as engine temperature between –40 °C...+150 °C).
- When additional information is available, the system runs plausibility checks (such as comparisons of crankshaft and camshaft rotation rates).
- Systems are engineered for redundancy in areas related to vital sensors (such as the accelerator-pedal travel sensor). This strategy provides multiple signals for mutual correlation and comparison.

Monitoring output signals

This function monitors the actuators as well as the wiring linking them to the electronic control unit. The corresponding performance checks can detect open wiring and short circuits as well as problems related to the actuators themselves. The system implements this functions by:

- Monitoring the progress of output signals through the output driver circuit. The system registers short circuits to battery voltage U_{Batt}, open circuits and shorts to ground.
- System data are then correlated with actuator control commands to determine whether the resulting conditions are plausible. On example is exhaust-gas recirculation, where the system checks to verify that intake-manifold pressure is responding to actuator triggering by moving into a specific range.

Monitoring data communications between ECUs

Communications with other electronic control units are generally expedited through the CAN bus. The CAN protocols include control mechanisms to allow recognition of malfunctions, allowing transmission errors to be detected before the signals leave the CAN chip. The electronic control unit also runs a variety of other test routines. Because each control unit normally transmits messages through the CAN at regular periodic intervals, the system can employ periodicity as a tool for detecting failure in any individual ECU.

In addition, when redundant information is held in the ECU, this information is used in checking the received signals in the same manner as all input signals.

Monitoring the ECU's internal operations

The control unit's internal monitoring apparatus includes hardware-based ("intelligent" output driver chips, etc.) as well as software-based functions to ensure consistently reliable operation.

The monitoring functions are tested by individual components within the control unit (such as the microcontroller, flash EPROMs, RAMs, etc.). A number of test routines start to run as soon as the control unit is activated. Other test routines run at regular intervals during normal vehicle operation to ensure recognition of any component failures that might occur on the road. Checks that require a considerable amount of processing capacity (EPROM tests, etc.) are implemented in the post-operative phase immediately after the engine is switched off (currently available for gasoline-engines only). This prevents the test routine from interfering with efficient implementation of normal operating processes. The immediate post-operational phase is employed to test the deactivation paths on diesels.

1 Monitoring vital input signals

Signal path	Monitoring
Accelerator-pedal travel sensor	Check on supply voltage and signal range
	Plausibility check against redundant signal
	Plausibility check against brake
Crankshaft rpm sensor	Check of signal range
	Plausibility check against camshaft rpm sensor
	Check on temporal shift (dynamic plausibility)
Engine-temperature sensor	Check of signal range
	Logical plausibility relative to rpm
	and injected fuel quantity/engine load factor
Brake-pedal switch	Plausibility check with redundant contact switch
Vehicle-speed signal	Check of signal range
	Plausibility check against engine rpm and injected fuel quantity/engine load factor
EGR valve return mechanism	Check for short and open circuits
	Closed-loop EGR control
	Check of system response to valve control signals
Battery voltage	Check of signal range
	Plausibility check against engine rpm (currently on gasoline engines only)
Fuel-temperature sensor	Check of signal range (currently on diesel engines only)
Boost-pressure sensor	Check of supply voltage and signal range
	Plausibility check against barometric-pressure sensor and/or other signals
Bypass-valve controller	Check for short and open circuits
	Boost-pressure control deviation
Air-mass meter	Check of supply voltage and signal range
	Logical plausibility check
Air-temperature sensor	Check of signal range
	Logical plausibility check against, e.g., engine-temperature sensor
Clutch-signal sensor	Plausibility check against vehicle speed
Barometric-pressure sensor	Check of signal range
	Logical plausibility check against manifold-pressure sensor

Table 1

Malfunction response

Problem recognition

Signal paths are classified as defective once an error state remains present for a specific amount of time. The last data recognized by the system as valid are inserted as defaults in the period until the problem is reclassified as a consistent error. Once the problem is assigned consistent error status, the system usually reverts to operation using substitute default data (such as a default temperature figure of $T = 90\,°C$).

A "restored-signal recognition" feature is available for most errors. This feature is available once the system recognizes the signal path as having regained "intact" status for a defined period.

Error-code storage

Each problem is stored as an error code in the malfunction log's non-volatile memory. When storing the error codes, the system simultaneously saves supplementary information in a "freeze frame" definition of the operating and environmental conditions present when the error occurred (such as engine speed, coolant temperature, etc.). Also stored are data portraying the error class (short circuit, open wire, etc.) and the error status (consistent error, sporadic error, etc.).

A number of error codes related to exhaust emissions are prescribed by official regulations. As an option the system can also store additional, problem information specifically intended to assist technicians in servicing individual vehicle models.

Following registration of the error code, the diagnosis process focusing on the specific system or components continues. Once compliance with defined criteria has been achieved, malfunctions that fail to recur (sporadic errors) can be deleted from the malfunction log.

Error-code access

Stored error codes can be accessed using a service tester provided by the manufacturer, a system tester (such as the Bosch KTS500) or a scan tool. The tester can also be used to delete the stored error codes following read-out and repair in the service facility.

Diagnostic interface

The data generated by the on-board diagnosis are relayed to the outside tester through the communications interface. The mandatory configuration for this interface is defined in ISO 9141 (diagnostic interface for access through communications cable). This serial interface operates at a data-transfer (baud) rate of between 10 and 10k baud. It can be in the form of a single-wire interface using one shared transceiver link, or a dual-wire socket using separate wires for data (COM wire) and triggering (L wire). A single diagnostic socket can provide access to numerous control units.

The tester transmits a triggering address to all control units, one of which recognizes the address and transmits an "acknowledge" code in return. Using the interval between pulse flanks as an index of baud rate, the tester adjusts its own communications rate accordingly. It then proceeds to establish communications with the control units.

In future applications, communications between the control units and the testers will be expedited via the CAN bus.

Laws and official regulations

Self-diagnosis was originally limited to an electrical component check. Increasingly complex diagnostic functions with new test processes (plausibility checks, etc.) combined with the official demand for diagnostic capabilities embracing emissions-relevant systems and components to make a single, standardised diagnostic system essential. In response to these developments, basic self-diagnosis evolved into the On-Board Diagnosis (OBD) system.

On-Board Diagnosis (OBD)

Successive model years have seen continuing reductions in the toxic emissions produced by each vehicle. In order that in everyday use vehicles continue to comply with the emissions limits stipulated by the manufacturer, it is necessary for the engine and the components to be consistently monitored. In response to this imperative, legislators have defined mandatory diagnosis processes for those components and systems with the potential to affect emissions.

OBD I appeared in 1988, marking the debut of on-board diagnosis designed to comply with CARB (California Air Resources Board) legislation. All new vehicles registered in California for the first time must meet the legal requirements. 1994 witnessed the introduction of OBD II, representing the second stage.

In the period since 1994 the corresponding regulations for the remaining 49 states have been defined by the EPA (Environmental Protection Agency). While the EPA stipulations governing diagnostic capabilities are essentially the same as the CARB rules (OBD II), the EPA regulations are less stringent in certain areas.

EOBD is a version of the OBD adapted to European conditions. It has been in force since the year 2000. The basic concept mirrors the EPA OBD system. The current EOBD requirements are, however, somewhat more lenient.

OBD I

The first stage of CARB OBD tests emissions-relevant components for short and open circuits. The electrical signals must lie within the defined plausibility limits.

When the system detects a problem or malfunction, it triggers a warning lamp in the instrument cluster to alert the driver. It must be possible to determine which component has failed using on-board resources (flash code through a connected diagnosis lamp, etc.).

OBD II

The diagnostic procedure for the second stage of CARB OBD extends well beyond that prescribed for OBD I. In addition to checking signal transmission paths, OBD II also monitors system operation. For example, it is no longer enough to run checks on the signals from the engine-temperature sensor to ensure that they do not violate the defined limits. OBD II also registers errors in response to excessively low (such as 10 °C) temperature data for a running engine (plausibility check).

OBD II demands that all the emissions-relevant systems and components are monitored which, in case of malfunction, could provoke a substantial increase in emissions. The system must also keep tabs on all components with the potential to affect the results of the diagnostic processes. All recognized malfunctions must be stored in the malfunction log. The warning lamp located within the instrument cluster must also indicate any and all malfunctions. The stored error codes can be accessed using the connected diagnosis testers.

The OBD II legislation prescribes standardized error protocols as defined by the SAE (Society of Automotive Engineers) in accordance with ISO 15 031. This stipulation ensures that stored error codes can be accessed using standard, commercially available scan tools.

Control of diagnostic sequences

As a rule, the diagnosis functions for all the systems and components which are to be tested, must be run through at least once during the emissions test cycle (e.g. FTP75, NEDC).

The diagnostic system's management can initiate dynamic adjustments in the processing sequence to accommodate different vehicle operating conditions. The ultimate objective is to ensure that the entire array of diagnostic functions runs with adequate frequency in everyday vehicle operation.

OBD – General requirements

The engine-management ECU is required to use OBD on-board diagnosis functions to monitor all of the systems and components within the vehicle whose failure could lead to substantial increases in pollutant emissions. An error is present once defined diagnostic thresholds (limits) are exceeded.

Application

The OBD regulations defined by the CARB and EPA apply to all passenger vehicles with up to 12 seats as well as small trucks up to 6.35 t. The EOBD stipulations, valid since 01.01.2000, apply to all gasoline-engine passenger vehicles and light commercial vehicles of up to 3.5 t and 9 seats. Starting in 2003 EOBD capabilities will also be mandatory for passenger vehicles and light commercial vehicles with diesel engines.

Limits

Table 1

The CARB OBD (OBD II) concept is based on relative limits. This means that the limits on acceptable pollutant concentrations within the exhaust gases vary according to the emissions category in which each individual vehicle is certified (LEV, ULEV, etc.). The European EOBD regulations are based on absolute limits (Table 1).

Operational requirements

One of the OBD requirements stipulates monitoring for all electrical wiring leading to the control unit. This means that "comprehensive components" (such as the air-mass meter) are monitored for signal plausibility (OBD II) as well as signs of electrical failure (EOBD). Complex OBD functions check the diagnostic system to verify that it is operational.

Table 2

The prescribed response to failures varies according to the problem's potential consequences. CARB OBD and EOBD use different criteria (Tables 2 and 3). The type of diagnosis is defined by the pollutants concentration which could be expected (empirical data) due to failure of a given component.

Table 3

Simple operational checks (black and white tests) only assess the basic operational status of the system or component (whether the secondary-air injection valve opens and closes, etc.). The qualitative operational check (flow check) provides more precise information on system performance. One example is the catalytic-converter check, where the monitored data are employed to assess ageing. The corresponding data are available for readouts through the diagnostic interface.

1	Exhaust-emissions limits
CARB:	relative limits 1.5 times limit in each emissions category
EOBD:	absolute limits CO: 3.2 g/km HC: 0.4 g/km NO_X: 0.6 g/km New EOBD limits are anticipated for implementation starting 01.01.2005

2	Diagnostic processes and malfunction response in CARB and EPA
Malfunction leads to pollutant concentration < 1.15 times limit	
•	Error status indicated only by service tester
Malfunction leads to pollutant concentration < 1.5 times limit	
•	Operation check (black and white test)
•	Error status indicated by MIL
•	Error status registered by scan tool
Malfunction leads to pollutant concentration ≥ 1.5 times limit	
•	Qualitative operation check
•	Error status indicated by MIL
•	Error status registered by scan tool

3	EOBD diagnosis and malfunction response
Malfunction leads to pollutant concentration < limit	
•	Monitoring of electrical wiring and min./max. plausibility checks adequate
•	Error status indicated by MIL
•	Error status registered by scan tool
Pollutant concentration ≥ limit	
•	Qualitative operation check
•	Error status indicated by MIL
•	Error status registered by scan tool

Increasingly stringent emissions controls have led to progressively more complex diagnostic routines. As a result, almost 50 % of the Motronic system's entire performance potential is devoted to on-board diagnosis processes.

Malfunction indicator lamp (warning lamp)

Die MIL (Malfunction Indicator Lamp) alerts the driver to operating problems. Systems designed for conformity with CARB and EPA regulations must trigger the MIL to indicate detected errors after no more than two operating cycles. Within the EOBD's range of application, the MIL must respond to detected problems by lighting up no later than in the third operating cycle (with waivers available for a maximum of up to ten operating cycles).

If the error disappears again (problem with an intermittent contact, etc.) the corresponding error code remains stored in the malfunction log throughout the subsequent 40 driving cycles. The MIL goes out again after three driving cycles with no detected malfunctions. The MIL responds to problems that could lead to catalytic-converter damage (combustion miss) by flashing.

Emergency operation

When the system detects an error it reverts to operation on substitute default data (for engine temperature, etc.) or to the emergency backup mode (e.g. limitation of engine output power). These strategies are intended to

- Maintain vehicle safety
- Avoid subsequent damage (from overheated catalytic converters, etc.)
- Minimize exhaust emissions

Activation conditions

The diagnostic routines run only after the activation requirements have been fulfilled. Among these are

- Torque thresholds
- Engine-temperature thresholds
- rpm limits

Inhibit conditions

The system cannot always run the engine-management and diagnostic functions simultaneously. Certain inhibit functions that prevent specific operations from being processed are also present. To cite one example, the fuel tank's ventilation system (evaporative-emissions control) cannot operate while the catalytic-converter diagnosis function is in progress.

Temporary interruption of diagnostic routines

Under specific conditions, diagnostic routines can also be suspended to prevent spurious malfunction alerts. These conditions include

- Extreme altitudes of more than 2,400 m (or 8,000 feet) above sea level (CARB OBD) or 2,500 m above sea level (EOBD)
- Fuel level ≤ 15 % (CARB OBD) or 20 % (EOBD) of nominal volume (the EOBD arrangement deviates from CARB OBD practice by not requiring plausibility checks on fuel level)
- Very low ambient temperatures during cold starts ($T < -7$ °C)
- Low battery voltage

Readiness code

Before proceeding to access the malfunction log, it is important for the technician to verify that the diagnostic routines have really been run during the proceeding driving cycle. Corresponding confirmation is available in the form of readiness codes available through the diagnostic interface. The system registers these codes to confirm that the essential diagnostic routines have been completed.

Recalls

The government can demand that manufacturers of vehicles that fail to comply with the OBD requirements recall these vehicles at their own expense.

4 OBD II and EOBD system and component monitoring

Diagnostic function	Special OBD components	Remark
Error response	MIL Malfunction Indicator Lamp	
"Freeze Frame" storage	Vehicle-speed sensor, etc.	Stores mileage when MIL is active, EOBD requirement
Scan-tool interface	DLC (standardized plug)	
Monitor input-signal plausibility [1]		EOBD recommendation for service support in field
Monitor catalytic converter	Cat-back O_2 sensor	
Monitor cat-forward and cat-back O_2 sensors [1]		Electric diagnosis for cat-back sensor required to support deactivation of catalytic-converter diagnosis
Heater diagnosis at cat-forward and cat-back O_2 sensors [1]		Electric diagnosis for cat-back O_2 sensor required, depends on catalytic-converter diagnosis
O_2 sensor signal transmission		Signal accessed through test interface
Monitor fuel system		Control of mixture adaptation and evaporative-emissions control
Engine-misfire detection	RPM sensor	
Rough-surface recognition	Wheel-speed sensor or acceleration sensor	Triggers deactivation of misfire detection Electrical diagnosis min. requirement for components, not stipulated by EOBD
Output-amplifier circuit diagnosis	Output-circuit components	Checks for open and short circuits in wiring to all emissions-relevant components
EGR diagnosis [1]	Manifold-pressure sensor	
EVAP valve (evaporative-emissions control system)		Only open-circuit check is required for EOBD
Tank-leak diagnosis	Tank-pressure sensor	Not an EOBD requirement
Other emissions-relevant systems: operation check [1]		
Fuel-level monitor	Fuel-gauge sensor	Electrical diagnosis is minimum EOBD requirement, for deactivation of diagnostic functions at fuel level < 20 % of capacity

Table 4

[1]) EOBD requirement, if failure could lead to violation of EOBD emissions limits, or electrical monitoring if other diagnostic utilities could be deactivated.

OBD – Diagnosis System Management (DSM)

Diagnosis System Management (DSM) controls the processing sequence of the on-board diagnosis facility. It consists of the following three components (Figure 1):

DFPM

The primary function of the Diagnostic Fault Path Management (DFPM) is to store data in response to detected system errors. The fault paths also contain information on ambient conditions and other data of value to the vehicle repairer and/or in locating problem sources.

Other functions discharged by the DFPM include triggering the MIL and managing data communications to allow the scan tool to access the malfunction log.

DSCHED

The Diagnostic Function Scheduler (DSCHED) is responsible for co-ordinating the assigned engine and diagnostic functions. It is supported in this process by information received from the DVAL and DFPM. Meanwhile, the functions that require DSCHED release transmit-readiness confirmation, initiating a check of current system status and function activation.

The scheduler incorporates a number of subsidiary components:

- The *inhibit handler* prevents a particular function from running if one of the components which is essential for correct implementation of that function is found to be defective
- The *priority handler* calculates a current priority for each function based on a range of parameters
- The *priority scheduler* accesses an exclusion table indicating which functions may not run simultaneously

DVAL

The Diagnosis Validator (DVAL) analyzes current stored error codes and various supplementary data related to each detected malfunction. These serve as the basis for determining whether the problem represents a primary malfunction or a subsequent fault stemming from another source. The diagnostic tester used to access the malfunction log is provided with information which has been stored by the validation utility.

This makes it possible to run diagnostic routines in any sequence, regardless of the secondary effects of the malfunctions they are intended to detect. All released diagnosis routines and their results are assessed subsequently. In a system without the DVAL, each diagnosis routine would have to await clearance from other diagnostic utilities before proceeding. This would limit diagnosis tools to operation under very specific conditions.

The validator is the key to rapid problem isolation and therefore to effective repair – even of complex systems.

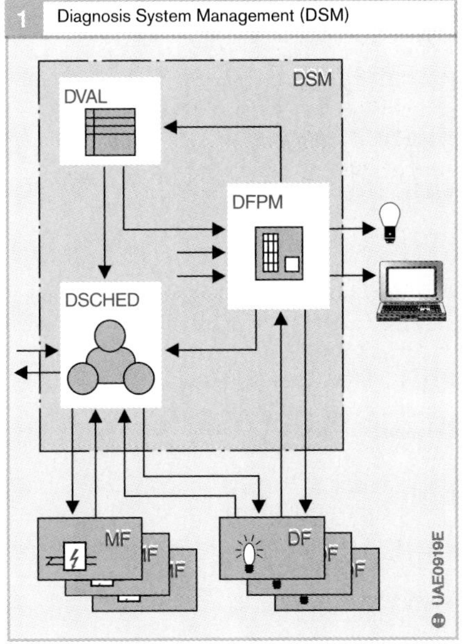

Fig. 1
DSM Diagnosis System Management
DFPM Diagnostic Fault Path Management with malfunction log, MIL control and interface for scan tool
DVAL Diagnosis validator
DSCHED Diagnostic function scheduler
MF Engine-management function
DF Diagnostic function

OBD – Individual diagnoses

Whereas EOBD and EPA OBD explicitly specify monitoring of only a few emission-control systems in detail, the specific requirements of CARB OBD II have continually been expanded. A further revision (OBD II Update) is due for model years 2004/2005. The list below details the CARB requirements for gasoline engines with effect from the beginning of 2003. The EOBD requirements are identified by (E).

- Catalytic converter (E), heated catalytic converter
- Combustion misses (E)
- Evaporative-emissions reduction system (fuel-tank leakage diagnosis)
- Secondary-air injection
- Fuel system
- Lambda (oxygen) sensors (E)
- Exhaust-gas recirculation
- Crankcase ventilation
- Engine cooling
- Cold-starting emission-control system
- Air conditioner (components)
- Variable valve timing
- Direct ozone-reduction system
- "Comprehensive components" (E)
- "Other emissions-related components" (E)

Some of the components and systems listed fall into the category of "comprehensive components" or "other emissions-related components" for EOBD and EPA.
They relate to:

- Other components or subsystems of the emission-control system or
- Emission-related components connected to a processor or
- Subsystems of the drivetrain whose failure or malfunction could result in the exhaust-gas emissions exceeding OBD limits

The sections that follow describe some of the diagnoses in more detail, while only the general objectives are outlined for other diagnoses.

Catalytic-converter fault diagnosis

The function of the three-way catalytic converter is to convert to safer substances the harmful emissions of carbon monoxide (CO), nitrogen oxides (NO_X) and hydrocarbons (HC) that are produced by the combustion process. Aging or damage (thermal, poisoning) can diminish the converter's efficiency. Therefore, the effectiveness of the catalytic converter has to be monitored.

One measure of catalytic-converter efficiency is its oxygen-storage capacity. To date it has been possible to demonstrate a correlation between oxygen-storage capacity and converter efficiency for all types of three-way catalytic-converter coating (wash coat with ceroxides as oxygen-storing component and noble metals as the actual catalysts).

Depending on the requirements regarding emission reduction, one or more main catalytic converters (usually underfloor catalytic converters) are used either alone or in combination with one or more primary catalytic converters close to the engine. Primary mixture control takes places using a lambda sensor upstream of the first catalytic converter in the exhaust-gas system. Present-day concepts incorporate secondary lambda sensors downstream of the primary catalytic converter and/or the main converter. They serve the purpose firstly, to fine-tune the primary lambda sensor(s) and secondly for the OBD functions. The basic principle of catalytic-converter fault diagnosis is to compare the sensor signals upstream and downstream of the converter under review.

Primary catalytic-converter fault diagnosis

In systems in which the secondary lambda sensor is located directly after the primary catalytic converter, the primary catalytic converter can be separately monitored. Fault diagnosis is based on the following principle: The setpoint value for lambda closed-loop control is modulated at a specific frequency and amplitude (Fig. 1). The variations in the oxygen content of the exhaust gas resulting from these fluctuations in the control settings are attenuated in the cat-

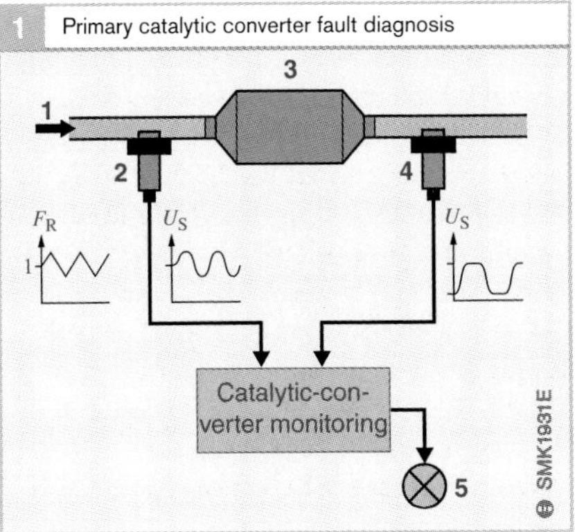

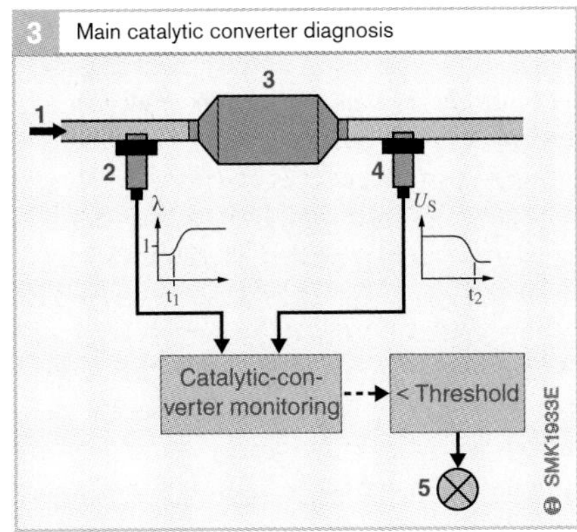

Fig. 1
1 Exhaust-gas mass flow from the engine
2 Broadband lambda sensor
3 Primary catalytic converter
4 Two-point lambda sensor
5 Diagnosis lamp

U_S Sensor voltage
F_R Lambda control factor (control setpoint)

Fig. 3
1 Exhaust-gas mass flow from the engine
2 Broadband lambda sensor
3 Catalytic converter (primary and main catalytic converters)
4 Two-point lambda sensor
5 Diagnosis lamp

U_S Sensor voltage

alytic converter by absorbing or releasing oxygen in the converter coating material, i.e. the rear sensor emits a signal with a very small amplitude (Fig. 2, top signal curve).

By contrast, a primary catalytic converter that has lost its oxygen-storage capacity as a result of aging or damage produces an oscillating two-point signal since almost no attenuation takes place (Fig. 2, bottom signal curve). Loss of oxygen-storage capacity can be calculated from the amplitude by filtering the signal in a special process. Conclusions can then be drawn as to converter efficiency.

Main catalytic converter diagnosis

The entire arrangement of primary catalytic converter and main converter is monitored (Fig. 3) on systems in which the secondary lambda sensor is positioned downstream of the main catalytic converter. However, the oxygen-storage capacity of the main catalytic converter is far greater than a smaller primary catalytic converter. This means that the regulation of the control setpoint is still significantly damped even if the converter is damaged. For this reason, a change in the oxygen concentration downstream of the main catalytic converter is too small for passive assessment, as in the method described above. This necessitates a fault-diagnosis process involving active intervention in the lambda closed-loop control system.

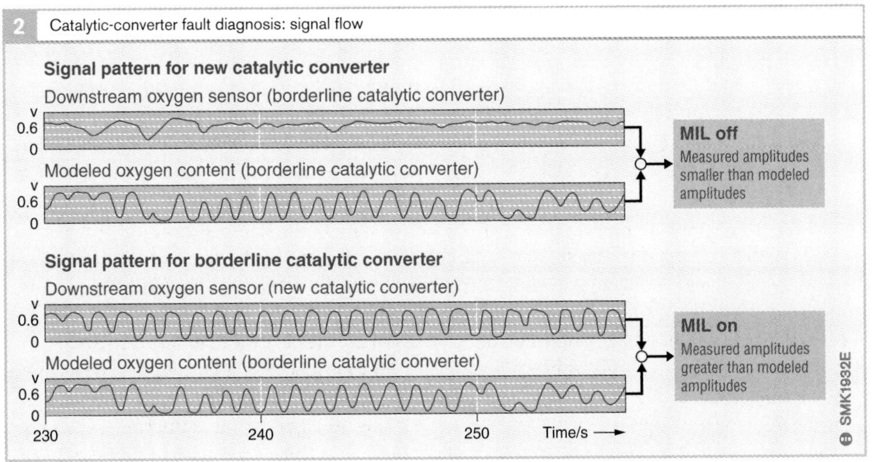

Main catalytic converter fault diagnosis is based on measuring oxygen storage directly on changeover from rich to lean mixture. A constant broadband lambda sensor is fitted upstream of the catalytic converter and measures the oxygen content in the exhaust gas. Downstream of the catalytic converter is a two-point lambda sensor which detects the condition of the oxygen accumulator. The reading is taken under static engine-operating conditions in the part-load range.

In the first stage of the process, the oxygen accumulator is completely emptied while the engine is running with a rich mixture ($\lambda < 1$). The signal from the rear lambda sensor indicates this by a voltage of 650 mV. In the next stage, while the engine is running with a lean mixture ($\lambda > 1$), the mass of oxygen absorbed to the point at which the accumulator reaches overflow is calculated with the aid of the mass-air flow and the primary lambda-sensor signal. The point of overflow is indicated by the signal voltage from the sensor downstream of the catalytic converter dropping to 200 mV. The calculated integral of the oxygen mass indicates the oxygen-storage capacity. That figure must exceed a reference figure, otherwise a fault is recorded.

Theoretically, the analyzis would also be possible by measuring the amount of oxygen released on changeover from lean to rich mixture. However, measuring the amount of oxygen absorbed on changeover from rich to lean mixture offers the following benefits:

- Less dependence on temperature and
- Less dependence on sulfurization

This method allows more accurate measurement of oxygen-storage capacity.

NO_X accumulator catalytic converter diagnosis

As well as functioning as a three-way catalytic converter, the NO_X accumulator catalytic converter required for gasoline direct-injection engines has the task of temporarily storing the nitrogen oxides that cannot be converted when the engine is running on a lean mixture ($\lambda > 1$). It then converts them at a later stage when the engine is running with a homogeneously distributed air-fuel mixture of ($\lambda < 1$). The NO_X storage capacity of this catalytic converter – indicated by the catalytic-converter quality factor – diminishes as a result of aging or contamination (e.g. sulfur absorption). Therefore, its functional capacity has to be monitored. This can be achieved with the aid of lambda sensors fitted upstream and downstream of the catalytic converter, or a NO_X sensor in place of the downstream lambda sensor.

In order to determine the catalytic-converter quality factor, the actual NO_X accumulator capacity is compared with the NO_X accumulator capacity of a new NO_X accumulator catalytic converter (new catalytic converter model, Fig. 4). The actual NO_X storage capacity is equal to the metered consumption of reduction agents (HC and CO) during catalytic-converter regeneration. The quantity of reduction agents is determined by integrating the reduction-agent mass flow during the regeneration phase when $\lambda < 1$. The end of the regeneration phase is indicated by an abrupt change in the signal voltage detected by the secondary oxygen sensor.

Alternatively, the actual NO_X accumulator capacity can be determined by means of an NO_X sensor.

4 Principle of determining catalytic-converter quality-factor

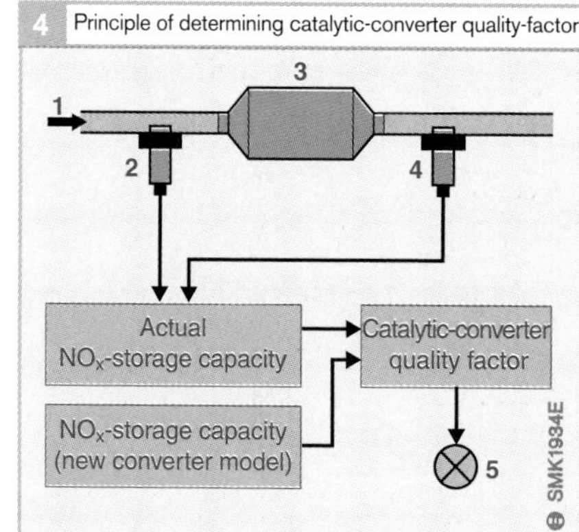

Fig. 4
1 Exhaust-gas mass flow from the engine
2 Broadband lambda sensor
3 NO_X accumulator catalytic converter
4 Two-point lambda sensor/NO_X sensor
5 Diagnosis lamp

Combustion miss detection

Current legislation demands detection of combustion misses that may be caused by worn-out spark plugs, for example. A miss occurs when the spark plug fails to produce an ignition spark. The result is that combustion of the air-fuel mixture does not take place and unburned fuel is emitted into the exhaust-gas system. Consequently, misses result in afterburning of unburned fuel in the catalytic converter and cause a temperature rise. This can lead to more rapid aging or even complete destruction of the catalytic converter. In addition, misses increase exhaust-gas emissions, particularly of HC and CO. Detection of misses is therefore a necessity.

The miss detection function measures the time elapsed between one combustion stroke and the next – the cycle time – for each cylinder. The time is calculated from the speed-sensor signal. The time it takes a certain number of teeth to pass the crankshaft sensor wheel is measured. If the engine has a combustion miss, it does not produce the amount of torque that would normally be expected. The result is that it rotates at a slower speed. A significant increase in the resulting cycle time is an indication of misfiring (Fig. 5).

At high engine speeds and low engine load, the increase in cycle time is only about 0.2 %. Consequently, precise monitoring of engine rotation and a complex calculation method are required in order to distinguish misfiring from other effects (e.g. judder caused by poor road surface).

Sensor-wheel compensation corrects for discrepancies arising from variations in manufacturing tolerances in the sensor wheel. This function is only active when the engine is overrunning since no acceleration torque is generated under these operating conditions. Sensor-wheel compensation supplies correction values for the cycle times.

If the combustion misfire rate exceeds a permissible level, fuel injection to the cylinder(s) concerned is deactivated in order to protect the catalytic converter.

5 Principle of combustion miss detection

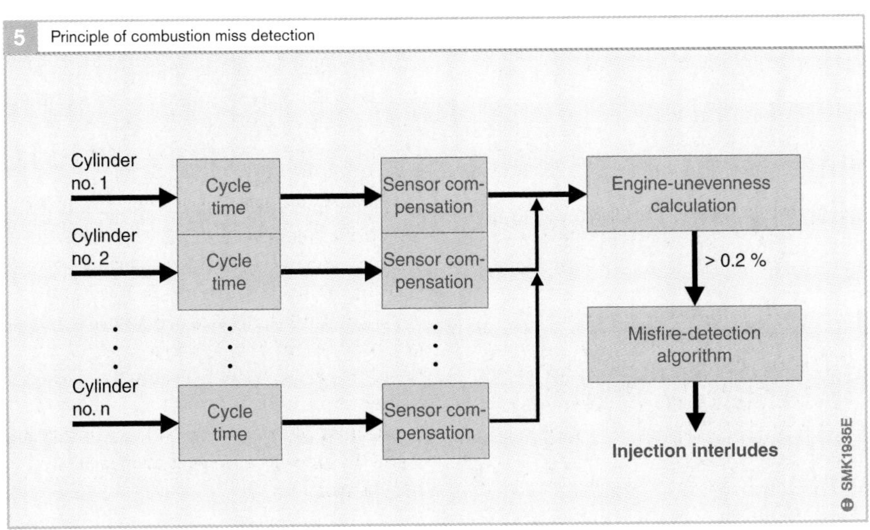

Fuel-tank leakage diagnosis

It is not only exhaust-gas emissions that are harmful to the environment. Evaporative emissions loss escaping from the fuel system – especially the fuel tank – are equally undesirable and consequently subject to emission limits as well. In order to limit evaporative emissions, evaporated fuel is collected by the activated charcoal canister in the evaporative-emissions control system. They are then released to the intake manifold via the canister-purge valve from where it joins the normal combustion process (Fig. 6). Monitoring the fuel-tank system is part of the on-board diagnosis functions.

Legislation for the European market is initially limited to straightforward checking of the electrical circuits for the fuel-tank pressure sensor and the canister-purge valve. In the U.S., on the other hand, detection of leaks from the fuel system is required. There are two different diagnosis methods for leakage detection. They can detect a major leak with a flow diameter of up to 1.0 mm and a minor leak with a flow diameter of up to 0.5 mm.

6 Fuel tank leakage detection using vacuum method

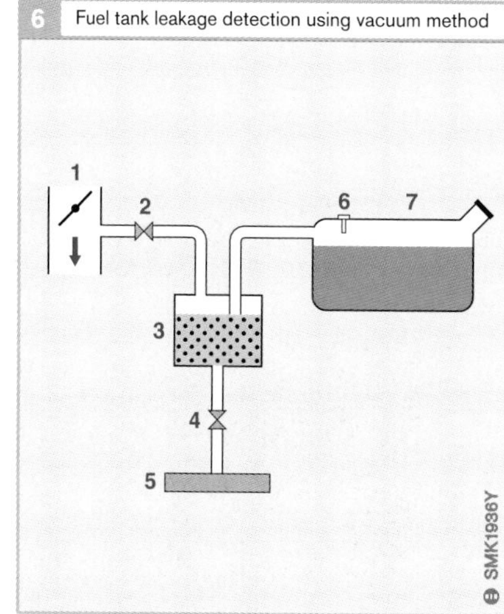

Fig. 6
1 Intake manifold with throttle valve
2 Canister-purge valve (regeneration valve)
3 Activated charcoal canister
4 Shutoff valve
5 Air filter
6 Fuel-tank pressure sensor
7 Fuel tank

Vacuum-relief diagnosis method

When the vehicle is stationary and the engine idling, the canister-purge valve (Fig. 6, Pos. 2) is closed. As the shutoff valve (4) is open, air is drawn into the tank by the vacuum inside the tank. The pressure in the tank system then increases. If the pressure measured by the pressure sensor (6) does not reach atmospheric pressure within a certain period of time, it is assumed that the shutoff valve has failed to open sufficiently or at all and is therefore faulty.

If a shutoff-valve fault is not detected, the valve is closed. Fuel evaporation may now be expected to cause a pressure increase. The resulting pressure may not exceed specific upper and lower limits.

If the pressure measured is below the specified lower limit, the canister-purge valve is faulty. In other words, the lack of pressure is caused by a leaking canister-purge valve which is allowing vapor to be drawn out of the fuel tank by the vacuum in the intake manifold.

If the pressure measured is above the specified upper limit, this indicates that too much fuel is evaporating (e.g. due to high ambient temperature) for fault diagnosis to be carried out.

If the pressure produced by fuel evaporation is within the specified limits, then the pressure rise is stored as the compensation gradient for minor-leak detection.

Only after testing the shutoff and canister-purge valves can fuel-tank leakage diagnosis be continued.

Major-leak detection

When the engine is idling, the canister-purge valve (2) is opened. As a result, the vacuum in the intake manifold (1) "spreads" to the fuel-tank system. If the pressure change detected by the fuel-tank pressure sensor (6) is too small because air is able to enter due to a leak to balance out the induced pressure drop, a major leak is detected and the fault-diagnosis sequence is terminated.

Minor-leak detection
The minor-leak diagnosis can start once the diagnosis system has failed to identify a major leak. This starts by closing the canister-purge valve (2) again. The pressure should then only increase by the fuel-evaporation rate (compensation gradient) previously stored, since the shutoff valve (4) is still closed. If the pressure increases at a higher rate, there must be a minor leak through which air is able to enter.

Overpressure method
If the diagnosis-activation conditions are satisfied and the ignition is switched off, the overpressure test is started as part of the ECU run-on sequence.

Major and minor-leak detection
For minor-leak detection, the electric vane pump (6) integrated in the diagnosis module (Fig. 7a, Pos. 4) pumps air through a "reference leak" (5) which has a diameter of 0.5 mm. As a result of the back pressure caused by the constriction at that point, the load on the pump increases, resulting in a drop in pump speed and an increase in electric current. That current level is measured and stored (4, Fig. 8).

Next, (Fig. 7b) the solenoid valve (7) is switched over and the pump pumps air into the fuel tank. If there are no leaks in the tank, the pressure increases and the pump current rises accordingly (Fig. 8) to a level (3) higher than the reference current. If there is a minor leak, the pump current will reach the reference current but not exceed it (2). If the pump current fails to reach the reference current after an extended period, a major leak is present (1).

8 Signal curves for overpressure method

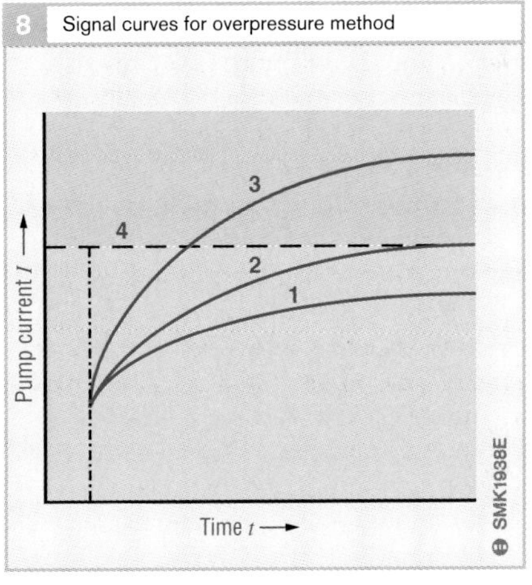

Fig. 8
1 Current curve for tank leak of d > 0.5 mm
2 Current curve for tank leak of d ≤ 0.5 mm
3 Current curve for tank without leaks
4 Reference current value

7 Fuel tank leakage detection using overpressure method

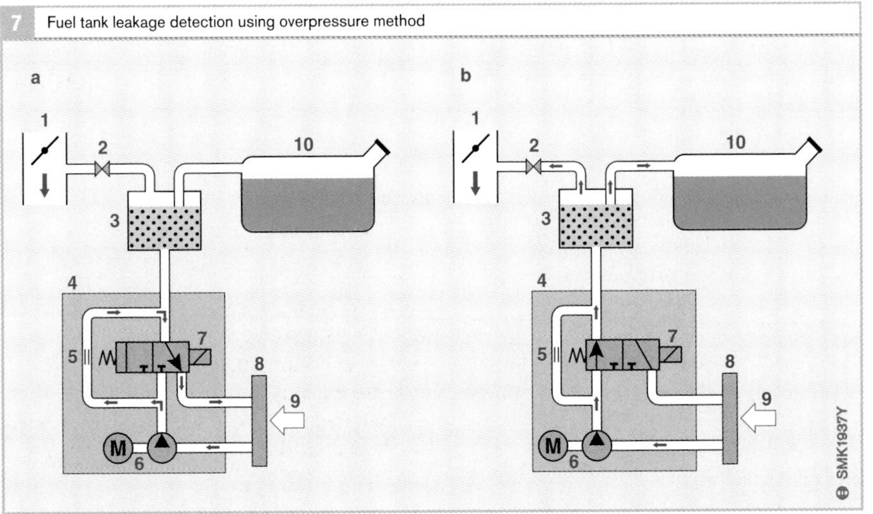

Fig. 7
a Measuring the reference-leak current
b Minor and major-leak test

1 Intake manifold with throttle valve
2 Canister-purge valve (regeneration valve)
3 Activated charcoal canister
4 Diagnosis module
5 Reference leak of d = 0.5 mm
6 Vane pump
7 Changeover valve
8 Air filter
9 Intake air
10 Fuel tank

Secondary-air injection fault diagnosis

Running the engine with a rich mixture ($\lambda < 1$) – which is necessary in cold weather, for example – leads to high concentrations of hydrocarbons and carbon monoxide in the exhaust gas. These emissions have to be re-oxidized in the exhaust-gas system, i.e. passed through an afterburning process. Therefore, there is a secondary-air injection facility which blows the oxygen required for catalytic afterburning into the exhaust gas directly downstream of the exhaust manifold (Fig. 9).

If this system fails, there will be an increase in exhaust-gas emissions when cold-starting the engine or before the catalytic converter has reached operating temperature. For this reason, a fault-diagnosis function is required.

Secondary-air injection fault diagnosis is a function check which establishes whether the pump is working properly and whether there are problems with the air pipe to the exhaust-gas system. The check can be performed in two ways.

The passive test takes place immediately after the engine is started and while the catalytic converter is being heated up. It involves the secondary-air system measuring the mass of secondary air injected. IT uses an active lambda closed-loop control system and compares the secondary-air mass with a reference figure. If the calculated air mass differs from the reference figure, a fault is detected.

The active test involves activating secondary-air injection for diagnostic purposes only during engine idle phase. The signals from the lambda sensors are used directly for calculating the secondary-air mass.
As with the passive test, the calculated secondary-air mass flow rate is compared with a reference value.

Although less accurate, the passive test is necessary because it is precisely during the period when the engine has only been running for a short time that secondary-air injection is active and its correct functioning must be ensured.

Fuel-system fault diagnosis

Faults in the fuel system (e.g. defective fuel valve, leaks in the intake manifold) can prevent the optimum mixture form being formed. For this reason, the OBD needs to monitor the system. The ECU therefore processes data such as the intake air mass (air-mass meter signal), the throttle-valve position, the air-fuel ratio (primary lambda-sensor signal) and engine operating data. This data is then compared with modeled calculations.

9 Principle of secondary-air injection

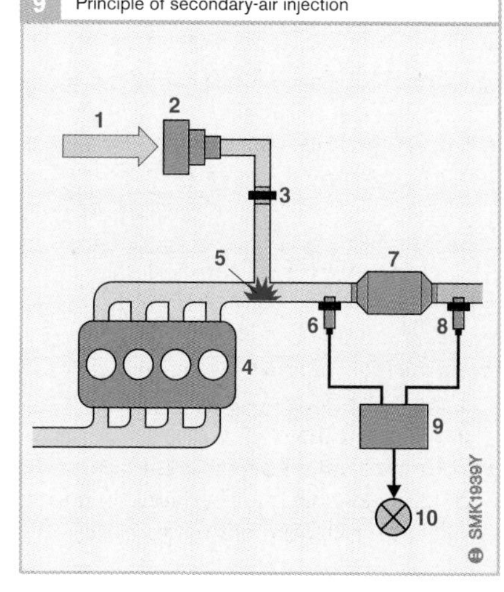

Fig. 9
1 Intake air
2 Secondary-air pump
3 Secondary-air valve
4 Engine
5 Injection point in exhaust pipe
6 Upstream lambda sensor
7 Catalytic converter
8 Lambda sensor downstream of catalytic converter
9 Engine ECU
10 Diagnosis lamp

Lambda-sensor fault diagnosis

The lambda-sensor system generally consists of two sensors (upstream and downstream of the catalytic converter) and the lambda closed-loop control circuit. Upstream of the catalytic converter, there is normally a broadband lambda sensor. It measures the λ level (i.e. the air-fuel mixture across the entire range from rich to lean) continuously by means of a variable voltage (Fig. 10a). It is also to controls this voltage. On older systems, a two-point lambda sensor was used upstream of the catalytic converter. The two-point sensor indicates only whether the mixture is lean ($\lambda > 1$) or rich ($\lambda < 1$) by an abrupt change in the voltage signal (Fig. 10b).

Present-day concepts incorporate a secondary lambda sensor – generally a two-point sensor – downstream of the primary catalytic converter and/or the main converter. It serves firstly to fine-tune the primary lambda sensor and secondly, for OBD functions. The lambda sensors not only measure the air-fuel mixture in the exhaust gas for the engine management system, they also monitor the function of the catalytic converter(s).

Possible lambda-sensor faults include:

- Breaks or short-circuits in the electrical circuit
- Aging (thermal, poisoning) of the sensor (leads to diminished sensor-signal dynamics)
- Corrupted signals caused by the sensor failing to reach operating temperature

Primary sensor

The sensor upstream of the catalytic converter is referred to as the primary or upstream sensor. It is checked for:

- Plausibility (internal resistance, output voltage – the actual signal – and other parameters) and
- Dynamics (rate of signal change at changeover from "rich" to "lean" and from "lean" to "rich", and period duration)

If the sensor has a heater, the heater function must also be checked. The tests are carried out while the vehicle is in motion at relatively constant operating conditions.

The broadband lambda sensor requires different diagnostic procedures from the two-point sensor since it can also record levels other than $\lambda = 1$.

Secondary sensor

The secondary or downstream sensor(s) is/are used to monitor the catalytic converter as well as for other functions. They check the conversion efficiency of the catalytic converter and supply the most important data for converter fault diagnosis. The secondary-sensor signals can also be used to check the data provided by the primary sensor. In addition, the secondary sensor can help to ensure the long-term stability of emission levels by correcting the primary-sensor signals.

Apart from the period duration, all of the characteristics and parameters listed for the primary sensor are also checked for the secondary sensor.

10 Lambda-sensor voltage curve

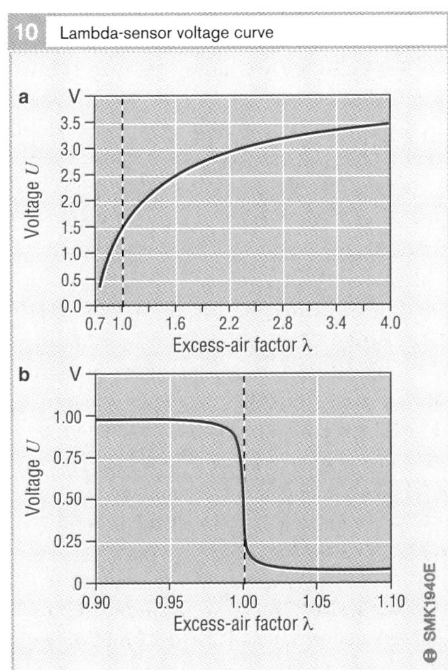

Fig. 10
a Broadband lambda sensor
b Two-point lambda sensor

Exhaust-gas recirculation system fault diagnosis

The exhaust-gas recirculation system is an effective means of reducing the emissions of nitrogen oxides. Adding recirculated exhaust gas to the air-fuel mixture entering the engine lowers the peak combustion temperature. In turn, this reduces the formation of nitrogen oxides. The functional efficiency of the exhaust-gas recirculation system therefore has to be monitored. There are two alternative methods that may be applied.

The method based on intake-manifold pressure involves closing the exhaust-gas recirculation valve briefly while the engine is operating at medium power, and measuring the pressure change. A comparison between the measured intake-manifold pressure and the pressure calculated using a model allows diagnosis of the closing function of the exhaust-gas recirculation valve.

The method based on uneven engine running is used in systems without an air-mass meter without an additional intake-manifold pressure sensor. The exhaust-gas recirculation valve is slightly opened while the engine is idling. If the exhaust-gas recirculation system is functioning correctly, the slight increase in residual-gas mass causes the engine to run slightly more unevenly. This is used by the engine-smoothness monitoring function to diagnose exhaust-gas recirculation valve problems.

Crankcase ventilation fault diagnosis

The so-called "blow-by gas" which enters the crankcase by escaping past the pistons, piston rings and cylinders has to be removed from the crankcase. This is the function of the positive crankcase ventilation (PCV) system. The exhaust-gas enriched air has the soot removed from it by a cyclone separator and is then passed through the PCV valve into the intake manifold so that hydrocarbons can be returned to the combustion cycle.

One possible fault diagnosis method is based on measuring the idle speed which, when the PCV valve is opened, should exhibit specific characteristics that are predicted by a model. If the observed change in the idle speed differs too greatly from the modeled response, a leak is assumed to exist.

Engine cooling system fault diagnosis

The engine cooling system consists of a large and a small circulation system which are connected by a thermostatic valve. The small circulation system is used during the starting phase to bring the engine quickly to operating temperature and is brought into action by closing the thermostatic valve. If the thermostat is defective or it has seized in the open position, the rate of coolant temperature rise is slowed down – particularly in cold weather conditions – and this results in higher emission levels. A thermostat monitoring function should therefore detect a slow rate of coolant temperature increase. The system's temperature sensor is first tested. Then the thermostatic valve is checked on this basis.

Monitoring catalytic-converter heating functions

In order to obtain a high conversion rate, the catalytic converter needs to operate at a temperature of 400...800 °C. Higher temperatures can damage the converter coating, however.

A catalytic converter operating at optimum temperature reduces engine emissions by more than 99 %. At low temperatures, its efficiency diminishes to the extent that a cold catalytic converter is almost totally ineffective. Therefore, in order to be able to meet emission-control requirements, a special strategy is required to bring the catalytic converter up to operating temperature as quickly as possible. The special converter-heating phase is terminated when the converter reaches a temperature of 200...250 °C (light-off temperature, approx. 50 % conversion efficiency). From this point on, the catalytic converter heats itself by exothermic conversion reactions.

When the engine is first started, two methods can be employed to heat up the catalytic converter more quickly:

- The ignition can be retarded to increase the exhaust-gas temperature
- The catalytic reactions produced in the catalytic converter by the incompletely burned fuel cause the converter to generate heat itself

The consequence of these effects is that the catalytic converter reaches its operating temperature more quickly so that exhaust-gas emissions are more rapidly reduced.

In order to ensure proper converter function, legislation requires monitoring of the temperature immediately upstream of the catalytic converter and monitoring of the converter warm-up phase. The warm-up phase can be monitored by checking and analyzing parameters such as ignition angle, engine speed and intake-air mass. In addition, other factors related to catalytic-converter warm-up are also specifically monitored during this phase (e.g. camshaft position/lock).

Air conditioner (components)

In order to cover the power requirements of the air conditioner, the engine is run in a different mode when given different requirements apply. If the optimized operating mode is not activated when the air conditioner is switched on (or if it is activated when the air conditioner is switched off), higher exhaust-gas emissions can result, which means that the function has to be monitored.

Variable valve timing (VVT) fault diagnosis

Variable valve timing is used under certain conditions to reduce fuel consumption and exhaust-gas emissions. Whereas previously VVT was indirectly subject to legislative requirements as a "comprehensive component", the OBD II Update introduces explicit demands. Present fault diagnosis involves measuring the camshaft position and performing a setpoint/actual comparison.

Direct ozone-reduction system

A particular feature of the Californian emission-control legislation is the possibility of reducing not only exhaust gases and evaporative-loss emissions, but also the atmospheric concentration of the air pollutant, ozone. This is the purpose of the catalytic coating applied to the vehicle radiator to create a direct ozone reduction (DOR) system. Ozone reduction is calculated on the basis of the surface area and the air throughput. A "credit" amount is then calculated. This can then be taken into account when calculating the total reduction of exhaust gases and evaporative-loss emissions (hydrocarbons only). The coated radiator is therefore an emissions-reducing component and will have to be monitored by the OBD system with effect from model year 2006 (legal requirement under OBD II).

As yet, no cost-efficient testing method has become widespread. The possible methods of diagnosis currently under discussion are listed below.

- *Pressure sensor:* Contamination reduces the amount of ozone passing through the radiator. A pressure sensor can detect the resulting drop in pressure.
- *Measurement of resistance:* The coating has a specific electrical resistance. Corrosion of the coating changes this resistance.
- *Photodetectors:* The catalytic coating is impermeable to light. Detectors can identify if there are gaps in the coating.
- *Ozone sensors:* They are used to measure the ozone concentration levels in front of and behind the radiator.

As an alternative, there is a possibility that CARB may accept a "presence test" for the DOR system, but only in return for half of the normal ozone credit.

Comprehensive components

In addition to the specific diagnoses described above, which are explicitly required by Californian legislation and described individually in separate sections, all sensors and actuators (e.g. throttle valve or high-pressure pump) must also be monitored as faults in these components could affect emission levels or prevent other diagnostic functions. The fault-diagnosis requirements for these components are described in this section – itemized separately for sensors and actuators.

Sensor fault diagnosis

Sensors have to be monitored for:

- Electrical faults
- Range errors and – where possible –
- Plausibility errors

Electrical faults

The legislation defines electrical faults as short-circuits to ground, power-supply short-circuits or circuit breaks.

Checking for range errors ("range checks")

Normally, sensors have a specified output characteristic, often with a lower and an upper limit, i.e. the physical detection range of the sensor is mapped to an output voltage in the range of 0.5...4.5 V, for example. If the output voltage produced by the sensor is outside this range, a range error is present. This means that the limits for this test (the "range check") are fixed limits specific to each sensor that are not dependent on the momentary operating status of the engine.

Where certain types of sensor do not permit a distinction between electrical faults and range errors, the legislation allows for this.

Plausibility errors ("rationality checks")

As a means of achieving greater fault-diagnosis sensitivity, the legislation demands that plausibility checks (so-called "rationality checks") are carried out in addition to the range checks. The characteristic feature of such plausibility checks is that the momentary sensor output voltage is not – as is the case with the range checks – compared with fixed limits but with narrower limits that are determined by the momentary operating status of the engine. That means that for this check current data from the engine management system must be referred to. These checks may be implemented as comparisons between the sensor output voltage and a model by cross-reference with another sensor. The model defines an expected range for the modeled variable for all engine operating conditions.

In order to make repairs as straightforward and effective as possible, the defective component has to be identified as accurately as possible. In addition, the categories of fault referred to should be distinguishable from one another and – in the case of range and plausibility errors – as to whether signal is outside the upper or lower limit. In the case of electrical faults, the problem can generally be assumed to be a wiring fault, whereas a plausibility error rather tends to indicate a fault in the component itself.

As testing for electrical faults and range errors must be continuous, checking for plausibility errors must take place at a specified minimum frequency during normal operation.

Among the sensors to be monitored in this way are the following:

- The air-mass meter
- Various pressure sensors (intake manifold, atmospheric pressure, fuel tank)
- The engine-speed sensor
- The phase sensor
- The intake-air temperature sensor
- The exhaust-gas temperature sensor

Example

The diagnostic process is described below using the air-mass meter as an example.

The air-mass meter, which is used to detect the amount of air drawn in by the engine to enable calculation of the amount of fuel to be injected, measures the air-mass rate and sends that information to the Motronic in the form of an output voltage signal. The air masses vary according to the throttle setting and/or engine speed.

The diagnostic function monitors whether the sensor output voltage is outside certain (specifiable, fixed) upper and lower limits, and if they output a range error.

By comparing the air mass indicated by the air-mass meter with the current position of the throttle valve – taking account of the engine operating status – the sensor signal can be taken to be implausible if the divergence between the two signals is greater that a certain tolerance. Example: The throttle valve is fully open but the air-mass meter indicates the contradiction of air mass equivalent to idle speed.

Actuator fault diagnosis

Actuators have to be monitored for electrical faults and – if technically possible – for correct function. Function monitoring in this case means that the execution of a given actuation command is monitored by observing whether the system responds in the appropriate manner. This means that – compared with sensor-signal plausibility checking – additional information has to be obtained from the system in order to assess component function.

Actuators include the following:

- All output stages
- The throttle valve
- The electronic throttle control system
- The canister-purge valve and
- The activated charcoal canister shutoff valve

However, the majority of these components are already taken into account in the system diagnoses.

Example

The throttle valve is responsible for controlling the amount of air that mixes with each injection of fuel. In the electronic throttle control system, it is electronically controlled. The throttle-valve angle (aperture) for adjusting the rate of air intake is controlled by a digital position controller. In order to diagnose throttle-valve faults, the controller is monitored for discrepancies between the specified and the actual throttle-valve angle. If the divergence is too great, a throttle-valve positioner fault is registered. The same fault is recorded if the throttle-valve positioner response is too small.

With the electronic throttle control system, there is no longer a mechanical link between the accelerator pedal and the throttle valve. Instead, the throttle setting desired by the driver and indicated by the position of the accelerator pedal is detected by two identical (for verification) potentiometers and processed by the engine ECU.

Data transfer between automotive electronic systems

Today's vehicles are being equipped with a constantly increasing number of electronic systems. Along with their need for extensive exchange of data and information in order to operate efficiently, the data quantities and speeds concerned are also increasing continuously.

For instance, in order to guarantee perfect driving stability, the Electronic Stability Program (ESP) must exchange data with the engine management system and the transmission-shift control.

System overview

Increasingly widespread application of electronic communications systems, and electronic open and closed-loop control systems, for automotive functions such as

- Electronic engine-management (EDC and Motronic)
- Electronic transmission-shift control (EGS)
- Antilock braking system (ABS)
- Traction control system (TCS)
- Electronic Stability Program (ESP)
- Adaptive Cruise Control (ACC) and
- Mobile multimedia systems together with their display instrumentation

has made it vital to interconnect the individual ECUs by means of networks.

The conventional point-to-point exchange of data through individual data lines has reached its practical limits (Fig. 1), and the complexity of current wiring harnesses and the sizes of the associated plugs are already very difficult to manage. The limited number of pins in the plug-in connectors has also slowed down ECU development work.

To underline this point:
Apart from being about 1 mile long, the wiring harness of an average medium-size vehicle already includes about 300 plugs and sockets with a total of 2,000 plug pins. The only solution to this predicament lies in the application of specific vehicle-compatible Bus systems. Here, CAN has established itself as the standard.

Serial data transfer (CAN)

Although CAN (Controller Area Network) is a linear bus system (Fig. 2) specifically designed for automotive applications, it has already been introduced in other sectors (for instance, in building installation engineering).

Data is relayed in serial form, that is, one after another on a common bus line. All CAN stations have access to this bus, and via a CAN interface in the ECUs they can receive and transmit data through the CAN bus line. Since a considerable amount of data can be exchanged and repeatedly accessed on a single bus line, this networking results in far fewer lines being needed.

1 Conventional data transfer

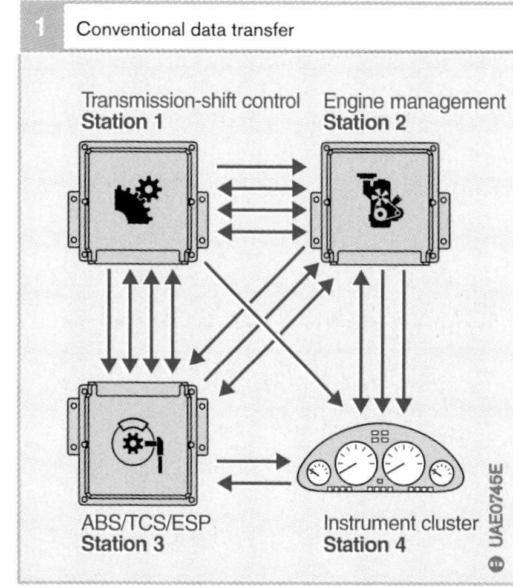

Applications in the vehicle

For CAN in the vehicle there are four areas of application each of which has different requirements. These are as follows:

Multiplex applications

Multiplex is suitable for use with applications controlling the open and closed-loop control of components in the sectors of body electronics, and comfort and convenience. These include climate control, central locking, and seat adjustment. Transfer rates are typically between 10 kbaud and 125 kbaud (1 kbaud = 1 kbit/s) (low-speed CAN).

Mobile communications applications

In the area of mobile communications, CAN networks such components as navigation system, telephone, and audio installations with the vehicle's central display and operating units. Networking here is aimed at standardizing operational sequences as far as possible, and at concentrating status information at one point so that driver distraction is reduced to a minimum. With this application, large quantities of data are transmitted, and data transfer rates are in the 125 kbaud range. It is impossible to directly transmit audio or video data here.

2 Linear bus topology

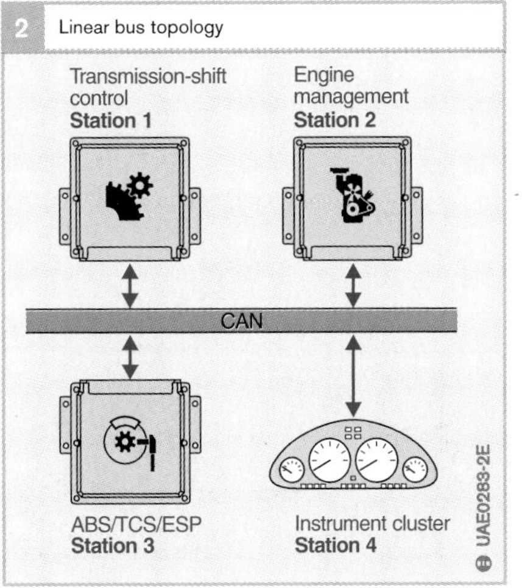

Diagnosis applications

The diagnosis applications using CAN are aimed at applying the already existing network for the diagnosis of the connected ECUs. The presently common form of diagnosis using the special K line (ISO 9141) then becomes invalid. Large quantities of data are also transferred in diagnostic applications, and data transfer rates of 250 kbaud and 500 kbaud are planned.

Real-time applications

Real-time applications serve for the open and closed-loop control of the vehicle's movements. Here, such electronic systems as engine management, transmission-shift control, and electronic stability program (ESP) are networked with each other. Commonly, data transfer rates of between 125 kbaud and 1 Mbaud (high-speed CAN) are needed to guarantee the required real-time response.

Bus configuration

Configuration is understood to be the layout and interaction between the components in a given system. The CAN bus has a linear bus topology (Fig. 2) which in comparison with other logical structures (ring bus and/or star bus) features a lower failure probability. If one of the stations fails, the bus still remains fully accessible to all the other stations. The stations connected to the bus can be either ECUs, display devices, sensors, or actuators. They operate using the Multi-Master principle, whereby the stations concerned all have equal priority regarding their access to the bus. It is not necessary to have a higher-order administration.

Content-based addressing

The CAN bus system does not address each station individually according to its features, but rather according to its message contents. It allocates each "message" a fixed *"identifier"* (message name) which identifies the contents of the message in question (for instance, engine speed). This identifier has a length of 11 bits (standard format) or 29 bits (extended format).

With content-based addressing each station must itself decide whether it is interested in the message or not ("message filtering" Fig. 3). This function can be performed by a special CAN module (Full-CAN), so that less load is placed on the ECU's central microcontroller. Basic CAN modules "read" all messages. Using content-based addressing, instead of allocating station addresses, makes the complete system highly flexible so that equipment variants are easier to install and operate. If one of the ECUs requires new information which is already on the bus, all it needs to do is call it up from the bus. Similarly, provided they are receivers, new stations can be connected (implemented) without it being necessary to modify the already existing stations.

Bus arbitration

The identifier not only indicates the data content, but also defines the message's priority rating. An identifier corresponding to a low binary number has high priority and vice versa. Message priorities are a function for instance of the speed at which their contents change, or their significance with respect to safety. There are never two (or more) messages of identical priority in the bus.

Each station can begin message transmission as soon as the bus is unoccupied. Conflict regarding bus access is avoided by applying bit-by-bit identifier arbitration (Fig. 4), whereby the message with the highest priority is granted first access without delay and without loss of data bits (nondestructive protocol).

The CAN protocol is based on the logical states "dominant" (logical 0) and "recessive" (logical 1). The "Wired And" arbitration principle permits the dominant bits transmitted by a given station to overwrite the recessive bits of the other stations. The station with the lowest identifier (that is, with the highest priority) is granted first access to the bus.

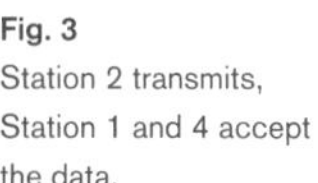

Fig. 3
Station 2 transmits, Station 1 and 4 accept the data.

Fig. 4
Station 2 gains first access
(Signal on the bus = signal from Station 2)

0 Dominant level
1 Recessive level

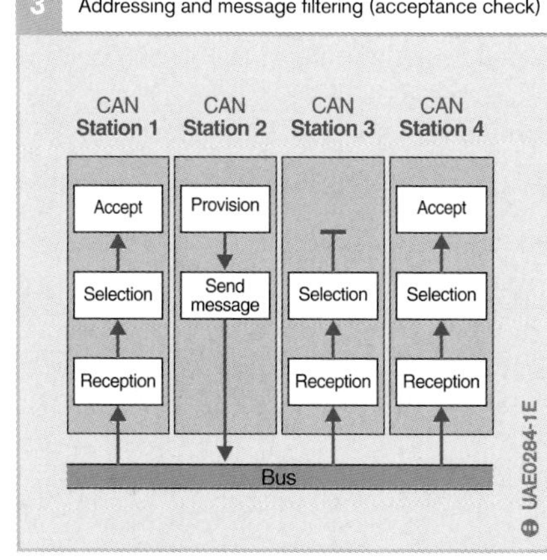

Addressing and message filtering (acceptance check)

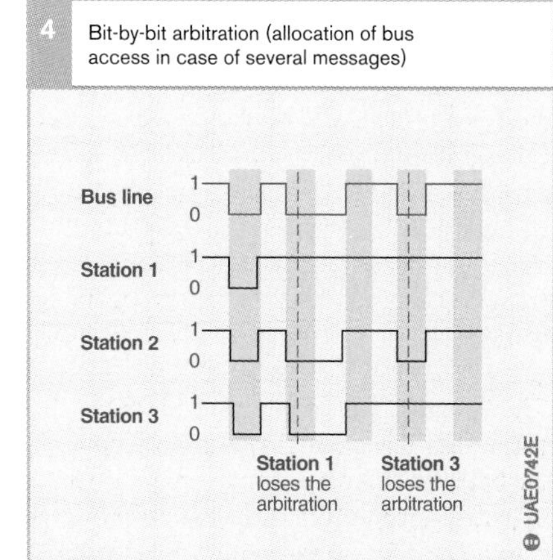

Bit-by-bit arbitration (allocation of bus access in case of several messages)

The transmitters with low-priority messages automatically become receivers, and repeat their transmission attempt as soon as the bus is vacant again.

In order that all messages have a chance of entering the bus, the bus speed must be appropriate to the number of stations participating in the bus. A cycle time is defined for those signals which fluctuate permanently (e.g. engine speed).

Message format

CAN permits two different formats which only differ with respect to the length of their identifiers. The standard-format identifier is 11 bits long, and the extended-format identifier 29 bits. Both formats are compatible with each other and can be used together in a network. The data frame comprises seven consecutive fields (Fig. 5) and is a maximum of 130 bits long (standard format) or 150 bits (extended format).

The bus is recessive at idle. With its dominant bit, the *"Start of frame"* indicates the beginning of a message and synchronises all stations.

The *"Arbitration field"* consists of the message's identifier (as described above) and an additional control bit. While this field is being transmitted, the transmitter accompanies the transmission of each bit with a check to ensure that it is still authorized to transmit or whether another station with a higher-priority message has accessed the Bus. The control bit following the identifier is designated the RTR-bit (**R**emote **T**ransmission **R**equest). It defines whether the message is a "Data frame" (message with data) for a receiver station, or a "Remote frame" (request for data) from a transmitter station.

The *"Control field"* contains the IDE bit (**Id**entifier **E**xtension **B**it) used to differentiate between standard format (IDE = 0) and extended format (IDE = 1), followed by a bit reserved for future extensions. The remaining 4 bits in this field define the number of data bytes in the next data field. This enables the receiver to determine whether all data has been received.

The *"Data field"* contains the actual message information comprised of between 0 and 8 bytes. A message with data length = 0 is used to synchronize distributed processes. A number of signals can be transmitted in a single message (e.g. engine rpm and engine temperature).

The *"CRC Field"* (**C**yclic **R**edundancy **C**heck) contains the frame check word for detecting possible transmission interference.

The *"ACK Field"* contains the **ack**nowledgement signals used by the receiver stations to confirm receipt of the message in non-corrupted form. This field comprises the ACK slot and the recessive ACK delimiter. The ACK slot is also transmitted recessively and overwritten "dominantly" by the receivers upon the message being correctly received. Here, it is irrelevant whether the message is of significance or not for the particular receiver in the sense of the message filtering or acceptance check. Only correct reception is confirmed.

5 CAN message format

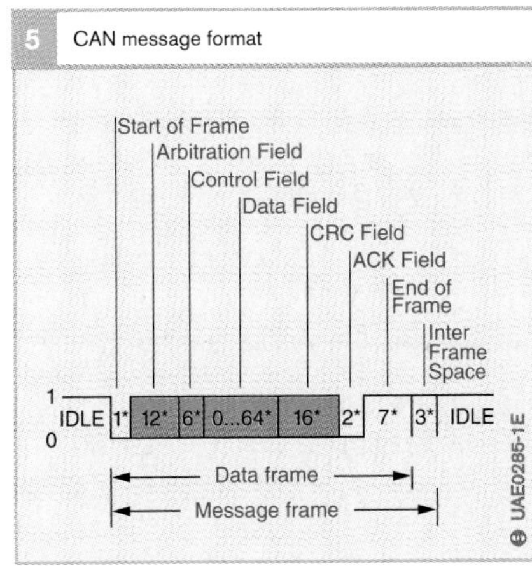

Fig. 5
0 Dominant level
1 Recessive level
* Number of bits

The *"End of frame"* marks the end of the message and comprises 7 recessive bits.

The *"Inter-frame space"* comprises three bits which serve to separate successive messages. This means that the bus remains in the recessive IDLE mode until a station starts a bus access.

As a rule, a sending station initiates data transmission by sending a "data frame". It is also possible for a receiving station to call in data from a sending station by transmitting a "remote frame".

Detecting errors

A number of control mechanisms for detecting errors are integrated in the CAN protocol.

In the *"CRC field"*, the receiving station compares the received CRC sequence with the sequence calculated from the message.

With the *"Frame check"*, frame errors are recognized by checking the frame structure. The CAN protocol contains a number of fixed-format bit fields which are checked by all stations.

The *"ACK check"* is the receiving stations' confirmation that a message frame has been received. Its absence signifies for instance that a transmission error has been detected.

"Monitoring" indicates that the sender observes (monitors) the bus level and compares the differences between the bit that has been sent and the bit that has been checked.

Compliance with *"Bitstuffing"* is checked by means of the "Code check". The stuffing rule stipulates that in every *"data frame"* or *"remote frame"*, a maximum of 5 successive equal-priority bits may be sent between the *"Start of frame"* and the end of the *"CRC field"*. As soon as five identical bits have been transmitted in succession, the sender inserts an opposite-priority bit. The receiving station erases these opposite-polarity bits after receiving the message. Line errors can be detected using the "bitstuffing" principle.

If one of the stations detects an error, it interrupts the actual transmission by sending an "Error frame" comprising six successive dominant bits. Its effect is based on the intended violation of the stuffing rule, and the object is to prevent other stations accepting the faulty message.

Defective stations could have a derogatory effect upon the bus system by sending an "error frame" and interrupting faultless messages. To prevent this, CAN is provided with a function which differentiates between sporadic errors and those which are permanent, and which is capable of identifying the faulty station. This takes place using statistical evaluation of the error situations.

Standardization

The International Organization for Standardization (ISO) and SAE (Society of Automotive Engineers) have issued CAN standards for data exchange in automotive applications:

- For low-speed applications up to 125 kbit/s: ISO 11519-2
- For high-speed applications above 125 kbit/s: ISO 11898 and SAE J 22584 (passenger cars) and SAE J 1939 (trucks and buses)
- Furthermore, an ISO Standard on CAN Diagnosis (ISO 15765 – Draft) is being prepared

Prospects

Along with the increasing levels of system-component performance and the rise in function integration, the demands made on the vehicle's communication system are also on the increase. And new systems are continually being introduced, for instance in the consumer-electronics sector. All in all, it is to be expected that a number of bus systems will establish themselves in the vehicle, each of which will be characterized by its own particular area of application.

In addition to electronic data transmission, optical transmission systems will also come into use in the multimedia area. These are very-high-speed bus systems and can transmit large quantities of data as needed for audio and video components.

Individual functions will be combined by networking to form a system alliance covering the complete vehicle, in which information can be exchanged via data buses. The implementation of such overlapping functions necessitates binding agreements covering interfaces and functional contents. The CARTRONIC® from Bosch is the answer to these stipulations, and has been developed as a priority-override and definition concept for all the vehicle's closed and open-loop control systems. The possible sub-division of the functions which are each controlled by a central coordinator can be seen in Fig. 1. The functions can be incorporated in various ECUs.

The combination of components and systems can result in completely novel functions. For instance, the exchange of data between the transmission-shift control and the navigation equipment can ensure that a change down is made in good time before a gradient is reached. With the help of the navigation facility, the headlamps will be able to adapt their beam of light to make it optimal for varying driving situations and for the route taken by the road (for instance at road intersections). Car radios, sound-carrier drives, TV, telephone, E-mail, Internet, as well as the navigation and terminal equipment for traffic telematics will be networked to form a multimedia system.

1 CARTRONIC®: Design schematic

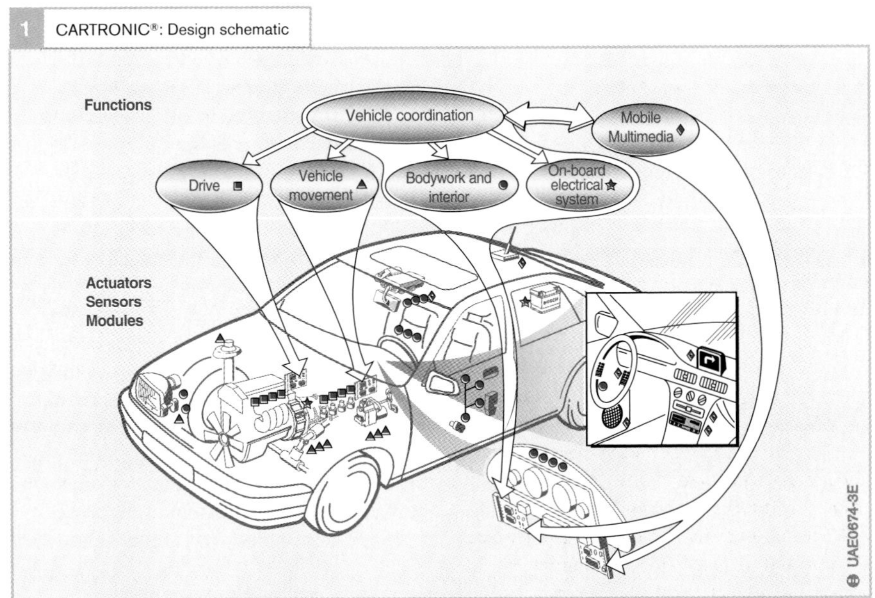

Exhaust emissions

Increasing energy consumption, and especially employment of the energy contained in fossil fuels, have transformed air quality into a critical concern. The quality of the air we breathe depends upon a wide range of factors. In addition to emissions from industry, homes and power plants, the exhaust generated by motor vehicles also plays a significant role (Figure 1).

Overview

The officially mandated limits restricting pollutant emissions from motor vehicles have been progressively tightened in recent years. In order to achieve compliance with these limits, vehicles have been equipped with supplementary emissions-control systems.

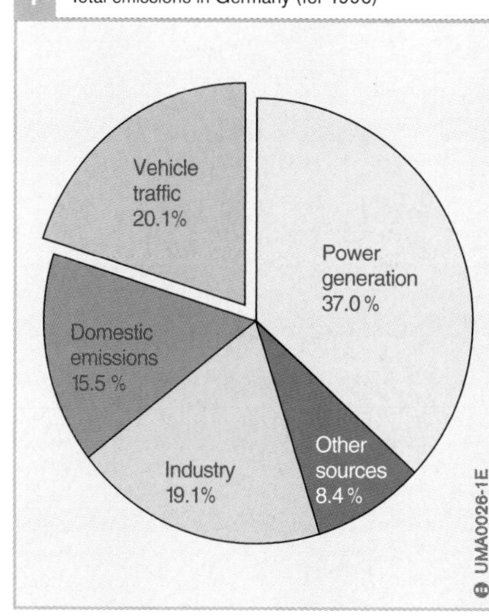

Fig. 1
Data in percent by weight, without emissions from natural sources

Total emissions: 935 Mt (megatons)

Source: Emissions protection report issued by German government in 1997

Combustion of the air/fuel mixture

A basic rule that applies to all internal-combustion engines is that absolutely complete combustion does not occur inside the engine's cylinders. This rule remains valid even when the combustion mixture contains excess air. Less efficient combustion leads to an increase in levels of toxic components within the exhaust gas. In addition to a high percentage of non-toxic elements, the internal-combustion engine's exhaust also contains secondary products which – at least when present in high concentrations – represent potential sources of environmental damage. These are classified as pollutants.

Positive crankcase ventilation

Additional emissions stem from the engine's crankcase ventilation system. Combustion gases travel along the cylinder walls and into the crankcase, whence they are returned to the intake manifold for renewed combustion within the engine.

Because nothing more than pure air is compressed in the diesel's compression stroke, diesels generate only regligible amounts of these bypass emissions. The gases that make their way into the crankcase during the power (combustion) stroke contain only about 10% as much pollution as the bypass gases in a gasoline engine. Despite this fact, closed crankcase-ventilation systems are now also mandatory on diesel engines.

Evaporative emissions

Additional emissions can escape from vehicles powered by gasoline engines when volatile components in the fuel evaporate and emerge from the fuel tank, regardless of whether the vehicle is moving or parked. These emissions consist primarily of hydrocarbons. To prevent these gases from evaporating directly into the atmosphere, vehicles must be equipped with an evaporative-emissions control system designed to store them for subsequent combustion in the engine.

Evaporative emissions from diesels are not a major concern, as diesel fuel possesses virtually no highly volatile components.

Major components

Assuming the presence of adequate oxygen, ideal, complete combustion of pure fuel could be portrayed in the following chemical reaction:

$$n_1\, C_xH_y + m_1\, O_2 \rightarrow n_2\, H_2O + m_2\, CO_2$$

The absence of ideal conditions for combustion combines with the composition of the fuel itself to produce a certain number of toxic components in addition to the primary combustion products water (H_2O) and carbon dioxide (CO_2) (Figure 2).

Water (H_2O)

During combustion, the water chemically bound within the fuel is transformed into water vapor, most of which subsequently condenses when its cools. This is the source of the exhaust plume visible on cold days. Water makes up about 13.1 % of the exhaust gas.

Carbon dioxide (CO_2)

In complete combustion, the hydrocarbons in the fuel's chemical bonds are transformed into carbon dioxide (CO_2), which makes up approximately 13.7 % of the exhaust gas. The amount of converted carbon dioxide in the exhaust is a direct index of fuel consumption. Thus the only way to reduce carbon-dioxide emissions is to reduce fuel consumption.

Carbon dioxide is a natural component of atmospheric air, and the CO_2 contained in automotive exhaust is not classified as a pollutant. However, it is one of the substances responsible for the greenhouse effect and the global climate change that this causes. In the period since 1920, atmospheric CO_2 has risen continually, from roughly 300 ppm to over 360 ppm in the year 1995. This renders efforts to reduce carbon-dioxide emissions and fuel consumption more important than ever.

Nitrogen (N_2)

Nitrogen is the primary constituent (78 %) of the air drawn in by the engine. Although it is not directly involved in the combustion process, it is the largest single component within the exhaust gas, at approximately 71.5 %.

2 Composition of exhaust gas from gasoline-engine during operation at $\lambda = 1$

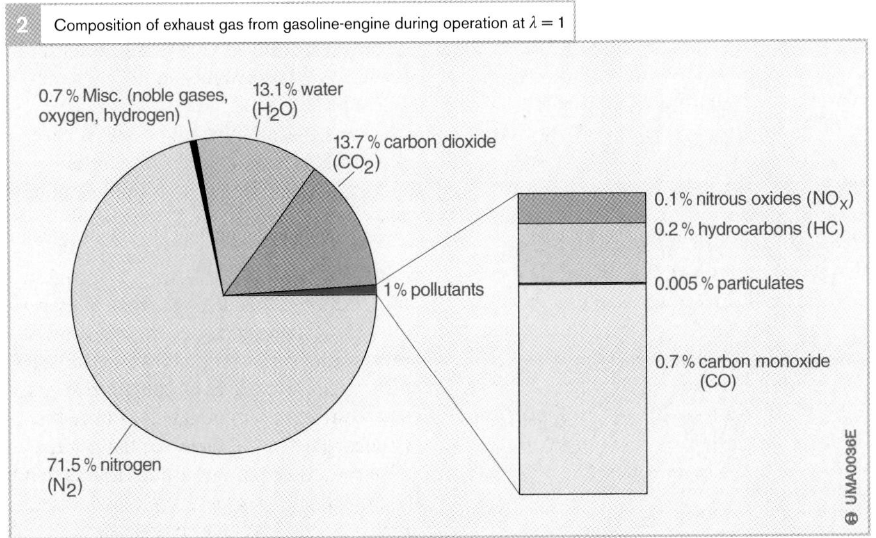

Fig. 2
Data in percent by volume

Actual concentrations of exhaust-gas components can vary in response to engine operating conditions, environmental factors (atmospheric humidity, etc.) and other parameters

Combustion by-products

During combustion, the air/fuel mixture generates a number of by-products. With the engine warmed to its normal operating temperature and running on a stoichiometric A/F ratio (λ=1), the proportion of these by-products in the engine's untreated emissions (exhaust gas after combustion, but before treatment) is about 1% of the total exhaust-gas quantity. The most significant of these combustion by-products are

- Carbon monoxide (CO)
- Hydrocarbons (HC) and
- Oxides of nitrogen (NO_X)

State-of-the-art catalytic converters are capable of converting more than 99 % of these gases once the engine has warmed to normal operating temperature.

Carbon monoxide (CO)

Carbon monoxide results from incomplete combustion in rich air/fuel mixtures under conditions characterized by an air deficiency.

Although carbon monoxide is also produced during operation with excess air, the concentrations are minimal, and stem from brief periods of rich operation or inconsistencies within the air/fuel mixture. Fuel droplets that fail to vaporize form pockets of rich mixture that do not combust completely.

Carbon monoxide is an odourless and tasteless gas. In humans it inhibits the ability of the blood to absorb oxygen, thus leading to asphyxiation.

Hydrocarbons (HC)

Hydrocarbons, or HC, is a generic designation for the entire range of chemical compounds uniting hydrogen H with carbons C. HC emissions are the result of inadequate oxygen being present to support complete combustion of the air/fuel mixture. The combustion process also produces new hydrocarbon compounds not initially present in the original fuel (by separating extended molecular chains, etc.).

Aliphatic hydrocarbons (alkanes, alkenes, alkines and their cyclical derivatives) are virtually odourless. Cyclic aromatic hydrocarbons (such as benzol, toluol and polycyclic hydrocarbons) emit a discernible odour.

Some hydrocarbons are considered to be carcinogenic in long-term exposure. Partially oxidized hydrocarbons (aldehydes, ketones, etc.) emit an unpleasant odour. The chemical products that result when these substances are exposed to sunlight are also considered to act as carcinogens under extended exposure to specified concentrations.

Nitrous oxides (NO_X)

Nitrous oxides, or oxides of nitrogen, is the generic term embracing chemical compounds consisting of nitrogen and oxygen. They result from secondary reactions that occur in all combustion processes where air containing nitrogen is burned. The primary forms encountered in the exhaust gases from internal-combustion engines are nitrogen oxide (NO) and nitrogen dioxide (NO_2), with dinitrogen monoxide (N_2O) also present in minute concentrations.

Nitrogen oxide (NO) is colourless and odourless. In atmospheric air it is gradually converted into nitrogen dioxide (NO_2). Pure NO_2 is a poisonous, reddish-brown gas with a penetrating odour. NO_2 can induce irritation of the mucous membranes when present in the concentrations found in highly-polluted air.

Nitrous oxides contribute to forest damage (acid rain) and also act in combination with hydrocarbons to generate photochemical smog.

Sulphur dioxide (SO_2)

Sulphurous compounds in exhaust gases – primarily sulphur dioxide – are produced by the sulphates contained in fuels. A relatively small proportion of these pollutant emissions stem from motor vehicles. These emissions are not restricted by official emissions limits.

It is not possible to use a catalyst to convert sulphur dioxide. Sulphur forms deposits within catalytic converters, reacting with the active chemical layer and inhibiting the catalyst's ability to remove other pollutants from the exhaust gases. While sulphur contamination can be reversed in the NO_X storage catalysts employed for emissions control with direct-injection gasoline engines, this process requires a considerable amount of energy, with consequent negative effects on the fuel-economy benefits afforded by direct injection.

The earlier limits on sulphur concentrations within fuel of 500 ppm (parts per million, 1,000 ppm = 0.1 %), valid until the end of 1999, have now been tightened by EU legislation. The new limits, valid from 2000 onward, are 150 ppm for gasoline and 350 ppm for diesel fuels. A further reduction to 50 ppm for both types of fuel is slated for 2005. Some countries will be reducing the limits prior to 2005.

Particulates

The problem of particulate emissions is primarily associated with diesel engines. Levels of particulate emissions from gasoline engines are regligible.

Particulates result from incomplete combustion. While exhaust composition varies as a function of combustion process and engine operating condition, these particulates basically consist of hydrocarbon chains (soot) with an extremely extended specific surface ratio. Uncombusted and partially combusted hydrocarbons form deposits on the soot, where they are joined by aldehydes, with their penetrating odour. Aerosol components (minutely dispersed solids or fluids in gases) and sulphates bond to the soot. The sulphates result from the sulphur content of the fuel.

Odour

The problem of odours from diesels has yet to be solved. The combinations of processes within the diesel engine that produce the distinctive olfactory sensation in its exhaust emissions are not fully understood. There is no standard test procedure for general use.

Ozone and smog

Exposure to the sun's radiation splits nitrogen-dioxide molecules (NO_2). The products are nitrogen oxide (NO) and atomic oxygen (O), which combines with the ambient air's atomic oxygen (O_2) to form ozone (O_3). Ozone formation is also promoted by volatile organic compounds. This is why higher ozone levels must be anticipated on hot, windless summer days when high levels of air pollution are present.

In normal concentrations ozone is essential for human life. However, in higher concentrations it leads to coughing, irritation to the throat and sinuses, and burning eyes. It adversely affects lung function, reducing performance potential.

There is no direct contact or mutual movement between the ozone formed in this way at ground level, and the stratospheric ozone that reduces the amount of ultraviolet radiation penetrating the earth's atmosphere.

Smog is not limited to the summer. It can also occur in winter in response to atmospheric layer inversions and low wind speeds. The temperature inversion in the air layers prevents the heavier, colder air containing the higher pollutant concentrations from rising and dispersing.

Smog leads to irritation of the mucous membranes, eyes and respiratory system. It can also impair visibility. This last factor explains the origin of the term smog, which combines "smoke" and "fog".

Factors affecting raw emissions

The primary by-products of combustion in the air/fuel mixture are the pollutants NO_X, CO and HC. The quantities of these pollutants present in raw exhaust gases (post-combustion gases prior to exhaust treatment) display major variations in response to different kinds of engine operation. Emissions-control systems featuring catalytic converters substantially reduce concentrations of these toxic substances in exhaust gases.

To achieve optimal transformation within the catalytic converter it is important to ensure that raw emissions be as low as possible.

Parameters

Engine speed

Higher engine speeds lead to greater friction losses within the engine as well as increased demand for energy to power ancillaries such as the water pump. Under these conditions the power output per consumed unit of energy decreases. The engine's operating efficiency falls as engine speed rises.

Generating a given level of power at high rpm equates with a higher level of fuel consumption than producing the same output at a lower engine speed. This leads to even higher levels of toxic emissions.

Mixture formation

In the interests of optimal combustion efficiency the fuel destined for combustion should be thoroughly dispersed to form the most homogeneous mixture possible with the air. In manifold-injection engines, this fuel is distributed throughout the entire combustion chamber. During operation in the stratified-charge mode, direct-injection engines employ a contrasting concept, concentrating fuel within the centre of the combustion chamber. A homogeneous mixture may be present within the concentrated mixture cloud.

Consistent distribution of uniform mixture to all cylinders is important for low emissions. Fuel-injection systems that employ their intake manifolds exclusively to transport air ensure consistent mixture distribution by discharging fuel into the intake port directly in front of the intake valve (manifold injection) or into the combustion chamber (direct gasoline injection). This type of consistency is less certain with systems relying on carburetors and central, throttle-body injection, as fuel tends to condense on the walls of the individual intake runners.

Engine load factor

The engine's load factor, as reflected in torque generation, is of major significance as a determinant of carbon monoxide (CO), unburned hydrocarbon (HC) and nitrous oxide (NO_X) levels. The various influences are described in more detail below.

Excess-air factor

Another primary factor defining the engine's toxic emissions is the air/fuel ratio (excess-air factor λ). To obtain maximum emissions reductions from the catalytic converter(s), manifold-injection engines run on a stoichiometric air/fuel mixture ($\lambda = 1$) under most operating conditions.

Direct-injection engines operate with both homogenous and stratified-charge mixtures, with selection varying according to operating conditions. In the homogeneous mode, the system injects fuel during the intake stroke to produce conditions comparable to those encountered with manifold injection. The system reverts to this mode of operation in response to demand for high torque and at high engine speeds. Under these conditions the system usually dials in an excess-air factor equal to or in the immediate vicinity of $\lambda = 1$.

The fuel is not distributed evenly throughout the entire combustion chamber during stratified-charge operation. The desired effect is achieved by waiting until the compression stroke to inject the fuel. The mixture cloud formed at the center of the combustion chamber should be as homogeneous as possible, with an excess-air factor of $\lambda = 1$. Only unadulterated air or an extremely lean mixture is present in the extremities of the combustion chamber. This results in a composite excess-air factor of $\lambda > 1$ (lean) for the overall charge within the chamber.

Ignition timing

The ignition of the air/fuel mixture that occurs within the time frame between the initial spark and the formation of a stable flame front is of decisive significance for the combustion sequence. The character of the ignition process is shaped by the ignition timing, which defines the point at which the spark is transferred to the mixture, the ignition energy, and the composition of the mixture immediately adjacent to the spark plug. Large quantities of surplus energy translate into stable ignition with positive effects, both for the consistency of the consecutive combustion cycles and the composition of the exhaust gases.

Aside from the excess-air factor λ it is the ignition timing that exercises the most pronounced effect on exhaust emissions.

1 Influence of excess-air factor λ and ignition timing α_Z on raw HC emissions

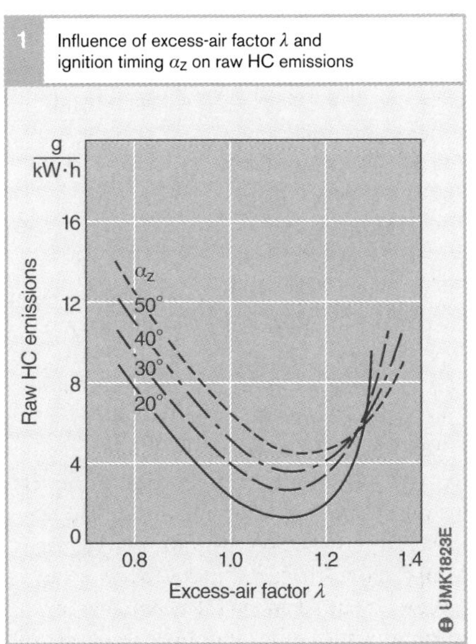

Raw HC emissions

The influence of torque

Temperatures within the combustion chamber rise as torque generation increases. As a result, the depth of the zone next to the cylinder walls in which the flame is extinguished shrinks as torque rises. This reduces the extent of the low-temperature zone where unburned hydrocarbons could be produced.

The high exhaust-gas temperatures that accompany higher combustion-chamber temperatures under high-torque operation promote secondary reactions in the unburned hydrocarbons during the ignition and exhaust strokes. Because high-torque operation equates with higher temperatures in combustion chambers and exhaust gases, it leads to reductions in quantities of unburned hydrocarbons relative to units of power generated.

The effects of engine speed

By reducing the time available for forming and then combusting the mixture, higher engine speeds lead to higher gasoline-engine HC emissions.

Effects of excess-air factor

During operation with excess fuel (air deficiency), incomplete combustion leads to generation of unburned hydrocarbons. Richer mixtures produce progressively greater HC concentrations (Figure 1). This is why richer mixtures (with progressively lower excess-air factor λ) are characterized by increased HC emissions throughout the rich-mixture range ($\lambda < 1$).

HC emissions also increase in the lean range ($\lambda > 1$). Minimum HC generation coincides with the range $\lambda = 1.1...1.2$. The rise within the lean range is caused by incomplete combustion at the extremities of the combustion chamber. Extremely lean mixtures, where combustion lag can ultimately lead to miss, aggravate this effect and produce a dramatic rise in HC emissions. This phenomenon results from the inadequate conditions for propagation of a secure flame front that accompany inconsistent mixture distribution within the combustion chamber. The gasoline-engine's lean-burn limit is primarily defined by the excess-air factor of the charge immediately adjacent to the spark plug at the instant of ignition, and by the composite excess-air factor. The flow pattern of the charge in the combustion chamber can be manipulated to obtain a more homogeneous mixture to ensure more reliable ignition, while at the same time accelerating propagation of the flame front.

The stratified-charge method used in conjunction with direct gasoline injection presents a contrasting picture. Instead of focusing on obtaining a homogeneous air/fuel mixture throughout the combustion chamber, this concept creates a highly ignitable mixture only in the area immediately adjacent to the tip of the spark plug. This concept thus allows substantially higher composite excess-air factors than would be available using a homogenous mixture. HC emissions in stratified-charge operation are essentially determined by the mixture formation process. It is vital to avoid depositing liquid fuel on the combustion-chamber walls and the piston crown, as the resulting surface film usually fails to combust completely, leading to high HC emissions.

Effects of ignition timing

Increasing the ignition advance (greater than α_Z) produces a rise in emissions of unburned hydrocarbons, as the resulting reduction in exhaust-gas temperatures has a negative effect on secondary reactions that occur during the ignition and exhaust phases (Figure 1). It is only during operation with extremely lean mixtures that this response pattern is inverted. These types of lean mixtures result in such a low flame-front propagation rate that the combustion process will still be in progress when the exhaust valve opens if ignition is late. With late ignition, the engine reaches its lean-burn limit early, at an excess-air factor of λ.

Raw CO emissions

The influence of torque

As with raw HC emissions, the high process temperatures that accompany high torque foster secondary reactions in CO during the ignition stroke. The CO oxidizes to form CO_2.

The effects of engine speed

CO emissions also mirror the pattern of HC emissions in their response to variations in engine speed.

Effects of excess-air factor

Within the rich range, CO emissions display a virtually linear correlation with the excess-air factor (Figure 2). This is the result of the incomplete carbon oxidation during operation with an air deficiency. In the lean range (air surplus) CO emissions remain at extremely low levels, and the influence of changes in the excess-air factor is minimal. Under these conditions the only source of CO generation is inefficient combustion stemming from inconsistencies in the air/fuel mixture.

Effects of ignition timing

Ignition timing has virtually no effects on CO emissions (Figure 2), which are almost entirely a function of the excess-air factor λ.

2 Influence of excess-air factor λ and ignition timing α_Z on raw CO emissions

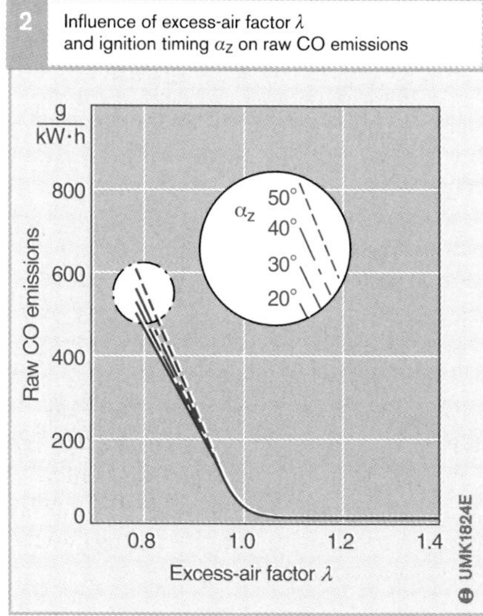

Raw NO_X emissions

The influence of torque

The higher combustion-chamber temperatures that accompany increased torque generation promote the formation of NO_X. As torque output rises, raw NO_X emissions display a disproportionate increase.

The effects of engine speed

Because there is less reaction time available for the formation of NO_X at high engine speeds, NO_X emissions fall along with increasing rpm. In addition, the residual gases in the combustion chamber must also be taken into account since these lead to lower peak temperatures. Because levels of residual gases tend to fall off as engine speed rises, this effect counteracts the response pattern described above.

Effects of excess-air factor

The way in which the excess-air factor affects NO_X production is entirely different from its influence on HC and CO emissions. Within the rich range ($\lambda < 1$) NO_X emissions respond to increases in the excess-air factor by rising (Figure 3). This phenomenon is caused by the progressively higher oxygen concentrations in the exhaust gas, which inhibit reduction of the nitrous oxides.

Within the lean range ($\lambda > 1$) emissions of NO_X respond to higher excess-air factors by falling as the decreasing density of the air/fuel mixture leads to progressively lower combustion-chamber temperatures. Maximum NO_X emissions occur with slightly lean mixtures in the range of $\lambda = 1.05...1.1$.

A characteristic of stratified-charge operation in direct-injected engines is a high excess-air factor. While NO_X emissions are low compared with those produced in operation at $\lambda = 1$, the lean mixture prevents the 3-way catalytic converter from reducing the nitrous oxides. The answer is to use an NO_X storage catalyst with these systems.

Effects of ignition timing

Throughout the range with excess-air factors of λ, NO_X emissions rise as ignition advance is increased (Figure 3). The higher combustion temperatures promoted by earlier ignition timing not only shift the chemical equilibrium toward greater NO_X formation, but – most significantly – they also accelerate the speed at which this formation takes place.

Soot emissions

Gasoline-engines do not produce substantial soot emissions during operation on mixtures in the vicinity of stoichiometric. However, soot can be generated on direct-injection engines during stratified-charge operation, when its formation can be fostered by localized areas with extremely rich mixtures or even fuel droplets. To ensure that adequate time remains available for efficient mixture formation, operation in the stratified-charge mode must therefore be restricted to low and moderate rpm.

3 Influence of excess-air factor λ and ignition timing α_Z on raw NO_x emissions

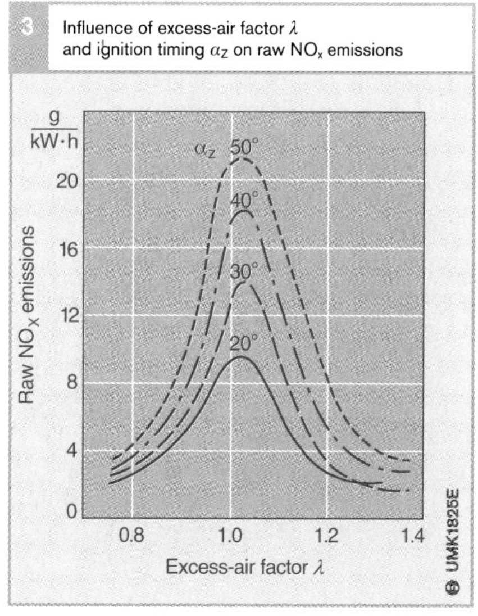

Reducing emissions

Legal limits on toxic emissions are defined in legislation. Engine design engineers strive to produce power plants offering optimal energy utilization, good fuel economy, and high levels of power and torque. These assets must be combined with yet another essential objective: levels of raw emissions generated when the air/fuel mixture combusts must remain as low as possible.

Overview

In recent years advances in power-plant technology have led to improved combustion processes producing lower raw emissions. Development of electronic engine-management systems has made it possible to provide precise control of injected fuel quantities and ignition timing, while also allowing optimal control of all components for ideal response under any given conditions (with the electronic throttle plate, etc.). Along with enhancements in engine performance, these advances also lead to substantial improvements in the quality of the exhaust gas.

Yet another factor that should not be forgotten is the improvement in fuel quality. Progressively higher levels of power-plant performance have resulted in increasing demands on fuel quality. Additives inhibit deposit formation in the combustion chamber during combustion, reduce the toxicity of the exhaust gases, and prevent damaging residue from impairing the efficiency of the fuel system. The conversion to unleaded fuels was an important milestone on the road to lower exhaust emissions.

These developments alone have reduced raw emissions by approximately 80 % since the 1970s. Yet it was the advent of the catalytic converter that made it possible to achieve compliance with the new legal requirements being mandated by legislators.

Engine design features

One contribution to reducing toxic emissions is furnished by camshaft control systems allowing variable valve timing, which supply greater latitude for governing combustion gases. Further reductions in pollutant emissions are furnished by advanced combustion-chamber configurations with

- Improved combustion-chamber geometry and
- Multi-valve technology
- Centrally positioned spark plugs
- Dual ignition with two spark plugs per cylinder on multi-valve engines
- Higher compression ratios
- Ideal placement of the high-pressure injectors used for direct injection

Outside the engine, other systems and components providing improved emissions control potential are installed. Examples include:

- Systems for post-combustion thermal treatment
- Exhaust-gas recirculation
- Evaporative-emissions control systems

1 Secondary-air injection system

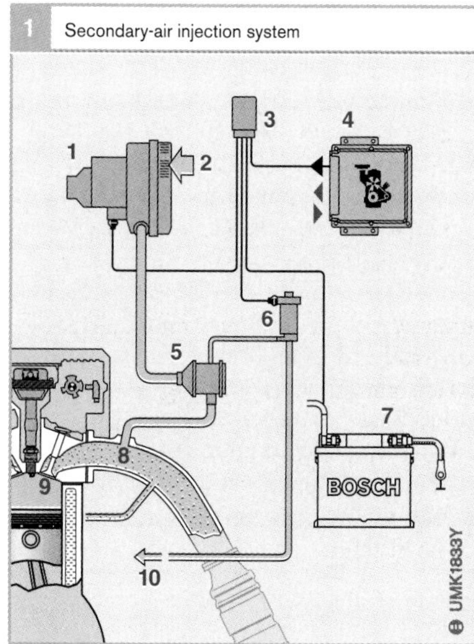

Fig. 1
1 Secondary-air pump
2 Induction air
3 Relay
4 Engine-management ECU
5 Secondary-air valve
6 Control valve
7 Battery
8 Exhaust input fitting
9 Exhaust valve
10 to intake manifold connection

Post-combustion thermal treatment

Combustion of the air/fuel mixture within the engine can never be 100 % complete. The fuel that fails to combust flows into the exhaust system during the exhaust stroke. This fuel does not contribute to torque generation. This is why combustion always produces raw HC and CO emissions.

In addition, when the engine is cold, fuel condenses on the cylinder walls. This fuel is subsequently discharged from the combustion chamber without burning. To compensate for these effects and to ensure smooth and stable engine operation in the warm-up phase, a richer air/fuel mixture is required. Raw HC, and above all raw CO emissions, rise dramatically when this unburned fuel is discharged. Yet another complication is the fact that the catalytic converter must attain a temperature of at least 300 °C before it can start to convert pollutant emissions.

This results in two imperatives. The first is to minimize levels of raw emissions produced during the warm-up phase. The second is to ensure that the catalytic converter warms to its operating temperature as quickly as possible. Strategies to reduce raw emissions in the period before the catalytic converter comes on line include:

- Improved starting processes (injection and ignition timing)
- Lean mixtures during warm-up phase (engine must be able to operate on lean mixtures to support this concept) and
- Secondary air-injection

Action designed to support rapid catalytic-converter response:

- Late ignition timing and high mass gas flow to induce high exhaust temperatures
- Catalytic converters mounted close to the engine
- Two-phase fuel delivery with direct gasoline injection

Secondary-air injection

The electric secondary-air pump (Figure 1, Position 1) draws in air (2) and then injects this air into the exhaust tract (8) in a process governed by the secondary-air valve (5). Since the valve prevents backflow of exhaust gases into the pump and air-control system, it must remain closed when the pump is not in active operation.

Because the secondary-air pump produces a substantial current draw during initial operation, a control relay (3) is required. An electric control valve (6) connected to a pneumatic system governs operation of the secondary air-valve. The control valve controls air flow between the secondary-air valve and the intake manifold (secondary-air valve opens) or to atmospheric pressure (valve closes). Operation of the pump and the control valve is controlled by the engine-management ECU (4) to inject secondary air under precisely defined operating conditions.

To exploit high exhaust-gas temperatures and promote efficient exothermic reactions, secondary air must be injected into the exhaust tract as close as possible to the exhaust valve (9). The secondary-air valve should not be placed too close to the exhaust manifold, as exposure to excessively high temperatures must be avoided. At the same time it is vital to prevent resonation (whistling) in the tube between the secondary-air valve and the air injection fitting in the exhaust manifold.

Supplementary air is required only in the post-start warm-up phase (at $\lambda < 1$). One function of the exothermic reaction is to reduce high HC and CO concentrations in the exhaust gas during this operational phase. Another function of this oxidation process is to generate heat and warm the exhaust gas, thus allowing the catalytic converter to reach its operating temperature more quickly. By promoting rapid heating of the catalytic converter, and bringing it up to temperature faster, secondary-air injection allows the catalyst to start converting NO_X emissions sooner.

Catalytic emissions control

Emission-control legislation defines the limits for the toxic agents generated during the combustion process in the spark-ignition engine. Catalytic treatment of the exhaust gas is necessary in order to comply with these limits.

Overview

Before leaving the exhaust pipe, the exhaust gas flows through the catalytic converter installed in the exhaust-gas tract (Figure 1, Pos. 3). Inside the converter, special coatings ensure that the toxic agents in the exhaust gas are chemically converted to harmless substances. Lambda oxygen sensors (2, 4) measure the residual-oxygen content in the exhaust gas. These measured values are then applied in adjusting the A/F mixture so that the catalytic converter can work at maximum efficiency.

A number of different catalytic-converter concepts were applied in the past years. The three-way catalytic converter represents the state-of-the-art for engines with homogeneous A/F mixture distribution and operation at $\lambda = 1$. Engines which run with a lean A/F mixture also require a NO_x accumulator-type catalytic converter.

Oxidation-type catalytic converter

In this type of catalytic converter, the hydrocarbons and the carbon monoxide in the exhaust gas are converted by oxidation (burning) into water vapor and carbon dioxide. The oxygen needed for the burning process is already present in the case of a lean A/F mixture ($\lambda > 1$) or by blowing air into the exhaust-gas tract upstream of the converter. The oxidation converter cannot convert the oxides of nitrogen (NO_x).

Oxidation-type catalytic converters were first introduced in 1975 in order to comply with the exhaust-gas legislation in force in the USA at that time. Today, catalytic converters which operate exclusively with oxidation principles are used only very rarely.

1 Exhaust-gas tract with Lambda oxygen sensors and a three-way catalytic converter installed in the immediate vicinity of the engine

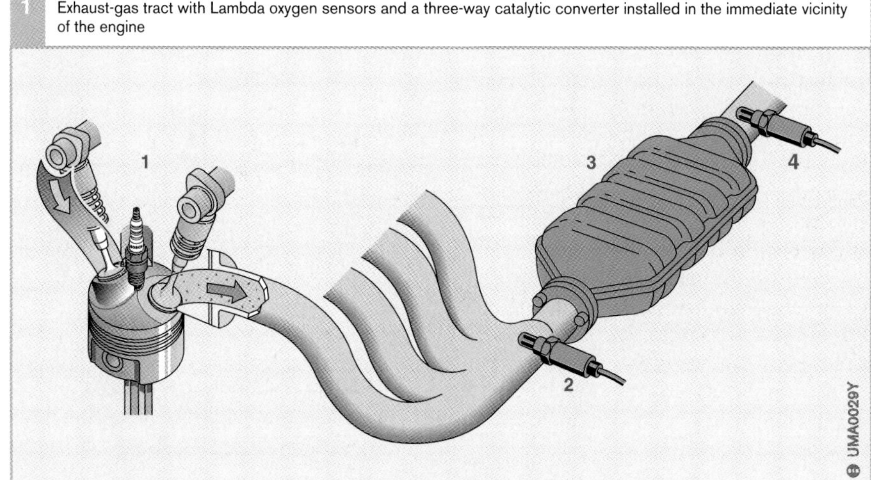

Fig. 1
1 Engine
2 Lambda oxygen sensor upstream of the catalytic converter (two-step sensor or broad-band sensor depending upon system)
3 Three-way catalytic converter
4 Two-step lambda oxygen sensor downstream of the catalytic converter (only on systems with lambda dual-sensor control)

Three-way catalytic converter

The three-way catalytic converter is installed in the exhaust-emission control systems of manifold-injection engines and gasoline direct-injection engines.

Assignment

Three toxic components are generated during the combustion of the A/F mixture: HC (hydrocarbons), CO (carbon monoxide), and oxides of nitrogen (NO_x). It is the job of the three-way catalytic converter to convert these into harmless components. The products which result from this conversion are H_2O (water vapor), CO_2 (carbon dioxide), and N_2 (nitrogen).

Operating concept

The toxic components are converted in two phases: Firstly, the carbon monoxide and the hydrocarbons are converted by oxidation (Figure G, Equations 1 and 2). The oxygen needed for the oxidation process is available in the exhaust gas in the form of the residual oxygen resulting from incomplete combustion, or it is taken from the oxides of nitrogen whereby these reduce as a result (Figure G, Equations 3 and 4).

The concentration of the toxic substances in the untreated exhaust gas is a function of the excess-air factor λ (Fig. 2a). For carbon monoxide and hydrocarbons (HC), the conversion level increases steadily along with increasing excess-air factor (Fig. 2b). At $\lambda = 1$, there is only a very low level of toxic components in the untreated exhaust gas. With high excess-air factors ($\lambda > 1$), the concentration of these toxic components remains at this low level.

Conversion of the oxides of nitrogen (NO_x) is good in the rich range ($\lambda < 1$). The lowest levels of NO_x are present during stoichiometric operation ($\lambda = 1$). Even a small increase in the exhaust-gas oxygen content as caused by operation at $\lambda > 1$ impedes the nitrogen reduction and causes a sharp increase in its concentration.

In order to maintain the three-way catalytic converter's conversion level for all three toxic substances at as high a level as possible, these must be present in a chemical balance in the exhaust gas. This means that the A/F mixture composition must have a stoichiometric ratio of $\lambda = 1$, so that the "window" for the A/F mixture ratio l is necessarily very restricted. A/F mixture formation must be controlled by a Lambda closed-loop control circuit.

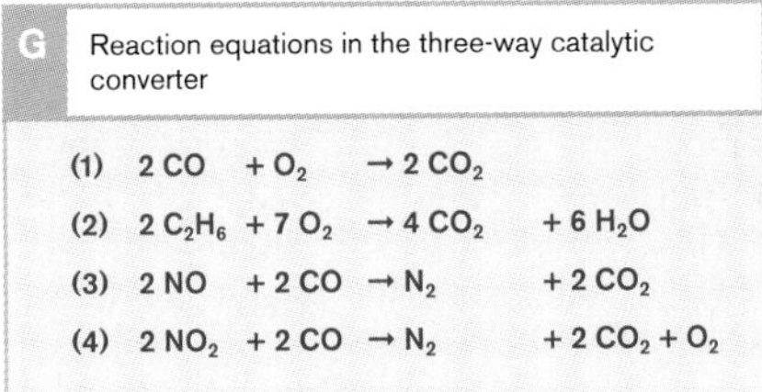
G Reaction equations in the three-way catalytic converter

$$(1)\quad 2\,CO + O_2 \rightarrow 2\,CO_2$$
$$(2)\quad 2\,C_2H_6 + 7\,O_2 \rightarrow 4\,CO_2 + 6\,H_2O$$
$$(3)\quad 2\,NO + 2\,CO \rightarrow N_2 + 2\,CO_2$$
$$(4)\quad 2\,NO_2 + 2\,CO \rightarrow N_2 + 2\,CO_2 + O_2$$

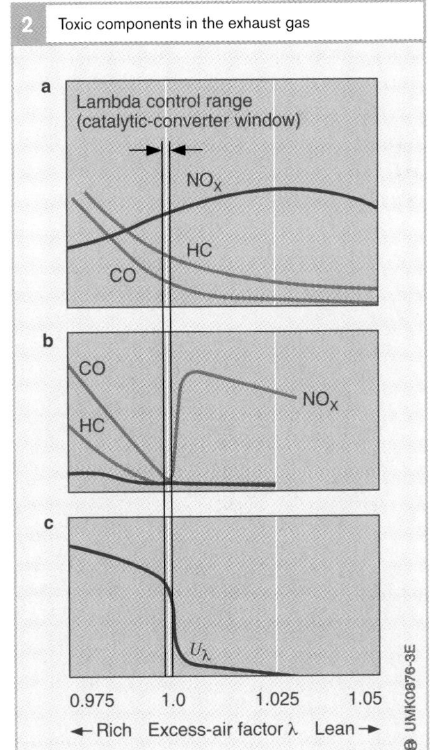

2 Toxic components in the exhaust gas

Fig. 2
a Before catalytic aftertreatment (raw exhaust gas)
b After catalytic after-treatment
c Voltage characteristic of the two-step Lambda sensor

Design and construction

The catalytic converter (Fig. 3) comprises a steel casing (6), a substrate (5), and the active catalytic noble-metal coating (4).

Substrates

Two substrate systems have come to the forefront

Ceramic monoliths

These ceramic monoliths are ceramic bodies containing thousands of narrow passages through which the exhaust gas flows. The ceramic is a high-temperature-resistant magnesium-aluminum silicate. The monolith, which is highly sensitive to mechanical tension, is fastened inside a sheet-steel housing by means of mineral swell matting (2) which expands the first time it is heated up and firmly fixes the monolith in position. At the same time the matting also ensures a 100 % gas seal.

Ceramic monoliths are at present the most commonly used catalyst substrates.

Metallic monoliths

The metallic monolith (metal catalytic converter) is an alternative to the ceramic monolith. It is made of finely corrugated, 0.05 mm thin metal foil which is wound and soldered in a high-temperature process. Thanks to its thin walls, more passages can be accomodated inside the same area, which means less resistance to exhaust-gas flow, a fact which is important in the case of high-performance engines.

Coating

The ceramic and metallic monoliths require an aluminum oxide (Al_2O_3) substrate coating, the so-called "Washcoat" (4). This coating serves to increase the converter's effective surface area by a factor of around 7,000. On the oxidation catalytic converter, the effective catalytic coating applied to the substrate contains the noble metals platinum and/or palladium. On the three-way converter, rhodium is also applied. Platinum and palladium accelerate the oxidation of the hydrocarbons (HC) and of the carbon monoxide. Rhodium accelerates the reduction of the oxides of nitrogen (NO_x).

Depending upon the engine's displacement, a catalytic converter contains about 1...3 g of noble metal.

3 Three-way catalytic converter with Lambda oxygen sensor

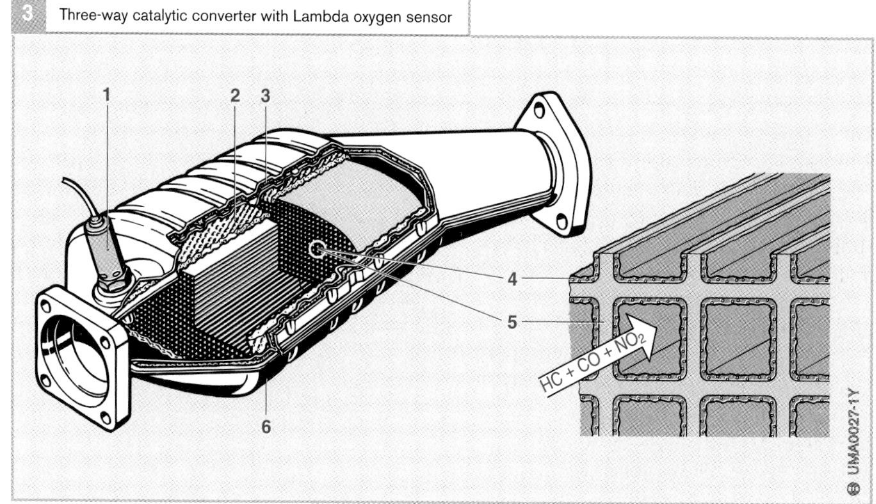

Fig. 3
1 Lambda oxygen sensor
2 Swell matting
3 Thermally insulated double shell
4 Washcoat (Al_2O_3 substrate coating) with noble-metal coating
5 Substrate (monolith)
6 Housing

Operating conditions

Operating temperature

The catalytic converter's temperature plays a decisive role in emission-control efficiency. Considering a three-way catalytic converter, no worthwhile conversion of toxic substances takes place until temperature exceeds 300 °C. Operation within a temperature range of 400...800 °C is ideal with regard to high conversion levels and a long service life.

At temperatures between 800...1,000 °C, thermal aging is accelerated due to the sintering of the noble metals and of the Al_2O_3 substrate layer, and this leads to a reduction of the effective surface. The time spent at 800...1,000 °C is of vital importance, and above 1,000 °C thermal aging increases drastically and leads to the catalytic converter becoming practically 100 % ineffective.

Engine malfunction (ignition misfire) can cause the temperature inside the catalytic converter to exceed 1,400 °C. Since such temperatures melt the substrate and completely destroy the catalyst, it is imperative that the ignition system is highly reliable and maintenance-free. Modern engine-management systems are able to detect ignition and combustion miss, and in such cases interrupt the fuel injection to the cylinder concerned so that unburned A/F mixture cannot enter the exhaust-gas tract.

Unleaded fuel

Another prerequisite for long-term operation is the use of unleaded fuel. Otherwise, lead compounds are deposited in the pores of the active surface and reduce their number. Residues from the engine oil can also "poison" the catalyst and damage it so far that it becomes ineffective.

Installation point

Strict emissions-control legislation demands special concepts for heating the catalytic converter when the engine is started. The catalytic converter's installation point is determined by such concepts (for instance, secondary-air injection, shift of the timing in the "retard" direction). The three-way catalytic converter's sensitivity regarding operating temperature limits the choice of installation point. The temperature conditions needed for a high conversion level make it absolutely imperative that the three-way converter is installed close to the engine.

In the case of the three-way catalytic converter, a configuration featuring a "pre-cat" near the engine followed by a second (main) underfloor catalytic converter has come to the forefront. Catalytic converters near the engine demand that their coating techniques be optimized to provide for high-temperature stability. Underfloor converters on the other hand, require optimization in the so-called "low light-off" direction (low start-up temperature) and good NO_x conversion characteristics.

An alternative is available with just one "overall" catalytic converter which is then installed close to the engine.

Effectiveness

For a spark-ignition engine with homogeneous mixture distribution operating at $\lambda = 1$, catalytic treatment of the exhaust gas using a three-way catalytic converter is at present the most effective emission-control method. Included in this system is the Lambda closed-loop control which monitors the composition of the A/F mixture. Using the three-way catalytic converter, the pollutant emissions of carbon monoxide, hydrocarbons, and oxides of nitrogen can be practically eliminated provided the engine operates with homogeneous A/F-mixture distribution and at stoichiometric A/F ratio. Notwithstanding the fact that it is not always possible to comply fully with these operating requirements, one can still presume an average pollutants reduction of more than 98 %.

NO_x accumulator-type catalytic converter

Assignment

During lean-burn operation, it is impossible for the three-way catalytic converter to completely convert all the oxides of nitrogen (NO_x) which have been generated during combustion. In such cases namely, the oxygen that is needed for the oxidation of the carbon monoxide and of the hydrocarbons is not split off from the oxides of nitrogen but instead is taken from the high level of residual oxygen in the exhaust gas. The NO_x accumulator catalytic converter reduces the oxides of nitrogen in a different manner.

Design and special coating

The NO_x accumulator-type catalytic converter is similar in design to the conventional three-way converter. In addition to the platinum, palladium and rhodium coatings, the NO_x converter is provided with special additives which are capable of accumulating oxides of nitrogen. Typical accumulator materials are the oxides of potassium, calcium, strontium, zirconium, lanthanum, and barium.

The coating for NO_x accumulation and for the 3-way catalytic converter can be applied on a common substrate.

Operating concept

At $\lambda = 1$, due to the noble-metal coating the NO_x converter operates the same as a three-way converter. In lean exhaust gases though it also converts the non-reduced oxides of nitrogen. This conversion is not a continuous process as it is with the hydrocarbons and the carbon monoxide, but instead takes place in three distinct phases:

1. NO_x accumulation (storage)
2. NO_x release and
3. Conversion

NO_x accumulation (storage)

On the surface of the platinum coating, the oxides of nitrogen (NO_x) are oxidized catalytically to form nitrogen dioxide (NO_2). The NO_2 then reacts with the special oxides on the catalyst surface and with oxygen (O_2) to form nitrates. For instance, NO_2 combines chemically with barium oxide (BaO) to form barium nitrate $(NO_3)_2$ (Fig. G, Equation 1). This enables the NO_x converter to accumulate the oxides of nitrogen which have been generated during engine operation with excess air.

There are two methods in use to determine when the NO_x converter is full and the accumulation phase has finished:

- Taking the catalyst temperature into account (Fig. 1, Pos. 4), the model-based method calculates the quantity of stored NO_x.
- An NO_x sensor (6) downstream of the NO_x converter continually measures the NO_x concentration in the exhaust gas.

NO_x removal and conversion

The more NO_x that is stored, the less the ability to chemically bind further nitrogens of oxide. This means that regeneration must take place as soon as a given level is exceeded, in other words the accumulated oxides of nitrogen must be released and converted. To this end, the engine is run briefly in the rich homogeneous mode ($\lambda < 0.8$). The processes for releasing the NO_x and converting it to nitrogen and carbon dioxide take place separately from each other. H_2, HC, and CO are used as reducing agents. Reduction is slowest with HC and most rapid with H_2. NO_x release takes place as follows, whereby the following description applies with carbon monoxide (CO) as the reducing agent: The carbon monoxide reduces the nitrate (e.g. barium nitrate $Ba(NO_3)_2$ to an oxide (e.g. barium oxide BaO). This leads to the generation of carbon dioxide (CO_2) and nitrogen monoxide (NO) (Fig. G, Equation 2). Subsequently, using the carbon monoxide (CO), the rhodium coating reduces the NO_x to nitrogen and carbon dioxide (CO_2) (Fig. G, Equation 3).

G Reaction equations for the NO_x accumulation phase (1), removal phase (2), and conversion phase (3)

(1) $2\,BaO + 4\,NO_2 + O_2 \rightarrow 2\,Ba(NO_3)_2$

(2) $Ba(NO_3)_2 + 3\,CO \rightarrow 3\,CO_2 + BaO + 2\,NO$

(3) $2\,NO + 2\,CO \rightarrow N_2 + 2\,CO_2$

There are two different methods for determining the end of the NO_x-release phase:

- The model-based method calculates the quantity of NO_x still held by the converter.
- A Lambda oxygen sensor (Fig. 1, Pos. 6) downstream of the converter measures the exhaust-gas oxygen concentration and outputs a voltage jump from "lean" to "rich" when conversion has finished.

Operating temperature and installation point

The NO_x converter's ability to accumulate/store NO_x is highly dependent upon temperature. Accumulation reaches its maximum between 300 and 400 °C, which means that the favorable operating-temperature range is much lower than that of the three-way catalytic converter. For catalytic emissions control, therefore, two separate catalytic converters must be installed – a three-way pre-cat near the engine (Fig. 1, Pos. 3), and an NO_x accumulator-type main converter (5) remote from the engine (underfloor cat).

Sulphur in the NOx accumulator-type catalytic converter

The sulphur in gasoline presents the accumulator-type catalytic converter with a problem. The sulphur contained in the exhaust gas reacts with the barium oxide (accumulator material) to form barium sulphate. The result is that, over time, the amount of accumulator material available for NO_x accumulation diminishes. Barium sulphate is extremely resistant to high temperatures, and for this reason is only degraded to a slight degree during NO_x regeneration. When sulphurized gasoline is used therefore, desulphurization must be carried out at regular intervals. Here, selective measures are applied to heat the converter to between 600 and 650 °C. For instance, the engine can be run in the "stratified-charge/cat-heating mode". Rich ($\lambda = 0.95$) and lean ($\lambda = 1.05$) exhaust gases are then passed through the cat one after the other. The barium sulphate reduces to barium oxide as a result.

1 Exhaust-gas system with three-way catalyic converter as pre-cat, and downstream NO_X accumulator-type converter and Lambda oxygen sensors

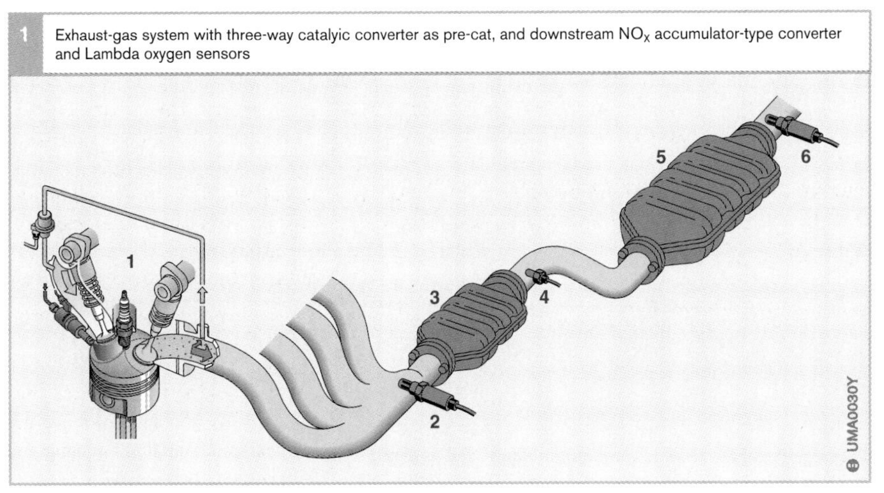

Fig. 1
1 Engine with EGR system
2 Lambda oxygen sensor upstream of the catalytic converter
3 Three-way catalytic converter (pre-cat)
4 Temperature sensor
5 NO_x accumulator-type catalytic converter (main cat)
6 Two-step Lambda oxygen sensor, optionally available with integral NO_x sensor

Lambda control loop

Assignment

For systems which operate with only a single three-way catalytic converter, the pollutants must be in a state of chemical balance in order that the conversion level for all three pollutant constituents is as high as possible. This necessitates a stoichiometric A/F-mixture composition with $\lambda = 1.0$, which means that the "window" in which the A/F ratio must be located is very narrow. The only solution is to apply closed-loop control to the adjustment of the A/F mixture ratio. Open-loop control of fuel metering is not accurate enough.

Direct-injection gasoline engines are run with A/F mixtures which deviate from stoichiometric. Closed-loop control can also be used on these systems for A/F-mixture adjustment.

Design and construction

A Lambda oxygen sensor (Figure 1, Pos. 3a) is located upstream of the pre-cat (4). The sensor signal U_{Sa} is inputted to the engine ECU (7). In order to do so, either a two-step Lambda sensor (two-step control) or a broad-band Lambda sensor (continuous-action Lambda control) must be used. A further Lamda oxygen sensor (3b) can be situated downstream of the main catalytic converter (5). This is always a two-step sensor, and it delivers the sensor signal U_{Sb}. This form of control is known as two-sensor control.

Operating concept

Using the Lambda control loop, deviations from a specific A/F-ratio can be detected and corrected. The control principle is based on the measurement of the residual oxygen in the exhaust gas. This is a measure for the composition of the A/F mixture supplied to the engine (2).

Two-step control

The sensor voltage U_{Sa} generated by the two-step Lambda oxygen sensor upstream of the pre-cat (4) is high in the rich range ($\lambda < 1$) and low in the lean range ($\lambda > 1$). Since the sensor voltage jumps abruptly at $\lambda = 1$, the two-step Lambda oxygen sensor can only differentiate between rich and lean A/F mixtures.

1 Functional diagram of the Lambda closed-loop control

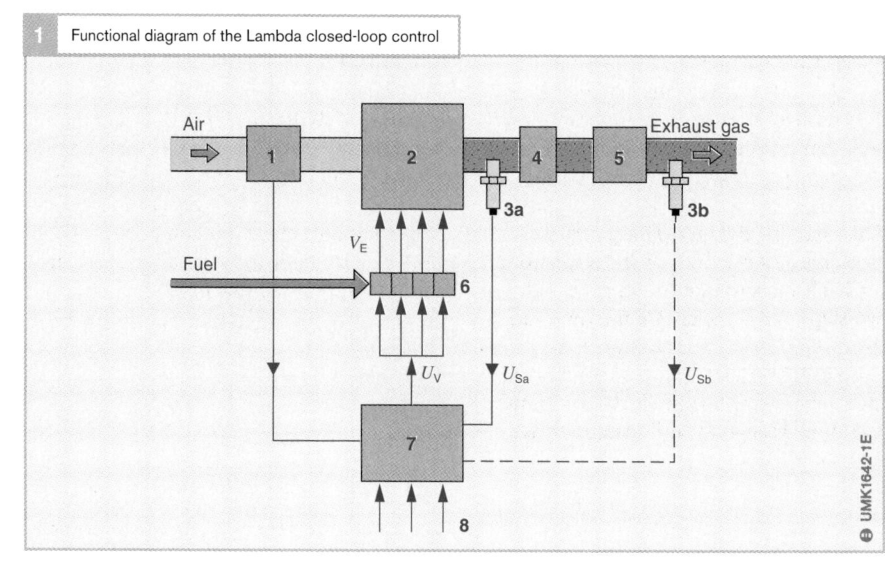

Fig. 1
1 Air-mass meter
2 Engine
3a Lambda oxygen sensor upstream of the pre-cat (two-step Lambda sensor, or broad-band Lambda sensor)
3b Two-step Lambda sensor downstream of the main catalytic converter (only if required; on gasoline direct injection with integral NO_x sensor)
4 Pre-cat (three-way catalytic converter)
5 Main cat (On manifold injection: three-way converter; on gasoline direct injection: NO_x accumulator-type converter)
6 Injectors
7 Engine ECU
8 Input signals

U_S Sensor voltage
U_V Injector-triggering voltage
V_E Injected fuel quantity

The sensor output signal is converted to a binary signal in the engine ECU and used as the input signal for the Lambda closed-loop control as implemented using software. The Lambda control has a direct influence on the A/F mixture formation and sets the correct A/F ratio by adapting the injected fuel quantity. The manipulated variable comprises a step change and a ramp, and its control direction changes with each jump of the sensor voltage. In other words, a jump of the manipulated variable causes the A/F mixture to change. This change is first of all very abrupt, and then it follows a ramp. With a high sensor voltage ("rich" A/F mixture), the manipulated variable adjusts in the "lean" direction, and for a low sensor voltage ("lean" A/F mixture) in the "rich" direction. This so-called two-step control enables A/F mixture to be closed-loop controlled to values around $\lambda = 1$.

Shaping the manipulated variable's characteristic curve asymmetrically compensates for the Lambda sensor's typical false signal caused by variations in A/F mixture formation (rich/lean shift).

Continuous-action Lambda control

The broad-band Lambda sensor outputs a continuous voltage signal U_{Sa}. This means that not only the Lambda area (rich or lean) can be measured, but also the deviation from $\lambda = 1$ so that the Lambda control can react more quickly to an A/F mixture deviation. This leads to better control behaviour with highly improved dynamic response.

The broad-band Lambda oxygen sensor can measure A/F mixtures which deviate from $\lambda = 1$. This means that (in contrast to the two-step control), such A/F mixtures can also be controlled. The control range covers $\lambda = 0.7...3.0$ so that continuous Lambda control is suitable for the "rich" and "lean" operation of engines with gasoline direct injection.

Two-sensor control

When it is situated upstream of the pre-cat, the Lambda oxygen sensor (3a) is heavily stressed by high temperatures and untreated exhaust gas, and this leads to limitations in accuracy. On the other hand, locating the sensor downstream of the main catalytic converter (3b) means that these influences are considerably reduced.

The only problem here though is that a single downstream sensor would be far too "sluggish" due to the exhaust gases taking so long to reach it. The principle of two-sensor control relies upon the upstream sensor controlling the "lean" and "rich" shift, while the downstream sensor is part of a "slow" corrective closed control loop responsible for additive changes.

Lambda closed-loop control of gasoline direct injection

The NO_x accumulator-type catalytic converter has two different functions. During lean-burn operation, NO_x accumulation and CO oxidation must take place. In addition, at $\lambda = 1$, a stable three-way function is needed which provides for a minimum level of oxygen-accumulation. The Lambda sensor upstream of the catalytic converter monitors the stoichiometric composition of the A/F mixture.

Together with the integrated NO_x sensor, the two-step Lambda sensor downstream of the NO_x accumulator converter not only takes part in the two-sensor control but also monitors the behaviour of the combination O_2 and NO_x accumulator (detection of the end of the NO_x release phase).

Catalytic-converter heating

Ignition timing towards "retard"

In order to keep the pollutant concentration in the exhaust gas down to a minimum, it is necessary that the catalytic converter reaches its operating temperature as soon as possible. One method is to adjust the ignition timing towards "retard".

This step lowers the engine efficiency, and in doing so leads to hotter exhaust gases which then heat-up the converter.

Secondary-air injection

The unburnt components of the A/F mixture still present in the exhaust gas are burnt in the thermal afterburning process. With "lean" A/F mixtures, the oxygen required for this afterburning process is available in the exhaust gas in the form of residual oxygen. With "rich" A/F mixtures, as often needed for an engine which has not yet reached operating temperature, extra air (secondary air) is injected into the exhaust-gas passage to speed-up the catalytic-converter heating.

On the one hand, this exothermic reaction reduces the hydrocarbons and the carbon monoxide. On the other, afterburning also heats up the catalytic converter so that it quickly reaches its operating temperature. During the warm-up phase, this process considerably increases the conversion rate so that the catalytic converter is quickly ready for operation. Fig. 1 shows the curves of the hydrocarbon and carbon-monoxide emissions in the first seconds of an emissions test, with and without secondary-air injection.

In line with present state-of-the-art, electric secondary-air pumps are used for secondary-air injection.

Post injection (POI)

On gasoline direct-injection engines, another method can be used for quickly bringing the catalytic converter up to temperature. In the "stratified-charge/cat-heating" operating mode, during stratified-charge operation with high levels of excess air a second injection of fuel takes place during the engine's power cycle. This fuel is combusted late and causes considerable heat-up of the engine's exhaust side and of the exhaust manifold. This means, that in those cases in which conventional measures (adjust ignition timing in the "retard" direction) do not suffice for complying with the stipulated exhaust-gas limits, the secondary-air pump used for manifold injection can be dispensed with.

1 Influence of secondary-air injection on CO and HC emissions

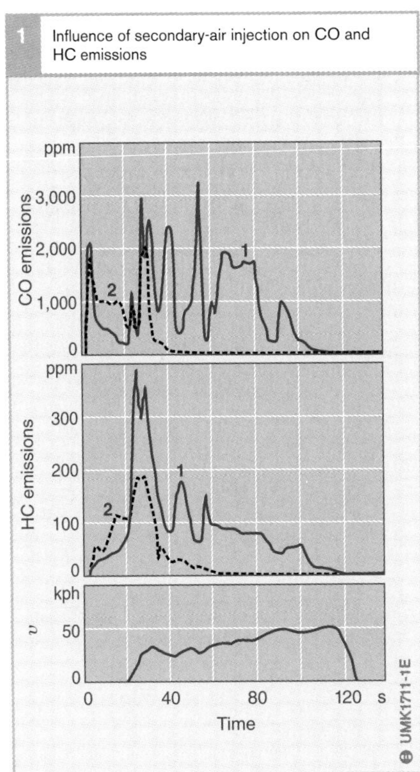

Fig. 1
1 Without secondary-air injection
2 With secondary-air injection

v Vehicle speed

Patents

Not only do new ideas and inventions have to be thought up in the first place, they also have to be protected against imitators. In cases where the copying of an invention is difficult to prove (e.g. production processes), the commercial advantage arising from the invention can be protected by secrecy. However, that method is not always possible or sensible. Better protection is afforded by patent law.

What is a patent?

A patent is a legal title that gives the patent holder (a private individual, a company or an organization) the exclusive rights within a specific geographical area (e.g. Europe, USA) and for a limited period (maximum of 20 years) to the manufacture, sale and use of the patented invention.

What can be patented?

Patents are granted for inventions that are new, original and have a commercial application. An invention is new if it was not publicly known of in any way prior to the date of the patent application (the "priority date"), i.e. was not already part of the state of the art. It is original if it is not obviously deducible from the state of the art by an expert in the field concerned. A patent can be granted for

- An object (e.g. a spark plug)
- A process (e.g. a special manufacturing method)
- A chemical substance (e.g. a medication)
- A computer program if it makes a technical contribution to the state of the art (e.g. ABS software)

A patent is therefore an instrument of intellectual property law. Other means of legal protection afforded by such legislation include the registered design, which is issued in respect of utility articles, copyright and trademarks, and the new semiconductor protection law.

Patenting an invention

A company's patents expert will normally first of all check whether an invention submitted by an employee is capable of being patented and whether a patent application is worthwhile from a commercial point of view. If that is the case, a patent application will be submitted to the appropriate patent office (national and/or international). Of course, not every patent application will be granted. Around 50 % of applications are rejected on the grounds that they are already obvious from the state of the art (e.g. existing patents).

The number of new patent applications is an indication of the inventiveness of the applicant (or the applicant's employees). Bosch, for example, applied for more than 2,400 patents in the year 2000.

Publication

Patent applications and granted patents are made public. They are not only a useful means of market analysis, they also provide an insight into the creative trends in all areas of science and technology. As such, they are an effective instrument in preventing duplication of research and development work.

Use of patents

Patents are important aids in the transfer of technology for the promotion of creative potential.

- The exclusive rights to a commercially applicable invention facilitate the financing of research and development costs by a business.
- By conferring exclusive rights, a patent strengthens the commercial position of a company.
- Patented inventions encourage researchers to find alternative solutions.
- The licensing of patented inventions promotes the spread of new technologies.

Emissions-control legislation

Over the years traffic density has displayed a marked increase, creating a corresponding negative array of consequences for the environment. The repercussions have been especially conspicuous in urban areas. As a result, it has became imperative to place legal limits on exhaust emissions from motor vehicles. Both component limits and the procedures for verifying compliance are defined in legislation. Each new vehicle model must comply with applicable regulations.

Overview

The state of California assumed a pioneering role in efforts to restrict toxic emissions emanating from motor vehicles. This development arises from the fact that the geography of cities like Los Angeles prevents wind from dispersing exhaust gases, fostering formation of smog layers that encompass the city. The resulting smog not only has substantial negative effects on the health of the residents, but also impairs visibility.

California introduced the first regulations restricting emissions levels from motor vehicles in the 1960s. These directives became progressively more stringent in the ensuing years. In the intervening period, regulations governing exhaust emissions have been adopted in all industrialised nations. These laws impose mandatory limits on emissions from gasoline and diesel engines while also defining the test procedures employed to confirm compliance. In some countries, regulations governing exhaust emissions are supplemented by limits on evaporative losses from the fuel system.

The most important legal restrictions on exhaust emissions are (Figure 1):

- CARB regulations
- EPA regulations
- EU regulations
- Japanese regulations

Test procedures

Japan and the European Union have followed the lead of the United States by defining test procedures for certifying compliance with emissions limits. These procedures have been adopted in modified or unrevised form by other countries.

Legal requirements prescribe any of three different test procedures according to vehicle class and the object of the test

- Type test for homologation approval
- Random testing of vehicles from series production conducted by the approval authorities and
- Field monitoring of specified exhaust-gas components from vehicles in highway operation

The most extensive test procedures are those used for type approval. The procedures employed for field monitoring are much simpler.

Classifications

Countries with legal limits on emissions from motor vehicles divide these vehicles into various classes:

- Passenger cars: Testing is conducted on a chassis dynamometer.
- Light commercial vehicles: The upper limit lies at an approved gross vehicle weight of between 3.5 and 3.8 tons, varying according to country. As with passenger cars, testing is carried out on a chassis dynamometer.
- Heavy commercial vehicles: approved gross vehicle weights in excess of 3.5...3.8 tons. Testing is performed on an engine dynamometer, with no provision for in-vehicle testing.

Type test

Vehicles must successfully absolve emissions testing as a condition for receiving homologation approval for each specific engine and vehicle type. This process entails proving compliance with stipulated emissions limits in defined test cycles. Different countries have defined individual test cycles and emissions limits (Figure 1).

Test cycles

Two types of test cycle are specified for passenger cars and light commercial vehicles. The differences between the two procedures are rooted in their respective origins:

- Test cycles designed to mirror conditions recorded in actual highway operation (FTP test cycle in the USA, etc.) and
- Synthetically generated test cycles consisting of phases with constant cruising speeds and acceleration rates (MNEDC in Europe, Japanese test cycles)

The mass of the toxic emissions from each vehicle is determined by operating it in conformity with speed curves precisely defined for the test cycle. During this test cycle the exhaust gases are collected for subsequent analysis to determine the mass of the pollutants emitted during testing.

Testing series-production vehicles

This testing is usually conducted by the vehicle manufacturer in the quality control checks that accompany the production process. The authorities responsible for granting homologation approval can demand confirmation testing as often as deemed necessary. EU and ECE directives[1]) take account of production tolerances by carrying out random testing on between 3 and a maximum of 32 vehicles. The most stringent requirements are encountered in the USA, and particularly in California, where the authorities require what is essentially comprehensive and total quality monitoring.

[1]) ECE: **E**conomic **C**ommission of **E**urope

On-Board Diagnosis

Emissions legislation also defines the processes to be employed in confirming conformity with the specified limits. The engine-management ECU incorporates diagnostic functions (software algorithms) designed to detect emissions-relevant malfunctions within the system. OBD functions (**On-Board Diagnosis**) monitor performance of all components in which malfunctions could lead to higher levels of exhaust emissions. Different countries have defined their own specific emissions limits. When the vehicle exceeds these limits the malfunction indicator lamp lights up to alert the driver.

1 Application areas for individual emissions regulations

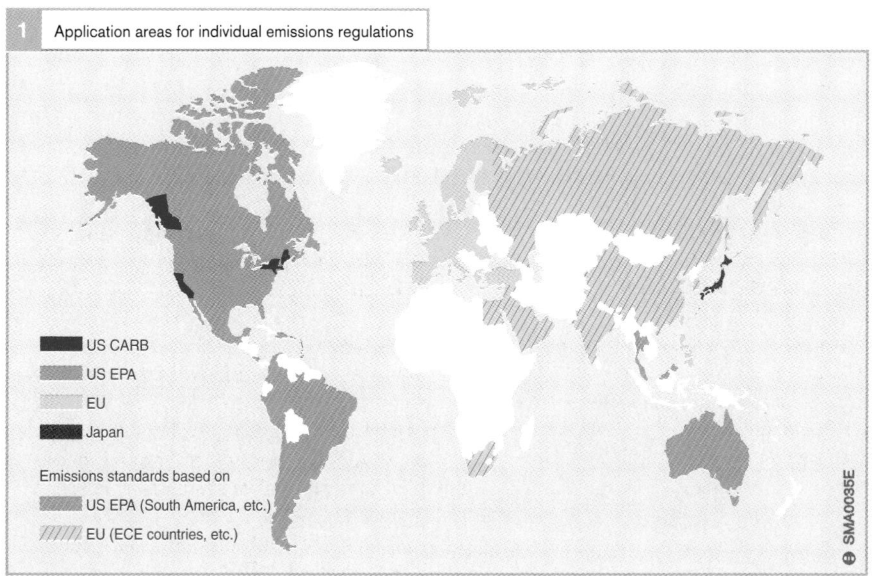

Fig. 1
Test cycles prescribed by various emissions regulations:

- FTP 75 cycle (CARB and EPA)
- Highway cycle (CARB and EPA, for determining fleet averages only)
- MNEDC (EU/ECE test cycle)
- 11-mode cycle and
- 10·15-mode cycle (both in Japan)

Other test cycles are in the introductory phase in the US:

- SC03 cycle and
- US06 cycle

CARB legislation

The CARB, or California Air Resources Board emissions limits for passenger cars and light trucks and vans (LDT, Light-Duty Trucks) are defined in statutes governing exhaust emissions.

- LEV I and
- LEV II

The LEV I standard applies to passenger cars and light commercial vehicles of up to 3,750 lbs (1.7 metric tons) manufactured in the model years 1994...2003. The LEV II standard will come into effect on 1.1.2004. It applies to all vehicles with an approved gross vehicle weight of up to 8,500 lbs (3.85 metric tons) from the 2004 model year onwards.

Emissions limits

The CARB regulations define limits on

- Carbon monoxide (CO)
- Nitrous oxides (NO_x)
- NMOG (non-methane organic gases)
- Formaldehyde (LEV II only) and
- Particulate emissions (LEV II only for gasoline-engines, also LEV I for diesels)

Actual emission levels are determined using the FTP 75 driving cycle (Federal Test Procedure). Limits are defined in relation to distance and specified in grams per mile.

Starting in 2004 the same limits will apply to both diesel and gasoline-engines.

Emissions categories

Automotive manufacturers can apply various vehicle concepts for classification in the following categories according to their respective emissions of NMOG, CO, NO_x and particulates:

- TLEV (Transitional Low-Emission Vehicle)
- LEV (Low-Emission Vehicle), applying to both exhaust and evaporative emissions
- ULEV (Ultra-Low-Emission Vehicle)
- SULEV (Super Ultra-Low-Emission Vehicle)
- ZEV (Zero-Emission Vehicle), vehicles without exhaust or evaporative emissions and
- PZEV (Partial ZEV), which is basically SULEV, but with more stringent limits on evaporative emissions and stricter long-term performance criteria

1 CARB emissions categories and limits

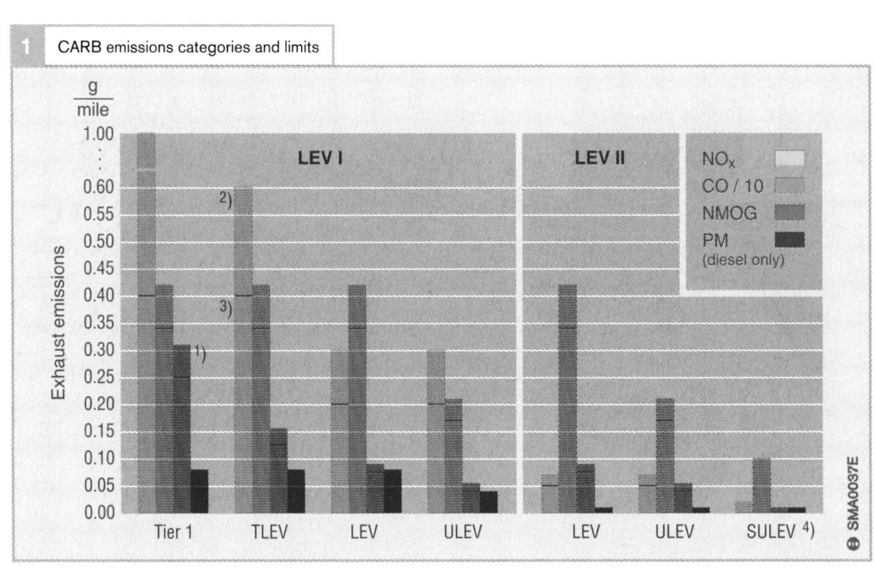

Fig. 1

1) For Tier 1, NMHC limit value applies instead of NMOG limit value

2) Limit value in each case for "full useful life" (10 years/ 100,000 miles with LEV I or 120,000 miles with LEV II)

3) Limit value in each case for "intermediate useful life" (5 years/ 50,000 miles)

4) Only limit values for "full useful life"

LEV I:
for passenger cars and light trucks/vans up to 3,750 lbs
Model years 1994 to 2003

LEV II:
for all vehicles up to 8,500 lbs
From 2004 model year only

The essential categories for LEV I are TLEV, LEV I and ULEV I. Figure 1 illustrates the limits on NO_x, CO and NMOG in the individual categories. Figure 2 is a graphic illustration of the limits on NO_x and NMOG in the various emissions categories.

The LEV II emissions standards come into effect on January 1, 2004. At the same time TLEV will be replaced by SULEV with its substantially lower limits. The LEV and ULEV classifications remain in place. The CO and NMOG limits from LEV I remain unchanged, but the NO_x limit is substantially lower for LEV II. To distinguish between the two categories the LEV II standard uses the designations LEV 2 and ULEV 2.

The LEV II standard also includes new, supplementary limits governing formaldehyde and particulate emissions.

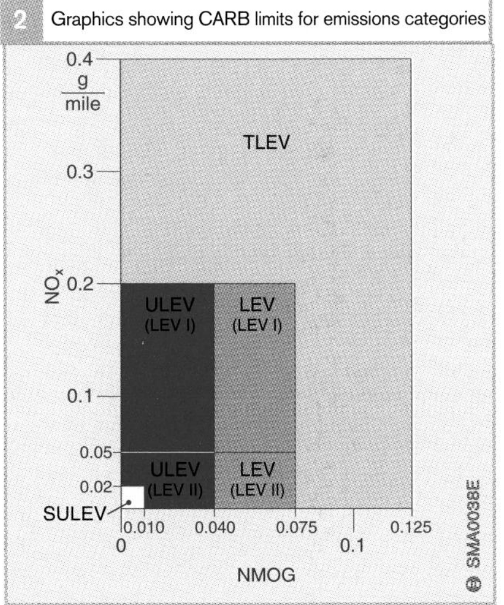

2 Graphics showing CARB limits for emissions categories

US type approval testing

To obtain approval for each vehicle model the manufacturer must prove compliance with the official emissions limits over a period of

- 50,000 or 100,000 miles ("full useful life") for LEV I (80,000 and 160,000 km in metric form) and 120,000 miles for LEV II (192,000 km) or
- 5 years (LEV I, 50,000 miles) or 10 years

The applicable figures for the PZEV emissions category are 150,000 miles and 15 years.

Manufacturers also have the option of certifying vehicles for 150,000 miles using the same limits that apply to 120,000 miles. The manufacturer then receives a bonus when the NMOG fleet average is defined (refer to section on "Fleet averages" on next page).

For this type of approval test the manufacturer must furnish two vehicle fleets from series production:

- One fleet in which each vehicle must cover 4,000 miles prior to testing.
- One fleet for long-term testing, in which the deterioration factors for individual components are defined.

Long-term testing entails subjecting the vehicles to specific driving programs over periods of 50,000 and 100,000 miles. Exhaust emissions are tested at intervals of 5,000 miles. Service inspections and maintenance are restricted to the standard prescribed intervals.

Countries that rely on the US test cycles (such as Switzerland) allow application of defined deterioration factors to simplify the certification process.

Phase-in

Following introduction of the LEV II standards in 2004 compliance will be mandatory for at least 25 % of new vehicles being registered for the first time in that year. The phase-in rule stipulates that an additional 25 % of the vehicles will then be required to conform to the LEV II standards in each consecutive year. All new vehicles will be required to meet the LEV II standards starting in 2007.

Fleet averages

Each vehicle manufacturer must ensure that exhaust emissions for its total vehicle fleet do not exceed a specified average. NMOG emissions serve as the reference category for assessing compliance with these averages. The fleet average is determined based on average emissions levels displayed by all of the manufacturer's vehicles in complying with the NMOG limits. Different fleet averages apply to passenger cars and light-duty trucks and vans.

The compliance limits for the NMOG fleet average are lowered in each subsequent year (Figure 3). To meet the lower limits, manufacturers must thus produce progressively more "clean" vehicles in the more stringent emissions categories in each consecutive year. This phase-in rule does not affect the fleet averages.

Average fleet fuel consumption

Each manufacturer's fleet must comply with US legislation specifying maximum average fuel consumption in miles per gallon (CAFE, Corporate Average Fuel Economy). The current figure for passenger cars is 27.5 miles per gallon. This corresponds to 8.55 litres per 100 kilometers in metric terms. At the end of each year the average fuel economy for each manufacturer is calculated based on the numbers of individual vehicle models that have been sold. The manufacturer must remit a penalty fee of $ 5.50 per vehicle for each 0.1 miles per gallon by which its fleet exceeds the target. Buyers also pay a "gas-guzzler" tax on vehicles with especially high fuel consumption. Here the limit is 22.5 miles per gallon (corresponding to 10.45 litres per 100 kilometers in metric terms).

These penalties are intended to spur development of vehicles offering high levels of fuel economy.

3 Fleet averages

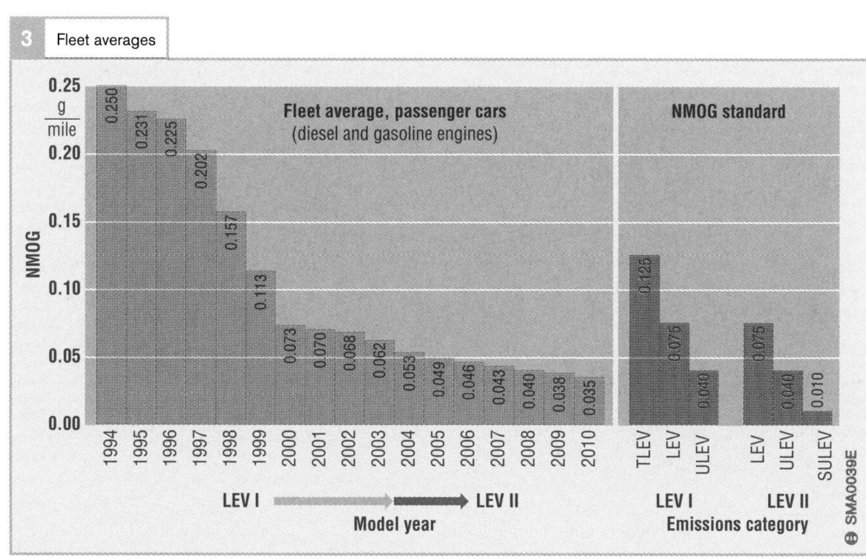

On-Board Diagnosis

With the introduction of OBD II, all new passenger cars and light-duty trucks and vans with an approved gross vehicle weight of up to 3.85 metric tons are now required to incorporate diagnostic capabilities that can detect all problems with the potential to affect vehicle emissions.

The system's response threshold is defined as 1.5 percent of the official emissions limit in each exhaust category. Once this threshold is crossed a warning lamp must be triggered after completion of two driving cycles at the absolute latest. This malfunction lamp can extinguish again should three subsequent driving cycles elapse with no detected errors.

Field monitoring

Unscheduled testing

Random emissions testing is conducted on in-use vehicles using the FTP 75 test procedure as well as an evaporation test. Only vehicles with milages of less than 50,000 or 75,000 miles (varies according to the certification status of the individual vehicle model) are selected for testing.

Vehicle monitoring by the manufacturer

Official reporting of problems and damage related to defined emissions-relevant components and systems has been mandatory for vehicle manufacturers since the 1990 model year. The reporting obligation remains in force for a period of 5 or 10 years, or 50,000 or 100,000 miles, depending on the length of the warranty applying to the component or assembly.

The reporting procedure consists of three stages

- Emissions Warranty Information Report (EWIR)
- Field Information Report (FIR) and
- Emission Information Report (EIR)

distinguished by progressively more detailed data at each consecutive level. Information concerning

- Problem reports
- Malfunction statistics
- Defect analysis and
- The effects on emissions

is then forwarded to the emissions authorities. The authorities use the FIR as the basis for issuing mandatory recall orders to the manufacturer.

Emissions-free vehicles

Starting in 2003 10 % of the new vehicles being registered for the first time will have to meet requirements for classification as ZEV, or Zero-Emission Vehicles. These vehicles must emit no exhaust or evaporative emissions during operation. This category essentially applies to electric cars.

While vehicles in the PZEV (Partial Zero-Emission Vehicles) category are not absolutely free of emissions, they do emit extremely low levels of pollutants. PZEV vehicles can also be used to comply with the standard mandating that 10 % of each vehicle fleet meet the ZEV standards. These vehicles are counted using a rating factor of 0.2...1 depending on the extent to which emissions have been reduced. The minimum weighting factor of 0.2 is granted when the following demands are met:

- SULEV certification indicating long-term compliance extending over 150,000 miles or 15 years
- Warranty coverage extending over 150,000 miles or 15 years on all emissions-relevant components
- No evaporative emissions from the fuel system (0 EVAP, Zero Evaporation), achieved through extensive encapsulation of tank and fuel system

Special regulations apply to hybrid vehicles with spark-ignition engines and electric motors. These vehicles can also contribute to achieving compliance with the 10 % limit.

EPA regulations

The EPA (Environment Protection Agency) regulations apply to the 49 states outside California, where the CARB stipulations are in force. The EPA regulations are not as strict as the CARB requirements. The individual states also have the option of adopting the CARB emissions regulations. This step has already been taken in some states, such as Maine, Massachusetts and New York. The legislative foundation is provided by the "Clean Air Act," which contains an action catalog with measures to protect the environment but does not specify actual limits.

The EPA regulations currently in force conform to the "Tier 1" standard. The next stage, "Tier 2", is slated to enter effect in 2004.

The NLEV (National Low Emission Vehicle) program is a voluntary undertaking aimed at reducing emissions in the 49 states (except California). Vehicles are classified in four emissions categories: Tier 1, TLEV, LEV and ULEV. As in California, average fleet fuel economy is then calculated based on NMOG emissions.

The NLEV program lapses upon introduction of the "Tier 2" emissions standards.

Limits

EPA regulations define limits on emissions of the following pollutants

- Carbon monoxide (CO)
- Nitrous oxides (NO_x)
- Non-methane organic gases (NMOG)
- Formaldehyde (HCHO) and
- Particulates

Pollutant emissions are determined using the FTP 75 driving cycle. Limits are defined in relation to distance and specified in grams per mile.

With the introduction of the Tier 2 standards, vehicles with diesel and spark-ignition engines will be subject to a single set of emissions standards.

Emissions categories

The Tier 1 standard defines limits on each regulated emissions component. Tier 2 classifies limits according to 10 emissions "bins" (Bin10...Bin1).

The transition to Tier 2 will produce the following changes:

- Introduction of fleet averages (analogous to the CARB regulations) for NO_x
- Formaldehydes (HCHO) will be subject to individual pollutant limits
- Passenger cars and light-duty trucks with AGVW up to 6,000 lbs (2.72 metric tons) will be combined in a single vehicle class
- The "full useful life" will be extended to 120,000 miles (192,000 kilometres)

Phase-in

At least 25 % of all new vehicles being registered for the first time will be required to conform to the Tier 2 standards once they take effect in 2004. The phase-in rule stipulates that an additional 25 % of the vehicles will then be required to conform to the Tier 2 standards in each consecutive year. All vehicles will be required to conform to the Tier 2 standard starting in 2007.

Fleet averages

NO_x emissions will be used in determining fleet averages for individual manufacturers under EPA regulations. This procedure is at variance with the CARB procedure, in which fleet averages are based on NMOG emissions.

Average fleet fuel consumption

The regulations defining average fleet fuel consumption in the 49 states are the same as those applied in California. Again, the limit applicable to passenger cars is 27.5 miles per gallon (8.55 litres per 100 kilometres). Beyond this figure manufacturers are required to pay a penalty. The purchaser also pays a penalty tax on vehicles providing less than 22.5 miles per gallon.

On-Board Diagnosis

The on-board diagnosis utility employed to detect emissions-relevant malfunctions under EPA regulations is basically the same as that prescribed by the CARB mandates.

Field monitoring

Unscheduled testing

The EPA regulations mirror the CARB laws by requiring random FTP 75 emissions inspections on vehicles in highway operation (in-use vehicles). Testing is conducted on low-milage vehicles (10,000 miles, roughly one year old) and higher milages (50,000 miles, and at least one vehicle per test group with 75,000/90,000 miles). The number of vehicles tested varies according to the number sold. At least one vehicle in each group is also tested for evaporative emissions.

Vehicle monitoring by the manufacturer

Since the 1972 model year, manufacturers have been obligated to submit mandatory reports detailing all known defects in defined emissions-relevant components and systems. Mandatory reports are required when defects occur in at least 25 similar emissions-relevant components in any model year. Reporting periods expire five years after the end of each model year. In addition to indicating the relevant components, the reports also contain descriptions of the defects and their effects on emissions as well as information concerning the remedial action undertaken by the manufacturer. The environmental authorities use this information as the basis for determining whether to issue recall orders to the manufacturer.

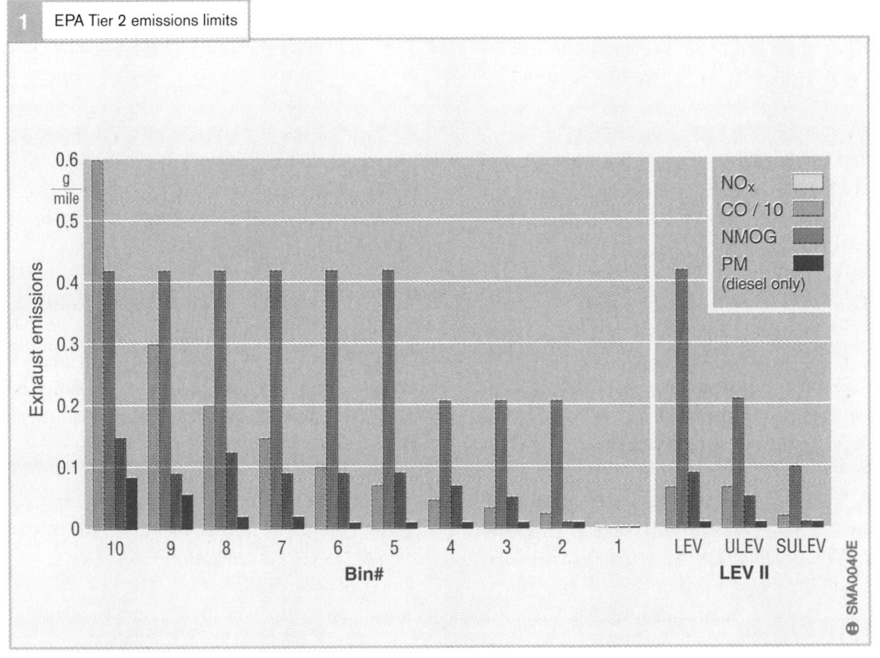

Fig. 1
The indicated figures apply to the "full useful life" (entire service life of 120,000 miles, 10 years)
The figures for the "intermediate useful life" of 50,000 miles or 5 years are lower for Bin 10...5, no data are specified for Bin 4...1
Bin 1 applies only for zero-emissions vehicles (electric vehicles, etc.)

EU regulations

The regulations contained in European Union directives are defined by the EU Commission. The emissions regulations defining emissions limits for passenger cars and light-duty trucks (LDT) are

- EU 1 (as from 1 July, 1992)
- EU 2 (as from 1 January, 1996)
- EU 3 (as from 1 January, 2000) and
- EU 4 (slated to come into effect on 1 January, 2005)

New emissions regulations are generally introduced in two stages. In the first stage, compliance with the newly defined emissions limits is required in vehicle models submitted for initial homologation approval certification (TA, Type Approval). In the second stage every new vehicle must comply with the new limits at initial registration (FR, First Registration). The authorities can also inspect vehicles from series production to verify compliance with emissions limits (COP, Conformity of Production).

The individual nations within the European Union can adopt the regulations defined in the EU 1 and EU 2 directives as national law. In Germany this proviso led to the creation of the D 3 and D 4 emissions levels. The earlier D 3 standards were stricter than the EU 2 regulations. Within the EU, Germany assumes the role of leader in implementation of new standards.

The EU 3 standards superseded the several national regulatory instruments then effective within the individual countries when it entered effect on January 1, 2000. The national regulations lapsed on this date. EU 4 assumes legal force in January, 2005.

Aside from the emissions standards, Germany also has vehicle tax rates based on emissions. EU directives allow "tax incentives" for vehicles that comply with upcoming standards before these actually become law.

Limits

The EU standards define limits for the following pollutants:

- Carbon monoxide (CO)
- Hydrocarbons (HC)
- Nitrous oxides NO_x and
- Particulates, although these limits are initially restricted to diesel vehicles

The limits are defined based on milage and indicated in grams per kilometer (g/km). Emissions are measured on a chassis dynamometer using the MNEDC (Modified New European Driving Cycle).

Levels EU 1 and EU 2 used a composite figure in assessing unburned hydrocarbons and nitrous oxides (HC+ NO_x). Separate limits, of the kind already employed with carbon monoxide (CO) for these two components, were introduced in EU 3.

The CO limit defined in EU 3 is actually somewhat higher than that in EU 2. This "worse" limit is explained by the fact that EU 3 also calls for exhaust emissions to be tested during starting. Earlier test procedures excluded the starting process, postponing actual monitoring to 40 seconds after the engine start. Because CO emissions are quite high in this phase, a direct comparison of the respective CO limits for EU 2 and EU 3 is impossible.

Although the limits in effect for diesels and spark-ignition engines currently differ, they are slated for harmonization at a future date.

Type approval testing

While type approval testing basically corresponds to the US procedures, deviations are encountered in the following areas: measurements of the pollutants HC, CO, NO_x are supplemented by particulate and exhaust-gas opacity measurements on diesel vehicles. Test vehicles absolve an initial break-in period of 3,000 kilometers before being subjected to testing. Deterioration factors for use in assessing test results are defined in the legislation; manufacturers are also allowed to present documentation confirming lower factors following specified long-term durability testing programs extending over 80,000 km (100,000 km starting with EU 4).

Compliance with the defined limits must be maintained over a distance of 80,000 km or 5 years. Verification of compliance is part of the certification test.

Directives

These emissions standards are based on EU Directive 70/220/EC dating from the year 1970. This directive placed the first official limits on exhaust emissions. The data have since been subjected to repeated updates.

Type tests

This directive defines six different test procedures:

The type I test evaluates exhaust emissions immediately following cold starts. Exhaust-gas opacity is also assessed on vehicles with diesel engines. While compliance with the EU 3 regulations is currently mandatory for new vehicles, many are already able to meet the limits defined in EU 4 (which comes into effect in 2005).

Type IV testing measures evaporative emissions from parked vehicles. These emissions consist primarily of the gases that evaporate from the fuel tank.

Type VI testing embraces hydrocarbon and carbon monoxide emissions immediately following cold starts at –7 °C. The first section of the MNEDC (urban portion) is employed for this test. This test assumed mandatory status in 2002.

1 Limits defined in EU legislation (vehicles with spark-ignition engine)

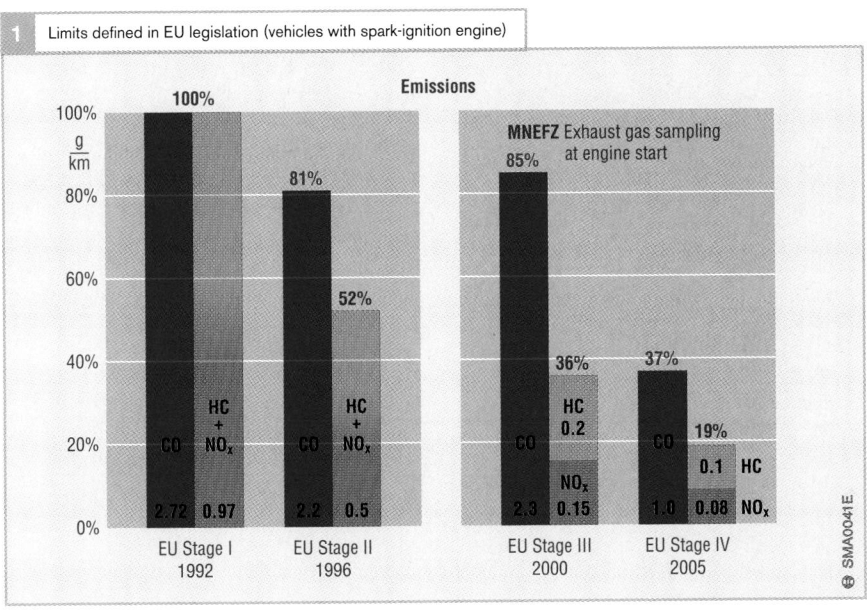

CO_2 emissions

Data indicating CO_2 emissions in grams per kilometer are specified for new vehicles being registered for the first time in the EU countries. Although no legal restrictions limit emissions of CO_2, which would also equate with a fuel-economy regulation, the vehicle manufacturers (ACEA, Association des Constructeurs Européen d'Automobiles) have united in promoting a voluntary program. By the year 2003, CO_2 emissions from Class M1 vehicles are not to exceed 165...170 grams per kilometer, corresponding to a fuel consumption rate of 6.8...7.0 l/100 km. The objective is to achieve CO_2 emissions of 140 g/km (5.8 l/100 km fuel consumption) by 2008.

Vehicles with extremely low CO_2 emissions currently enjoy tax advantages within Germany.

On-Board Diagnosis

Introduction of the EU 3 emissions standards has been accompanied by the advent of EOBD (European On-Board Diagnosis) for spark-ignition engines. These regulations call for a diagnosis system capable of detecting all malfunctions with the potential to affect emissions in all new passenger cars and light-duty trucks and vans with up to 9 seats and an approved gross vehicle weight of up to 3.5 metric tons. The EOBD requirement is being extended to embrace vehicles with diesel power plants starting on 1. 1. 2003.

The following absolute emissions limits are defined as error thresholds for pollutant concentrations:

- Carbon monoxide CO: 3.2 g/km
- Hydrocarbons HC: 0.4 g/km
- Nitrous oxides NO_x: 0.6 g/km

Greenhouse effect

Shortwave solar radiation penetrates the earth's atmosphere and continues to the ground, where it is absorbed. This process promotes warming in the ground, which then radiates long-wave heat, or infrared energy. A portion of this radiation is reflected by the atmosphere, causing the earth to warm.

Without this natural greenhouse effect the earth would be an inhospitable planet with an average temperature of −18 °C. Greenhouse gases within the atmosphere (water vapor, carbon dioxide, methane, ozone, dinitrogen oxide, aerosols and particulate mist) raise average temperatures to approximately +15 °C. Water vapor, in particular, retains substantial amounts of heat.

Carbon dioxide has risen substantially since the dawn of the industrial age more than 100 years ago. The primary source of this increase has been combustion of coal and petroleum products. In this process the carbon bound within the fuels is released in the form of carbon dioxide.

The processes that influence the greenhouse effect within the earth's atmosphere are extremely complex. While some scientists maintain that anthropogenic (of human origin) emissions are the primary source of climate change, this theory is challenged by other experts, who believe that the warming of the earth's atmosphere is being caused by increased solar activity.

There is, however, a large degree of unanimity in calling for reductions in energy use to lower carbon-dioxide emissions and combat the greenhouse effect.

The EOBD system must respond to detection of malfunctions causing the vehicle to exceed the specified limits by triggering the error lamp after no more than three driving cycles. The system also starts to record milage when the alert is issued.

This malfunction lamp can extinguish again should three subsequent driving cycles elapse with no detected errors.

Field monitoring

EU legislation also calls for conformity-verification testing on in-use vehicles as part of the Type I test regimen. The minimum number of vehicles to be tested is three, while the maximum number varies according to the test procedure.

Vehicles selected for testing must meet specific criteria:

- The vehicle model must have been granted previous type approval in accordance with applicable regulations, and certification of conformity must be present.
- The milage and vehicle age must lie between 15,000 km/6 months and 80,000 km/5 years (starting with Euro 3 in 2000) or 100,000 km (Euro 4, 2005).
- Proof of regular periodic service inspections as specified by the manufacturer must be available.
- The vehicle is to display no indications of non-standard use (manipulation, major repairs, etc.).

If emissions from an individual vehicle fail substantially to comply with the standards, the source of the high emissions must be determined. If more than one vehicle from a series displays excessive emissions in random testing, then the results of the test are classified as negative. As long as the maximum number of vehicles specified for the random testing series has not been exceeded, an additional vehicle may be subjected to testing in response to various scenarios.

If the authorities arrive at the conclusion that a particular vehicle model does not conform with the legal requirements, they can respond by demanding a remedial-action plan from the manufacturer. The action catalog must be applicable to all vehicles displaying the same defect. Implementation of the action plan can also entail of a vehicle recall.

Periodic emissions inspections (AU)

Within the Federal Republic of Germany, all passenger cars and light-duty trucks and vans are required to undergo emissions inspections (AU) three years after their initial registration, and then at subsequent intervals of two years. This test consists essentially of a measurement of emissions and calculation of the corresponding lambda figure.

Operation of the lambda control system is also examined on vehicles equipped with closed-loop-controlled catalytic converter. Testing of such vehicles which also feature on-board diagnosis (OBD) also includes a readout of error codes stored in the malfunction log and examination of the readiness code as one of several supplements to the CO measurements.

US test cycles

FTP 75 test cycle

The FTP 75 test cycle (Federal Test Procedure) consists of three phases, and represents the speeds and conditions actually recorded in morning commuter traffic in the US city of Los Angeles (Figure 1a):

Preconditioning

The vehicle to be tested is first conditioned (allowed to stand with engine off for 12 hours at a room temperature of 20...30 °C), then started and run through the prescribed test cycle:

Collecting emissions

The emitted toxic emissions are collected separately during various phases.

Phase ct: During the cold transition phase, diluted exhaust gases are collected in bag 1 for the CVS test (refer to section on "emissions testing").

Phase s: Exhaust gases are diverted to sample bag 2 at the beginning of the stabilized phase (after 505 s) without any interruption in the driving cycle. Upon termination of phase s, after a total of 1,365 seconds, the engine is switched off for a period of 600 seconds.

Phase ht: The engine is restarted for hot testing, which employs the speed curve used for the cold transition phase (Phase ct) in unmodified form. Exhaust gases are collected in a third sample bag.

Analysis

The bag samples from the previous phases are analysed during the pause before the hot test, as samples should not remain in the bags for longer than 20 minutes.

The sample exhaust gases contained in the third bag are also subjected to analysis following completion of the driving cycle. The results of the three individual phases are added using the weighting factors 0.43 (ct phase), 1 (s phase) and 0.57 (ht phase). The test distance is then incorporated in the calculations, and the weighted sums of the emissions (HC, CO and NO_x) from all three bags are converted into emissions per mile.

Outside the USA and California this test is also employed in various other countries (in South America, etc.).

SFTP schedules

The tests defined by the SFTP standard are being introduced in stages between 2001 and 2004. These are composites including the following driving cycles:

- FTP 75
- SC03 and
- US06

This extended test routine allows assessment of the following vehicle operating conditions (Figure 1b, c):

- Aggressive driving
- Radical changes in vehicle speed
- Engine start and acceleration from a standing start
- Operation with frequent minor variations in speed
- Periods with vehicle parked and
- Operation with air conditioner on

Following preconditioning, the SC03 and US06 cycles proceed through the ct phase from FTP 75 without exhaust-gas collection. Other conditioning procedures may also be used.

The SC03 cycle is carried out at a temperature of 35 °C and 40 % relative humidity (vehicles with air conditioning only). The individual driving schedules are weighted as follows:

- Vehicles with air conditioning: 35 % FTP 75 + 37 % SC03 + 28 % US06
- Vehicles without air conditioning: 72 % FTP 75 + 28 % US06

The SFTP and FTP 75 test cycles must be successfully completed on an individual basis.

Cold-start enrichment, which is necessary when a vehicle is started at low temperatures, produces particularly high emissions. These cannot be measured in current emissions testing, which is conducted at ambient temperatures of 20...30 °C. A supplementary emissions test is performed at –7 °C to support enforcement of limits on these emissions. However, only carbon monoxide is subject to specified limits in this test.

Test cycles for determining fleet averages

Each vehicle manufacturer is required to provide data on fleet averages. Manufacturers that fail to comply with the specified limits are required to pay penalties. A bonus is awarded for figures that lie below specified levels. Fuel consumption is determined based on exhaust emissions produced during two test cycles: the FTP 75 test cycle (55 %) and the highway test cycle (45 %).

An unmeasured highway test cycle (Figure 3d) is conducted once after preconditioning (vehicle allowed to stand with engine off for 12 hours at 20...30 °C). The exhaust emissions from a second test run are then collected. The emissions can be used to calculate fuel consumption.

Compliance with emissions limits

Each new vehicle being registered for the first time must comply with the emissions limits over a defined distance. The same limits apply regardless of vehicle weight and engine displacement.

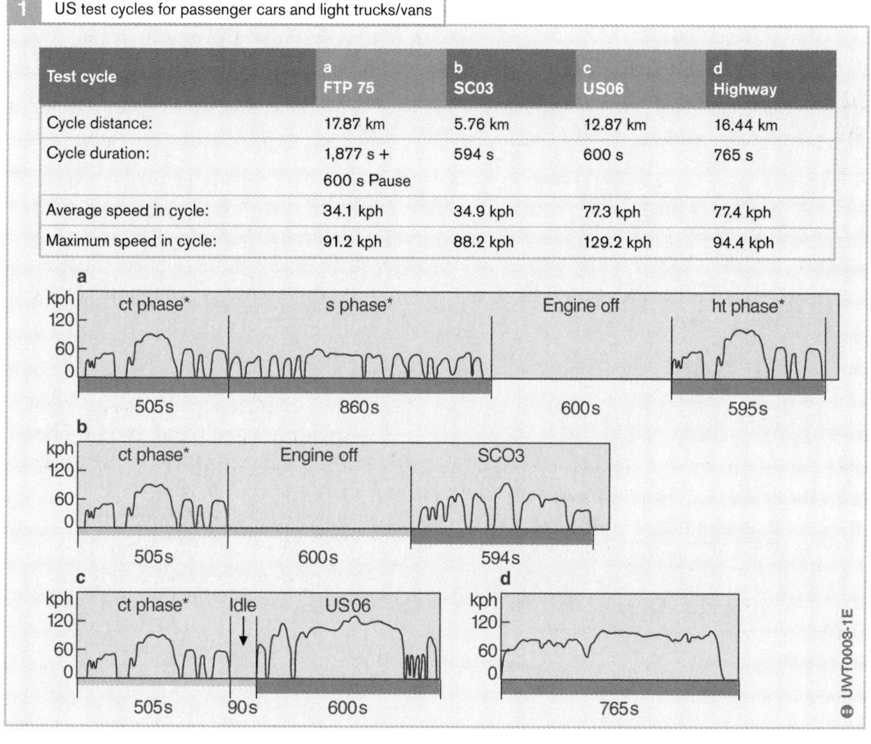

1 US test cycles for passenger cars and light trucks/vans

Test cycle	a FTP 75	b SC03	c US06	d Highway
Cycle distance:	17.87 km	5.76 km	12.87 km	16.44 km
Cycle duration:	1,877 s + 600 s Pause	594 s	600 s	765 s
Average speed in cycle:	34.1 kph	34.9 kph	77.3 kph	77.4 kph
Maximum speed in cycle:	91.2 kph	88.2 kph	129.2 kph	94.4 kph

Fig. 1
* ct transitional phase
s stabilized phase
ht hot test
Exhaust-gas collection phases
Preconditioning (may consist of other driving cycles)

The USA grants exemptions for various model years under specified conditions. The law distinguishes between limits for 50,000 and 100,000 miles. The limits for the 100,000 mile mark are somewhat more lenient (deterioration factor).

Other test cycles

FTP 72

The FTP 72 test routine – also known as the UDDS (Urban Dynamometer Driving Schedule) – corresponds to the FTP 75 test, but does not include the *ht* test component (hot test). This cycle is completed during the running loss test.

New York City Cycle (NYCC)

This cycle is also carried out during the running loss test.

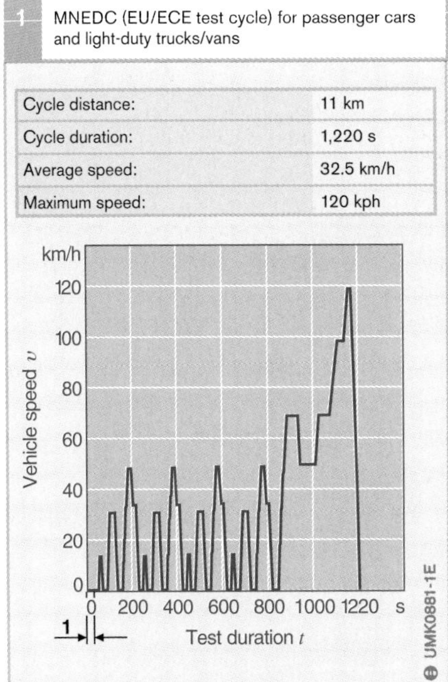

Cycle distance:	11 km
Cycle duration:	1,220 s
Average speed:	32.5 km/h
Maximum speed:	120 kph

1 MNEDC (EU/ECE test cycle) for passenger cars and light-duty trucks/vans

Fig. 1

1 Pretest running phase (no gas collection); formerly 40 s, abolished starting with EU stage III

European test cycle

The EU/ECE test cycle (Economic Commission of Europe) – also known as the European test cycle – employs a driving curve (Figure 1) representing a close approximation of urban operation (UDC, Urban Driving Cycle). In 1993 the cycle was extended to include a rural component with speeds of up to 120 km/h (EUDC, Extra-Urban Driving Cycle). The composite test cycle uniting these two tests is referred to as the NEDC (New European Driving Cycle).

The 40 second waiting period prior to the start of actual measurements was deleted for EU 3 testing (2000) (MNEDC, Modified New European Driving Cycle). It thus includes measurements of emissions immediately following starting.

Preconditioning

Prior to emissions testing, vehicles must remain parked for a period of at least 6 hours at a defined temperature. This temperature is currently 20...30 °C. Starting in 2002 the starting temperature for Type VI testing was lowered to –7 °C.

Urban cycle

The urban cycle consists of four sections, all lasting 195 seconds, performed in immediate and uninterrupted sequence. The distance is 4.052 km, which produces an average speed of 18.7 kph. The maximum speed is 50 km/h.

Extra-urban cycle

The urban cycle is followed by operation at speeds of up to 120 kph. This section lasts 400 seconds, and extends over a distance of 6.955 kilometers.

Analysis

During measurement, exhaust gas is collected in a sample bag using the CVS method. The mass of the pollutants ascertained in analysis of the bags' contents is converted to mass per unit of distance.

While hydrocarbons and nitrous oxides were formerly combined in a composite figure ($HC + NO_x$), these components have been subject to individual assessment since the introduction of EU 3.

1 Japanese test cycles for passenger cars and light-duty trucks/vans

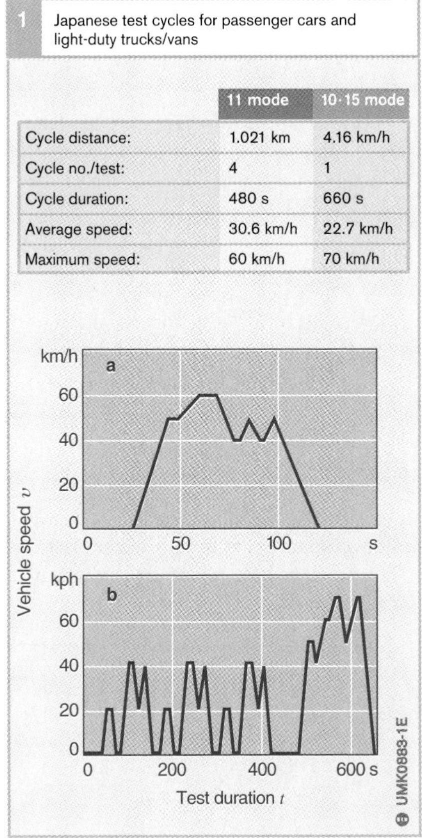

	11 mode	10·15 mode
Cycle distance:	1.021 km	4.16 km/h
Cycle no./test:	4	1
Cycle duration:	480 s	660 s
Average speed:	30.6 km/h	22.7 km/h
Maximum speed:	60 km/h	70 km/h

Fig. 1
a 11-mode cycle (cold test)
b 10·15-mode cycle (hot test)

Japanese test cycles

The overall test is composed of two test cycles based on different, synthetically generated driving curves. Following a cold start, the 11-mode cycle is run four times, with evaluation of all four cycles. The 10·15-mode test cycle is conducted once with a hot start. This test cycle, simulating typical operating conditions in Tokyo, has been expanded to include a high-speed component. However, the top speed is lower than that used in the European test cycle, as the higher traffic densities in Japan result in generally lower driving speeds.

The preconditioning procedure for the hot test includes the mandatory idle emissions test. The routine is as follows: After the vehicle is allowed to warm up during approximately 15 minutes of operation at 60 kph, the concentrations of HC, CO and CO_2 are measured in the exhaust pipe. The 10·15-mode hot test commences after a second warm-up phase consisting of 5 minutes at 60 km/h. Both the 11-mode test and the 10·15-mode test rely on CVS equipment for exhaust-gas analysis. Exhaust gases are diluted and collected in bags during each test. While the cold test examines pollutants in grams per test, the hot test's results are defined relative to distance, and are indicated in grams per kilometer.

The exhaust gas regulations in Japan include limits on evaporative emissions, which are measured using the SHED method.

Emissions testing

Vehicle testing must reflect real-world conditions to allow precise assessment of vehicle emissions. Compared to highway driving, operation in test cells offers the advantage of allowing tests to be conducted at precisely predefined speeds, without distortions arising from variations in traffic flow patterns. This is essential in obtaining reproducible test results suitable for comparison purposes.

Test structure

The test vehicle is parked on a chassis dynamometer with its drive wheels on the rollers (Figure 1, Pos. 15).

The various forces applied to the vehicle – such as inertia, rolling resistance and aerodynamic drag – must be simulated to ensure that the emissions generated on the chassis dynamometer correspond to those produced in actual highway operation. Asynchronous motors, DC generators and eddy-current brakes (1) can be used to generate loads that reflect those encountered at various vehicle speeds. These forces act upon the rollers and must be overcome by the vehicle. On more modern systems, electronic simulation of the centrifugal or flywheel mass is used for the inertia simulation. Older test stands use actual oscillating masses (2) which can be connected to vehicles with rapid-action couplings to obtain simulated vehicle mass.

Precise adherence to the curve for load over vehicle speed, and maintenance of required inertial masses, are vital. Deviations lead to inaccurate test results. Environmental factors such as humidity, temperature and barometric pressure also affect the results. A fan placed immediately forward of the vehicle provides the required engine cooling.

1 CVS test methods for passenger cars and light-duty trucks/vans

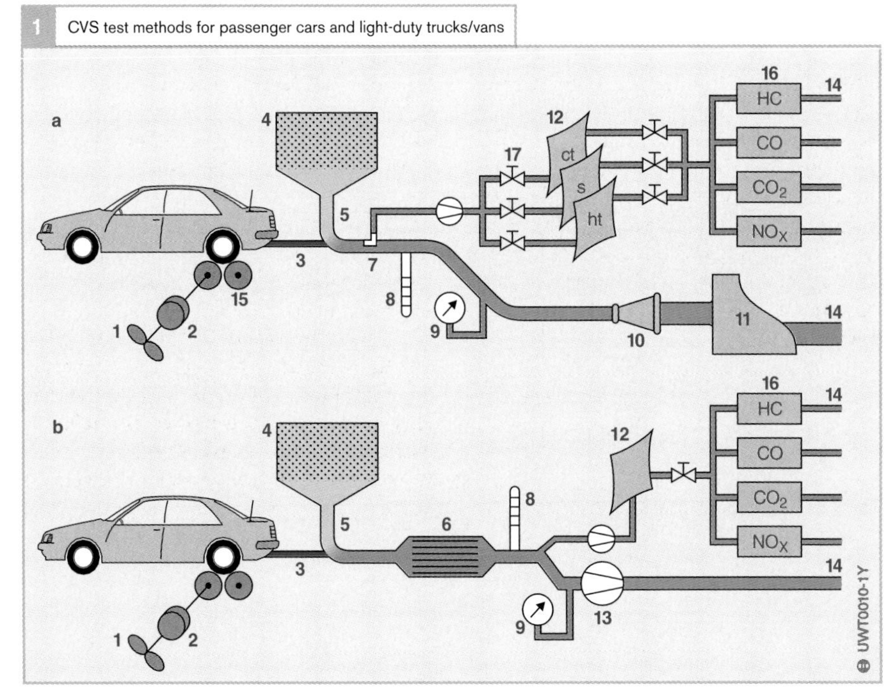

Fig. 1
a for US Federal test (here with venturi system)
b for European test (here with rotary-piston fan)

1 Brake
2 Inertial mass
3 Exhaust gas
4 Air filter
5 Dilution air
6 Radiator
7 Sample venturi nozzle
8 Gas temperature
9 Pressure
10 Venturi nozzle
11 Fan
12 Collection bag
13 Rotary-piston fan
14 to discharge
15 Rollers
16 Analyser
17 Changeover valves

ct Exhaust gases from transitional phase
s Exhaust gases from stabilized phase
ht Exhaust gases from hot test

Dilution procedure (CVS)

The CVS dilution method (Constant Volume Sampling) represents one procedure for collecting the exhaust gas emitted by the engine. Originally introduced for passenger cars and light-duty trucks and vans in the US in 1972, CVS technology has since evolved through several stages. Europe converted to the CVS method in 1982. This means that a single concept for collecting exhaust-gas samples is now in global use.

Concept

The concept employed in the CVS method is as follows: the exhaust gases emerging from the test vehicle are diluted at a mean ratio of 1:5 ... 1:10. The gases are then extracted by a special system of pumps designed to maintain the respective flow rates of exhaust gases and fresh air at a specific volumetric ratio. The system thus regulates air-feed rate in line with the vehicle's instantaneous exhaust volume. Throughout the test, a constant proportion of the diluted exhaust gas is extracted for storage in several sample bags.

The pollutant concentration in the sample bags at the end of the test cycle corresponds precisely to the mean concentration in the total quantity of fresh-air/exhaust mixture which has been extracted. Because the total volume of the fresh-air/exhaust mixture can be defined, pollutant concentrations can be used as the basis for calculating the pollutant masses produced during the course of the test.

Advantages of the CVS method

Due to the dilution, condensation of the water vapor contained in the exhaust gases is avoided. This provides a substantial reduction in the rate of nitrous-oxide loss during the residence time in the bag. In addition, dilution greatly inhibits the tendency of the exhaust components (especially hydrocarbons) to support mutual secondary reactions.

However, dilution does mean that pollutant concentrations decrease proportionally as a function of the mean dilution ratio, necessitating the use of more sensitive analysis equipment. Standardized equipment is available for analysis of the pollutants in the bags.

Properties of the CVS test method

The CVS method is distinguished by the following:

- Results based on actual exhaust-gas volumes generated by the engine during testing
- Accurate registration of all stationary and non-stationary vehicle operating conditions
- Avoids condensation of water vapor and unburned hydrocarbons and
- Provides technically precise measurements of particulate emissions

Dilution systems

One of two different but equally acceptable pump arrangements is generally used to maintain a constant flow volume during the test. In the first, a standard blower extracts a mixture of exhaust gas and fresh air through a venturi nozzle. In the second, a special rotary-piston fan (Roots blower) is employed. Either method is capable of measuring the flow volume with an acceptable degree of accuracy.

Testing diesel vehicles

In the USA the CVS method has also been applied in testing diesel-powered vehicles since 1975. Modifications in both sampling modalities and analysis equipment used in measuring hydrocarbons had to be introduced for this application. The entire system used to extract the samples must be heated to roughly 190 °C to prevent condensation of heavy, low-volatility hydrocarbons in the gas samples. Heating also inhibits recondensation of the condensed hydrocarbons initially present in the diesel's exhaust.

Inclusion of particulate limits in the emissions legislation also led to modification of the CVS method. A dilution tunnel with increased internal flow turbulence (Reynolds number > 40,000) and the required filter test points were integrated in the system to allow collection of particulate emissions.

Measurement concepts

All of the nations that have adopted the CVS method for verifying compliance with their emissions legislation, employ standard test concepts for analysis of exhaust and other emissions:

- Concentrations of CO and CO_2 using non-dispersive infrared (NDIR) analysers
- Concentrations of NO_x based on the CLD chemiluminescence principle
- Gravimetric quantification of particulate emissions (particulate filters conditioned and weighed before and after collection of sample charges)
- Concentrations of total hydrocarbons using the flame ionisation method (FID)

Evaporative-emissions testing

Apart from the emissions stemming from the gasoline-engine's combustion processes, vehicles also emit hydrocarbons (HC) when fuel evaporates in the tank and within the fuel system. The amount of fuel that evaporates varies according to the fuel temperature and the individual vehicle's design configuration. Some nations (such as the US and European countries) have adopted regulations limiting these evaporative losses.

Test principle

These evaporative emissions are usually quantified with the aid of a hermetically sealed SHED (**S**ealed **H**ousing for **E**vaporative **D**etermination) chamber. For the test, HC concentrations are measured at the beginning and the end of the test, with the difference representing the evaporative losses.

Evaporative emissions are measured under some or all of the following conditions – depending upon individual country – and must comply with the stipulated limits:

- Evaporation emerging from the fuel system in the course of the day: "tank ventilation test" or "diurnal test" (EU and USA).
- Evaporation that emerges from the fuel system when the vehicle is parked with the engine warm following operation: "hot parking test" or "hot soak" (EU and USA).
- Evaporation during on-the-road operation: "running-loss test" (USA).

Evaporative emissions are measured in a precisely defined test sequence embracing a number of phases. Prior to testing, the vehicle undergoes preconditioning in a process including the activated charcoal canister. With the tank filled to the stipulated level of 40%, testing starts.

Tests

1st test: Hot-soak losses

Before testing to determine evaporative emissions in this phase, the vehicle is first warmed to normal operating temperature using the test cycle valid in the particular country. It is then parked in the SHED chamber. The increase in HC concentration within a period of 1 hour is measured during the vehicle's cooling period.

The vehicle's windows and trunk lid must remain open throughout the test. This makes it possible for the test to include evaporative losses from the vehicle's interior in its results.

2nd test: Tank-ventilation losses

For this test, a typical temperature profile for a warm summer day (maxima of 35 °C in the EU, 35.5 °C in EPA testing and 40.6 °C for CARB, respectively) is simulated within the hermetically sealed climate chamber. The hydrocarbons emitted by the vehicle under these conditions are then collected.

The USA requires testing in both 2-day diurnal (48-hour) and 3-day diurnal (72-hour) procedures. EU legislation prescribes a 24-hour test.

Running-loss test

The running-loss test is conducted prior to the hot-soak test. It is used to assess the hydrocarbon emissions generated during vehicle operation in the prescribed test cycles (1 FTP 72 cycle, 2 NYCC cycles, 1 FTP 72 cycles; refer to section on "US test cycles").

Emissions limits

EU regulations

The sum of the results from the first and second tests provides the evaporative losses. This sum must remain below the currently required limit specifying 2 grams of evaporated hydrocarbons for the entire test series.

USA

In the USA (CARB and EPA Tier 1), the evaporative losses monitored in the running-loss test must remain below 0.05 g per mile. Other limits are:

- 2-day diurnal: 2.5 g (sum from the first and second test)
- 3-day diurnal: 2.0 g (sum from first and second tests)

Compliance with these limits must be proved over a distance of 100,000 miles.

The EPA has promulgated even tighter limits in the Tier 2 regulations:

- 2-day diurnal: 1.2 g (sum from the first and second test)
- 3-day diurnal: 0.95 g (sum from first and second tests)

Compliance with these limits must be proved over a distance of 120,000 miles. They are to be introduced in stages, starting with the 2004 model year, and 100 % compliance will be required for the 2007 model year.

Other tests

Refuelling test

The refuelling test monitors evaporation of fuel vapors emitted during refuelling by measuring HC emissions (limit: 0.053 g HC per litre of fuel supplied to the tank).

In the US this test is used in both CARB and EPA procedures.

Spit-back test

The spit-back test monitors the amount of fuel spray generated during each refuelling process. The tank must be refuelled to at least 85 % of its total volume.

This test is conducted only in response to failure to successfully pass the refuelling test (limit: 1 g HC per test).

Service technology

Important!
This chapter provides general descriptions of service technology, and is *not* intended to replace repair and instruction manuals! Repairs should always be performed by qualified professional technicians!

Over 10,000 Bosch service centres in 132 countries are standing by to provide motorists with assistance. And, because Bosch centres do not represent the interests of any one vehicle manufacturer, this help is neutral and unbiased. Fast assistance is always available, even in sparsely populated countries in South America and Africa. And the same quality standards apply everywhere. It is thus no wonder that the Bosch service warranty is valid throughout the world.

Overview

The specifications and performance data of Bosch components and systems are precisely matched to the requirements of each individual vehicle. Bosch also develops and designs the test equipment, special tools and diagnosis technology needed for tests and inspections.

General-application test equipment from Bosch – extending from basic battery testers to comprehensive vehicle inspection bays – is used by vehicle service facilities and official inspection agencies throughout the world.

Service personnel receive training in the efficient use of this test technology as well as information focusing on a range of automotive systems. Meanwhile, feedback from our customers flows into the development of new products.

The service AWN

Test technology

It is still possible to test mechanical systems in motor vehicles using relatively basic equipment, but mastering the increasingly complex electronic systems found in mod-

[1] Bosch service technology stems from development activities carried out by the Bosch AWN service network. The "asanetwork GmbH" is responsible for advanced development and marketing under the "AWN" name.

1 The AWN service network [1]

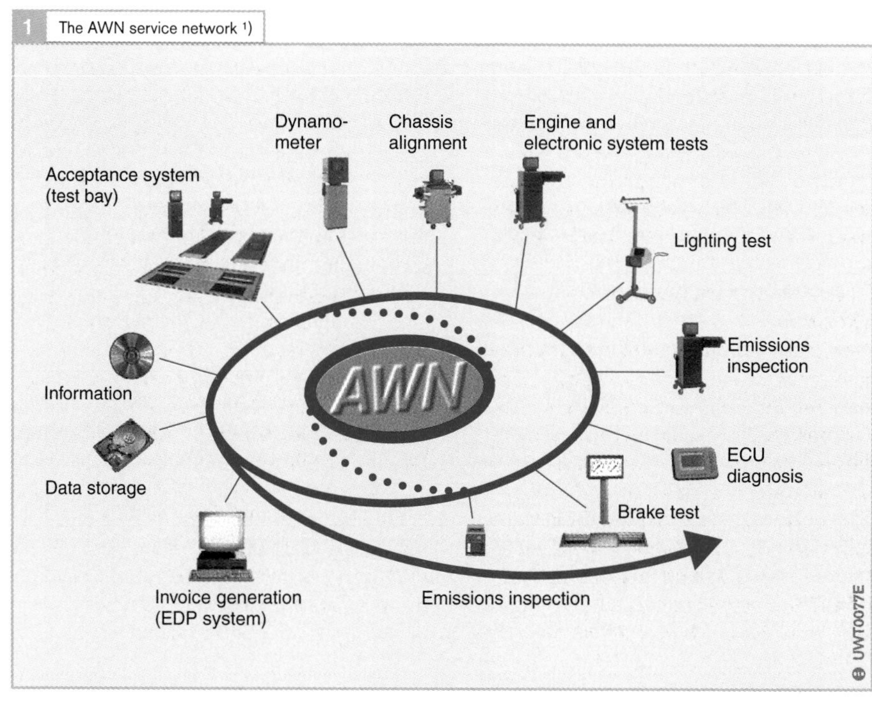

ern vehicles means using new test methods relying on electronic data processing. Tomorrow's technology is represented by the AWN service. This was conceived to link the entire range of workshop data-processing systems within a single integrated network (Figure 1). This concept earned Bosch the 1998 Automechanika Innovation Prize in the "service" category.

Test sequence

When a vehicle arrives for a service inspection the job order processing system's database furnishes immediate access to all available information on the vehicle. Immediately the vehicle enters the shop, the system provides access to its entire history. This includes all service work and repairs carried out on the vehicle up to that point.

Individual diagnostic testers provide the data needed for direct comparisons of specified results and current readings, without the need for supplementary entries. All service procedures and replacement components are recorded to support the invoicing process. Following the final road test the invoice can be generated with nothing more than a few key entries. The system also provides a clear and concise printout with the results of the vehicle diagnosis. This offers the customer a clear and precise protocol detailing all of the service operations and material that went into the vehicle's repair.

Electronic Service Information ESI[tronic]

Even in the past, the wide variety of vehicle makes and models made data control systems essential (with part numbers, test specifications, etc.). Large data records, such as those containing information on spare parts, are contained on microfiche cards. Microfiche readers provide access to these microfiche libraries, and are still standard equipment in every automotive service facility.

In 1991, Bosch introduced its ESI[tronic] (Electronic Service Information) system on CD ROM for use with PCs. ESI[tronic] vastly increases data-storage potential. The system offers additional application options extending beyond those available from microfiche cards. It can also be incorporated in electronic data processing networks.

Application

The ESI[tronic] software package supports service personnel throughout the entire vehicle repair process by providing the following information:

- Spare-part identification (correlating spare-part numbers with specific vehicles, etc.)
- Work units
- Repair instructions
- Vehicle circuit diagrams
- Test specifications and
- Vehicle diagnosis

Service technicians can select from two available options to diagnosis problems and malfunctions: the KTS500 is a high-performance portable system tester, while the KTS500C has been designed to run on the PCs used in service areas (diagnosis stations). The latter consists of a PC adapter card, a slot card (KTS) and a test module for measuring voltage, current and resistance. The interface allows ESI[tronic] to communicate with the electronic systems within the vehicle, such as the engine-management ECU. Working at the PC, the technician starts by selecting the SIS (Service Information System) utility to initiate diagnosis of on-board control units and access the engine-management ECU's malfunction log. ESI[tronic] uses the results of the diagnosis as the basis for generating specific repair instructions. The system also provides displays with other information, such as component locations, exploded views of assemblies, diagrams showing the layouts of electrical, pneumatic and hydraulic systems, etc. Using the PC, the technician can then proceed directly from the exploded view to the parts list with part numbers, and order the required replacement components.

Testing on-board control units

Each of the electronic control systems in the vehicle (Motronic, etc.) is equipped with an ECU diagnosis feature for the extensive and comprehensive testing of the complete electronic system.

Test equipment

Specialised test equipment is essential for effective system analysis. While earlier electronic systems could be tested with basic equipment such as a multimeter, ongoing advances have resulted in electronic systems that can only be diagnosed with complex testers.

1 KTS500C diagnosis station

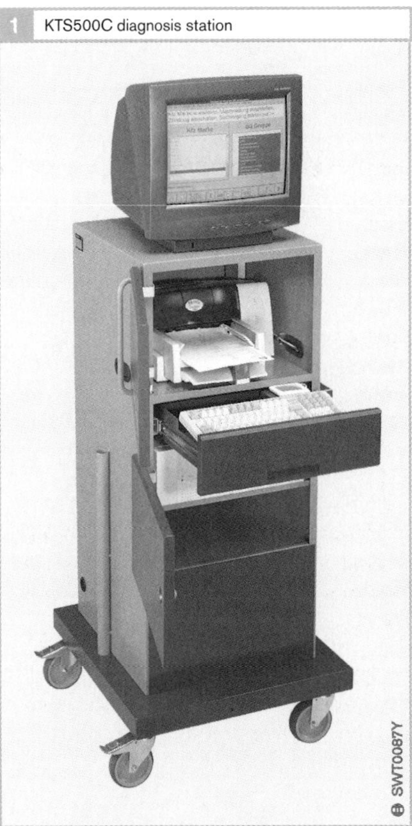

Test equipment from the KTS Series is in widespread use in service operations. Examples include the portable KTS500 and the KTS500C diagnosis station for use with ESI[tronic] (Figure 1). With its extensive diagnosis and test features, including graphics (to portray test results, etc.), this equipment offers a range of options for service use. In its descriptions of system test equipment, the following text also uses the alternate term "system tester."

KTS500 functions

KTS500 offers an extended range of functions. User access is through menus, and the selection process is guided and controlled using entry keys and a large-format display. The following is a list of the most important functions offered by the KTS500.

Stored error-ode readout: The KTS500 can be used to read out the error codes stored during the vehicle control unit's diagnostic processes. The KTS500 accesses the malfunction log to provide plain-text error descriptions on the display screen.
Instantaneous data readouts: Monitored data, such as the results of processing operations in the engine-management ECU, are converted to physical units for user viewing (engine speed in rpm, etc.).
Multimeter function: Measurements of current, voltage and resistance levels as with a conventional multimeter.
Performance curves: Monitored data (O2-sensor voltages, etc.) can be shown in graphic displays as signal curves similar to those available from an oscilloscope.
Supplementary information: Special supplementary information related to components and indicated problems can also be selected for viewing (component locations, component test specifications, electrical circuit diagrams, etc.).
Printouts: All data can be printed out on a standard DIN A4 PC printer (with a list of the test results, as documentation for the customer, etc.).

In the workshop, the capabilities of the KTS500 are utilized to the extent demanded by the particular system being tested. Not all the vehicle's systems are able to support the complete functionality.

Yet another advantage is the ability to link ECU diagnosis with the ESI[tronic] Electronic Service Information system using CAS (Computer-Aided Service). As part of the problem-diagnosis process, control-unit diagnosis is initiated from the ESI[tronic] interface. ESI[tronic] displays possible error codes for use in the focused repair processes it supports.

Standard service procedure

The basic diagnostic procedure is the same for all electronic systems. The most important tool is the diagnostic tester, which is connected to the vehicle's electronic control units at the diagnostic interface socket.

Vehicle identification

The first step is to select the vehicle model. This information must be entered in the system tester so that it can access the data required to test the specific vehicle.

Error-code readout

Most systems in the vehicle are equipped with ECU diagnosis to test electrical components for malfunctions. Codes identifying detected problems are stored in the malfunction log, accompanied by information concerning

- Error path (engine-temperature sensor, etc.)
- Problem class (short to ground, implausible signal, etc.)
- Error status (consistent problem, sporadic malfunction, etc.)
- Environmental conditions (data on engine speed, temperatures, etc., monitored when error code is stored)

"Malfunction log" can be selected to initiate a data transmission from the control unit to the system tester. The corresponding problem descriptions appear on the display screen in plain text along with the error path, malfunction location, error status, etc.

Trouble-shooting

Not all sources of engine malfunctions can be identified using control-unit diagnosis. Service technicians must also be able to respond to this type of problem with fast and effective diagnosis and repair.

The ESI[tronic] electronic service information system provides assistance with the diagnostic process in both cases, with or without stored error codes. The diagnosis instructions furnished by the system cover all imaginable problems (engine surge, for instance) as well as specific malfunctions (short circuit in engine-temperature sensor, etc.)

Repair

After localizing the problem source using the information provided by ESI[tronic], the user proceeds to the repair phase.

Deleting stored error codes

Once the defect has been repaired, the corresponding error code(s) must be cleared from the malfunction log. The user proceeds by selecting the system tester's "delete stored error codes" function.

Road test

The next stage is a road test intended to confirm that the problem has been resolved. During the road test the control unit diagnosis utility runs a system check, and responds to any detected problems by storing a new error code.

Checking malfunction log

The malfunction log is accessed again following the road test. It should be empty at this point. The repair has now been successfully completed.

Other test methods

In addition to its standard functions, the KTS500 system tester also provides supplementary tools for use in diagnosing electronic systems. The system tester triggers the functions and the control unit executes them.

Actuator diagnosis

Many control-unit functions (such as fuel-tank purge, etc.) are activated only under specific vehicle operating conditions. This means that it is not possible to trigger every actuator (servo device) individually for operation testing (canister-purge valve, etc.) without using special equipment.

The system tester makes it possible to trigger actuators for diagnosis in the workshop (Figure 2a). Correct operation can then be verified using visual or acoustic feedback.

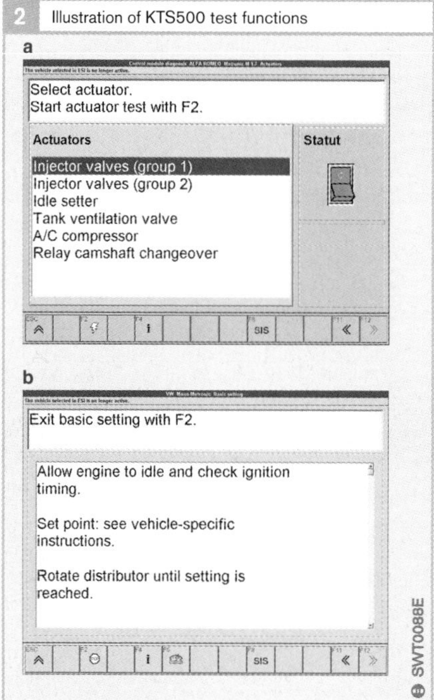

Fig. 2
a Triggering actuator diagnosis
b Check ignition timing

This actuator-diagnosis facility tests the entire electrical circuit from the engine-management ECU and through the wiring harness to the actuator as well as checking operation of the actuator per se.

Actuator-diagnosis processes are usually performed for specifically limited periods with the vehicle stationary. The limitation on test duration is intended to prevent damage to actuators and motors. Injectors are also triggered for a minimum period to avoid catalyst damage due to fuel discharge.

Signal testing

In the event of malfunction, the technician can use the oscilloscope to examine the patterns of the trigger signals. This is especially important in dealing with actuators that are not included in the actuator-diagnosis process.

New test functions with modern control units

ECU initialisation
In the ECU initialisation the engine-management ECU is initialised to recognise the throttle plate's closed position (following component replacements during repairs).

Figure 2b uses the KTS500 display to show how this function is employed.

Control-unit release activation (version code)
When a new engine-management ECU is installed the correct version codes for the vehicle must be entered.

Immobilizer (drive-away protection)
Following replacement or repair of an engine-management ECU, the immobilizer must be reinitialised for operation with the new unit.

Kickdown point
If the vehicle is equipped with an automatic transmission, the kickdown shift point must be tested and adjusted through the control unit.

Global service

"When you ride in a motorcar you will discover that horses are really incredibly boring (...). But the car needs a conscientious mechanic (...)."

Robert Bosch wrote these lines to his friend Paul Reusch in 1906. In those days it was indeed possible for the hired chauffeur or mechanic to repair problems. Later on, however, in the period following the First World War, rising numbers of motorists driving their own vehicles led to a corresponding increase in the demand for service facilities. In the 1920's Robert Bosch launched a campaign aimed at creating a comprehensive service organisation. Within Germany these service centres received the uniform designation "Bosch-Dienst" in 1926

Today's Bosch operations bear the name "Bosch Car Service". They feature state-of-the-art electronic service equipment to meet the demands defined by the automotive technology and the customer expectations of today.

1 A repair operation in the year 1925 (Photo: Bosch)

2 Bosch Car Service in 2001, with state-of-the-art electronic test equipment

Testing the ignition system

Caution!
Universal rule: During operation, dangerous voltages are present in the ignition system's primary and secondary circuits. Always avoid all contact with these components when ignition is on!
Service operations on the ignition system should be performed exclusively by individuals with the required technical background.
Always observe all prescribed safety precautions!
Always observe all precautions and recommendations for working on high-voltage systems!

In case of gasoline-engine malfunction, trouble-shooting to determine whether the problem stems from the ignition system is not always a simple matter. The purpose of the ignition system is to ignite an air-fuel mixture that should be as homogeneous as possible. Within the system as a whole therefore, it is always possible that malfunctions can occur due to the interplay between A/F-mixture preparation and the ignition process triggered by the ignition system.

The ignition system is subject to immense dynamic demands throughout the spark-ignition engine's operating range. The voltage levels required to produce flashover at the spark plug vary in response to changes in ignition timing, engine speed, charge density, boost pressure and mixture composition. This makes diagnosis of sporadic errors extremely difficult.

The gasoline engine's ignition system should always be viewed as an overall system comprising ignition control (Motronic), ignition driver stage, ignition coil, spark plugs, and all the hardware connecting them.

The repair guides incorporated in ESI[tronic] are of inestimable value in trouble-shooting. Engine testers are employed to perform elementary diagnosis of basic ignition functions (check of basic ignition timing, etc.). The range of available test equipment encompasses everything from pocket testers to complete diagnosis systems incorporating numerous functions, including emissions analysis, oscilloscope, etc. Stationary testers are equipped to monitor signals from the primary and secondary circuits in several ignition coils at once. The oscillogramme illustrates operation of a six-cylinder engine with dual-spark coils in grid form (Fig. 1). The unit can be switched to its selective mode to monitor operation at an individual cylinder. The two-channel oscilloscope also allows simultaneous observation of the primary and secondary circuits.

Another trouble-shooting aid is a function designed to search for irregularities in the primary and secondary signal patterns. In this mode the unit analyzes the signal patterns recorded in the seconds before the "store" key is pressed. This mode supports problem detection by allowing direct comparisons of ignition voltage and spark duration in individual cylinders, etc.

1 Ignition pattern (secondary) in 3D grid display for a six-cylinder engine with dual-spark ignition coils at 760 rpm

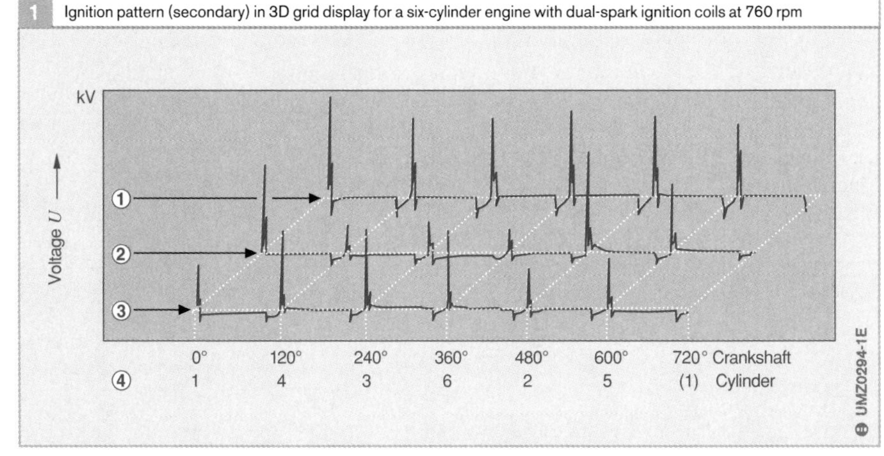

Fig. 1

1 Sequential display: primary and secondary sparks overlap
2 Positive display: Primary sparks on cylinders 1, 2 and 5; secondary sparks on cylinders 4, 3 and 6
3 Negative display: Primary sparks on cylinders 4, 3 and 6; secondary sparks on cylinders 1, 2 and 5
4 Firing order 1–4–3–6–2–5

Ignition coils in service

The ignition coil is a maintenance-free component, and will usually provide reliable service throughout the lifetime of the vehicle.

Problem diagnosis

However, should a problem arise, it can lead to combustion miss, producing rough and uneven running. The first response should be to examine the wiring from the wiring harness to the coil, and the high-tension cables from the coil to the spark plugs, to determine if any of the contacts are loose or intermittent (visual inspection).

The engine-management system's (OBD) diagnostic utility will not necessarily be able to detect malfunctions arising in an ignition coil. When very high levels of ignition voltage are required, a defect in the coil's internal insulation for instance, could lead to an arc inside the coil itself instead of across the spark-plug electrodes. Neither the OBD nor a visual inspection will detect this type of problem.

Effects of defective ignition coils

Defects can be divided into various categories:

Total coil failure

On vehicles with a rotating distributor coil, failure will completely immobilize the vehicle. If the vehicle is equipped with a distributorless ignition system, the mixture in the cylinder with the defective coil will fail to ignite, and its reciprocating masses will be maintained in motion by the remaining cylinders. Modern engine-management systems respond to this scenario by deactivating the fuel supply to the affected cylinder in order to protect the catalyst. Although operating smoothness will suffer, the vehicle can proceed to a service facility under its own power.

Reduced coil performance

Then engine's requirements for energy and ignition voltage rise to extreme levels under some conditions. These circumstances lead to sporadic combustion miss when the system fails to furnish adequate ignition voltage and/or a spark of sufficient duration. Because various interrelationships arising from component wear and the operating environment are factors in this type of problem, it can be extremely difficult to localize, both with rotating and distributorless high-voltage distribution.

Under these conditions modern systems are at an advantage, as the engine's diagnostic functions are capable of detecting the problem.

Sporadic coil failure

This defect produces rough engine running and ignition miss throughout the entire range of engine speeds and load factors, with pronounced miss occuring in engines with rotating high-voltage distribution. As a rule, the diagnostic utilities incorporated in modern engine-management systems can detect this problem and identify the cylinder in which it arises.

Symptoms of a defective coil

A defective coil can be identified by the following symptoms:

No ignition spark

This indicates an open circuit in the primary winding, the connection cables or the ignition driver stage, or an internal short circuit between the high-voltage circuit and ground.

➔ No generation of high-voltage.

Coil heats up with engine off

Indicates a short circuit in the ignition driver stage or in the ground wire between Terminal 1 and the driver stage.

Caution: This condition can lead to burned wiring and/or destruction of the ignition coil!

The ignition coil becomes unusually hot during operation

This can be caused by shorts between individual loops in one of the windings as well as incorrect triggering.

Caution: This condition can lead to burned wiring and/or destruction of the ignition coil!

Ignition miss and rough running in certain engine-speed and load ranges.

Reduced high voltage:

➔ The ignition coil fails to provide the specified performance.

Damaged insulation on high-voltage components and worn spark plugs can also lead to this condition.

Ignition-coil replacement

The coil is one element within an ignition system in which the electrical specifications of each component have been matched to those of all other system elements to provide optimal performance. When replacing the ignition coil it is thus vital to ensure that its electrical specifications make it compatible with the rest of the system.

Installation of an ignition coil of incorrect specifications can lead to the following malfunctions and/or system damage:

- Excessive current draw resulting in destruction of the engine-management ECU's driver stages
- Destruction of the ignition coil owing to excessive current
- Premature spark-plug wear
- Engine malfunction and ignition miss
- Damage or destruction of the catalytic converter
- Engine damage, from missing EFU diode, etc.

Handling regulations

General information

Handling regulations are specified for all plastic ignition coils from Bosch. Failure to observe these regulations can result in damage to the ignition coil and malfunctions or damage in other ignition-system components, ultimately producing ignition miss or complete failure of the ignition system. Consequential damage to the engine, injection and exhaust systems is possible. Under exceptional conditions the problem can even result in complete destruction of the plastic coil and the primary connection wires with a corresponding *fire hazard.*

This is why ignition coils that have been damaged or are not correctly connected should never be used!

Bosch plastic ignition coils have been designed and tested for motor vehicles. Other applications (in aviation and marine applications, etc.) are the exclusive responsibility of the user.

Shipping and handling

Plastic coils must remain dry while being carefully transported in the prescribed packaging. The packaging does not provide protection against impact damage (when a carton drops to the floor, etc.). When extracting the coil from the package it is important to verify that the rubber cap, protective casing and the plug terminal socket display no cuts or fissures.

Plastic ignition coils that have been dropped should be separated from the good units.

Installation, removal

Severe impact (hammer blow, etc.) and other use of force during installation and removal (excessive leverage, etc.) must be avoided. Never use anything other than the prescribed installation and removal tools. Always comply with all torque specifications when tightening connections, and observe all instructions regarding the installation surface to avoid damage to the plastic coil and to prevent it from loosening during vehicle operation.

Lubricants (talcum or special grease) in the rubber cap, the casing and the spark-plug socket are intended to prevent their surfaces from sticking to the spark-plug insulator and should thus not be removed. The lubricant coating may be refreshed with factory-recommended lubricants only.

Modifications to the primary connection wiring (wiring harness) that change the weight, rigidity, open length, etc., can lead to overtension and damage or rupture the plug connections.

Storage

Whenever possible, ignition coils should be stored in their original factory packaging at room temperature in a cool, dry place.

Large variations in temperature may foster formation of surface condensation when atmospheric humidity is high. For this reason an adequate period to allow the coils to dry and their temperatures to stabilize should be provided prior to installation. This precaution prevents moisture from being trapped in the high-tension connections and primary-circuit plugs and producing contact corrosion.

Installation area

Sharp objects and protruding metallic edges close to the coil's high-tension components in the installation area can induce excessive corona discharge and must thus be avoided. A distance of at least 2 mm should always be maintained between the high-voltage components and the surfaces of any adjacent metallic engine surfaces.

The yoke plate on pencil coils guards the secondary winding and provides electrical shielding on these units. No minimum distance to the yoke plate is specified.

Electrical connections and operation

Durable connections in the correct positions are required. Loose and intermittent contacts in high-tension connections lead to tracking, which reduces ignition voltage and can destroy the connections themselves.

Always ensure that the wires are connected to the correct terminals on the primary winding. Incorrect polarity can produce short circuits in the driver stages and cause complete destruction of the coil and the wiring to the primary circuit.

Never connect the ignition coil directly to the battery for testing or any other purpose.

Never allow the ignition coil to operate with the spark-plug cable removed (without secondary draw): the resulting, extremely high voltages would break through the coil's insulation and produce permanent damage. Contact with exposed surfaces and with the exposed rubber cap can be dangerous, as can contact with the open protective shield or when the spark-plug connection is pulled back; the arcs that occur under these conditions pose a potentially *lethal hazard!*

The external iron core or the yoke plate should usually be connected to ground. Missing or poor ground contact results in accumulation of an excessively high electrical charge with uncontrolled discharge in the iron core or yoke plate. Result: Hazard of personal injury in the event of contact, malfunctions in or destruction of components in the ignition or other vehicle systems, malfunctions in electronic communications equipment. Failure to provide a satisfactory ground can also lead to destruction of internal components within the coil, with destruction of the complete coil occuring in extreme cases. A coil that has been operated without a ground connection must be replaced.

Handling spark plugs

Spark-plug installation

Correct selection and installation will ensure that the spark plug continues to serve as a reliable component within the overall ignition system.

Gapping prior to installation is recommended only on spark plugs with front electrodes. Because it would affect the operating characteristics, the gaps of the ground electrodes on surface-gap and semi-surface-gap spark plugs should never be adjusted.

Removal

The first step is to screw out the spark plug by several threads. The spark-plug well is then cleaned using compressed air or a brush to prevent contaminants from becoming lodged in the cylinder-head threads or entering the combustion chamber. It is only after this operation that the spark plug should be completely unscrewed and removed.

To avoid damaging the threads in the cylinder head, respond to stiction and any tendency to seize in spark plugs by unscrewing them by only a small amount. Then apply oil or a penetrating solvent to the threads and screw the spark plug back in. Wait for the penetrating oil to work, then screw the plug back out all the way.

Installation

Please observe the following when installing the spark plug in the engine:

- The contact surfaces between spark plug and engine must be clean and free of all contamination.
- Bosch spark plugs are coated with anti-corrosion oil, making application of any other lubricant unnecessary. Because the threads are nickel-coated, they will not seize in response to heat.

Spark plugs should be tightened to the specified torques using a torque wrench. The torque applied to the spark plug's hexagon fitting is transfered to the seat and the socket's threads.

Application of excessive torque or failure to keep the socket attachment correctly aligned within the spark-plug well can place stress on the shell and loosen the insulator. This destroys the spark plug's thermal response properties and can lead to engine damage. This is one reason why torque should never be applied beyond the specified level.

The torque specifications apply to new spark plugs, with a light coating of oil.

Under actual field conditions, spark plugs are often installed without a torque wrench, and as a result far too much torque is usually applied. Bosch recommends the following procedure:

First: Screw the spark plug into the clean spark-plug well by hand until it is too tight to continue. Then apply the socket wrench. At this point we distinguish between:

- New spark plugs with a flat seat, which are tightened through an angle of approximately 90° after initial resistance to turning (Fig. 1) and
- Used spark plugs with a flat seat, which should be tightened further by an angle corresponding to roughly 5 minutes on a clock, or an angle of approximately 30° and

Table 1

1 Torque specifications

Spark-plug seat	Threads	Torque for cast-iron cylinder heads (N·m)	Torque for light-alloy cylinder heads (N·m)
Spark plug with Flat seat	M 10 x 1	10...15	10...15
	M 12 x 1.25	15...25	15...25
	M 14 x 1.25	20...40	20...30
	M 18 x 1.5	30...45	20...35
Spark plug with Tapered seat	M 14 x 1.25	20...25	15...25
	M 18 x 1.5	20...30	15...23

- Spark plugs with a conical seat, which should be turned further by an increment corresponding to 2 to 3 minutes on a clock, or an angle of approximately 15° (Fig. 2).

Second: Do not allow the socket attachment to tilt to an angle relative to the plug while either tightening or loosening; this would apply excessive vertical or lateral force to the insulator, making the plug unsuitable for use.

Third: When using a separate socket attachment ensure that its insertion opening is all the way above the top of the spark plug to allow it to snap into place securely. If the opening is too low on the plug, preventing the extension from seating correctly, spark-plug damage can result.

1 Tightening with flat seat

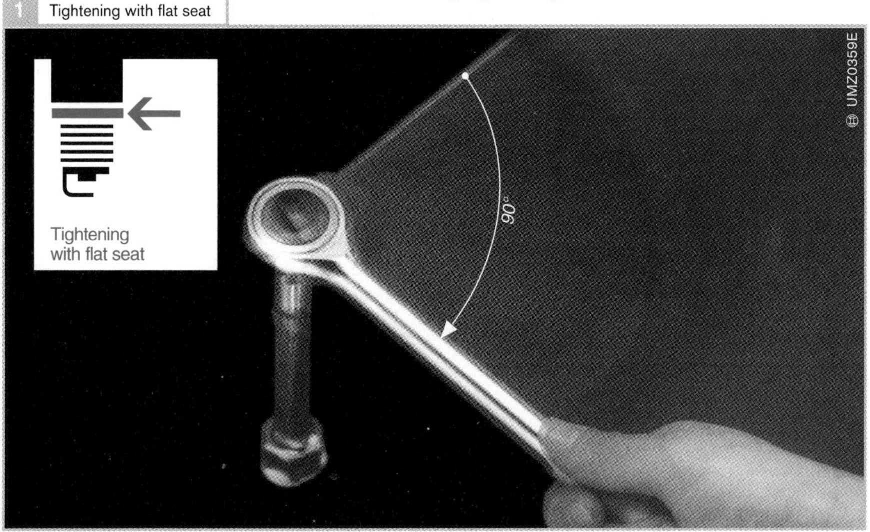

2 Tightening with conical seat

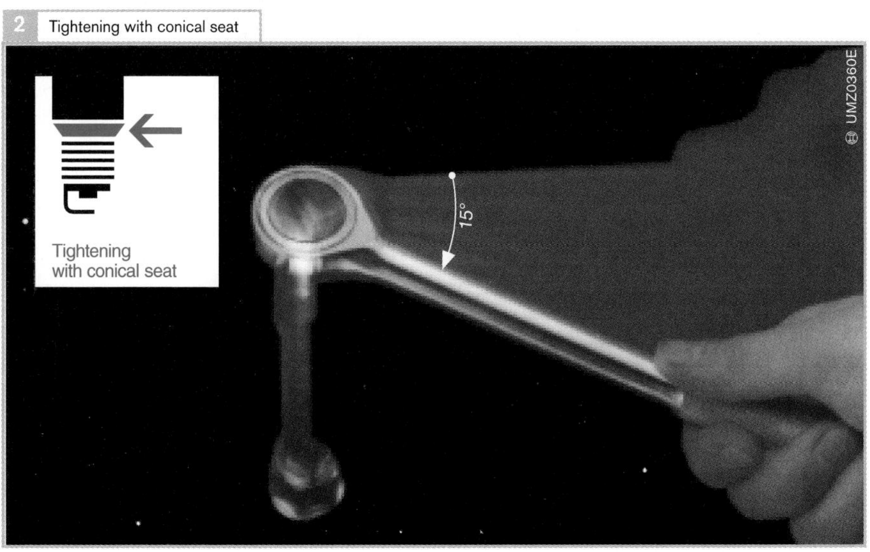

Mistakes and their consequences

Only spark plugs specified by the engine manufacturer or as recommended by Bosch should be installed. Motorists should consult the professionals at a Bosch service center to avoid the possibility of incorrect spark-plug selection. Sales and orientation assistance are available from catalogs, sales displays with reference charts and application guides available on the premises.

Use of the wrong spark plug type can lead to engine damage. The most frequently encountered mistakes are:

- Incorrect heat range
- Wrong thread length and
- Modifications to the seat

Incorrect heat range

It is vital to ensure that the spark-plug's heat range corresponds to the engine manufacturer's specifications and/or Bosch recommendations. Use of spark plugs with a heat range other than that specified for the specific engine can cause auto-ignition.

Wrong thread length

The length of the threads on the spark plug must correspond precisely to the depth of the socket in the cylinder head. If the threads are too long, the spark plug will protrude too far into the combustion chamber. Possible consequences:

- Piston damage
- Carbon residue baked onto the spark plug threads can make it impossible to remove the plug or
- Overheated spark plugs

A threaded section that is too short will prevent the spark plug from reaching far enough into the combustion chamber. Possible consequences:

- Poor ignition and flame propagation to the mixture
- The spark plug fails to reach burn-off temperature and
- The lower threads in the cylinder-head socket become coated with baked-on carbon residue

Modifications to the seat

Never install a seal ring, shim or washer with a spark plug featuring a conical seat. On spark plugs with a flat seat use only the captive gasket installed on the plug at the factory. Never remove this gasket, and do not replace it with another shim or washer of any kind.

The gasket prevents the spark plug from protruding too far into the combustion chamber. This reduces the efficiency of thermal transfer from the spark-plug shell to the cylinder head while also preventing an effective seal at the mating surfaces.

Installation of an additional washer prevents the spark plug from threading all the way into its socket, which also reduces thermal transfer between the spark plug shell and the cylinder head.

Spark-plug faces

Reading the spark plugs provides information on the condition of both engine and plugs. The appearance of the spark plug's electrodes and insulator – "the spark-plug face" – furnishes insights into how the spark plug is performing. It also furnishes information on the composition of the induction mixture and the combustion process within the engine (Fig. 3 to 5, following pages).

Reading the spark plugs is thus an important part of the engine-diagnosis procedure. It is vital to observe the following procedure in order to obtain accurate results: the vehicle must be driven before the spark plugs are examined. If the engine is run at idle, and especially after cold starts, carbon residue will form, preventing an accurate assessment of the spark plug's condition. The vehicle should first be driven a distance of 6 miles (10 kilometres) at various engine speeds and under moderate load. Avoid extended idling before switching off the engine.

3 Spark-plug faces, Part 1

① **Normal.**
Insulator nose with color between whitish or yellowish gray and russet. Engine satisfactory. Correct heat range. Mixture adjustment and ignition timing are good, no ignition miss, cold-start enrichment functioning properly.
No residue from leaded fuel additives or engine oil additives. No overheating.

② **Carbon fouling.**
Insulator nose, electrodes and spark plug shell covered with a felt-textured, matt-black coating of soot.
Cause: Incorrect mixture adjustment (carburetor, injection): rich mixture, extremely dirty air filter, automatic choke or choke cable defective, vehicle used only for extremely short hauls, spark plug too cold, heat range too low.
Effects: Ignition miss, poor cold starts.
Corrective action: Adjust mixture and cold-start system, check air filter.

③ **Oil fouling.**
Insulator nose, electrodes and spark-plug shell covered with shiney, oily layer of soot or carbon.
Cause: Excessive oil in combustion chamber. Oil level too high, severe wear on piston rings, cylinder walls or valve seals/guides.
Two-stroke engines: too much oil in fuel mixture.
Effects: Ignition miss, poor starting.
Corrective action: Overhaul engine, use correct oil/fuel mixture, replace spark plugs.

④ **Lead deposits.**
A brownish-yellow glaze, possibly with a greenish tint, forms on the insulator nose.
Cause: Fuel additives containing lead. The glaze forms when the engine is operated under high loads after extended part-throttle operation.
Effects: At high loads the coating becomes electrically conductive, leading to ignition miss.
Corrective action: New spark plugs, cleaning is not possible.

4 Spark-plug faces, Part 2

⑤ Severe lead deposits.
Thick, brownish-yellow glaze with possible green tint forms on the insulator nose.
Cause: Fuel additives containing lead: the glaze forms during operation under heavy loads following an extended period of part-throttle operation.
Effects: At high loads the coating becomes electrically conductive, leading to ignition miss.
Corrective action: New spark plugs. Effective cleaning is not possible.

⑥ Ash deposits.
Serious ash residue from oil and fuel additives on the insulator nose, in the insulator recess and on the ground electrode. Loose or cinder-flake deposits.
Cause: Substances from additives, especially those used for oil, can leave these ash deposits in the combustion chamber and on the spark plug.
Effect: Can produce auto-ignition with power loss as well as engine damage.
Corrective action: Repair engine. Replace spark plugs, use different oil as indicated.

⑦ Melting on center electrode.
Melted center electrode, insulator nose is soft, porous and spongey.
Cause: Overheating due to auto-ignition. Can stem from overadvanced ignition timing, deposits in the combustion chamber, defective valves, malfunctioning ignition distributor and low-quality fuel. May also possibly be caused by heat range that is too low.
Effects: Ignition miss, lost power (engine damage).
Corrective action: Check engine, ignition and injection/induction systems. Install new spark plugs with correct heat range.

⑧ Melted center electrode
Severely melted center electrode with serious damage to ground electrode.
Cause: Overheating from auto-ignition. Can stem from over-advanced ignition timing, deposits in the combustion chamber, defective valves, malfunctioning ignition distributor and low-quality fuel.
Effects: Ignition miss, low power, possible engine damage. Insulator nose may rupture due to overheated center electrode.
Corrective action: Check engine, ignition and injection/induction systems. Replace spark plugs.

5 Spark-plug faces, Part 3

⑨ **Melted electrodes.**
Electrodes melted to form a cauliflower pattern. Possibly with deposits from other sources.
Cause: Overheating owing to auto-ignition. Can stem from over-advanced ignition timing, deposits in the combustion chamber, defective valves, malfunctioning ignition distributor and low-quality fuel.
Effect: Low power followed by complete engine failure.
Corrective action: Check engine, ignition and injection/induction systems. Replace spark plugs.

⑩ **Severely eroded center electrode.**
Cause: Failure to observe spark-plug replacement intervals.
Effects: Ignition miss, especially during acceleration (ignition voltage not adequate for bridging wider electrode gap). Poor starting.
Corrective action: New spark plugs.

⑪ **Severely eroded ground electrode.**
Cause: Aggressive fuel and oil additives.
Deposits or other factors interfering with flow patterns in combustion chamber. Engine knock. No overheating.
Effects: Ignition miss, especially during acceleration (ignition voltage not adequate for bridging wider electrode gap). Poor starting.
Corrective action: New spark plugs.

⑫ **Cracked insulator nose.**
Cause: Mechanical damage (impact, fall or pressure on the center electrode from incorrect handling). In extreme cases the insulator nose may be split by deposits between the center electrode and the insulator nose, or by corrosion in the center electrode (especially when replacement intervals are neglected).
Effects: Ignition miss. Flashover occurs in locations with poor access to the combustion mixture.
Corrective action: New spark plugs.

Emissions inspections (AU)

In their efforts to reduce levels of harmful pollutants in exhaust gases legislators have extended emissions-control regulations to include vehicles currently in highway use (periodic emissions inspections). In Germany, the regular periodic test procedure (AU) consists of measurements to verify compliance with specific limits for individual components (such as CO). All vehicles are required to report for an initial emissions inspection three years after initial registration, and then at two-year intervals. Inspections are intended to determine whether the emissions of each vehicle can be classified "satisfactory" based on contemporary technology.

The introduction of on-board diagnosis marked the advent of a system suitable for monitoring all emissions-relevant systems and components on a continuous basis. Periodic emissions inspections (AU) determine whether the OBD system is operating as designed, and verify that it continues to meet the legal requirements throughout the vehicle's service life.

Regulatory basis

The general conditions that apply to emissions inspections have been defined in a directive issued by the European Union (EU). The individual member states have then incorporated the directive's stipulations into their own national codes. The results are specific individual test sequences in the various countries, such as the AU, or *Abgasuntersuchung*, in Germany (1993).

1 Emissions testing with the BEA250 Bosch Emissions Analyser

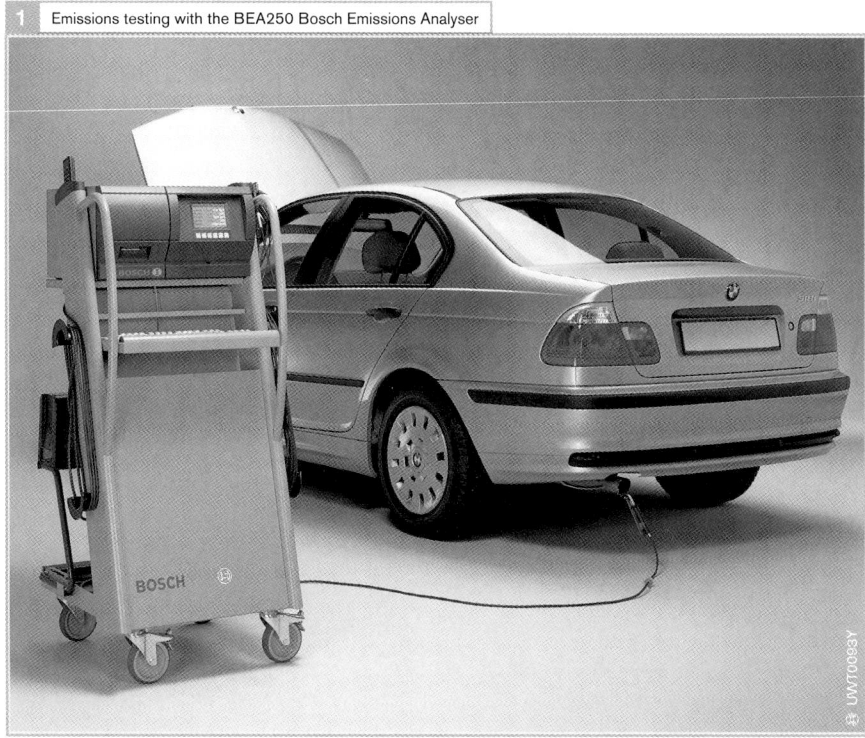

Among the requirements of yet another directive are on-board diagnosis systems for the type certification of all new passenger cars and light-duty trucks starting in the year 2000, and on all new vehicles being registered for the first time in 2001. Within Germany, the emissions inspection procedure was revised to reflect the new demands with the addition of OBD testing.

Test sequence

The test sequence for each vehicle with OBD is as follows:

- Entry of vehicle identification data (license-plate number, vehicle manufacturer, odometer reading, etc.). The test specifications for the vehicle are downloaded from the database or entered manually (optional database available in Germany only).
- Visual inspection of the exhaust system: examination to confirm that all components are present, check for damage and leakage.
- Visual examination of the MI lamp when the ignition is switched on and the engine started. The OBD interface is used to read out MI lamp status from the control unit. The malfunction log and the readiness codes are then accessed.
  An error code in the malfunction log will result in the vehicle failing to pass the emissions test. If one or several readiness tests are not performed, a supplementary test of the O_2 sensor voltage will be carried out following the "idle CO" test (consisting of voltage check on step-function sensor, voltage, current or lambda measurement on broadband probe). The vehicle fails to pass the emissions inspection if the results are outside the tolerance range.
- Engine speed and temperature check.
- CO and lambda data are monitored within a defined rpm window (high idle).
- Test passed? Yes/No.
  Inspection officer's entries.
- Automatic printout of test protocol.

Tester

Bosch offers equipment such as the modular BEA 250 Bosch Emissions Analyser for conducting emissions inspections. The most important features of this 5-component tester are

- Extremely precise measurements, designated Class 0 by the OIML (Organisation Internationale de Métrologie Légale)
- Comparison of test data and specifications from database
- Stable performance over extended periods (calibration required only once annually)
- OBD function
- Colour TFT display screen
- Clear operator guidance
- Various sensors for monitoring engine speed (B+, B–, Terminal 1, Terminal 15, TN/TD, TDC)
- Monitors O_2-sensor voltage and ignition timing
- Suitable for expansion into complete emissions-inspection station (gasoline and diesel engines)
- Optional equipment available for NO testing
- Optional AWN network connection

Emissions measurement concept

Emissions testers are employed to examine exhaust-gas composition in the field (AU emissions inspection) and ensure compliance with legally mandated emissions limits. They also represent essential service tools. The emissions tester provides the data needed for optimal adjustment of induction and injection systems, and is an essential diagnosis tool.

Test procedure

Required is the ability to measure specific exhaust components with a high degree of accuracy. Laboratories employ complex and extensive procedures for these measurements. Over the course of time, the infrared process has advanced to assume the status of the method of choice in vehicle service operations. The concept exploits the fact that individual substances absorb infrared light at different rates determined by each gas' characteristic wavelength.

Both single (for instance, for CO) and multiple-component (for CO/HC, CO/CO_2, CO/HC/CO_2, etc.) testers are available.

Emissions-tester measurement chamber

Infrared radiation is discharged from an emitter (Figure 1, Pos. 5) which has been heated to approx. 700 °C. The radiation then passes through a measuring cell (3) and into the receiver chamber (1).

Example of CO measurement

For CO measurements the sealed receiver chamber is charged with a gas of a defined CO content. This gas absorbs a portion of the CO-specific radiation. The absorption process heats the gas, and generates a current of gas which flows from volume V_1 through the flow sensor (2), and into the compensation volume V_2. A rotating chopper disk (4) induces a rhythmic interruption in the beam to produce an alternating flow pattern between the two volumes V_1 and V_2. The flow sensor converts this motion into an alternating electrical signal.

When a test gas with a variable CO content flows through the measuring cell, it absorbs radiant energy in quantities proportional to its CO content; this energy is then no longer available in the receiver chamber. The result is a reduction of the base flow in the receiver chamber. The deviation from the alternating base signal serves as an index of the CO content in the test gas.

1 Test chamber using infrared concept (schematic)

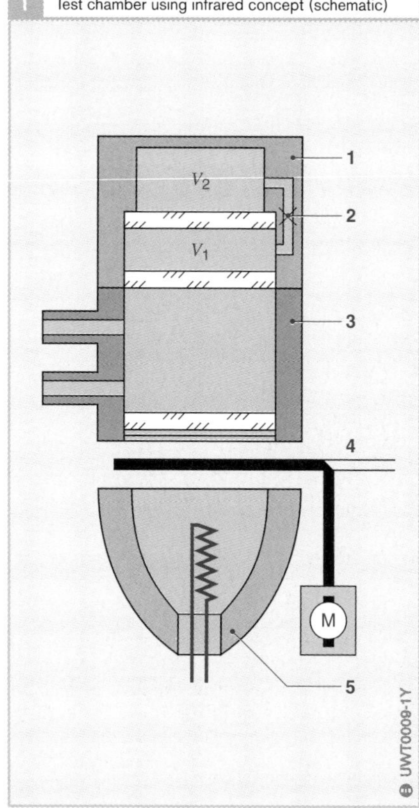

Fig. 1
1 Receiver chamber with compensation volumes V_1 and V_2
2 Flow sensor
3 Measuring cell
4 Rotating chopper with motor
5 Infrared emitter

Gas flow in the tester

A probe (Figure 2, Pos. 1) is used to extract exhaust gases from the test vehicle. The tester's integral diaphragm pump (7) draws the gas through a coarse-mesh filter (2) and into the condensation separator (3). The separator filters coarse particulates and ingested condensation from the test gas before it proceeds to yet another filter for further cleaning. A second diaphragm pump (8) conducts the condensation to the condensation discharge (16).

The next stop for the test gas is gas-analysis chamber GA1 (10), where CO_2 and CO levels are measured. The gas then proceeds to gas-analysis chamber GA2 (11) for an assessment of HC content. Before leaving the tester through the gas discharge (15), the test gas passes over electrochemical sensors (13 and 14) for measurements of oxygen (O_2) and nitrous oxide (NO) content.

A solenoid valve (6) upstream from the gas pump (7) switches the test chamber's supply from exhaust gas to fresh air for automatic zero-balance calibrations.

The activated-charcoal filter (5) in the air-inlet passage (4) prevents hydrocarbons in the surrounding air from entering the tester.

A pressure sensor (9) detects leakage at any point in the gas passage. The remaining pressure sensor (12) monitors atmospheric pressure, which is also included as a parameter in the unit's calculations.

2 Gas flow path in the multi-component tester

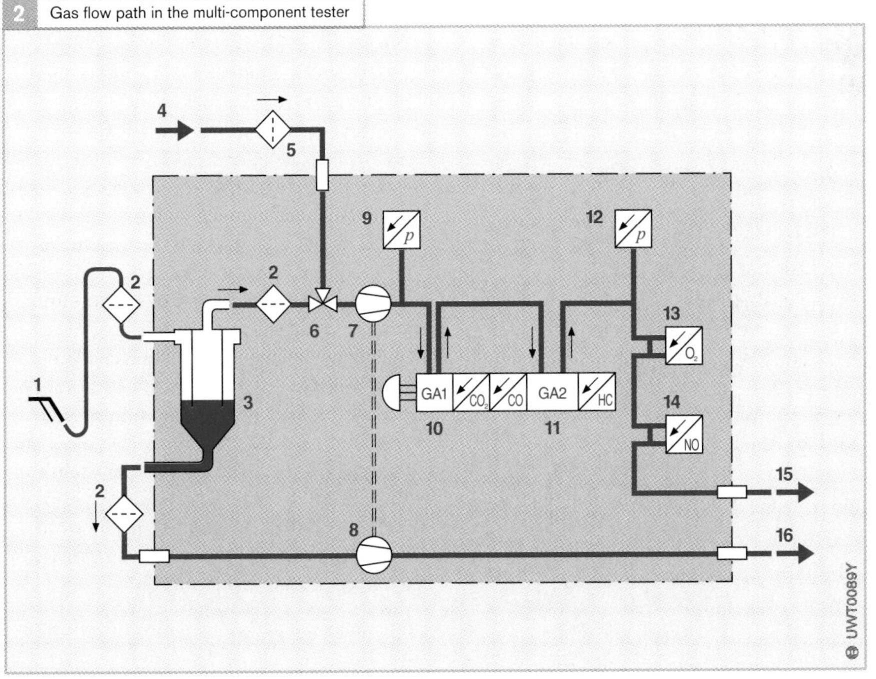

Fig. 2

1 Extraction probe
2 Coarse-mesh filter (wet filter)
3 Condensation separator
4 Air
5 Activated-charcoal filter
6 Solenoid valve
7 Diaphragm pump (gas pump)
8 Diaphragm pump (condensation pump)
9 Pressure sensor (internal pressure/leaks)
10 GA1 gas analyser (CO_2, CO test chamber)
11 GA2 gas analyser (HC test chamber)
12 Pressure sensor (barometric pressure)
13 Electrochemical sensor (O_2 sensor)
14 Electrochemical sensor (NO sensor)
15 Gas discharge
16 Condensation discharge

ECU development

The control unit (ECU) is the central point from which the functions of an electronic system in a motor vehicle are controlled. For that reason, extremely high demands in respect of quality and reliability are placed on ECU development.

Overview

An electronic system consists of sensors and setpoint generators, an ECU and actuators (Fig. 1).

The sensors detect the operating parameters of the electronic system (e. g. wheel speed, engine temperature, ambient pressure). The setpoint generators register the settings that the driver has specified with his/her operating controls (e.g. by means of the air-conditioner switches). The sensors and setpoint generators thus supply the input signals that are analyzed and processed by the ECU.

Actuators (e.g. ignition coils, fuel injectors) convert the electrical output signals into physical variables.

1 Components of an electronic system

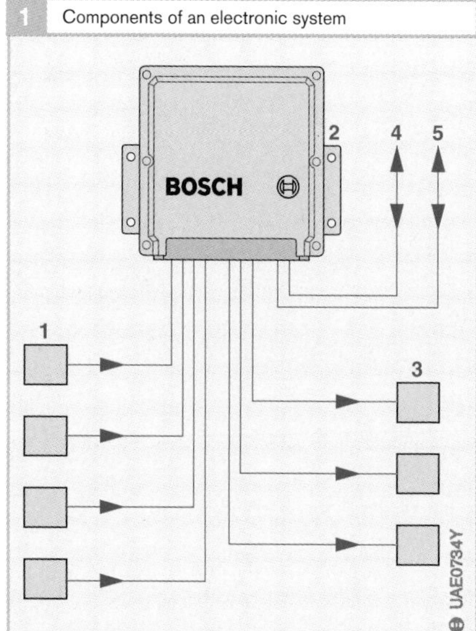

Fig. 1
1 Sensors and setpoint generators (input signals)
2 ECU
3 Actuators
4 Interface with other systems
5 Diagnosis interface

The process of developing an electronic system is made up of a number of stages, in which the ECU development stage plays a decisive role. The following tasks are involved in the development of the ECU (Fig. 2):

- Hardware development
- Function development
- Software development
- Application

Requirements

The product specifications and the development specifications document the requirements that a particular electronic system has to meet. Those two documents form the basis for the development process.

Product specifications

The product specifications define the requirements from the point of view of the vehicle manufacturer. They describe the functions that the product concerned must perform. They detail all requirements on the part of the vehicle manufacturer with regard to the products and services to be supplied. The requirements specified should be quantifiable and measurable. The product specifications thus define *what means* of performing *what task* is to be provided.

The product specifications are not revised during the course of the development process.

Development specifications

From the requirements set down in the product specifications, the ECU manufacturer draws up the development specifications. The development specifications define *how* and *by what means* the requirements are to be implemented (implementation specifications).

The development specifications are the basis for practical development of the ECU. They have to be regularly reviewed and updated in consultation with the vehicle manufacturer during the course of the development process.

2 ECU development: Interrelationship of development tasks

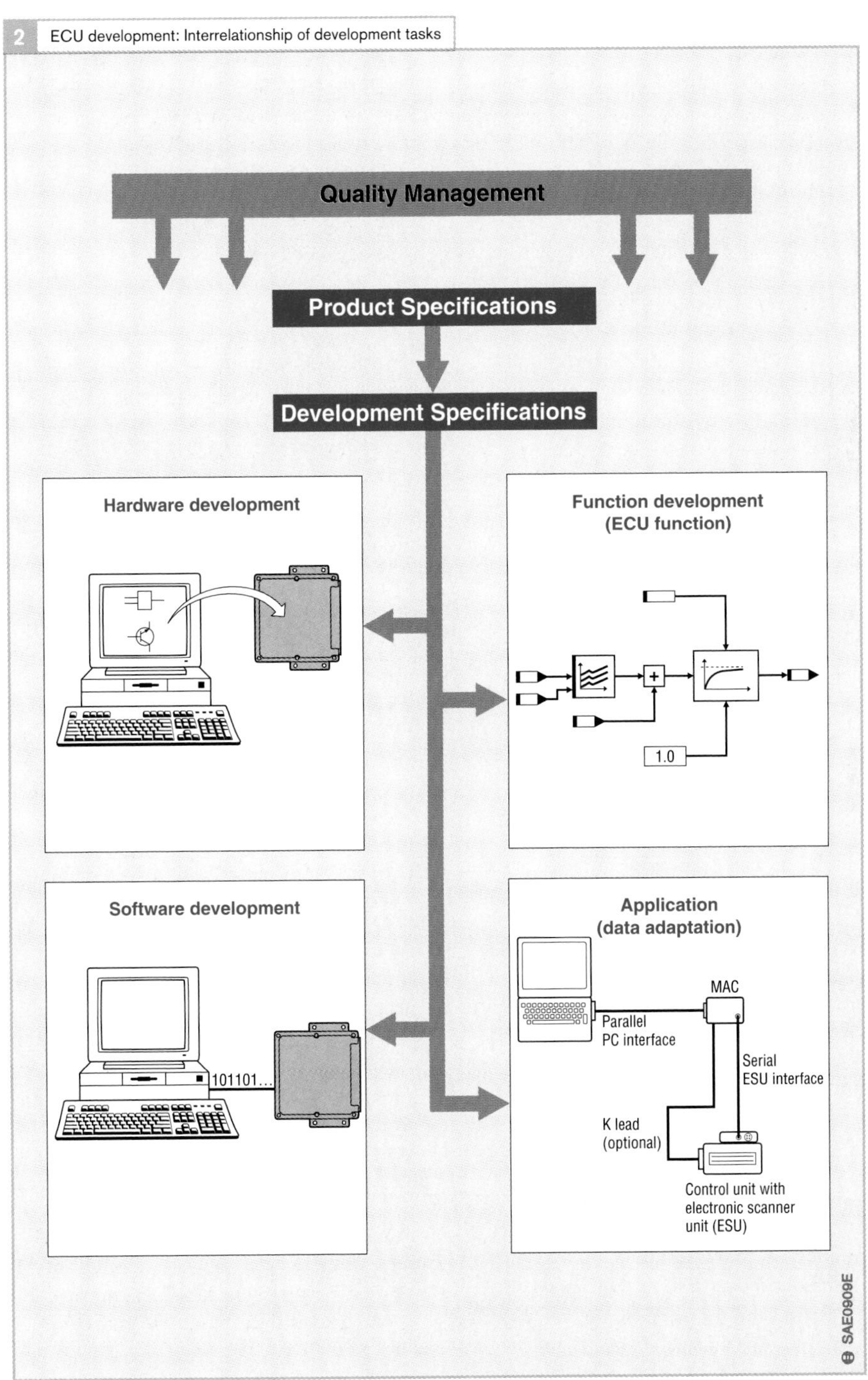

Hardware

Hardware is a generic term for the physical (i.e. tangible) components of a system whether mechanical (e.g. heat sink, casing) or electronic (e.g. microcontroller, memory modules, output modules). This applies to the components both individually and collectively.

The hardware of the ECU electrically processes the signals received from the sensors and passes them onto the "processor core" of the control unit for further processing. Output modules amplify the control signals so that the actuators can be operated with the required electrical power.

The task of *hardware development* (Fig. 2) is to design and produce an ECU that meets the requirements arising from the development specifications.

ECU functions

The control unit's job is to analyze the input signals and to control the actuators so that the system operates according to its intended purpose. The overall behaviour of the system can be broken down into a number of separate functions (e.g. for gasoline-engine management system: control of engine idling speed, exhaust emission levels, radiator fan, etc.). And even with today's state-of-the-art ECUs, the apparently straightforward control of the radiator fan is dependent upon a whole range of input variables. It is not sufficient merely to switch the fan on when the engine is hot, and off again when it has cooled down. Furthermore, every vehicle manufacturer has its own ideas as to exactly how this unit should operate.

Function development (Fig. 2) involves the implementation of the engine manufacturer's specifications, and the drawing up of the function descriptions which describe the ECU functions down to the very last detail. Those documents form part of the development specifications.

Software

Software is a generic term for the total of the programs and data stored in the memory of a computer-controlled system.

The central element of an ECU that performs a function in a motor vehicle is the microcontroller. It runs the program stored in the program memory. That program represents the functions of the ECU in the form of program code.

The process of *software development* (Fig. 2) converts the requirements arising from the function descriptions into a program. The machine code thus produced is entered in the program memory of the ECU. To simplify the process of writing a program, programming languages are used. The increasing complexity of electronic systems makes the use of high-level programming languages (such as the language C) absolutely essential. Software development is also assisted by simulation tools.

Data adaptation

The behaviour of an electronic system does not depend solely on the ECU program. A decisive role is also played by the data stored in the memory along with the program code. An example of such data in the case of an engine ECU would be the ignition timing map, which specifies the optimum ignition timing from the point of view of fuel consumption and emission levels for a range of engine operating conditions (engine speeds and loads/torques). Such data is engine-specific. For that reason, it has to be calculated and adapted during the process of development of the vehicle. Adaptation of the data to the engine in this way is the task of the *application* stage (Fig. 2).

Quality management

Quality assurance measures accompany the entire development process, and subsequently the production process as well. Only in that way can consistent quality of the end product be guaranteed. The quality requirements placed on safety-related systems (e. g. ABS) are particularly strict.

Quality assurance systems

All elements of a quality management system and all quality assurance measures have to be systematically planned. The various tasks, authorities and responsibilities are defined in writing in the quality management handbook. International standards such as ISO 9001 to 9004 are also adopted.

In order to regularly monitor all elements of a quality management system, quality audits are carried out. Their purpose is to assess the extent to which the requirements of the quality management system are being followed and the effectiveness with which the quality requirements and objectives are being met.

Quality assessment

On completion of specific stages in the development process, all information available up to that point about quality and reliability is subjected to a quality assessment and any necessary remedial action initiated.

FMEA

FMEA (Failure Mode and Effects Analyzis) is an analytical method for identifying potential weaknesses and assessing their significance. Systematic optimisation results in risk and fault cost reduction and leads to improved reliability. FMEA is suitable for analyzing the types of fault occurring on system components and their effects on the system as a whole. The effect of a fault can be described by a causal chain from point of origin (e.g. sensor) to system (e.g. vehicle).

The following types of FMEA are distinguished:

- Design FMEA: assessment of the design of systems for compliance with the specifications. It also tests how the system reacts in the event of design faults.
- Process FMEA: assessment of the production process.
- System FMEA: assessment of the interaction of system components.

FMEA assessments are based on theoretical principles and practical experience.

Example: a direction indicator fails. The effects in terms of road safety are serious. The likelihood of discovery by the driver is small, however, since the indicator is not visible from inside the vehicle. As a means of making the fault obvious, the rate at which the indicators flash must be made to change if an indicator fails. The higher flashing rate is discernible both visually on the instrument cluster and audibly. As a result of this modification, the effect of the fault can be reduced.

Review

The review is an effective quality assurance tool in software development in particular. Reviewers check the compliance of the work produced with the applicable requirements and objectives.

The review can be usefully employed as a means of checking progress made even at early stages of the development process. Its aim is to identify and eliminate any faults at as early a point as possible.

Hardware development

The complexity of electronic systems has seen a continuing increase in the past years. This tendency will continue in the future, and it will only be possible to master such developments by applying highly integrated circuits. The call for small dimensions for all system components places further exacting demands on the hardware development process.

Efficient and cost-effective hardware development is now only possible by using standard modules that are produced in large numbers.

Project starting point

A schematic diagram sets out all the functions that are to be performed by the ECU being developed. From that point, the following aspects can be clarified:

- Definition of hardware required
- Cost estimate for the hardware
- Extent of development work required
- Cost of tools

Hardware design prototypes

Once the project is underway, hardware design prototypes are produced and subjected to quality tests. These design prototypes may be one of four categories representing successive stages along the road to the final production ECU. Each prototype category is based on its predecessor and is designed for a particular purpose in each case.

'A' prototype

The 'A' prototype is derived from an existing or modified ECU or a development circuit board. Its range of functions is limited. Its technical function is largely in place but the 'A' prototype is not suitable for continuous testing. It is a *function prototype* that is used for initial trials and to confirm the basic viability of the design.

'B' prototype

The 'B' prototype includes all circuit components. It is a *trial prototype* that is used to test out the full range of functions and the technical requirements. At this stage it is ready for continuous testing in prototype vehicles.

The connection sizes and ECU dimensions are as required for final production. It may be, however, that not all of the vehicle manufacturer's specifications are satisfied at this stage, perhaps because different materials have been used, for instance.

'C' prototype

The 'C' prototype is the *approval prototype* on which the vehicle manufacturer's tests for "technical approval" are performed. This version of the ECU must reliably conform to all specifications. On successful approval of the product, the development process is complete. As far as possible, 'C' prototypes are produced using full-production tools and production methods as close as possible to full series production.

'D' prototype

The 'D' prototype is the *pilot-series prototype* which also carries the full-production identification plate showing the version number. 'D' prototypes are fitted in pilot-series vehicles for large-scale vehicle trials. This version of the ECU is produced using volume-production methods and is fitted and tested under volume-production conditions. It is the version with which the reliability of production is verified.

Preparations

The preparations for the 'B' prototype start from the beginning of the project and involve

- Definition of connector pin assignment
- Definition of the casing design
- Ordering and developing new circuit modules (function groups)
- Producing the circuit diagram
- Defining the components (only approved components may be included, or inclusion must be subject to approval)

When selecting the circuit modules (e.g. knock-sensor analyzer circuits integrated in an IC), developers check whether existing circuits – with modifications if necessary – can be used. If not, new modules have to be developed.

Circuit diagram and components list

A CAD (Computer-Aided Design) system is used to create the circuit diagram (Fig. 1 a) and the list of all components used. The components list also details the following information for each component:

- Its size
- Its pin assignment
- Its casing design and
- The supplier and terms of supply

Layout

A circuit-board layout (Fig. 1 b) is required for production of the circuit board. It shows the positions of the conductor tracks and the connector pins.

The layout is produced on a CAD system. The process starts by taking the circuit-diagram data and converting it. The component connection schedule (list of connections between components) then produced shows how the various components are connected to each other. From this component connection schedule and the CAD data for the components (size and connector pin assignment) the layout can then be produced.

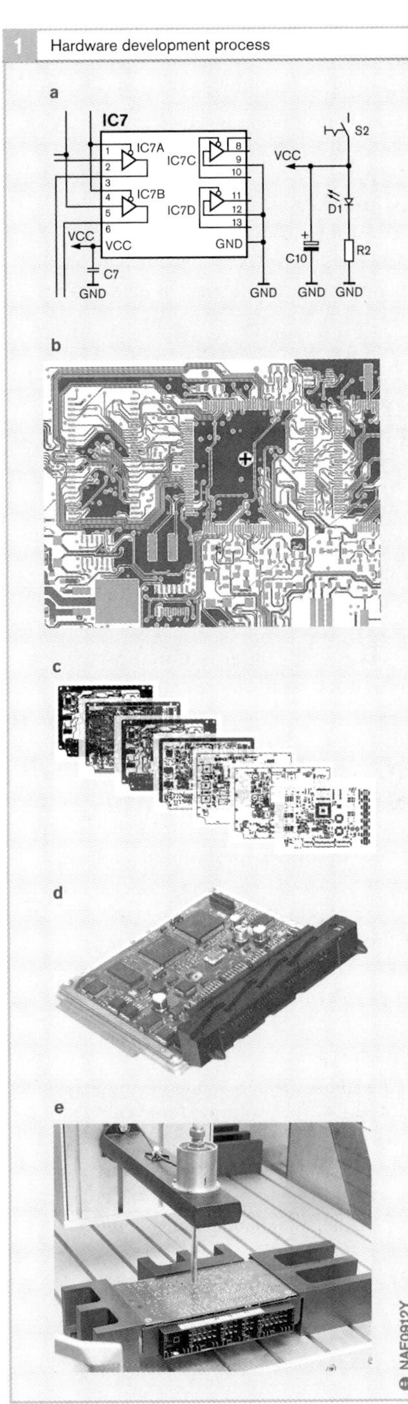

Fig. 1
a Circuit diagram
b Layout
c Circuit-board production
d Prototype construction
e Testing

There are specific criteria that have to be taken into account in the creation of the layout. In connection with the positioning of components these can include the following considerations:

- The power loss from specific components (possibilities for heat dissipation)
- EMC (electromagnetic compatibility) effects
- Convenient positioning of components in relation to connectors
- Observance of restriction zones (component size)
- Ease of fitting of components by automated production machinery
- Accessibility of testing points and
- Space requirements for testing adaptors

Circuit board

The layout data can be used to produce films for manufacture of the printed-circuit board (Fig. 1 c). The films are used to photographically expose the blanks (which are covered with a light-sensitive coating) and then develop and etch them. The individual layers of multilayer printed-circuit boards are placed on top of one another and hardened.

Finally, the component print, the solder resist and a carbon lacquer are applied to the circuit board.

Prototype construction

The finished circuit board has to have the components mounted on it (Fig. 1 d). In the case of prototype products, this is part of the process of prototype manufacture. Because of the miniaturisation of components and the high degree of integration on the printed circuit, even prototypes have to have their components fitted by machine. The machine is controlled by the CAD layout data.

Following fitting of the components, their connections are soldered. There are two alternative soldering methods:

- Wave soldering and
- Reflow soldering

Testing the finished circuit board

Electrical testing

Once all components have been fitted and all connections soldered, the circuit board must be tested. To this end, electrical testing sequences are defined which run on a computer. These automatic tests check that all components are fitted and that the circuit functions properly.

Thermographics

Thermographic images of the printed circuit show the heat generated by the components during operation (Fig. 2). Different temperature ranges appear on the film as different colours. In this way, components that are too hot can be identified. The information obtained is incorporated in the list of modifications between the 'B' and 'C' prototypes. Changes to the circuit layout (e.g. heat-dissipation through-contacting) can reduce the amount of heat generated.

2 Thermographic image of a printed-circuit board

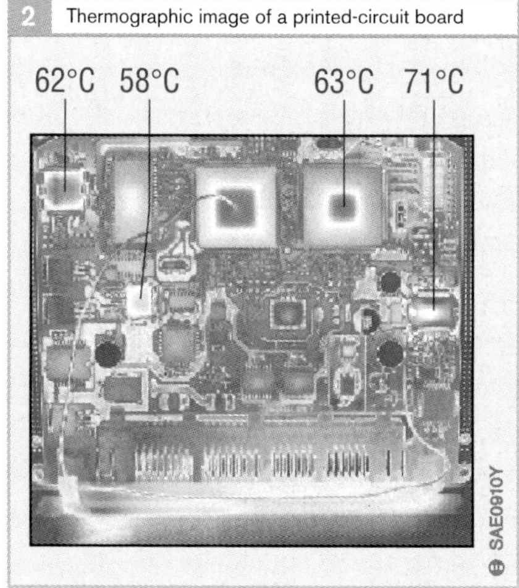

Fig. 2
Engine control unit
Power supply voltage
U = 14 V
Idling speed
n = 1,000 rpm

Electromagnetic testing

The electromagnetic fields created on the circuit board can be scanned with a magnetic field detector (Fig. 1 e). The readings are analyzed on a PC. Different field strengths show up as different colours. If necessary, modifications to the circuit layout have to be made and additional components fitted that reduce the magnetic field or make the ECU immune to interference.

These tests are performed on the 'B' prototype so that the necessary modifications can be incorporated in the 'C' prototype.

EMC tests

Readings taken in the EMC testing cell or testing chamber (Fig. 3) test the ECU response to external and internal sources of electromagnetic interference. Measurements are taken both with the ECU fitted in the vehicle (in-situ tests) and in the laboratory (e. g. stripline method).

The difficulty with the *in-situ tests* is that they cannot be carried out until development of the vehicle and all its electronic systems is at a very advanced stage. If the EMC characteristics are found to be unsatisfactory at this stage, the scope for modifications is then severely limited. For this reason, *laboratory testing* is extremely important because it permits potential problems to be identified on the hardware prototypes at an early stage in the process.

The EMC readings are taken at a variety of frequencies and with a range of electric field strengths. The immunity of the output signals (e.g. ignition signals, fuel injection signals) to external interference is examined as well as the level of interference generated by the unit itself.

3 Vehicle in the EMC testing chamber

Function development

Highly exacting demands regarding driver convenience and fuel consumption are placed on modern motor-vehicle engines. In order that these demands can be met economically at the same time as compliance with increasingly strict emission-control legislation, the prime requirement is for optimized coordination between the engine management system and the entire drivetrain, numerous sensors and actuators and the fault-diagnosis functions for them.

As, in contrast with earlier solutions, direct mechanical adjustment of the actuators in response to changes in ambient conditions is not possible, the engine electronic control unit– and therefore the control software running on it – is the only link between the sensors and the actuators. Consequently, the fuel consumption characteristics of a vehicle are determined to a great extent by the algorithms implemented on the engine-management system and the quality of them.

It is the purpose of function development to provide these algorithms.

Requirements placed on functions

Modularity

Modern control systems have to be modular in terms of the structure of their functions and hardware in order to manage the enormous variety of engine-configuration parameters, such as number of cylinders, fuel-injection method, sensors used and type of exhaust-gas system. A hierarchical structure divided into subsystems (air system, fuel system, exhaust-gas system) with stable interfaces enables parallel and therefore rapid development which, at the rate of innovation in today's automotive industry, is becoming an increasingly important consideration.

Component packages

For every class of sensors and actuators, a generally applicable interface with a physical representation is created. Higher-level control functions connect to this interface without knowledge of the way in which the component is structured and leave the detailed handling of special component properties to the lower function level, the so-called component package. The latter provides for optimum interaction between the mechanical component – i.e. the sensor or actuator –, the processing or control hardware and the hardware-related software. In addition to essential functions such as protection against excessive temperatures or operating voltages, this is also where the correction of nonlinear characteristics and conversion into the selected physically based component-package interface format take place. It means, for example, that the changeover from a pneumatic to an electric actuator can be implemented without altering the higher-level control functions.

This concept is the basic requirement for the use of similar components of differing generations and manufacturers, with minor consequences for the overall system.

Development process

Requirement

When new requirements emerge in relation to a customer or platform project, for example with regard to the use of new components, a possible solution has to be found and the functions that are affected have to be identified. When doing so, the existing function structure should be retained as far as possible in order that the interfaces, and therefore the interaction with other functions, are not endangered. Once the necessary algorithms are known and capable of implementation on the ECU, an estimate is produced of the amount of development work, the cost and the completion date. Following discussions with internal and external clients, the project is commissioned by the customer and a definite delivery deadline agreed.

Concept

If the available algorithms are inadequate for performing the required task and if the task is sufficiently significant, a new development process is initiated involving basic calibration and testing on the vehicle or engine test bench. The resulting concept is tested for conformity with physical principles, absence of conflicts, capability of implementation in the available engine-management system and anticipated applicability in the process of a concept review involving representatives from Function Development, System Development and Application.

Function definition

There then follows the implementation of the concept by the creation and specification of each function. After offline simulation of critical components using measured vehicle data, these functions are entered as function definitions in a central database together with the associated documentation which encompasses the written descriptions and application instructions.

Function reviews with other function developers prevent the repetition of known errors.

Encoding

Next, the software developers convert the functions into program code either by automatic code generation or manually in the case of critical functions. As with function reviews, a code review minimizes error frequency.

Function test

On completion, all new software modules are integrated in a program version for a specific vehicle project. Only then can the system be tested under real conditions on the vehicle. This test involves three stages:

- The function developer checks that the specifications and the software implementation are consistent with one another.
- The function developer compares the implemented solution with the customer/project requirements.
- In the course of initial commissioning, Function Development and Application jointly assess whether the chosen solution is adequately applicable and also usable for other projects.

These tests can lead to more or less extensive modifications to the selected solution. However, the aim is always to identify errors as early as possible in the development process in order to manage their impact on deadlines, costs and quality within acceptable limits.

Program version delivery

After the testing stage, the functions are delivered to the customer in the form of a program version. In the case of new concepts or extensive modifications, the function developer and the Application department assist the customer with commissioning by presenting the selected concept and discussing the calibration procedure on the engine test bench or in the course of summer or winter trials.

Software development

From Assembler...

When the first microcontrollers were used in control units, the programs they used required only 4 kilobytes of space or even less. At that time, it was all the memory that was available. For this reason, the programs had to be written in a space-saving language. The most commonly used programming language was Assembler. The commands in this language are mnemonics. They generally correspond directly to the instructions in the microcontroller's machine code. However, Assembler programs are generally difficult to read and maintain.

Over the course of time, the capacity of memory chips grew larger and larger, and the range of functions performed by the engine-management system became more and more complex. The expanding variety of functions made software modularization unavoidable. The ECU program is divided into modules, each of which incorporates a specific group of functions (e.g. lambda closed-loop control, idle-speed control). Of course, these modules have to be usable not just for one particular project but for large numbers of similar projects. For this reason, defined interfaces for the input and output variables of the functions are important. At this point, assembly language is reaching the limits of its capabilities.

...to high-level programming language

For the demands faced by present-day software developers, high-level programming languages are indispensable. The entire software for an engine-management system is nowadays written in a high-level language – the preferred choice being the C programming language. Programming in a high-level language ensures:

- Updatability of the software
- Modularity
- Interchangeability of software packages
- Independence of the software from the microcontroller used in the control unit

Software quality

Most innovations in automotive technology today are based on the use of electronics. In the past, the software was seen as an "appendage" to the hardware. Over time, however, the importance of the software has steadily grown. With the increasing complexity of microcontroller-controlled electronic systems, the quality of the software became a central pillar of software development – because problems caused by imperfections in the software damage the image of a manufacturer and drive up warranty-claim costs.

Software process improvement

The model used for improving processes in software development is CMM (Capability Maturity Model). It provides a framework for highlighting the elements of an effective software-development process. It describes an evolutionary path from a disorganized to a perfected, disciplined process and supports:

- Characterization of the degree of process maturity
- Definition of targets for process improvement and
- Setting of priorities for action to be taken

Distributed development

Bosch develops software not only in Germany but at a number of locations around the world. The same development process is applied throughout this international development network. Consequently, the same high standards of software quality are produced by Bosch worldwide.

Software sharing

As a result of software modularization and the use of defined interfaces, "external" software modules can also be incorporated in the ECU programs. This means that a vehicle manufacturer can use its own software for different vehicle models. The software then becomes a "distinguishing factor" on the competitive marketplace.

Creating program code

The basis for software development are the function definitions that are produced by the function developers. These documents describe the ECU functions (e.g. lambda closed-loop control) that are to be converted into a program by the software developers and then combined into an executable ECU program.

Creating source code

For every function, a separate module – a component of the overall program – is produced. The source code for the modules is written on a PC using a text editor (Fig. 1). The source code essentially contains the actual program instructions as well as documentation that facilitates the "readability" of the program (for program upgrading).

Compiler

Once created, the source code has to be translated into machine code so that it can be understood by the microcontroller. This involves the use of a compiler. It produces "object code" which contains relative rather than absolute memory addresses.

1 From FDEF to programmed ECU

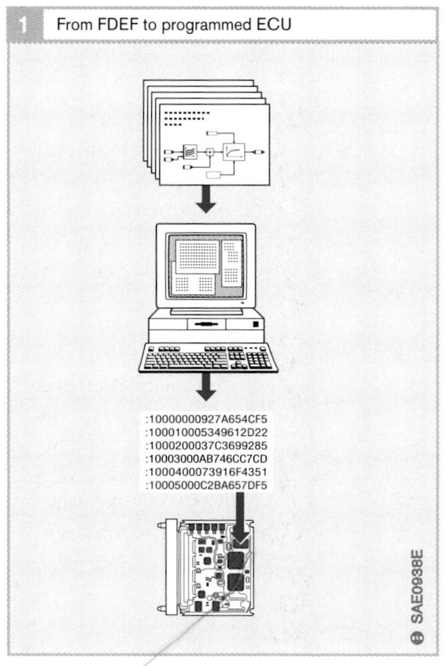

Linker

Once all the modules that make up the overall program have been created and compiled, all the object-code modules can be combined to form an executable program. This job is performed by a linker. The linker makes use of a file in which all the modules to be linked are listed. The file also details the memory addresses for the data and program memories. So all the relative addresses quoted in the object-code modules can be replaced by absolute addresses.

The result of the linking process is a machine-code program capable of running on the target system – the electronic control unit.

Module archiving

Software is subject to a rapid process of change. So that program versions supplied to customers can be reliably reproduced, it is essential to archive the modules. Archiving programs allow the tracking of every modification down to each individual module. For each archived program version, all the modules used can be listed and retrieved.

Testing station

The program code produced by the linker has to be tested in the laboratory before the ECU programmed with this code can be used on a vehicle. First of all, the new modules have to be individually tested in every detail. After this, the ability of all the modules to operate perfectly in combination with one another has to be thoroughly tested.

The tests are carried out on a testing station which is made up of a diverse range of testing equipment (Fig. 2 overleaf).

ECU with emulator module
A specially constructed laboratory version of the ECU is used for laboratory tests. It differs from the production version by virtue of an IC socket in place of the permanently soldered flash EPROM. Plugged into this socket is an emulator module which simulates the EPROM by means of a RAM. This makes it possible to modify data and program code "online". The control operations are performed on a PC.

LabCar
In a real operating environment, the control unit receives input signals from sensors and desired-value generators. It then produces output signals to control the physical actuators on the vehicle. The sensor signals are simulated in a laboratory environment. The necessary signal generators (e.g. inductive speed sensor) or hardware circuits (e.g. resistor sequence for simulating the temperature sensor) are accommodated in a "black box" known as the LabCar.

The LabCar also emulates the connections with all of the actuators controlled by the control unit. One of the most important is the electronic throttle-valve device since feedback from this device is continuously monitored by the ECU. Without the throttle-valve link, it is not possible to operate the vehicle.

The immobilizer also has to be connected to the ECU in order to allow vehicle operation. The electrical signals are simulated.

The LabCar thus provides the means for simulating the vehicle for the purposes of testing the ECU program.

Connection adaptor
In the wiring harness which connects the LabCar to the ECU, there is a connection adaptor. Every lead in the wiring harness is plugged into a socket on the connection adaptor. This means this every signal traveling to or from the control unit is accessible for testing purposes (e.g. tracking the voltage curve of a control signal using an oscilloscope).

2 Software testing station

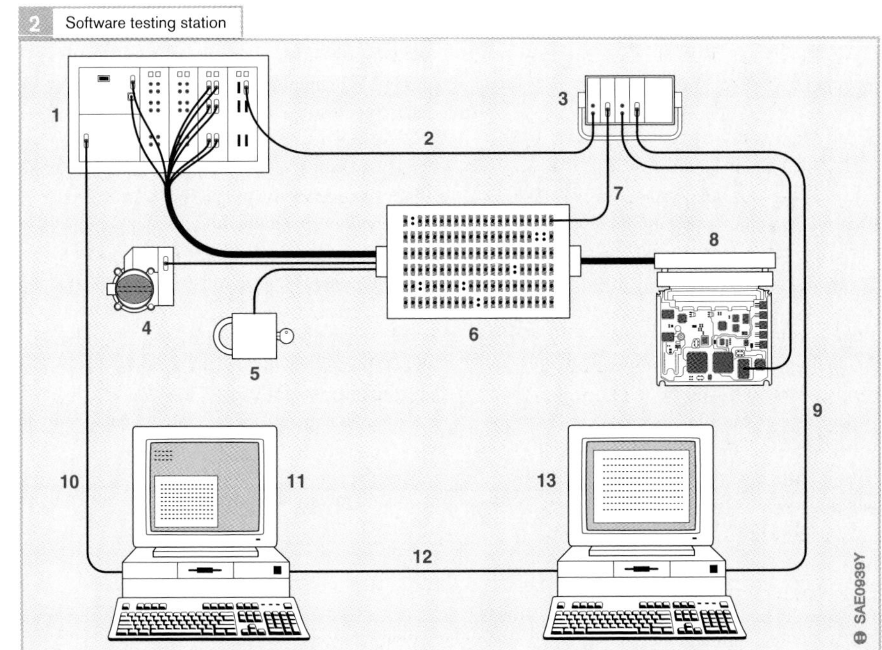

Fig. 2
1 LabCar
2 TRS 4.22 interface
3 INCA VME (calibration tool)
4 Throttle device (electronic throttle control)
5 Immobilizer
6 Connection adaptor
7 'K' wire (serial 1-wire interface)
8 Engine ECU with emulator module
9 Serial interface (RS232)
10 Parallel interface (Centronics link cable with additional fiber-optic cores)
11 PC (engine model, LabCar control)
12 Serial interface (RS232)
13 PC (testing computer, auxiliary computer for control of automatic testing)

Emulator mode

The control unit (with microcontroller socket) has an emulator plugged into it. The emulator replaces the ECU's microcontroller. The ECU program then runs on the emulator.

The program code is loaded to the emulator from a PC. The emulator is also controlled from this PC. This means that:

- The program can be started from a specific memory address.
- "Breakpoints" can be set (so this the program stops at defined points).
- The program sequence can be retraced from the breakpoint and the memory contents can be read and modified at each individual stage.
- Trigger conditions can be defined and the program sequence can be analyzed before and after the trigger point.
- Internal signals and registers (processor registers) that otherwise would not be accessible can be read.
- In single-step mode, program sequences can be processed one step at a time and the processing sequence tracked.
- Data and program code can be modified in order to manipulate the program sequence for testing purposes.

The ability to set trigger conditions and record results makes it possible to analyze the program run in relation to specific input signals. As a result, every branch of the program can be tested against the function definitions.

Logic analyzer

Another means of tracking the program sequence is provided by the logic analyzer. It is connected via an adaptor in such a way that it can "listen in" to data traffic on the address and data buses. In this way, it can record the program sequence and also track read-and-write access to the external data memory. However, access to the data memory integrated in the microcontroller is not possible.

Trigger conditions (e.g. reaching a specific address, storage of a specific value in a memory cell) can be set on the logic analyzer. When the trigger condition has been met, the program sequence can be retraced.

The emulator and the logic analyzer offer similar capabilities. The advantage of the logic analyzer is that it does not interfere with the program sequence (real time) and can therefore be used in a real operating environment on the vehicle.

Automatic test

The LabCar testing station not only provides the means for manual testing of ECU programs, it also offers the facility for substantially reducing testing times, particularly where repeat tests are concerned, by using an automatic testing sequence. For such purposes, the ECU is operated in a closed-loop control circuit. At present, there are four different tests that can be used:

1. The most frequently used plausibility test is a "rough check" of the most important ECU functions. It performs an electrical and physical check on all input variables, and tests fuel injection and ignition, throttle-valve response, and maximum runtime load on the ECU.
2. The OBD (on-board diagnosis) test checks out the most important ECU diagnostic functions and fault management by means of fault simulation.
3. The CAN test checks communication signals with reference to range of values and reference signal.
4. The start/stop test analyzes fuel-injection and ignition response to ECU start-and-stop processes. Among other things, this involves varying the battery voltage.

Application-related adaptation

So that a car is able to meet the driver's expectations, extensive development work is required, particularly in respect of the engine.

As a rule, the vehicle manufacturer will center the development of a new vehicle around a basic engine. The most important operating parameters for this engine are known quantities, such as:

- Compression ratio and
- Valve timing (on engines with variable valve timing, this can be altered while the engine is running)

The engine peripherals have to be modified to meet installation space constraints. This applies in particular to:

- The air-intake system and
- The exhaust-gas system

Other important aspects (e.g. fitting location of knock sensors) have to be defined in consultation with the Application department at Bosch.

Parameter definition

The next stage involves adapting the electronic control system – the Motronic – to the engine. This is the function of application-specific adaptation. Application-specific adaptation means adapting an engine to suit a particular vehicle.

The ECU program consists of the program itself plus a large amount of data. The program meets the requirements set out in the specifications document (function framework), but the data still has to be adapted to the particular engine and vehicle-model variant.

In the adaptation phase, all data – also referred to as parameters – has to be adjusted to achieve optimum efficiency of operation. The main evaluation criteria include:

- Low-emission exhaust gas (compliance with prevailing emission standards)
- High torque and power output
- Low fuel consumption and
- High level of user-friendliness

1 Calibration tools

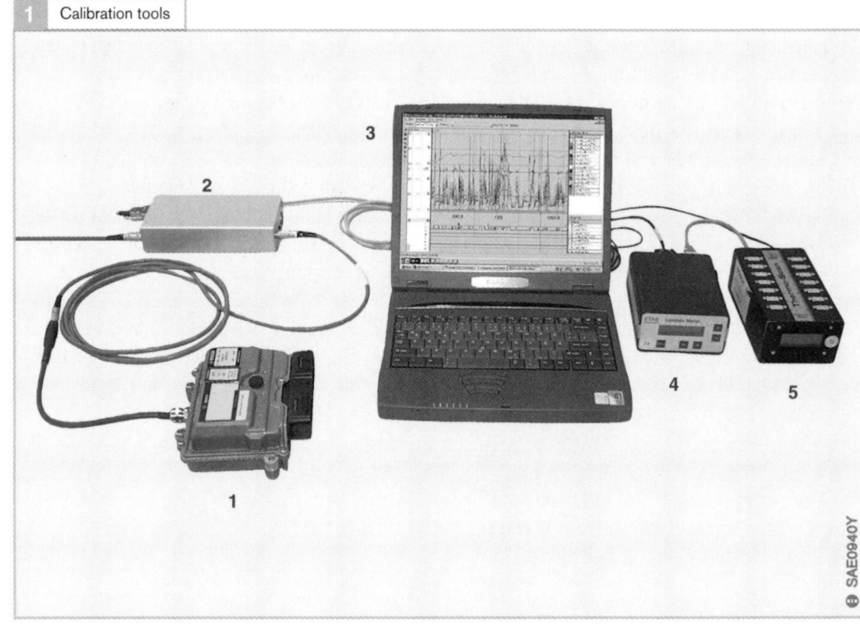

Fig. 1
1 Engine ECU
2 MAC (compact testing and calibration tool)
3 Laptop computer
4 Lambda Scan (test interface for broadband lambda sensors)
5 Thermo Scan (test interface for temperature sensors)

The aim of adaptation is to ensure that the objectives outlined above are achieved as fully as possible, i.e. the best possible compromise is reached between competing demands.

To this end, there is a data record with up to 5,500 "labels" that are capable of adjustment. These labels are subdivided into:

- Individual parameter values (e.g. temperature thresholds at which specific functions are activated)
- Data curves (e.g. engine speed vs. temperature curve for threshold at which a function is activated, temperature vs. ignition-timing curve) and
- Data maps (e.g. ignition timing as a function of engine load and speed)

Initially, work has to be carried out on the engine test bench. In this basic adaptation phase, parameters such as the ignition-timing map are defined. When basic adaptation has established the foundation for the initial vehicle trials, all parameters that affect engine response and dynamic characteristics are adapted. This work is for the greater part performed with the engine fitted in the vehicle.

The scope for optimization on Motronic systems has become so extensive and complex that many of the functions are now only possible with the aid of automated optimization methods and powerful tools.

Calibration tools

A large part of the adaptation work is carried out using PC-based calibration tools. Such programs allow developers to modify the engine-management software. One such calibration tool is the INCA (**In**tegrated **Ca**libration and Acquisition System) program. INCA is an integrated suite of several tools. It is made up of the following components:

- The Core System incorporates all measurement and adjustment functions.
- The Offline Tools (standard specification) comprise the software for analyzis of measured data and management of adjustment data, and the programming tool for the flash EPROM.

The ECU used for adaptation purposes has an emulator module instead of the program memory (EPROM) which emulates the ECU's EPROM and RAM. The INCA system has access to the data in these memories.

This memory emulator represents the most powerful ECU interface currently available for connecting calibration tools.

A simpler method of linking calibration tools (laptop) to the ECU is offered by the MAC (compact testing and calibration device). It connects via the 'K' wire of the diagnosis interface or – if present – the emulator module (Fig. 1).

The use and function of the calibration tools can be illustrated by the description below of a typical calibration process (Fig. 2 overleaf).

Software calibration process

Defining the desired characteristics

The desired characteristics (e.g. dynamic response, noise output, exhaust-gas composition) are defined by the engine manufacturer and the (exhaust-gas emissions) legislation. The aim of calibration is to alter the characteristics of the engine to meet these requirements. This requires testing on the engine test bench and in the vehicle.

Preparations

Special engine ECUs with emulator modules are used for calibration. Compared with the ECUs used on production models, they allow parameters that are fixed for normal operation to be altered. An important aspect of preparations is choosing and setting up the appropriate hardware and/or software interface.

Additional testing equipment (e.g. temperature sensors, flow meters) permit detecting other physical variables for special tests using INCA hardware components such as:

- Thermo Scan (for measuring temperatures)
- Lambda Scan (for measuring exhaust-gas oxygen content)
- Baro Scan (for measuring pressures) and
- A/D Scan (for other analog signals)

Determining and documenting the actual system responses

The detection of specific measured data is carried out using the INCA kernel system. The information concerned can be displayed on screen and analyzed in the form of numerical values or graphs (Fig. 3).

The measured data can not only be viewed after the measurements have been taken but while measurement is still in progress. In this way, engine response to changes (e.g. in the exhaust-gas recirculation rate) can be investigated. The data can also be recorded for subsequent analyzis of transient processes (e.g. engine starting).

2 Stages of calibration process

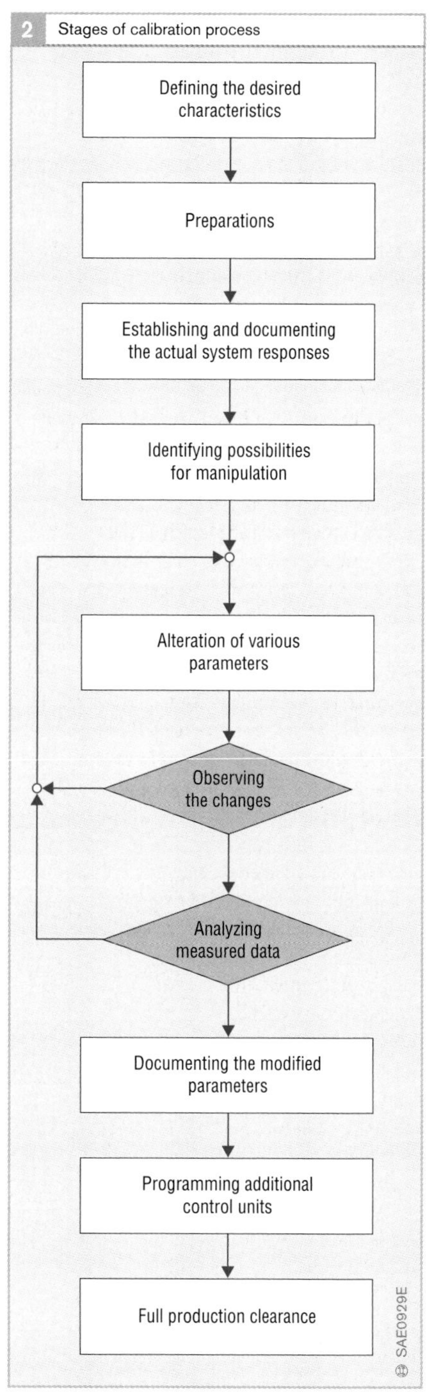

Identifying possibilities for manipulation
With the help of the ECU software documentation (function framework), it is possible to identify which parameters are best suited to altering system response in the manner desired.

Altering selected parameters
The parameters stored in the ECU software can be displayed as numerical values (in tables) or as graphs (curves) on the PC and altered. Each time an alteration is made, system response is observed.

All parameters can be altered while the engine is running so that the impacts are immediately observable and measurable.

In the case of short-lived or transient processes (e.g. engine starting), it is effectively impossible to alter the parameters while the process is in progress. In such cases, therefore, the process has to be recorded during the course of a test. The measured data is saved to file and then the parameters that are to be altered are identified by analyzing the recorded data.

Further tests are performed in order to evaluate the success of the adjustments made or to learn more about the process.

Analyzing measured data
Analyzis and documentation of the measured data is performed with the aid of the offline tool MDA (Measured Data Analyzer). This stage of the calibration process involves comparing and documenting system response before and after parameter alteration. Such documentation encompasses improvements as well as problems and malfunctions. Documentation is important because several people will be involved in the process of engine optimization at different times.

Documenting the modified parameters
The changes to the parameters are also compared and documented. This is done with the offline tool ADM (**A**pplication **D**ata **M**anager), sometimes also called CDM (**C**alibration **D**ata **M**anager).

The calibration data obtained by various technicians is compared and merged into a single data record.

Programming additional control units
The new parameter settings obtained can also be used on other engine ECUs for further calibration. This requires reprogramming of the flash EPROMs in the ECUs. This is carried out using the INCA kernel system tool PROF (Programming of Flash EPROM).

Depending on the extent of the calibration and design innovations, multiple looping of the steps described above may take place.

3 Software calibration monitor (example)

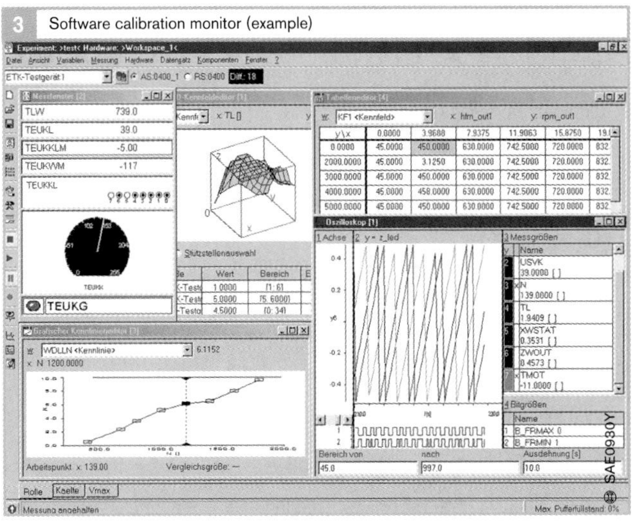

Example of calibration

Exhaust-gas temperature control

The following aspects play an essential role in operating a modern gasoline engine:

- Exhaust-gas emissions
- Fuel consumption and
- Thermal stresses on engine components

In order to minimize the stresses to which the engine is subjected, a low exhaust-gas temperature has to be aimed at (e.g. peak temperature below 1,050 °C and continuous temperature below 970 °C). This is a particularly important consideration in the case of turbocharged engines because the turbocharger has to be protected against thermal damage.

The exhaust-gas temperature can be reduced by enriching the air-fuel mixture by manipulating the optimization parameters, for example. Unfortunately, the negative side of enriching the mixture is that both fuel consumption and exhaust-gas emissions (CO and HC) are raised. This means that the mixture need only be enriched by the absolute minimum amount required.

Engine operation generally follows a dynamic pattern, which is why the exhaust-gas temperature is subject to fluctuations. Therefore, a physical model to take account of heat capacities, heat transfer and response times is required to determine exhaust-gas temperature. Such complex models can generally only be configured with the aid of optimization tools. This involves performing tests under all relevant operating conditions in which all essential input and output variables are recorded. In the case of exhaust-gas temperature control, the optimizer then adapts the optimization parameters until the modeled temperature matches the measured temperature as closely as possible. The accuracy of the modeled temperature is equivalent to that achieved with a temperature sensor if the choice of parameters is appropriate.

The advantages are obvious – the temperature sensor, sensor wire and installation position layout can be dispensed with on the production model. In addition, this removes the risk of the component failing over the life of the vehicle. It means that the fault-diagnosis function for the component is not required either.

Other adjustments

Safety-related adaptation

As well as the functions that determine emission levels, performance and user-friendliness, there are also numerous safety functions that require adaptation (e.g. response to failure of a sensor or actuator).

Such safety functions are primarily intended to restore the vehicle to a safe operating condition for the driver and/or to ensure the safe operation of the engine (e.g. to prevent engine damage).

Communication

The engine ECU is normally part of a network of several ECUs. The exchange of data between vehicle, transmission, and other systems takes place via a data bus (usually a CAN). Correct interaction between the various ECUs involved cannot be fully tested and optimized until they are installed in the vehicle, as the process of basic configuration on the engine test bench usually involves only the engine-management module on its own.

A typical example of the interaction between two vehicle ECUs is the process of changing gear with an automatic transmission. The transmission ECU sends a request via the data bus to reduce engine torque at the optimum point in the gear shifting operation. The engine ECU then initiates actions independently of the driver to reduce engine torque output and thus facilitate a smooth and judder-free gear change. The data that result in torque reduction has to be adapted.

Electromagnetic compatibility

The large number of electronic vehicle systems and the wide use of other electronic communications equipment (e.g. radio telephones, two-way radios, GPS navigation systems) make it necessary to optimize the electromagnetic compatibility (EMC) of the engine ECU and all its connecting leads in terms of both immunity to external interference and of emission of interference signals. A large proportion of this optimization work is carried out during the development of ECUs and the sensors concerned, of course. However, the dimensioning (e.g. length of cable runs, type of shielding) and routing of the wiring harnesses in the actual vehicle has a major impact on immunity to and creation of interference. As a result, testing and, if necessary, optimization of the complete vehicle inside an EMC room is absolutely essential.

Fault diagnosis

Due to legal requirements, the capabilities demanded of fault-diagnosis systems are very extensive. The engine ECU constantly checks that the signals from all connected sensors and actuators are within specified limits. It also tests for loose contacts, short circuits to ground or to the battery terminal, and for plausibility with other signals. The signal range limits and plausibility criteria must be defined by the application developer. These limits must firstly be sufficiently broad to ensure that extreme conditions (e.g. hot or cold weather, high altitudes) do not produce false diagnoses, but secondly, sufficiently narrow to provide adequate sensitivity to real faults. In addition, fault response procedures must be defined to specify whether and in what way the engine may continue to be operated if a specific fault is detected. Finally, detected faults have to be stored in a fault memory so that service technicians can quickly locate and rectify the problem.

Testing under extreme climatic conditions

Testing procedures include trials under extreme climatic conditions that are normally only encountered under exceptional circumstances during the service life of the vehicle. The conditions that are encountered during these trials can only be simulated to a limited degree on a test bench because the subjective judgement of the test driver and long experience play an important part in such tests. Temperature itself can easily be simulated on a test bench, but using a chassis dynamometer to assess a vehicle's response when pulling away, for example, is very difficult compared to making the judgement under real driving conditions.

In addition, road tests generally involve longer distances and several vehicles. This enables testing of calibration parameters across the spread of the vehicles tested and, therefore, allows wider conclusions to be drawn than with calibration based on a single test subject.

Another essential aspect is the impact of variations in fuel grade from one part of the world to another. The chief effect of such variations in fuel grade is on the engine's starting characteristics and warm-up phase. Vehicle manufacturers go to great lengths to ensure this a vehicle will run properly on all the fuels on the market.

Cold-weather trials

Cold-weather trials cover the temperature range from approx. 0 °C to –30 °C. Preferred locations for cold-weather trials are places such as northern Sweden and Canada. The primary function is to assess starting and pulling away.

During the starting sequence, every individual combustion process is analyzed and the appropriate parameters optimized where necessary. Correct configuration of parameters for every individual injection sequence is a decisive factor in engine start time and smooth increase of engine speed from starter speed to idle speed. Even a single imperfect combustion process with resulting reduced torque development during the startup phase is perceived as a deficiency – even by inexperienced customers.

Hot-weather trials

Hot-weather trials cover the temperature range from approx. +15 °C to +40 °C. These trials are carried out at locations such as southern France, Spain, Italy, the U.S., South Africa and Australia. Despite the great distances involved and the corresponding high cost of equipment transportation, South Africa and Australia are of interest because they offer hot-weather conditions during the European winter. Due to the ever increasing demands to shorten development times, such possibilities have to be considered. Hot-weather trials test such things as hot starting, tank ventilation, tank leakage detection, knock control, exhaust-gas temperature control and a wide variety of diagnostic functions.

Altitude trials

Altitude trials involve testing at altitudes between 0 and approx. 4,000 meters. It is not only the absolute altitude that is of importance to the tests but, in many cases, a rapid change in altitude within a short space of time. Altitude trials are generally carried out in combination with hot or cold-weather trials.

Once again, an important component is testing the start characteristics. Other aspects examined include mixture adaptation, tank ventilation, knock control and a range of diagnostic functions.

4 Recording cold-starting response in the cold room

The Bosch Boxberg Test Center

The practical tests which are performed at an early stage by the system supplier are an important factor in the development of modern automotive systems. But not all of these tests can be performed on public roads. The new Bosch Test Center in Boxberg between Heilbronn and Würzburg (southern Germany) has been in operation since mid-1998. Here, the most varied automotive systems and their components, no matter whether for driving, safety, comfort, or convenience, can be thoroughly tested on proving grounds covering an area of 92 ha (1 ha = 10^4 m^2). There are seven different track modules available for putting innovative systems through their paces under all possible driving situations and up to the limits of their physical capabilities. And all this, while still ensuring maximum-possible safety for the driver and for the vehicle.

The **rough-road tracks** (1) are designed for speeds up to max. 50 km/h (30 mph) or max. 100 km/h (60 mph). The following special types of track have been built:

- Pothole type
- Washboard type
- Shake and vibration type
- Belgian-block type and
- Tracks with varying degrees of unevenness

The asphalted **hill-climbing tracks** (2) for drive-off and acceleration tests on a hill with gradients of 5 %, 10 %, 15 %, and 20 %, also incorporate tile strips of varying widths that can be flooded with water.

Two **water-wading points** (3) with lengths of 100 m and 30 m (with depths of 0.3 and 1 m respectively) are also available.

Also at the disposal of the testers are **special, floodable tracks** (4) paved as follows:

- Chess-board-pattern pavement (with asphalt and tiles)
- Asphalt
- Tiles
- Blue basalt
- Concrete
- Aquaplaning tracks and
- Trapezoidal blue-basalt track

The 300 m diameter **vehicle-dynamic test-pad** (5) for special cornering tests is paved with asphalt, and can be partially flooded in order to simulate ice and aquaplaning conditions. The test pad is surrounded by a tire barrier as a precautionary measure for the safety of the drivers and vehicles.

The **high-speed oval** (6) is provided with three lanes and can be used by passenger cars as well as commercial vehicles. Design is such that tests can be performed at speeds of up to 200 km/h (120 mph.)

The **handling track** (9) is comprised of two different stretches of road: One for max. 50 km/h (30 mph) and the other for max. 80 km/h (50 mph). Both tracks feature curves with differing radiuses and inclinations. This track is mainly used for testing vehicle-dynamics systems (ESP).

View of the test-track section modules

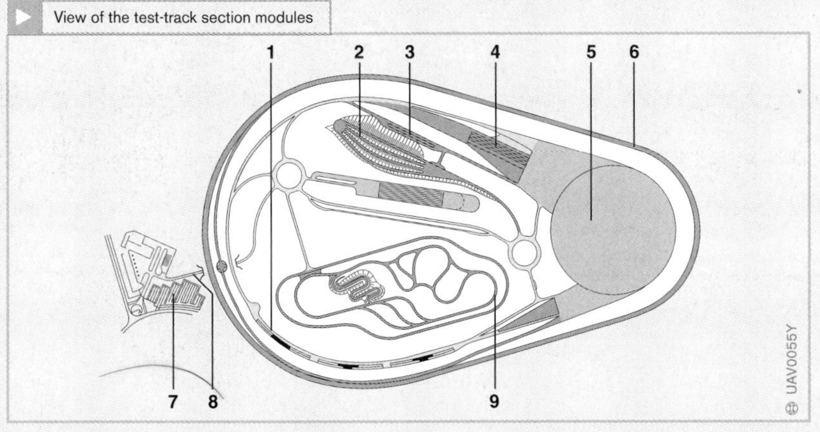

Fig. 1

1 Rough-road track
2 Hill-climbing track
3 Water-wading points
4 Special, floodable tracks
5 Vehicle-dynamics test pad
6 High-speed oval
7 Buildings
– Workshops
– Offices
– Test benches
– Laboratories
– Filling station and
– Welfare and medical facilities
8 Access road
9 Handling track

Index of technical terms

Technical terms

A
A/F-mixture cloud, 155
A/F-mixture formation, 39, 155
Acceleration response, 90
Accelerator-pedal sensors, 262f
Accessory Control, 303
Activation arc, 179
Activation conditions, 309
Actuator diagnosis, 374
Additives, 32
Air bypass actuator, 50
Air charge, 36
Air system, 298
Air-flow sensor, 85, 240
Air-flow sensor, L3-Jetronic, 104
Air-gap concept, 211
Air-mass meter, 141, 240
Air/fuel (A/F) ratio, 39
Air/fuel mixture, 18
Air/fuel ratio, 18
Aircraft engine, 72
Alcohol fuels, 33
Alternate fuels, 33
Altitude compensation, 79
Altitude trials, 410
Anti-ageing additives, 32
Application-related adaptation, 404
ASIC, 291
Asphalt ignition coil, 197
Assembler, 400
Auto-ignition, 43, 217, 221
Auxiliary systems, 245
Auxiliary-air device, 90, 103
Average fleet fuel consumption, 354, 356

B
Back-up spark, 192
Basic injection duration, 98
Basic mixture adaptation, 87
Battery ignition, 167ff
Benz, 161
Benz, Carl Friedrich, 13
Boiling curve, 31
Boost pressure, 60
Bosch, Robert f, 14
Bottom dead center, 16
"Boxer" (reciprocal) configuration of the mechanical fuel-injection pump, 72
Brake-booster vacuum control, 255
Breaker-triggered transistorized ignition, 172
Broad-band Lambda oxygen sensors, 286f
Brush carburetor, 68
Bus configuration, 325

C
Calibration tools, 405
Camshaft changeover, 53
Camshaft phase adjustment, 52
CAN, 324
CAN interface, 245
Canister-purge valve, 133
Capability Maturity Model, 400
Capacitance, 200
CARB legislation, 352ff
Carbon canister, 133
Carbon dioxide, 331
Carbon monoxide, 332
Carburetor, 68-75
CARTRONIC, 329
Catalytic converter, 340ff
Catalytic emissions control, 340ff
Catalytic-converter fault diagnosis, 312
Catalytic-converter heating, 348
Catalytic-converter quality factor, 314
Center electrode, 209
Central injection unit, 109
Centrifugal advance adjustment, 171
Centrifugal turbo-compressors, 60
Ceramic monoliths, 342
Charge-flow control flap, 55
Classifications, 350
Coating, 342
Coil designations, 192
Coil on Plug, 194
Cold-start enrichment, 88, 101
Cold-start valve, 88
Cold-weather trials, 409
Combustion by-products, 332f
Combustion knock, 20f, 221
Combustion miss detection, 315
Combustion process, 154
Common rail, 147
Communication, 302
Communications interface, 306
Compact coil, 193
Compiler, 401
Compound electrodes, 210
Comprehensive components, 322
Compression ratio, 18
Compression-stroke, 16
Compressor, 60, 62
Connecting devices, 182
Contact-breaker points, 170
Continuous-action Lambda control, 347
Continuous-delivery system, 131
Control pressure, 86
Conventional coil ignition, 159, 170
Corrosion, 220
Corrosion protection, 32
Crankcase ventilation, 320
Current control, 173
Cylinder charge, 36
Cylinder-individual fuel injection, 145

D
D-Jetronic, 76
Daimler, 162
Daimler, Gottlieb, 12
Data memory, 290
Data transfer, 324
Demand-controlled system, 129, 131
Development specifications, 390
Diagnosis System Management, 311
Diagnosis Validator, 311
Diagnostic Fault Path Management, 311
Diagnostic Function Scheduler, 311
Diagnostic interface, 245, 306
Diagnostic System, 303
Diesel, Rudolf Christian Karl, 13
Differential-pressure valves, 87
Dilution procedure, 367
Distributor, 170
Distributorless (stationary) voltage distribution, 180
Distributorless semiconductor ignition, 159
Division control multivibrator, 97
Downsizing, 65
Driver stage, 198
Dual injection, 157
Dual spray, 144
Dual-spark coils, 180
Dual-spark ignition coil, 178, 192
Dwell-angle control, 173
Dynamic internal resistance, 201
Dynamic supercharging, 57ff

E
Ecotronic carburetor, 75
ECU development, 390
EEPROM, 291
Efficiency, 24
EFU diode, 179
EGR valve, 56
Eisemann, 165
Electric fuel pump, 134f

Electro-hydraulic pressure actuator, 92
Electrode gap, 212
Electrode materials, 210
Electrode wear, 220
Electrodes, 209
Electromagnetic compatibility, 408
Electromagnetic fuel injectors, 142ff
Electromagnetic testing, 397
Electronic control systems, 294
Electronic control unit (ECU), 288
Electronic diagnosis, 234, 304
Electronic ignition, 159, 174
Electronic Service Information ESI[tronic], 371
Electronic Throttle Control, 51
EMC tests, 397
Emergency operation, 309
Emissions categories, 352, 356
Emissions inspections, 386f
Emissions measurement concept, 388f
Emissions testing, 366ff
Emissions-control legislation, 350ff
Emissions-free vehicles, 355
Emulator mode, 403
Emulator module, 405
Engine cooling system, 320
Engine efficiency, 24f
Engine response data, 111
Environmental conditions, 306
Environmentally compatible gasolines, 32
EOBD, 307
EoL programming, 292
EPA, 307
EPA regulations, 356
EPROM, 291
Error class, 306
Error code, 306
Error status, 306
Error-code storage, 306
EU regulations, 358
EU/ECE test cycle, 364
European test cycle, 364f
Evaporative emissions, 330
Evaporative-emissions control system, 133, 243
Evaporative-emissions testing, 368
Excess-air factor, 18, 39
Exhaust, 17
Exhaust emissions, 330ff
Exhaust System, 301
Exhaust valve, 17
Exhaust-gas recirculation, 56, 241
Exhaust-gas recirculation system, 320
Exhaust-gas temperature control, 408
Exhaust-gas turbine, 62
Exhaust-gas turbocharging, 62ff, 242
External EGR, 56

F
Field monitoring, 355, 357, 361
Finite Element Method, 202, 229
Flame front's propagation rate, 19
Flame propagation, 19
Flash-EPROM, 291
Fleet averages, 354, 356
Flow check, 308
Flux density, 189
FMEA, 393
Ford, Henry, 13
Four-stroke principle, 16
Freeze frame, 306
Fresh gas, 36
Frictional losses, 25
FTP 75 test cycle, 362
Fuel consumption, 40
Fuel distributor, 85
Fuel economy, 22
Fuel filter, 136
Fuel injector, 117, 243
Fuel lines, 138
Fuel metering, 85, 97
Fuel rail, 137
Fuel standards, 29
Fuel supply, 83, 110, 128
Fuel System, 299
Fuel tank, 138
Fuel-consumption map, 27
Fuel-consumption signal, 245
Fuel-injection valves, 84
Fuel-level sensor, 259
Fuel-pressure damper, 138
Fuel-pressure regulator, 137, 243
Fuel-quantity control, 150
Fuel-system components, 252
Fuel-system, diagnosis, 318
Fuel-tank leakage diagnosis, 316
Fuels, 28ff
Full-load enrichment, 90
Fully variable valve timing, 54
Function development, 398f

G
Gas-exchange valves, 16
Gasoline direct injection, 146ff
Gasoline direct-injection system, 73
Gasoline-engine management, 34ff
Goliath GP700E, 73
Greenhouse effect, 360
Gross calorific value, 29
Ground electrodes, 209
Group injection, 145
Gutbrod Superior, 73

H
Hall-effect phase sensors, 264f
Hall-effect sensors for transistorized ignition, 267
Handling spark plugs, 380ff
Hardware, 392
Hardware design prototypes, 394
Hardware development, 394
Heat-range code number, 215
High voltage, 190
High-level programming language, 400
High-pressure injector, 152f
High-pressure pump, 148
High-pressure sensors, 275
High-tension magneto ignition, 164
High-tension vibrator ignition, 161
High-voltage dome, 193
High-voltage properties, 201
History of the automobile, 10-15
Homogeneous and lean-burn mode, 157
Homogeneous and stratified-charge mode, 157
Homogeneous mode, 156
Homogeneous/anti-knock mode, 157
Honold, 164
Hot-film air-mass meter, 107
Hot-film air-mass meter HFM2, 279
Hot-film air-mass meter HFM5, 280ff
Hot-soak losses, 369
Hot-tube ignition, 162
Hot-weather trials, 410
Hot-wire air-mass meter, 106
Hot-wire air-mass meter HLM, 278
Huygens, 160
Hydrocarbons, 332
Hysteresis, 189

I
I core, 193
Idle stabilization, 89
Ignition, 19, 42
Ignition angle, 42
Ignition coil, 178f, 186ff
Ignition coils in service, 377ff
Ignition driver stage, 178
Ignition energy, 183ff
Ignition map, 42
Ignition point, 42
Ignition System, 300
Ignition timing, 185

Ignition voltage, 182
Ignition-advance map, 175
Ignition-coil development, 202
Ignition-coil modules, 195
Ignition-system, 158ff
In-tank unit, 132
INCA system, 405
Inductance, 200
Induction-mixture distribution, 18
Induction-stroke, 16
Induction-type sensors for transistorized ignition, 266
Inductive engine-speed sensors, 268
Inductive ignition system, 176ff
Inert gas, 37
Inhibit conditions, 309
Inhibit handler, 311
Injection diagram, 78
Injection duration, 77
Injection trigger, 77
Injection-orifice plate, 142
In-line fuel-injection pump, 12-cylinder, 72
Insulator, 207, 230
Intake manifold, 57
Intake valve, 16
Intake-manifold pressure, 78
Intake-manifold pressure sensor, 241
Intercooling, 65
Interference-suppression resistors, 182
Interference-suppressor equipment, 182
Internal EGR, 38, 53
Internal-gear pumps, 134
Ionic-current measurement, 199, 216

J

Japanese test cycles, 365
Jet-nozzle carburetor, 70

K

K-Jetronic, 82
KE-Jetronic, 92
KE-Motronic, 49
Knock, 20, 43
Knock control, 21, 43
Knock resistance, 30

L

L-Jetronic, 94-107
L3-Jetronic, 104f
LabCar, 402
Lambda, 18
Lambda closed-loop control, 91
Lambda control loop, 346f
Lambda program map, 115
Lambda sensors, 244
Lambda-sensor, diagnosis, 319
Lean-misfire limit, 39
Lenoir, 161
LH-Jetronic, 106f
Linker, 401
Liquified petroleum gas, 33
Logic analyzer, 403
Low-tension magneto ignition, 163

M

M-Motronic, 49, 238-245
MAC, 405
Magnetic circuit, 189
Magnetic field strength, 189
Magnetic-flux field, 188
Magnetization curve, 189
Magneto ignition, 163ff
Main functions, 296
Major-leak detection, 316
Malfunction indicator lamp, 245, 309
Malfunction response, 306
Manifold chamber, 57
Manifold fuel injection, 140ff
Manufacturing ignition coils, 203
Marcus, Siegfried, 68-75
Maybach, Wilhelm, 12
Maybach, Wilhelm , 70-75
ME-Motronic, 49, 246-249
Measuring the air flow, L-Jetronic, 96
Mechanical supercharging, 60
MED-Motronic, 49, 250-255
Mercedes-Benz 300 SL, 74
Message format, 327
Metallic monoliths, 342
Microcontroller, 290
Micromechanical pressure sensors, 270ff
Micromechanics, 257, 274
MIL, 309
Minor-leak detection, 317
Miss detection function, 315
Mixture adaptation, 121
Mixture formation, 68-75
Module archiving, 401
Monitoring, 303, 304
Monitoring concept, 249
Monitoring module, 291
Monitoring operational data, 232
Mono-Jetronic, 108-127
Mono-Motronic, 49
Monoliths, 342
Motor Octane Number, 30
Motronic, 49, 232
Motronic engine management, 232
Motronic versions, 235
Multiplex applications, 325
Multipoint fuel-injection systems, 66
Multispark ignition, 199

N

Natural gas, 33
Net calorific value, 29
Nitrous oxides, 332
NO_x accumulation, 344
NO_x accumulator-type catalytic converter, 344f
NO_x removal, 344

O

O core, 193
OBD I, 307
OBD II, 307
On-Board Diagnosis, 307, 355, 357, 360
Operating data, 302
Operating modes, 156
Operating-mode coordination, 254
Otto, Nikolaus August, 12, 69
Overpressure method, 317
Overrun, 41
Overrun fuel cutoff, 41, 91
Oxidation-type catalytic converter, 340
Ozone, 333
Ozone-reduction system, 321

P

p-V diagram, 24
Parameter definition, 404
Particulates, 333
Patents, 349
Pencil coil, 195
Pencil spray, 144
Periodic emissions inspections, 361
Peripheral pump, 135
Permanent magnet, 190
Petroleum, 28
Phase-in, 354, 356
Piezoelectric knock sensors, 269
Pioneers of automotive technology, 12f
Planar Lambda oxygen sensors, 284
Platinum +4 spark plug, 224
Platinum electrodes, 210
Plausibility check, 322
Plug core, 230
Positive crankcase ventilation, 330
Positive-displacement pumps, 134
Positive-displacement superchargers, 60
Post-combustion thermal treatment, 339
Post-ignition, 218

Post-start, 118
Powder machine, 160
Power, 23, 40
Power (combustion) stroke, 17
Power loss, 200
Pre-cat, 343
Pre-ignition, 218
Pre-supply pump, 134
Pressure regulator, 110
Pressure sensor, 79
Pressure-control valve, 147, 150
Primary pressure, 86
Primary winding, 188
Primary-current limitation, 178
Primary-pressure regulator, 84
Primary-voltage limitation, 178
Priority handler, 311
Priority scheduler, 311
Problem recognition, 306
Processing operational data, 234
Product specifications, 390
Program memory, 290
Pumping losses, 25

Q
Quality management, 393

R
Radial engine, 73
Rail, 147, 148
RAM, 291
Ram-tube supercharging, 57
Ram-tube systems, 59
Range check, 322
Rationality check, 322
Raw emissions, 334ff
Readiness code, 308, 309
Reducing emissions, 338ff
Refuelling test, 369
Research Octane Number, 30
Residual gas, 36
Residual mixture, 43
Review, 393
Rivaz, 160
Roller-cell pumps, 134
ROM, 290
Running-loss test, 369

S
Schematic pulse-timing diagram, L-Jetronic, 99
Seal seat, 208
Secondary winding, 188
Secondary-air injection, 339, 348
Secondary-air injection, diagnosis, 318
Secondary-air system, 244
Self-diagnosis, 304
Self-induction, 188
Semi-surface gap concepts, 211
Sensor plate, 85
Sensor-flap (impact-pressure) air-flow sensor LMM, 276
Sensor-plate potentiometer, 260
Sensor-wheel compensation, 315
Sensors, 256
Sequential fuel injection, 145
Serial data transfer, 324
Service technology, 370ff
SFTP schedules, 362
Shell, 208
Shrink-fitting process, 208
Shunt current, 214
Shunt losses, 184
Side-channel pump, 135
Silicone insulation layer, 193
Silver center electrodes, 210
Simultaneous fuel injection, 145
Single-barrel pump, 149
Single-point injection, 67
Single-spark coils, 180
Single-spark ignition, 199
Single-spark ignition coil, 178, 192
Smog, 333
Software, 392
Software development, 400
Software quality, 400
Software sharing, 400
Soot emissions, 337
Spark energy, 191
Spark erosion, 220
Spark head, 183
Spark plug, 181, 204ff
Spark position, 213
Spark tail, 183
Spark-plug assembly, 231
Spark-plug cables, 182
Spark-plug concepts, 211
Spark-plug heat ranges, 214f
Spark-plug manufacture, 230
Spark-plug performance, 220f
Spark-plug type designations, 228
Spark-plug wear, 181
Special-purpose spark plugs, 227
Specific fuel consumption, 26f
Spit-back test, 369
Spray formation, 144
Standard system, 129
Start of injection, 77
Stoichiometric ratio, 39
Stratified-charge concept, 18
Stratified-charge mode, 156
Stratified-charge/cat-heating, 157
Stripline method, 397
Subsystems, 296
Sulphur, 345
Sulphur content, 32
Sulphur dioxide, 332
SUPER 4 spark plug, 223
SUPER spark plug, 222
Surface carburetor, 68, 70
Surface-gap concept, 211
Swirl air flow, 154
Switch-on arcs (activation arcs), 190
System control, 296
System documentation, 296
System structure, 236, 295

T
Tank-ventilation losses, 369
Tapered spray, 144
Temperature measurements, 216
Temperature sensors, 258
Terminal post, 207
Test cycle, 362ff
Test procedures, 350
Testing on-board control units, 372ff
Testing series-production vehicles, 351
Testing station, 401
Testing the ignition system, 376
Thermal auto-ignition, 217
Thermal efficiency, 24
Thermal losses, 24
Thermo-time switch, 88
Thermographics, 396
Thick-film pressure sensors, 273
Three-barrel pump, 149
Three-way catalytic converter, 341ff
Throttle device, 51
Throttle valve, 50
Throttle-valve actuator, 123
Throttle-valve sensor, 241, 261
Throttle-valve switch, 80, 102
Throttling losses, 53
Top dead center, 16
Torque, 23
Torque demand, 296
Torque structure, 249, 297
Transistorized ignition, 159, 172
Transition compensation, 119
Tube-type (finger) sensors, 282
Tumble air flow, 154
Tuned-intake-tube charging, 58
Turbine pumps, 135
Turbo flat spot, 65
Turns ratio, 201
Twin-spark ignition, 192

Two-cylinder fuel-injection pump, 73
Two-step control, 346
Two-step Lambda oxygen sensors, 282ff
Type approval testing, 359
Type test, 350
Type tests, 359

U
Underfloor catalytic converter, 343
US test cycles, 362f
US type approval testing, 353

V
Vacuum advance adjustment, 171
Vacuum method, 316
Valve overlap, 38
Valve timing, 17
Vapor Lock Index, 32
Vapor pressure, 31
Vapor/liquid ratio, 31
Variable valve timing, 52ff, 242
Variable valve timing, diagnosis, 321
Variable-geometry intake manifold, 58, 241
Vehicle management, 235
Vibrator ignition, 161
Volatility, 31
Volta, 160
Voltage correction, 100
Voltage distribution, 180
Volumetric efficiency, 38
VST supercharger, 64
VTG turbocharger, 63

W
Wall film, 41
Wall wetting, 41
Warm-up enrichment, 89
Warm-up phase, 118
Warm-up regulator, 89
Wastegate supercharger, 63
Wick carburetor, 68
Wood-gas generator, 75

Y
Yoke plate, 196

Z
ZENITH carburetor, 71

Abbreviations

A

ABS: Antilock Braking System
AC: Accessory Control
ACC: Adaptive Cruise Control
ADC: Analog/Digital Converter
ADM: Application Data Manager (see also CDM)
AS: Air System
ASIC: Application Specific Integrated Circuit
ATL: Exhaust-gas turbocharger (German: Abgasturbolader)
AU: German emissions inspection (German: Abgasuntersuchung)

B

BDC: Bottom Dead Center
BDE: Gasoline direct injection (German: Benzin-Direkt-einspritzung)
BIP: Bosch Integrated Power (bipolar technology in transistors)

C

CAD: Computer Aided Design
CAE: Computer Aided Engineering
CAN: Controller Area Network
CARB: California Air Resources Board
CDM: Calibration Data Manager (Application data manager, ADM)
CI: Conventional Coil Ignition
CIFI: Cylinder Individual Fuel Injection
CMM: Capability Maturity Model
CO: Communication
COP: Coil on Plug
CPU: Central Processing Unit
CRC: Cyclic Redundancy Check
CVS: Constant Volume Sampling

D

DFPM: Diagnosis Fault Path Management
DOR: Direct Ozone Reduction
DS: Diagnostic System
DSCHED: Diagnostic Function Scheduler
DSM: Diagnosis System Management
DVAL: Diagnosis Validator

E

EA: Electrode gap (German: Elektrodenabstand)
ECU: Electronic Control Unit
EU/ECE: Economic Commission of Europe
EDC: Electronic Diesel Control
EEPROM: Electrically Erasable Programmable Read Only Memory
EFU: Activation arc (German: Einschaltfunkenunterdrückung)
EGAS: Electronic throttle control (German: Elektronisches Gaspedal)
EGR: Exhaust-Gas Recirculation
EGS: Electronic transmission control (German: Elektronische Getriebesteuerung)
EMC: Electro-Magnetic Compatibility
EMS: Electronic engine-performance control system (German: Elektronische Motorleistungs-steuerung)
EOBD: European On-Board-Diagnosis
EOL: End of Line
EPA: Environment Protection Agency
EPROM: Erasable Programmable Read Only Memory
ES: Exhaust System
ESI[tronic]: Electronic Service Information
ESP: Electronic Stability Program
ESU: Electronic Scanner Unit (Emulator module)
ETC: Electronic Throttle Control
EUDC: Extra Urban Driving Cycle
EV: Injector (German: Einspritzventil)
EZ: Electronic ignition system (German: Elektronische Zündung)

F

FEM: Finite Element Method
FID: Flame Ionisation Detector
FMEA: Failure Mode and Effects Analysis
FS: Fuel System
FTP: Federal Test Procedure

G

GS: Electronic transmission-shift control (German: Elektronische Getriebesteuerung)

H

HDEV: High-pressure injector (German: Hochdruck-Einspritzventil)
HDP: High-pressure pump (German: Hochdruckpumpe)
HFM: Hot-film air-mass meter
HLM: Hot-wire air-mass meter (German: Hitzdraht-Luftmassen messer)

I

IDE: Identifier Extension Bit
IGBT: Insulated Gate Bipolar Transistor (incorporates elements of field-effect and bipolar transistor)
INCA: Integrated Calibration and Aquisition System
IS: Ignition System
ISO: International Organization for Standardization
IZP: Inner-gear pump (German: Innenzahnradpumpe)

L

LDT: Light-Duty Trucks
LEV: Low Emission Vehicle
LML: Lean-Misfire Limit
LMM: Air-flow sensor (German: Luftmengenmesser)

M

MAC: Compact testing and calibration device (German: Mess- und Applikationsgerät Compact Serie)
MDA: Measure Data Analyzer
MFZ: Multispark ignition (German: Mehrfunkenzündung)
MIL: Malfunction Indicator Lamp
MNEDC: Modified New European Driving Cycle
MO: Monitoring
MON: Motor Octane Number
MSV: Delivery-quantity control valve (German: Mengensteuerventil)

N

NDIR: Non-Dispersive Infrared Analyser
NE: Post-ignition (German: Nachentflammung)
NEDC: New European Driving Cycle
NMOG: Non-Methane Organic Gases

O

OBD: On-Board-Diagnosis
OD: Operating Data

P

PCB: Printed-Circuit Board
PP: Peripheral Pump
PWM: Pulse-Width Modulation
PZEV: Partial Zero-Emission Vehicle
PZZ: Multispark ignition (German: Pulszugzündung)

R

RAM: Random Access Memory
RLFS: Returnless Fuel System
ROM: Read Only Memory
RON: Research Octane Number
ROV: Rotating high-voltage distribution
RTR: Remote Transmission Request
RUV: Stationary voltage distribution (German: Ruhende Spannungs-verteilung
RZP: Roller-cell pump (German: Rollenzellenpumpe)

S

SAE: Society of Automotive Engineers
SC: System Control
SD: System Documentation
SEFI: Sequential Fuel Injection
SHED: Sealed House for Evaporation Determination
SMD: Surface Mounted Device
SULEV: Super Ultra-Low-Emission Vehicle
SZ: Auto-ignition (German: Selbstzündung)

T

TCS: Traction Control System
TD: Torque Demand
TDC: Top Dead Center
TLEV: Transitional Low-Emission Vehicle
TS: Torque Structure
TZ: Transistorized ignition (German: Transistorzündung)

U

UDC: Urban Driving Cycle
ULEV: Ultra Low Emission Vehicle

V

VE: Pre-ignition (German: Vorentflammung)
VLI: Vapour-Lock Index
VST: Variable Sleeve Turbine
VTG: Variable Turbine Geometry
VVT: Variable Valve Timing
VZ: Distributorless ignition (German: Vollelektronische Zündung)

W

WOT: Wide-Open Throttle
WWR: Heat-range reserve in °Crankshaft (German: Wärme-wertreserve)

Z

ZEV: Zero-Emission Vehicle
ZZP: Ignition point in °Crankshaft BTDC (German: Zündzeitpunkt)

Bosch reference books – First-hand technical knowledge

Diesel-Engine Management

There is a lot of movement – also in a figurative sense – when it comes to the diesel engine and diesel-fuel injection, in particular. These developments are now described in the completely revised and updated 3rd Edition of the "Diesel-Engine Management" reference book. The electronics that control the diesel engine are explained in easy detail. It provides a comprehensive description of all conventional diesel fuel-injection systems. It also contains a competent and detailed introduction to the modern common rail system, Unit Injector System (UIS) and Unit Pump System (UPS), including the radial-piston distributor injection pump.

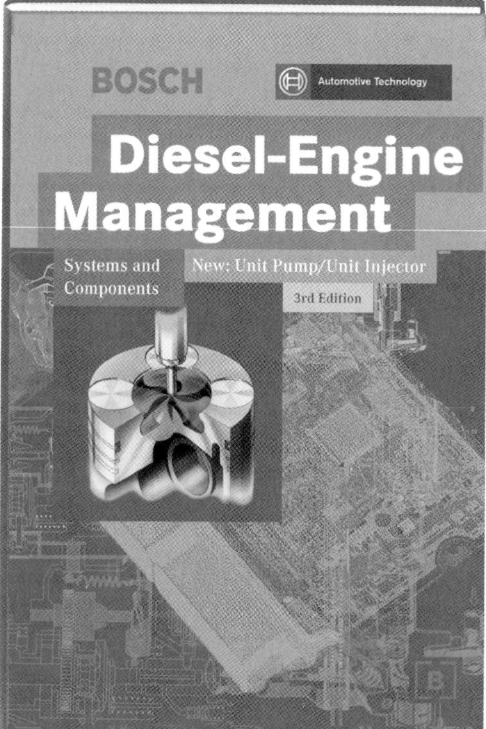

Contents

- Basics
- Cylinder-charge control systems
- Diesel fuel-injection systems
- In-line fuel-injection pumps
- Distributor injection pumps
- Single-plunger fuel-injection pumps
- Unit Injector System
- Unit Pump System
- Common-rail Electronic Diesel Control (EDC)
- Sensors
- Electronic control unit
- Electronic control and regulation
- Service (Overview)
- Emissions-control engineering and legislation
- Technical terms
  - Abbreviations

Hardcover,
17 x 24 cm format,
3rd Edition, completely revised and extended,
489 pages,
hardback,
with numerous illustrations.

ISBN
0-8376-1051-6

Automotive Electrics / Automotive Electronics

The rapid pace of development in automotive electrics and electronics has had a major impact on the equipment fitted to motor vehicles. This simple fact necessitated a complete revision and amendment of this authoritative technical reference work. The 4th Edition goes into greater detail on electronics and their application in the motor vehicle. The book was amended by adding sections on "Microelectronics" and "Sensors". As a result, the basics and the components used in electronics and microelectronics are now part of this book. It also includes a review of the measured quantities, measuring principles, a presentation of the typical sensors, and finally a description of sensor-signal processing.

Contents

- Automotive electrical systems, including calculation of wire dimensions, plug-in connections, circuit diagrams and symbols
- Electromagnetic compatibility and interference suppression
- Batteries
- Alternators
- Starters
- Lighting technology
- Windshield and rear-window cleaning
- Microelectronics
- Sensors
- Data processing and transmission in motor vehicles.

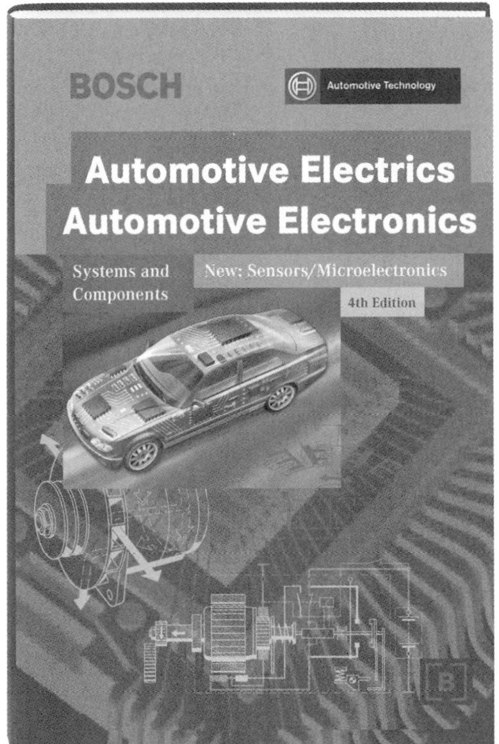

Hardcover,
17 x 24 cm format,
4th Edition, completely revised and extended,
503 pages,
hardback,
with numerous illustrations.

ISBN
0-8376-1050-8